The Ecology of North America

THE ECOLOGY OF
NORTH AMERICA

by Victor E. Shelford

UNIVERSITY OF ILLINOIS PRESS *Urbana Chicago London*

First paperback printing, 1978

ISBN 0-252-00707-7

Dedication

To the Memory of
STEPHEN ALFRED FORBES
Illinois State Entomologist

"The scientific investigator has no business to consider the practical value of his findings. It is his business to follow up the leads that open up to him."

and of
CHARLES MANNING CHILD
University of Chicago

"Do experimental work but keep in mind that other investigators in the same field will consider your discoveries as less than one fourth as important as they seem to you."

Preface

The author's first book, entitled *Animal Communities in Temperate America,* was based largely upon a region within about 100 miles (160 km) of Chicago. Soon after it was published in 1913, the idea of writing a similar treatise covering all North America was conceived. This was announced to a seminar group at the Puget Sound Marine Station in the summer of 1914. When work was begun on the subject, the need for locating and studying various relict areas of natural communities soon became evident. This gave the idea of the *Naturalist's Guide to the Americas,* which was published in 1926. Since invertebrate animals had been neglected in most studies of natural areas, quantitative collections were made and processed for all areas visited. The introduction of animals into the community classification based on plants required considerable reorganization of current community concepts.

Experimental studies were discontinued by the author after 1928 in order to concentrate on the study program of field communities. To further *Ecology of North America,* doctors' theses were assigned to graduate students on the hemlock–cedar forest in Oregon, the cold desert in Utah, the Rocky Mountain forest and pinyon–juniper in Arizona, the mountain forest and oak bush in Utah, the mountain forest in Maine, the deciduous forest in Illinois, and the tall-grass grassland in Illinois. These were biotic studies and all have been published. The author's personal studies included the tundra in northern Manitoba, the Mississippi River floodplain in western Tennessee, and 14 years of work with forest populations in central Illinois. These also have been published.

Trips with graduate students, with research assistants, or with small parties were made repeatedly over all of the United States, except Delaware, Maryland, New Jersey, and New England, between 1928 and 1945. Studies were also made in the western provinces of Canada, and a rather thorough series of inspections of community types in Quebec were taken under the guidance of Dr. Pierre Dansereau. Trips and studies extended into Mexico to include the states of Chihuahua, Coahuila, Nuevo León, Tamaulipas, San Luis Potosí, Veracruz, and Chiapas. Parts of two summers were spent at Barro Colorado Island in the Panama Canal Zone. Air travel to and from the island and auto travel from the principal airports en route gave a look at a number of vegetation types. Notes and records from all these trips, as well as unpublished theses of the University of Illinois and other institutions and numerous facts supplied from scientific observers as personal communications, have been utilized in this book. Unpublished information probably amounts to between 5 and 10 per cent of this treatise.

The writer has had the unstinted cooperation of his institution and a host of national, state, provincial, and local government agencies. The services and cooperation of many individuals have played an essential role in bringing together the scattered information available in many areas. The United States National Park Service, the Fish and Wildlife Service, and the Forest Service provided scientific guidance on numerous occasions for studies on lands under their control, threw open their records for use, copied data, and supplied me with their publications.

Several branches of the Canadian government responsible for biological resources and problems gave extended assistance. The Science Service of the Department of Agriculture made photographs of the original colored drawings of forest insect pests available for halftones. The Department of Resources and Development through the National Museum of Canada and the National Parks Branch supplied photographs without which the sections concerned would have been incomplete. The museum also identified various specimens taken in Canada.

The American Museum of Natural History, the Harvard Museum of Comparative Zoology, the Philadelphia Academy of Natural Sciences, the Chicago Museum of Natural History, and the University of Michigan Museum of Zoology allowed the writer to make full use of their specimens and records. These constituted important contributions.

The University of Illinois Research Board has granted every request for funds for this project, beginning in 1921. This early help enabled the author to have a considerable number of works of travel abstracted. In the past 15 years the board has assisted with travel expense to points in Panama, Central America, Mexico, the southwestern and Gulf Coast states, and Quebec. They provided for the large expense connected with preparing illustrations, especially where photography was used. The University of Illinois Foundation provided a substantial subsidy for the publication of this book.

The Department of Zoology of the University of Illinois has assisted in the development of the book in many ways, especially through the drawings by the departmental artists Katherine H. Paul, Charles A. McLaughlin, and Alice A. Boatright. Dr. S. Charles Kendeigh, with the assistance of Dr. Jean W. Graber, Ruth F. Bruckner and Frances E. Thompson, edited the manuscript and saw the book through the press. The long manuscript was retyped several times by patient and obliging stenographers.

There is no service of greater value to a field student than that which arises when a resident scientist shows a visitor communities with which he is unfamiliar. Such guidance from the following persons is appreciated: F. W. Albertson, Arni Arnison, Ralph D. Bird, Irving Blake, Victor H. Cahalane, E. P. Creaser, Pierre Dansereau, L. R. Dice, R. C. Donaldson, Samuel Eddy, Ira George, J. S. Horton, K. M. King, E. V. Komarek, R. V. Komarek, J. M. Linsdale, J. A. Macnab, R. L. Mayhew, C. E. Mickel, T. T. Munger, R. K. Nabours, Curtis L. Newcombe, Eugene P. Odum, A. S. Pearse, G. A. Pearson, Francis Ramaley, Martha W. Shackleford, A. J. Sharp, Forest Shreve, Doyle Stevens, H. L. Stoddard, W. P. Taylor, B. C. Tharp, Glenn Tunnehill, J. R. Watson, A. O. Weese, A. H. Wright, James Zetek, and other personnel of the Canadian Department of Railways and Canals, United States Fish and Wildlife Service, the Hudson Bay Railway, and the Urania Lumber Company.

A large number of specimens of plants and animals were identified by the staff of the National Museum and by individual specialists. The names of the scientists whose services are not otherwise acknowledged in published papers are as follows: AMPHIBIANS, S. C. Bishop, H. M. Smith; ANNELIDS (Enchytraeidae), L. L. Neave; (Lumbricidae), H. J. Van Cleave, W. J. Harman; APHIDS, L. M. Russell; ARACHNIDS (Arthrogastida), Clarence and Marie Goodnight; BIRDS, S. C. Kendeigh; CERCOPIDS, Louise M. Russell; CHILOPODS, R. V. Chamberlin; CICADAS, Tom Moore; COLEOPTERONS, N. H. Anderson, L. L. Buchanan, H. S. Barbour, W. S. Fisher, R. E. Blackwelder, J. M. Valentine, E. A. Chapin, M. W. Sanderson; CRUSTACEANS, Samuel Eddy, Melville H. Hatch (also Coleoptera), C. R. Shoemaker, J. O. Maloney, H. Hobbs, Jr., Fenner A. Chance; DIPLOPODS, H. J. Loomis, R. J. Chamberlin, R. A. Hefner; DIPTERONS, C. T. Green, M. T. James, Allen Stone, C. H. Curran, C. W. Sabrosky, W. W. Worth, R. H. Foot; FISHES, Hurst Shoemaker; FLOWERING PLANTS, G. N. Jones, B. C. Tharp, Mrs. Agnes Chase, Joseph Ewan; HEMIPTERONS, R. L. Usinger, R. I. Sailer, H. M. Harris; HOMOPTERONS, J. S. Caldwell, R. H. Beamer, D. M. DeLong; HYMENOPTERONS, A. G. Gahan, J. C. Crawford, L. H. Weld, C. F. W. Muesebeck, K. V. Krombein, H. K. Townes, R. A. Cushman, L. L. Buchanan, C. W. Sabrosky, M. R. Smith (Ants); LEPIDOPTERONS, J. R. Watson, J. C. Downey; MAMMALS, H. H. T. Jackson, Donald Hoffmeister; MITES, H. E. Ewing; MOLLUSKS, Henry Van der Schalie Wm. J Clench; NEUROPTERONS, H. K. Townes; ORTHOPTERONS, J. A. G. Rehn, A. B. Gurney; REPTILES, H. M. Smith; SPIDERS, Sarah E. Jones, W. J. Gertsch, C. R. Crosby, W. M. Barrows; TERMITES, A. E. Emerson; THYSANU-

RONS and COLLEMBOLONS, J. W. Folson, H. B. Mills; TIPULIDS, C. P. Alexander.

Scientific names have been omitted from the text as far as possible, but are given in the index after the common names used. Subspecific designations have not been given for mammals or birds and are cited for other forms only where they may have some significance. The nomenclature of the authors concerned has usually been adopted without change when the publication was not more than 20 or 25 years old, but help from specialists was frequently sought for the nomenclature in earlier papers. Authorities generally followed for classification and nomenclature are the following:

PLANTS: Fernald (*Gray's Manual*, 1950); Little (1953), trees, United States; Small (1933), flora, southeastern; McMinn (1939), shrubs, California; Standley (1920-26), trees and shrubs, Mexico; Tidestrum (1925), plants, Utah and Nevada; Rydberg (1954), Rocky Mountains; Frasier and Russell (1954), Saskatchewan; Hitchcock (1935), grasses; Brown, Clair A. (1945), trees and shrubs, Louisiana. MAMMALS: Miller and Kellogg (1955), North America. BIRDS: American Ornithologists Union (1957), checklist; Blake, E. R. (1953), Mexico; Sturgis (1928), Canal Zone. REPTILES and AMPHIBIANS: Schmidt (1950), checklist; Schmidt and Davis (1941), snakes; Smith and Taylor (1945), snakes, Mexico; Smith and Taylor (1948), amphibians, Mexico; Smith and Taylor (1950), reptiles, Mexico. INSECTS IN GENERAL: Smith, J. B. (1909), eastern; Smith, R. C., *et al.* (1943), midwestern; Essig (1934), western. BEETLES: Leng *et al.* (1920-39), catalog; Blatchley (1910), Indiana. ORTHOPTERONS: Blatchley (1920), eastern North America. HETEROPTERONS: Blatchley (1926), eastern North America; Banks (1910a), Nearctic region. HOMOPTERONS: DeLong and Knull (1945), leafhoppers, North America; Metcalf (1923) (Fulgoridae), eastern North America. HYMENOPTERONS: Muesebeck *et al.* (1951), United States. DIPTERONS: Aldrich, J. M. (1905), United States; Curran (1934), North America. SPIDERS: Comstock (1948), North America; Chamberlin and Ivie (1944), Georgia region; Banks (1910b), North America. MOLLUSKS: Pilsbry (1939-48), North America; Baker, F. C. (1939-48), snails, Illinois.

The author wishes to thank the University of Chicago Press for the use of Goode's outline maps in preparing a number of figures. Thanks are due Max M. Paris of the University of Illinois Visual Aids Service for air brush work on unduplicatable photographs, and to the staff of the University of Illinois Photographic Laboratory, especially Ray Harmeson.

VICTOR E. SHELFORD

Contents

Chapter 1

The Scope and Meaning of Ecology

The purpose of this volume is to describe North America from an ecological viewpoint as it appeared in the period A.D. 1500 to 1600 before European settlement. Unfortunately, plant and animal communities were in shambles before scientific study began. Thus the ecology of North America must be largely reconstructed from the observations of travelers and studies made from other viewpoints; hence it will, of necessity, be incomplete. A knowledge of the habitats, biotic communities, and the distribution and abundance of plants and animals in primeval North America should be a useful background for all ecologists to have, in order better to interpret present day conditions. This chapter deals with the concepts and terminology that are essential for an understanding of the descriptive material that makes up most of the book.

Habitat. A habitat is an area of surface with its minerals and climate. Primary habitats are bare areas of volcanic rock and ash, silt and sand of rivers, ground left by receding waters and glaciers, wind-deposited sand and loess, etc. Essential features of the habitat are also sunshine, rainfall, and temperature. Animals and plants quickly invade primary habitats, and as a result of their reactions, the habitats become changed. Animal droppings add organic matter to the substratum. Seeds of plants are blown or carried in by animals. Burrowing forms, such as tube-weaving spiders and tiger beetle larvae, carry fragments of their prey to considerable depths in the soil. The roots of plants also commonly penetrate the substratum to depths measurable in feet, and when they die, add organic matter to the soil. The top parts of plants reduce the extremes of weather at the ground surface. A long series of reactions such as these may change a primary habitat into one of such fertility that it will support a forest (Clements 1905).

Community. Plants and animals have lived together with interdependence

1

since the beginning of life. The term "biota" is commonly used to include all plants and animals that occupy an area. The term "community" has almost the same meaning but involves a dynamic rather than a taxonomic point of view. There is little or no justification for terms such as plant community, animal community, bird community, except for convenience in referring to particular groups; these groups are merely constituents of the biotic communities in which they occur. The term "biotic" is redundant since all communities are necessarily biotic. The dismissal of all the interactions between plants and animals, between plants and plants, or between animals and animals in the analysis of biota as "biological or biotic factors" is difficult to defend. This viewpoint fails to emphasize the many adaptations of life histories, structures, and physiological responses which have developed as a result of the close relation of organisms in communities.

Knowledge of the dynamic character of communities has been acquired slowly and with considerable difficulty through attempts at wildlife management in forests to prevent extensive loss of seedlings of valuable trees eaten or destroyed by animals (Ruff 1938, Trippensee 1948), through predator control to protect game, with the result that the game species has overmultiplied and destroyed its food supply (Rasmussen 1941), and through attempts to manage and increase forage in grasslands for livestock and game, first by poisoning predators and then, when rodents multiplied as a consequence, by poisoning the rodents (Merriam 1901, Vorhies and Taylor 1933). More recently, much has been learned through spraying to control insect pests of trees and tree diseases. This spraying has resulted in the killing of parasites and predators of otherwise unimportant arthropods which, as a result, have then increased in such numbers as to damage the trees (English 1954). Again, knowledge of both terrestrial and aquatic communities has been increased through attempts to control both aquatic and terrestrial organisms in the interest of public health. Ecke and Johnson (1952) found that both the mammal and bird populations were involved in the spread of plague in Colorado and that a relationship existed with the vegetation. Various attempts to maintain sport and commercial fishing in lakes have contributed much information of biotic interrelations (Bennett 1947).

These practical operations have demonstrated that the so-called animal communities do not merely live in or on the so-called plant communities (Tansley 1935). On the other hand, the combined plant and animal assemblages are socially interacting. This follows the pioneer viewpoint of Möbius (1877), who considered the oyster bed such an interacting community, which he called a biocoenose. This concept was also adopted by Gams (1918), a botanist, who introduced the term "biocoenology." Möbius emphasized social relations more than the habits of individual species. Clements (1905) adopted similar views, using quite different terms.

Haeckel (1869), who introduced the term "ecology," dealt mainly with species rather than communities. The title of a Russian publication, "Ecology and Biocenology," depicts the natural history of the individual (ecology) and the community (biocoenology) in their original meanings (Bukovsky 1935). Both Gams (1918) and Clements (Clements and Shelford 1939), however, advocated the continued recognition of both plant and animal communities.

Ecosystem. The term ecosystem is used when referring to habitat and community as an interacting unit (Tansley 1935, Odum 1955). Such chemicals as nitrogen, carbon, and oxygen cycle continuously between the habitat, plants, and animals. There is loss of nitrogen to the atmosphere through the action of the denitrifying bacteria, but this is offset by the capture of free nitrogen by the nitrogen-fixing bacteria. When the addition of nitrogen is greater than the loss, there is a gain in fertility of the habitat. This makes possible an increase in the density of organisms or the change of species so that development or succession of communities results.

Energy flows through the ecosystem; it does not cycle continuously. Green plants, called producers, capture energy from the sun by photosynthesis. Herbivorous and carnivorous animals are consumer organisms, since they must feed on plants or other animals to acquire energy. Bacteria, fungi, and some kinds of animals derive energy in the process of decomposition of dead organisms; hence they are called reducers, decomposers, or transformers. Each group of organisms uses energy to carry on its activities and dissipates heat energy back to the atmosphere. Since energy is continuously being lost, new energy must continuously be secured, hence the importance of solar radiation for the community. The total energy captured is called the "gross productivity" of the community. Any energy that remains as living protoplasm or organic matter in the substratum after all activities are completed is "net productivity." These terms may also be applied to each different group of organisms or trophic levels.

Food chains and pyramid of numbers. The flow of energy through plants, herbivores, and up to three links of carnivores makes a food chain. There are many different food chains in the community, and they anastomose in various ways to form the web-of-life. The total number or biomass of producer organisms must obviously be greater than the number or biomass of consumers, if the producers are to maintain their own existence and provide food for the consumers. Likewise, each link in the chain of consumers must be larger than the links that follow. The position of a species in the food chain thus provides the primary basis that determines its abundance in the community.

Abundance is also correlated with the size of organisms (Fig. 1-1). A pristine biota usually consists of a large number of small organisms associated with a small number of large ones. For example, in a deciduous forest many minute fungi and a few tall maples are associated with an infinite number of small collembolons and a few black bears. If the large number of small organisms is represented by a horizontal line to scale and each smaller number is drawn above to the same scale, a rough cross-section of a pyramid will be formed, and this is called a pyramid of numbers.

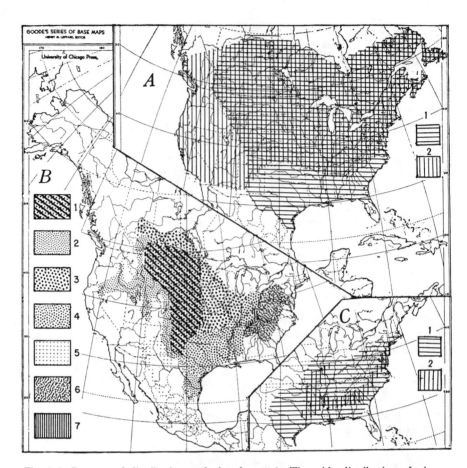

Fig. 1-1. Patterns of distribution and abundance. A. The wide distribution of pioneer species in early seral stages, (1) sandbar willow, (2) chokecherry (after Preston 1948, courtesy Iowa State College Press). B. The distribution of bison, a characteristic species of climax grassland: (1) as the principal animal dominant; (2) as an influent in the far west along with the pronghorn; (3) as an influent in the north and east along with the wapiti; (4) as an influent in an area of mixed grassland and forest; (5) small herds in the predominantly pronghorn country of Mexico and New Mexico; (6) in forests, its influence little known; (7) areas in which bison and wapiti probably maintained a grass and shrub vegetation (Anderson 1946, Roe 1951). C. The distribution of post oak: (1) general range; (2) area of climax dominance.

Mohr's (1940) composite diagram of the populations of mammals (Fig. 1-2) shows some facts of the pyramid of numbers. The 450 pound (240 kg) black bear *Euarctos (Ursus) americanus* has a population of approximately 7 per 36 square miles (7.4/100 km^2), while the jumping mouse *Zapus*, weighing 0.045 pound (20 g), has a population of 700,000 in the same area (740,000/100 km^2). The diagram also shows that the herbivorous wood rat *Neotoma* has a population of about 1280 per square mile (500/km^2), while the predatory weasel *Mustela frenata* has less than 3 (1/km^2).

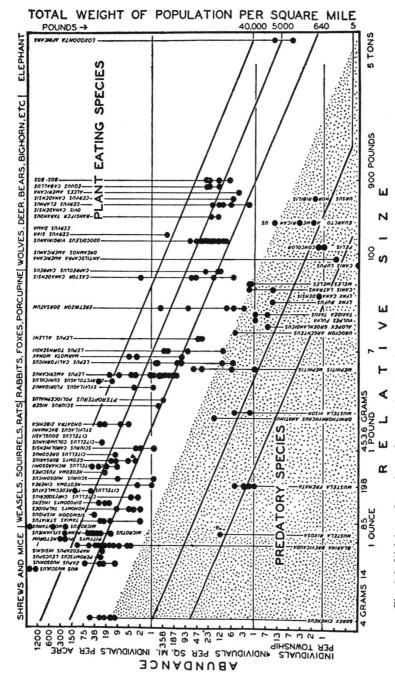

Fig. 1-2. Weight and numbers of mammals per unit area. Nomenclature as given by author (Mohr 1940).

Dominance. Dominant organisms are species which exercise a large measure of control over the composition of the community. The dominants of terrestrial communities are commonly plants and occasionally animals. Plants may control the composition of communities by reducing light intensity, affecting moisture content of the soil, crowding, or modifying the habitat in other ways to which a given species may or may not be tolerant. Animals may exert dominance through feeding or other actions if these are sufficiently intense to suppress or eliminate other species in the community.

Influence. One species of animal commonly affects the abundance and activities of other species without actually determining their presence in the community. The amount of such influence that an individual can exert depends in large part on its size.

Size in a plant is roughly indicated by its life forms, such as herb, shrub, or tree, but comparable categories are not in general use for animals. The important mammals range from one-seventh ounce (4 g) in the shrew *Sorex cinereus* to 800 pounds (363 kg) or more in the California grizzly bear *Ursus horribilis*. The terms "ungulates" and "rodents" convey some idea of size. Shelford and Olson (1935) used the term "major influent" to apply to the large mammals and birds in the coniferous forest. The smallest mammal so designated was the red fox, and the only bird was the great horned owl. Although the basic idea was size, speed and agility were secondary considerations. The largest "minor influent" was the slow-moving porcupine. There were 17 mammals and 12 resident birds in the list of minor influents. All arthropods are relatively small, and individually they exert little effect in the community. However, when combined with large numbers and destructive feeding habits, a species may be classified as an influent or even a dominant.

Community development. When the habitat changes as a result of the actions of organisms so that there is an increase of fertility in the soil, decrease in light intensity, or some other modification in the environment, conditions may become more favorable for some other species than for those already present. There may then be a replacement of one species by another or of one community by another, and this second community may be replaced by a third, and so on. A sere is a succession of easily recognized stages in the development of a climax community. The final stage or climax tends to persist as long as there is no change in climate. Two kinds of actions bring about this succession, reactions between the organisms and the habitat and coactions among the community constituents.

Within a particular climatic region, all seres on different habitats converge into the same general climax community (Figs. 3-9, 7-13). For example, maple–beech forests occur on limestone, granite and shale residuals, glacial till of various types, floodplain deposits of various types, wind-blown sand, and peat.

The substratum, however, often imposes limitations on the vigor of individual organisms, on productivity, and on community composition. There are soils with deficient minerals, notably serpentine soils, on which the community differs in composition and vigor of the constituents from adjacent communities that occur on a minerally balanced soil. A climax occurring on the optimal soil that is characteristic for a region is called the climatic climax, while the community that develops on deficient soils is called an edapho-climatic climax, since soil and climate are of nearly equal importance.

Population densities and distribution. Some species of plants and animals are primarily dependent on soil and water and relatively independent of climate. They generally occur in early seral stages of community development and have wide geographic distribution (Fig. 1-3). The sandbar willow, for example, grows at the water's edge on stream deposits. The chokecherry survives in a variety of primary and secondary habitats and follows fire. Most plants and animals, however, are closely dependent upon climate and have more restricted distributions. This type of distribution is well shown by constituents of late seral or climax communities such as grassland and decidu-ous forest (Fig. 1-3).

Plant and animal "preferences" for certain communities or habitats are generally indicated by higher population densities in those areas than else-where. Careful trapping on the desert floodplain of the Colorado River showed that a pocket mouse occurred in eight different communities but with

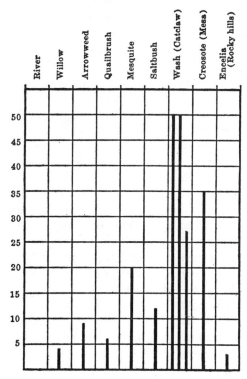

Fig. 1-3. Differences in number of desert pocket mice captured in seral stages on the floodplain of the Colorado River between Needles and Yuma, in the catclaw ecotone climax, in the creosote bush desert climax, and in seral communities on the rocky hills of the Colorado Desert in California (from Grinnell 1914).

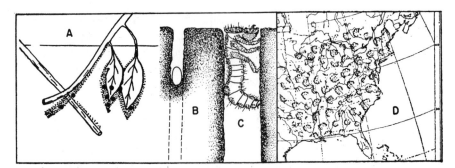

Fig. 1-4. Niches and distribution of the green tiger beetle. A. Position of egg burrows in the shade of leaves on paths of forest mammals. B. An egg in its burrow. C. A larva in its burrow from which it may reach out and snatch prey. D. Distribution of the species which coincides with the distribution of deciduous forest.

definitely greatest numbers in the catclaw ecotone climax (Fig. 1-3). The preferred habitat and community constitutes the typical niche of the species.

The habitat distribution of a species often depends on various delicate relations with their environment. The green tiger beetle of the deciduous forest deposits its eggs in deer trails and apparently deposits them with the ovipositor in the shade (Fig. 1-4). Taken with other forest conditions, this may be a factor in limiting the species to the deciduous forest.

Life span. In every community there are time elements of great importance. These are, first, the life span of the community constituents and, second, the time to reproductive maturity of new individual constituents. We must consider that a community includes everything from short-lived soil bacteria to trees which are the longest-living organisms. The time from one generation to the next generation is commonly stated to be minutes among bacteria, hours among protozoa, and days among aphids. At the other extreme, the interval is years or even decades for many vertebrates and centuries for many, especially climax, trees.

Individuals may live and continue to reproduce for three, four, or more times their age at sexual maturity and may live for a considerable time after reproduction ceases. This makes for large populations and severe intraspecies competition throughout the life of the individual. The populations of dominant and influent animals may, however, change very rapidly from maxima to minima levels in from one and one-half to six years or longer. In the case of the dominant plants, individual trees commonly stay in the same place from 50 to several hundred years. Trees live 20 to 100 times as long as such forest mammals as the deer. On the other hand, the length of time from birth to reproductive maturity is only one and one-half years for the deer but 10 to 15 years for climax trees.

Life cycles and sensitive periods. In order to understand community functions in a thorough manner, it is necessary to know about the constituent organisms. This involves a knowledge of their behavior and physiological responses to environmental factors. The activity or vital process which takes place within narrowest environmental limits is usually the most important ecological feature of a life cycle. This activity may be mating, egg laying, overwintering, or something else, but varies widely between species (Merriam 1890, Shelford 1907, 1911c, Adams 1909). Sensitive periods sometimes last only a few days and may be difficult to detect. However, the identification of these sensitive periods in the life cycles of both plants and animals and the factors that affect them is very important for understanding their distribution, seasonal occurrence, and abundance.

We can often see the effects of the interplay between environment and the physiology of organisms without knowing what factors are involved. This is, for instance, often evident in variations of fecundity. Female chinch bugs deposited an average of 500 eggs during the large population or outbreak period of 1864-65 (Shimer 1867). During the low population period of 1894, the average number recorded by Johnson (Forbes 1894) was 164 eggs. At the time of the large population of 1934-36 in eastern Iowa, the average fecundity was probably above 600 eggs (Janes 1937). The yield of Illinois redtop grass seed varied between 36 and 78 pounds per acre from 1919 to 1934 (Burlison *et al.* 1934).

A few physiological life histories of plants have been published (Pelton 1953). Annual plants develop from seed and mature a new crop of seed in one season. Some seeds may lie dormant and not germinate for several years. Perennial plants, such as grasses or trees, likewise begin with seed germination

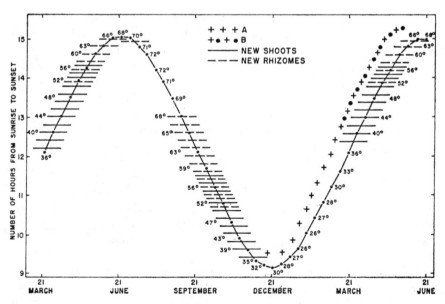

Fig. 1-5. The physiological life history of bluegrass (modified from Evans 1949). A. Initiation of inflorescence. B. Its development. Note rising temperature (°F.) and increasing daylength. The development of shoots and rhizomes occurs principally under declining daylength and falling temperatures.

but require several years to reach sexual maturity.   The mature plant then manifests annual cycles of physiological states and produces seed crops each year (Fig. 1-5).

The life histories of animals follow similar patterns.   For instance, the tiger beetle *Cicindela hirticollis* near Lake Michigan appears from hibernation in June.   Sexual maturity is attained in a few days and eggs are laid in July.   The larvae feed on other insects and molt once, or more often twice, during the summer.   When cold weather approaches, the larvae close their burrows and remain below ground during the winter.   They begin feeding again the following April, pass to the pupa stage, and emerge as adults in mid-July.   In August, after feeding for two or three weeks, the adults burrow into the sand and hibernate until the following June when they begin the cycle over again.   Development depends on the occurrence of proper temperatures and abundant food, the site of egg laying is determined by soil texture and moisture, sexual maturity is attained only after two winter dormant periods, and so forth (Shelford 1907).   Only by such detailed studies as these can the role of environmental factors be properly evaluated.

Solar radiation.   As discussed above, solar radiation is the source of all the energy required for carrying on the activities of plants and animals, and the community is organized to capture and transmit this energy to all constituent species.   Variations in the intensity of light available and the particular requirements of the organisms involved influence the position of the species in the seral development of climax communities.

The daily duration of light and the rate that it changes have been demonstrated to affect the development of gonads and season of reproduction.   Birds nest in the spring, apparently in response not so much to rising temperatures as to increasing daily photoperiods.   This was first shown in the slate-colored junco (Rowan 1925, 1938) and later in many other species (Burger 1949). Bird migration northward in the northern hemisphere is in part stimulated by increasing length of day.   After nesting, birds return south as the day's length decreases.   Tundra birds sometimes leave while the food supply for young and old birds is still plentiful.

Other animals and many plants have also been shown to respond to changing daily photoperiods.   For instance, sideoats and slender grama grasses in southern Arizona are typical short-day species. They fail or are delayed in flowering when placed on a 16-hour photoperiod.   They flower more profusely on a 12-hour photoperiod than on an 8-hour period.   Their normal time for flowering in Arizona comes when the days are 12 hours long (Olmsted 1943, 1945).

The orange and red portion of the spectrum gave most rapid growth of gonads in birds.   Ultraviolet and infrared gave slowest growth in certain experiments, but this was probably due to excess intensities being used.   Marshall and Bowden (1934) found ultraviolet most effective for shortening the anestrous period of female ferrets.   According to Bailey (1950), ultraviolet increases egg production in poultry.   Sabrosky, Larson, and Nabours (1933)

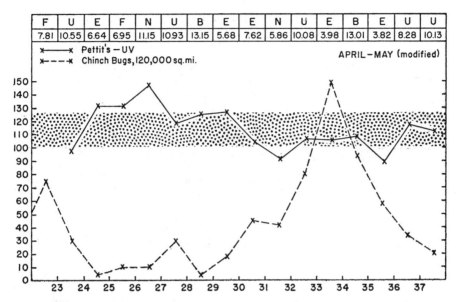

F	U	E	F	N	U	B	E	E	N	U	E	B	E	U	U
7.81	10.55	6.64	6.95	11.15	10.93	13.15	5.68	7.62	5.86	10.08	3.98	13.01	3.82	8.28	10.13

Fig. 1-6. Relative numbers of chinch bugs in Illinois and neighboring states. The impact of May and June rainfall on population is indicated: E, excellent; F, favorable; N, neutral; U, unfavorable; B, bad; rainfall is given below the letters in inches; less than 8 inches is usually favorable. Ultraviolet (UV) intensity, Mount Wilson, California, (Pettit, International Astronomical Union 1924-38) is shown for the sensitive period of April and May. The stippled band is an assumed optimum range for large populations (modified from Shelford 1951a).

caused a grouse locust to reproduce by the application of weak ultraviolet light over a long period. The young were more numerous and more vigorous than those produced under prolonged white light.

Solar ultraviolet varies more from year to year and decade to decade than other portions of the spectrum. In chinch bug populations an optimum intensity between 100 and 128 units is indicated (Fig. 1-6). Studies of this species suggest that increased fecundity is induced by favorable radiation conditions during the development of the reproductive cells in April or May. Ultraviolet appears to be more important in the control of some populations than either temperature or rainfall.

Much work has been done on the relations of plants to various portions of the spectrum, but it has not been possible to find results comparable to these for animals. We know little about sensitive periods in the life histories of plants.

Temperature and moisture. Temperature and moisture have not ordinarily been evaluated with reference to sensitive periods in life histories of plants and animals or to the aggregate of community activity. The early spring development of plant and animals in the temperate region responds similarly to temperature (Fig. 1-7). In a University of Illinois forest, low rainfall

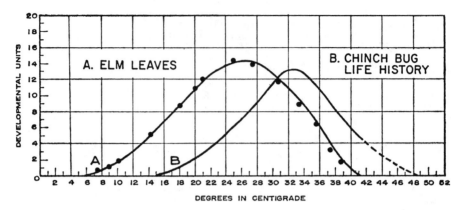

Fig. 1-7. Rate of development of elm leaves in developmental units per hour (after Jones 1938) and chinch bugs from eggs to adults (after Shelford 1932). The total developmental units required for the complete development of elm leaves is 2000 and of chinch bugs is 6800.

from November 1, 1933, through October 31, 1934, was associated with a thin tree leaf canopy, sparse herbs which died down by early June, and unusually small populations of invertebrates (Rice 1946). A native cutworm or larva of the moth *Agrotis orthogonia,* occurring in grassland, increased in years following less than ten "wet" days in May and June (Seamans 1926). May-June rainfall of more than six inches caused sharp reduction of populations of chinch bugs (Shelford 1951a). Temperature and moisture usually operate as paired factors.

Master factors and paired factors. Larsen (1943a) has pointed out that at different times a different factor assumes a dominating role in the activity of noctuid moths. The master factor may be either internal or external to the organism. The urge to secure food, for instance, may dominate all other activities at particular times. On the other hand, either optimum ultraviolet or rainfall may control the size of the population of the next generation (Seamans 1926).

When two factors exert effects simultaneously on an organism, each factor may modify the effect produced by the other over a large series of combinations of the two factors (Fig. 1-8). For instance, temperatures from below 90° to 100° F. (32° to 38° C.) bring the largest transformation of nymphs into adults in the African locust, but only if relative humidity is between 60 and 70 per cent. Combinations of higher and lower values of these two factors bring lower rates of transformation.

It is apparent that the manner in which environmental factors affect individual species and whole communities is complex and little understood. In this book we will give information concerning the climate, soil, topography, and other environmental factors that appear correlated with each community, but specific detail as to how these factors are effective cannot be thoroughly explored.

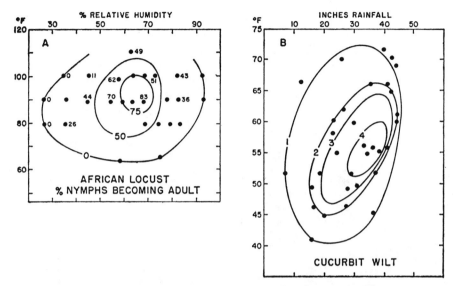

Fig. 1-8. A. The response of an animal to combinations of temperature and relative humidity or rainfall. B. The response of a fungus to temperature and relative humidity or rainfall combinations. The lines pass through equal survival rates for the locust nymphs and equal growth conditions for the cucurbit wilt. Survival rates above 75 per cent for locust nymphs occurred only at limited temperatures and humidities (Hamilton 1936). Maximum growth and destructiveness of the cucurbit wilt (inside line 4) occurred only in a narrow range of temperature and rainfall (after Tehon 1928).

Natural communities of North America. Figure 1-9 is a diagrammatic map of the large and medium sized communities north of 21° N. Lat. To the south are tropical rain and tropical deciduous forest communities. These major communities or biomes are seldom uniform in character throughout and may be subdivided into plant associations and faciations, as indicated for the temperate deciduous forest biome in Figure 2-9.

There are intensive biotic studies on typical areas over one or more annual cycles in parts of the deciduous forest, tundra, temperate grassland, chaparral, western hemlock forest, montane forest, woodland ecotone, and cold desert. The smaller invertebrates and plants are usually omitted from such studies. There are few or no biotic studies of major communities in the tropics, although there are numerous separate studies of fauna, flora, and vegetation, particularly in the tropical rain forest of the Canal Zone, the Yucatan Peninsula, northern Honduras, and near Palenque, Chiapas, Mexico, and in subtropical Florida. The scattered and fragmentary information that has been found has been brought together in the hope of stimulating further investigation of the pristine character of these biotic communities.

Nearly all studies of biotic communities at the present time are handicapped by the disappearance of large animals. Even the pioneer studies made at the University of Chicago around the beginning of the present century came after the wapiti, moose, deer, bear, bison, wolf, and coyote were gone. The early studies at the University of Nebraska came after the bison, pronghorn, wolf, kit fox, badger, and the river margin wapiti, deer, and bear had been greatly reduced in numbers.

In the original communities on land, animal groups in the probable order of their ecological importance were rated with mammals first, insects second, birds third, and the lower vertebrates and invertebrates other than insects,

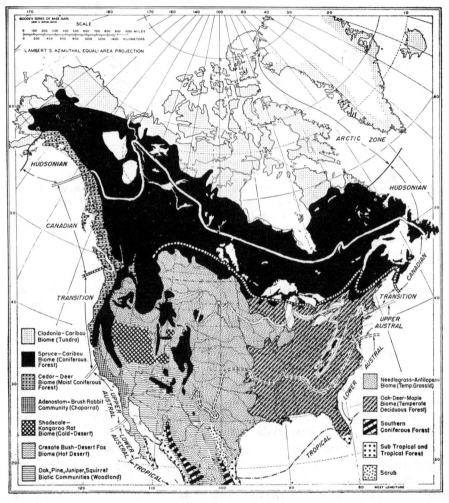

Fig. 1-9. North America, north of 21° N. Lat., showing the transcontinental life zones drawn on a diagrammatic map of biotic communities. The zones are not marked in the western mountains but tend to turn southward. The Arctic Zone is the same as the tundra community; the Hudsonian Life Zone is the northern portion of the coniferous forest community, consisting largely of muskeg and savanna; and the Canadian Zone is the southern part of the coniferous forest. All the other zones cut across natural communities as indicated.

fourth. Herbivorous insects and mammals have produced striking effects on vegetation, sometimes continually over long periods, at other times only cyclically. The importance of small birds is difficult to appraise. Thus the enormous toll of arthropods taken to support the parents and feed the young is to a considerable degree nullified when fast-breeding insect predators and parasites are also eaten, which if left alone would have been able to control the herbivorous prey species. In plants, the outstanding community constituents are usually spermatophytes. Lichens are characteristic of the tundra.

Community terminology. It may be useful to bring together here the principal terms used repeatedly in this treatise and to define them as they are interpreted by the author.

Habitat is originally a newly exposed bare area with its climate. Habitats become modified when occupied by plants and animals but still refer to the combination of physical factors that affect organisms.

A *biota* is the complete flora and fauna of an area from the taxonomic point of view.

Community is a general term applicable to aggregations of plants and animals of all sizes that have dynamic interrelations and dependencies. Communities are usually named for their most abundant and conspicuous constituents.

Dominants are community constituents which exercise a large measure of control over the composition of the community. Control by plants is brought about by effects on the habitat, such as shade production and lessened impact of weather conditions. Animal dominants act directly on the other organisms, especially plants, eliminating some and suppressing others.

An *influent* is a community constituent which modifies the abundance or well-being of other constituents but does not control community composition. *Major influents* are large animals; *minor influents* are small ones.

A *characteristic* species is one which is nearly always found in a community, regardless of its abundance or influence.

Permeants are animals that invade the climax and all or many of the seral stages and faciations in their day-to-day or week-to-week activities.

Coactions are interactions between individuals and species, usually involving physical contact of parts or of whole organisms.

Reactions are the effects of organisms on the habitat.

Community development is the process of change leading to a climax. It is essentially the same as "succession."

A *sere* is the series of stages passed through in the development of the climax community.

A *climax* community is the last stage in a sere and the most stable community of which we have knowledge. It is also applied as an adjective to plants and animals which occur in climax communities.

The *biome* is the largest community, for instance tundra, desert, deciduous forest. It is recognized by the character of the climax but includes seral stages as well. Biomes are plant formations with the animal constituents integrated. This is justified (a) when some of the animals are essentially dominants as shown by their influencing the composition of the vegetation, (b) when they show behavior or social characteristics suited to the environment, including contacts with other constituents of the community, (c) when they are physiologically suited to the extremes of the weather and other characteristics of the habitat, and (d) when some important animal species are distinctive and present throughout the community. This last condition is the most practical criterion.

An *association* is the largest distinct subdivision of a biome, identified by the presence of characteristic climax dominant or index species. It is used herein wherever it can be clearly delimited in the biotic sense, otherwise the term "region" or "faciation" is employed. This differs from Kendeigh (1954), who limits the concept of the association to plants and has introduced the term "biociation" for distinctive aggregations of animals and plants.

An *associes* is a developmental community equivalent to the association. Associes are usually small in size.

A *faciation* is a general term for a community which varies in composition and appearance from other communities adjacent to it but which is part of the same biome. Faciations possess considerable uniformity as to dominants. *Facies* is applied to developmental communities in the same manner as faciation is used with the climax.

An *ecotone* is a community which is transitional between two biomes or other large units. It is not applied here to seral transitional communities.

Aspection refers to seasonal changes in the appearance, composition, and activities of organisms in a community. The yearly cycle is commonly divided into prevernal, vernal, aestival, serotinal, autumnal, and hiemal *aspects*.

Chapter **2**

The Temperate Deciduous Forest Biome
(Northern and Upland Regions)

The temperate deciduous forest, or the *oak–deer–maple biome,* occupies North America from the center of the Great Lakes region south to the Gulf of Mexico. It covers the northern two-thirds of the Florida Peninsula and extends west beyond the Mississippi River to the Ozark Mountains (Fig. 2-1). The chief characteristic of the temperate deciduous forest is the predominance of trees with broad leaves which are shed each autumn. An understory of small trees and shrubs is usually also deciduous. The shedding of the leaves brings a striking change in light conditions and shelter for animals. The forest floor is covered with a dense layer of leaves in various stages of decay. The extreme southern part of the forest also contains evergreen species.

The white oak, white-tailed deer, and turkey are important throughout practically all of the biome. Likewise, the ranges of various less important species are nearly coextensive with the biome.

The annual rainfall ranges from 28 to 60 inches (70 to 150 cm) and is well distributed throughout the year. Rainfall is lowest in some of the northern areas and greatest near the Gulf of Mexico. The western boundary, except for the river-skirting forests, coincides approximately with a line marking rainfall equal to 80 per cent of evaporation (Transeau 1905). The northern boundary of the biome has a mean January temperature of 14° F. (−10° C.).

In 1600 the forest was intact except for some tree removal in the northeastern states and in Virginia where populations of Indians were large. Fire was sometimes used to remove trees, and fires also resulted from lightning. However, fires were generally unimportant, since the shady forest held considerable moisture and the many streams prevented their spread.

Since settlement by Europeans, the forest has been almost completely de-

17

stroyed by agriculture or modified by logging. Areas from which no trees have been removed are probably not over 0.1 per cent of the entire forest. These areas are too small to be frequented by all the native animals that were formerly present. The wolf, mountain lion, and wapiti have been essentially extirpated from the entire biome. In areas that have become reforested because cultivation proved unprofitable, white-tailed deer, black bear, and turkey have returned in numbers.

DECIDUOUS FOREST REGIONS

Three large subdivisions of the deciduous forest may be recognized: (a) northern and upland forests, (b) the southern and lowland forests, and (c) stream-skirting forests. Braun (1950) has discussed and mapped plant associations in the deciduous forest, but in the Indiana–Michigan area, the studies of Cain (1935), Potzger (1951), and others have emphasized the mixed character of the floral composition and the lack of sharp boundaries. We will, therefore, describe the subdivisions in terms of regions rather than associations.

The Northern and Upland Regions

The northern and upland regions usually have beech or sugar maple in the climax stands and wapiti and deer as permeant dominant animals. There are five regions that may be recognized (Fig. 2-1) :

(a)   Tulip–oak region. The "mixed mesophytic" and "western mesophytic" forests of Braun are included in this region. The term "mesic" is preferred over "mesophytic" when animals are included. The dense, mixed mesic forest contains a fair abundance of the two indicator species, white basswood and yellow buckeye, in a total group of 15 to 20 dominant species. The western mesic forest is less dense, has fewer dominants, and usually lacks the two indicator species. The greater part of the tulip–oak forest lies between 500 and 1000 feet (150 and 300 m) altitude but in places ranges above 3000 feet (900 m). It becomes scattered in the valleys of the Blue Ridge Mountains.

(b)   Oak–chestnut region. The oak–chestnut region occurs from Cape Ann, Massachusetts, and the Mohawk River Valley of New York to the southern end of the Appalachian highland. It begins in central West Virginia below 2000 feet (600 m), reaches to 3500 feet (1060 m) west of the Cheat Range, and becomes the predominant forest farther east. It covers mountains higher than those shown in the transect. It also occurs in many valleys west and north of the region mapped, in areas where other forest types predominate.

(c)   Maple–basswood–birch region. The maple–basswood–birch region consists of scattered areas on the mountain sides in the Appalachians, ranging from 2500 to 4200 feet (760 to 1275 m). It occurs in two or three faciations.

(d)   Maple–beech–hemlock region. Pure stands of maple–beech are limited to southern Michigan, northern Ohio, Indiana, and an area on the north side of Lake Erie and the south side of Lake Ontario. Mixed with hemlock and/or white pine, it predominates in the northern edge of the forest and southward into the Appalachians.

(e) Maple–basswood region. The maple–basswood region occupies northern Illinois, southern Wisconsin, and eastern Minnesota.

The Southern and Lowland Regions

The regions of the southern and lowland regions are (Fig. 2-1) :

(a) Oak–hickory region. As interpreted by Oosting (1942), the oak–hickory forest occupies a strip along the eastern edge of oak–chestnut forest

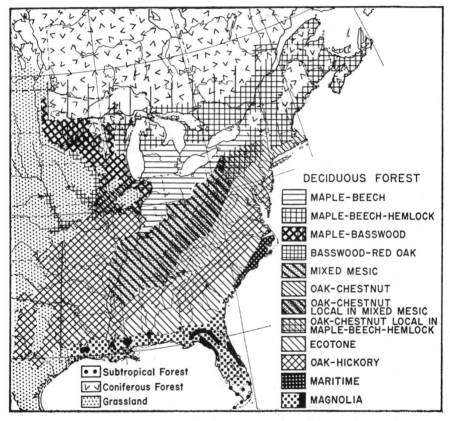

Fig. 2-1. The deciduous forest regions. The vertical lines through the maple–beech and maple–basswood areas indicate the presence of hemlock. The single lines extending northward into Canada indicate the presence of scattered deciduous forest trees. The oak–hickory region represents two faciations (Braun 1950). The long vertical line in eastern Texas and Oklahoma is the approximate western limit of *good* deciduous forest. The ecotone area is the pinelands which under primeval conditions showed transitions to oak–hickory, maritime, and magnolia forests. In the magnolia forest, black areas are climax and late subclimax.

from Sandy Hook, New Jersey, to Alabama, then extends westward across the Mississippi into Arkansas and Texas and northward to central Illinois.

(b)   The magnolia–maritime region.   Beginning at the southeastern corner of Virginia, the maritime forest extends southward along the coast and meets the magnolia forest near Charleston, South Carolina.   The magnolia forest continues along the coast to near Houston and covers the northern two-thirds of Florida.

Stream-skirting Forest Region

There are small strips of characteristic climax deciduous forest that extend westward along streams into the Great Plains.   This forest will be discussed in Chapters 11 and 12.

GENERAL FEATURES OF THE BIOME

The climate in various parts of the biome is shown by climographs (Fig. 2-2).   All regions have quite similar limits of temperature and rainfall, except the magnolia forest.   The arrangement of communities in the mountains is shown in Figure 2-3.

Most of the climax deciduous trees tend to be widely distributed through the deciduous forest.   There is an admixture of coniferous trees in the climax in the uplands toward the northern boundary of the biome and in seral sandy areas in both the north and the south.   Within the biome are also fresh-water climaxes (Clements and Shelford 1939), large lake climaxes (V. E. Shelford 1913), and possibly large river climaxes (Shelford 1954a).

Most of the large and medium sized animals are permeant through both climax and seral stages; only the gray squirrel shows a preference for the climax.   The dominant and influent animals of the biome, such as the deer, wolf, turkey, mountain lion, gray squirrel, bobcat, and others, were present throughout the biome under pristine conditions but often differentiated regionally into subspecies.

Plants

Tree species usually recognized as climax dominants on the basis of their range, frequency in the areas in which they occur, and their abundance follow.

A.   Covering 85 per cent or more of the entire climax: white oak, black oak, shagbark hickory, and bitternut hickory.

B.   Covering less than 75 per cent, chiefly in the northern one-half of the forest: northern red oak and American chestnut.

C.   Covering less than 75 per cent, chiefly in the southern one-half of the forest: southern red oak and laurel oak.

D.   Chiefly in the northern two-thirds of the forest: post oak and sugar maple.

E.   Chiefly in the southern two-thirds of the forest: mockernut hickory, redbay, and American holly.

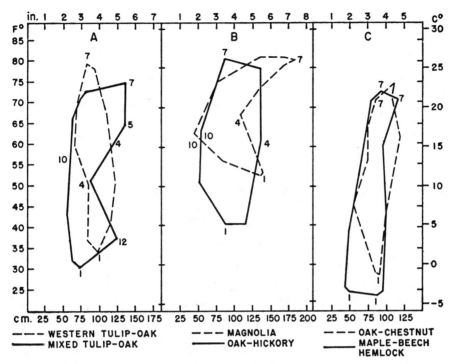

Fig. 2-2. Limit climographs of two U.S. weather stations near the center of each of six forest subregions. The mean monthly temperatures and rainfall for the area fall within the limits shown in each case. The numerals are near the month which they indicate. The weather stations used for mixed tulip–oak are Pittsburgh, Pennsylvania, and Hazard, Kentucky. The heavy summer rainfall (4, 5, 7) is in Kentucky; the low autumn rainfall (10), Pittsburgh, Pennsylvania. The stations used for the western tulip–oak are Louisville, Kentucky, and Jackson, Tennessee. The two climographs are quite similar. The stations for magnolia are Baton Rouge, Louisiana, and Gainesville, Florida; for the oak–hickory, Raleigh, North Carolina, and Little Rock, Arkansas. The limits for this pair of stations are quite different; particularly important are heavy summer rains in Florida (7). The stations used for oak–chestnut are Asheville, North Carolina, and Reading, Pennsylvania, and for maple–beech–hemlock, Elmira, New York, and Fort Wayne, Indiana. At Elmira, the rain is heaviest in the autumn (7 to 1, right side) as the temperature is falling. At Fort Wayne, rainfall is lightest during the autumn (7 to 1, left side).

    F.   Others: chestnut oak, black walnut, and American beech.

There is another large group of trees which fall into seral communities and are much more widely distributed. These include the willows, cottonwoods, common chokecherry, red maple, American elm, slippery elm, hackberry (north), sugarberry, bur oak (north), and swamp chestnut oak (south).

There are about 50 deciduous shrubs and understory trees that are important in the forest along with about 15 evergreen shrubs and a dozen vines. There are about twice as many herbs as shrubs and vines. These subordinate species support a rich fauna of insects and spiders. The understory trees in-

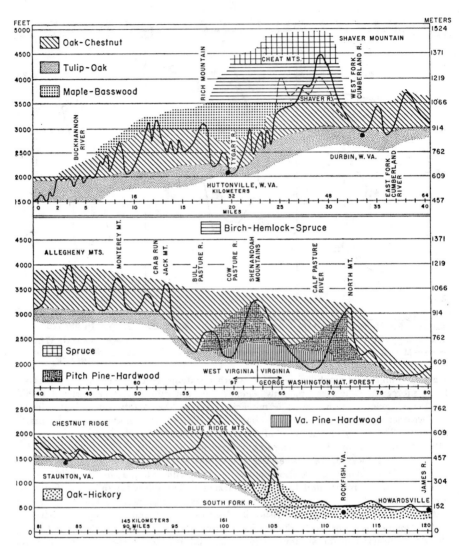

Fig. 2-3. A diagrammatic northwest–southeast cross-section of the Appalachian Mountains from a point nine miles south of Buckhannon, West Virginia, to Howardsville, Virginia (120 miles, 192 km). The vertical scale is exaggerated 26 times. A transect strip, four miles wide, drawn between the two localities on U.S. topographic maps is the basis for the line indicating the topography. The legend is spread across the mountains just as a layered fog would be, to indicate that the legend applies to all surfaces in the transect strip which are at the same altitude and with the same exposure. Collections at six study stations between the top of Shaver Mountain and Howardsville supplied the data on lower vertebrates and invertebrates noted in the text. Spruce and maple–basswood approach zero at the Cheat Range rainfall barrier; maple–basswood continues locally in valleys and on northeast slopes above 3500 feet (1066 m) for a few miles east of Durbin. The tulip–oak or mixed mesic continues in deep valleys well into the Blue Ridge Mountains up to 3500 feet. This figure was prepared with the assistance of Maurice Brooks.

clude American hornbeam, hophornbeam, sassafras, eastern redbud, flowering dogwood, and striped maple, all of which are widely distributed. Important shrubs are pawpaw, spicebush, arrow-wood, black huckleberry, blueberry, witch-hazel, and Virginia creeper.

Animals

Important animals are deer, wapiti (north), wolves (*Canis lupus* in the north, *Canis niger* in the south), mountain lion, black bear, bobcat, gray fox, raccoon, fox squirrel, eastern chipmunk, white-footed mouse, pine vole, short-tailed shrew, and others.

Most of the small birds do not stay in the forest throughout the year. The tufted titmouse is one of the few species ranging through nearly all the sub-divisions of the biome and present throughout the year. The blue jay extends its range northward into the boreal coniferous forest. This may have come with man's modification of the northern community. Other important year around resident species are the red-bellied, hairy, and downy wood-peckers, the white-breasted nuthatch, two species of owls and two of hawks. The eastern wood pewee, wood thrush, and Acadian flycatcher carry on their warm season activities within the deciduous forest but winter in Central America. Other common summer resident species in the forest are yellow-billed cuckoo, great crested flycatcher, red-eyed vireo, yellow-throated vireo, cerulean warbler, ovenbird, and scarlet tanager.

The box turtle, common garter snake, timber rattlesnake, and northern black racer are all characteristic reptiles of the forest. Lizards do not occur in the climax of northern areas but may be found in sandy habitats. The amphibians are represented by the slimy salamander through most of the deciduous forest.

GENERAL FEATURES OF THE NORTHERN AND UPLAND REGIONS

The northern and upland regions include the center of dispersal of the deciduous forest biota (Adams 1905) and the location of undifferentiated deciduous forest (Braun 1941). In its general appearance, the forest as a whole does not differ greatly from that of the maple–beech (Fig. 2-10).

Small areas of forest preserved by individuals or universities or incorporated into parks and sanctuaries provide the best locations for ecological research into natural conditions. Reservations are sometimes in contact or surrounded by extensive second-growth areas which help to protect them from disturbance at the present time. However, nearly all areas have been stripped of one or more species of trees that were widely used by the early settlers for equipping their households or farms.

The native Indians utilized and modified the forest in the northeastern United States to a greater extent than is commonly recognized. Investigators have found hundreds of village sites in New England, New York, and New Jersey. Land cleared for some sites covered 150 acres or more. Wood was used for weapons, utensils, canoes, houses, and especially firewood. When firewood was exhausted, the village was moved. Maize cultivation was extensive. Sullivan in 1779 is estimated to have destroyed 160,000 bushels of corn that had been raised and stored away by Indians (Day 1953).

The natives also utilized deer for food and their hides for clothing, shelter, and thongs. The pelts of many small animals were used for fur clothing. The Indians were conservationists, however, and frequently became angered by excessive killing of game by white men.

Climate and Physical Conditions

The mean annual rainfall ranges from 28 to 40 inches (70 to 100 cm) in the lowland north and west of the Ohio River and in the St. Lawrence and upper Hudson river valleys. The April to September rainfall ranges 18 to 24 inches (45 to 60 cm) in the northern lowlands. Figure 2-2 shows the mean monthly rainfall and temperature for six selected stations.

The reaction of the forest on physical conditions is striking. When the air movement is 885 feet (270 m) per minute outside the woods, it may be only 360 feet (110 m) per minute among the shrubs along the forest edge. When the sunlight between 12:00 noon and 2:00 P.M. is 5500 footcandles outside the woods, it may be about 500 footcandles under the canopy, 66 feet (20 m) inside the woods. At the same time, 660 feet (200 m) in from the forest edge, the illumination may be only 130 footcandles above the shrubs, 20 below the tops of the shrubs, and 2 under the herbs. In midsummer the temperature decreases downward in the soil and increases upward above the litter. The mean weekly temperature between 12:00 noon and 2:00 P.M. during September was 1.3° C. higher at a height of 33 feet (10 m) above the ground than at a height of 2 feet (0.6 m); the mean weekly humidity was 5.7 per cent lower at 33 feet than at 2 feet. In winter the temperature relations of the soil are reversed, increasing downward, and the gradient above the ground fluctuates (Weese 1924).

Layer Communities Within the Forest

The fragments of remaining forest available for study commonly have trees 75 to 100 feet (23 to 30 m) in height and 23 to 30 inches (58 to 75 cm) in diameter, with the bottoms of the crowns at 32 to 40 feet (10 to 12 m) above the ground. The canopies shade about 90 per cent of the ground under the forest. A number of distinct layer communities can be recognized.

The *canopy layer* includes the bark of the limbs and upper trunk as well as the leaves which are heavily inhabited by invertebrates. Smith-Davidson (1930) found the spiders *Eustala anastera, Bathyphantes alboventris, Araneus displicatus,* and a number of other species. She estimated the population of invertebrates of the canopy layer at 14,700 per ten square meters and of all lower layers, including the soil to a depth of seven centimeters, at 18,810.

Tree cavities, produced by decay where limbs have broken off, are often in the lower part of the canopy. The downy woodpecker, red-headed woodpecker, and tufted titmouse utilize cavities in trees and defend territories of 11 to 14 acres (4.4 to 5.5 hectares). The white-breasted nuthatch, another cavity-nesting species, has a very large territory of 55 acres (22 hectares). The barred owl also nests in tree cavities, as do such mammals as the raccoon, gray squirrel, and flying squirrel.

The *understory tree layer* is made up of young and suppressed individuals of large species and normally small species, mainly 30 to 34 feet (9 to 10 m) in height. Lower tree trunks of the sugar maple support small brown moths, spiders, snails, and centipedes at the rate of 75 per square meter of bark at heights of 0 to 3 meters. At heights of 3 to 9 meters, still larger numbers sometimes occur (Smith-Davidson 1930).

The *high shrub layer* (2 to 6 m) possesses the pawpaw as its most important shrub or treelet. A few nests of eastern wood pewee and red-eyed vireo occur at this level. Their defended nesting and feeding territories average about 7.5 acres (3 hectares). The density of their nests was one per 15 acres (6 hectares) (Twomey 1945).

The *low shrub layer* (1 to 2 m) includes spicebush as its principal shrub, but many tree seedlings occur. The mid-July invertebrate population is about one-third of that of the herbs but is composed of the same species. Tree cricket nymphs, the lace bug nymphs *Gargaphia tiliae,* with populations of 24 per square meter, and a second lace bug *Corythucha aesculi* occur. There is approximately one wood thrush nest at this level per 17.5 acres (7 hectares) with their territories averaging 6.75 acres (2.7 hectares).

The *upper herb layer* (15 to 100 cm) is made up of common nettle ($0.4/m^2$), waterleaf (6.5), sanicle (1.2), and jewelweed (0.3). At this season the nettles are badly punctured by wedge beetles (20) which also occur on the other plants. The small mirid bug *Dicyphus gracilentus* and its nymphs are abundant (19).

The *lower herb layer* (0 to 15 cm) is characterized by the stemless wild ginger ($1.2/m^2$) and the violet *Viola pensylvanica* ($0.1/m^2$). They support some of the same insect species found in the higher layers. The millipedes *Pseudopolydesmus serratus* (10 to $100/m^2$) (Fig. 2-4) and *Fontaria* ($10/m^2$) move and mix the soil. The predatory snail *Haplotrema concavum* ($1/m^2$), and fly larvae ($10/m^2$) are present. Fallen and standing dead tree trunks support shelf fungi and their beetles and the universally present flat underbark beetle larvae (Fig. 2-4). Collembolons are very numerous. Layering also occurs underground with the root systems of plants. According to McDougall (1922) many species of plants are able to live together largely because the main parts of their absorbing systems are placed at different levels in the soil and because different species carry on their more important activities at different times of the year. As an example of the first, *Circaea latifolia* has its rhizomes only about 2.5 centimeters beneath the surface of the soil; the rhizomes of *Asarum canadense, Sanguinaria canadensis,* and

Thalictrum dioicum are about 5 centimeters deep.   Trees and shrubs have their roots penetrating to depths of several meters.

The presence of mycorrhiza fungus around the roots of trees and shrubs is often necessary for their proper functioning.   McDougall (1922) and McDougall and Liebtag (1928) examined 145 species of mycorrhiza and found either ectotrophic or endotrophic mycorrhizas on 64 per cent of the possible host species.   There are some trees, such as hophornbeam, that rarely grow when transplanted unless the proper mycorrhiza is present.

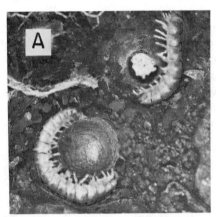

Fig. 2-4. Forest invertebrates. A. The millipede *Pseudopolydesmus serratus,* a ground animal, engaged in making its egg nests. B. The beetle larva *Dendroides canadensis,* which lives under the bark of dead trees (twice natural size).

Plants and Animals in 10 Square Miles (26 km²) of Forest

The picture of an area of deciduous forest can be only crudely drawn. A hypothetical circle 3.57 miles (5.7 km) in diameter enclosing an area of 10 square miles (26 km²) may be taken as a basis for calculating the density of plants and population size of animals.   This is approximately the area occupied by a city of 80,000 to 100,000 inhabitants.   An area the size of an ordinary building lot would have about 16 large trees, a few small ones, 75 shrubs, 76,000 herbs, 192,000 insects and other arthropods, and 8 mice.   The area would be too small for the territory of a nesting bird or the home range of a gray squirrel.   The areas from which this information is drawn are the oak–chestnut region in Snyder County, Pennsylvania; the maple–beech region in Cuyahoga County, Ohio; the maple–basswood region in Champaign County, Illinois (Trelease and Brownfield Woods of the University of Illinois), and the Pisgah National Game Preserve in North Carolina.

Ten square miles has approximately 750,000 trees, 3 inches (7.5 cm) or more in diameter, breast high, and reaching into the canopy (Vestal and Heermans 1945).   About a dozen species of trees are important, and about a dozen more are of lesser significance.   Oaks, maples, basswood, hickories, walnut, and formerly chestnut are numerous.

Below the 750,000 trees are about 786,000 tree seedlings or a little more than 1 per tree. There are about 3.7 shrubs per tree, making 2,810,000 shrubs per 10 square miles. The seedlings are sugar maple, basswood, buckeye, and ash; the shrubs are spicebush, pawpaw, strawberry-bush, and Virginia creeper (Shelford 1951b).

Among and below the shrubs and seedlings are 230 to 460 million herbaceous plants such as bloodroot, wild ginger, squirrel-corn, violets, false Solomon's seal, and jewelweed, with nettle in moist places and in wet years. This is about 300 to 600 plants per tree. These plants usually have their leaves nipped, skeletonized, perforated, and deformed and their sap sucked by a myriad of insects belonging to 10 to 20 species. Meanwhile a rather large percentage of both herbs and shrubs are trampled and broken by the activities of the larger vertebrates.

During the 60 days of early summer when birds are breeding, there are about 26,880 million invertebrates, mainly insects and spiders on the 10 square miles. These are about equally divided between the tree tops, shrubs, herbs, and the soil and litter (Rice 1946).

The tree tops have an estimated 3720 million invertebrates (Smith-Davidson 1930). Gall wasps (Cynipidae), gall flies (Cecidomyiidae), and plant lice (Aphididae) are responsible for the production of a multiplicity of gall deformities of leaf and petiole. Acorns become infested with larvae. Twigs are clipped and leaves removed by squirrels and birds and bored into by beetle larvae.

Some 8960 million arthropods large enough to be counted with the naked eye occur in the litter and soil (Rice 1946). This group includes snails, millipedes, centipedes, earthworms, and larvae. These animals and others too small to count consume roots of plants, grind and bury plant debris, and reduce leaves and litter to soil.

In this period of maximum foliage and insect population, there are about 7680 pairs of small nesting birds per 10 square miles, or about 1 pair for every 90 to 95 trees, although they do not all nest in trees. The adults and young normally consume about 386 million insects during the 60-day period. This is equivalent to more than 500 insects per nesting pair of birds per day (Rice 1946). This inroad does not greatly reduce the total insect population because there is continued replacement of species and development of immature stages.

The important predatory birds are the great horned owl, barred owl, redtailed hawk, red-shouldered hawk, and others averaging 2 to 5 per 75,000 trees. They feed largely on mice but include other vertebrate and invertebrate groups in their diet.

The number of mice, mostly white-footed mice, varies from year to year from about 160,000 to 320,000 and averages about one mouse per two trees. They take a very large number of seeds, nuts, and snails and are preyed upon by three species of shrews (Williams 1936).

The gray squirrel with its root-digging and nut-burying habits is an important influent or dominant.   Where the gray squirrel is absent or present in small numbers, the fox squirrel occurs, especially near the western edge of the forest.   A fair population is 2 to 3 per 25 to 40 trees.   Associated with them is the southern flying squirrel which may have a similarly sized population, or from 10,000 to 20,000 per 10 square miles.   These squirrels take an enormous toll of roots, nuts, fruits, and seeds, and may be an important factor in controlling forest composition during critical periods.

The turkey is one of the important animals from the standpoint of plant–animal coactions.   It utilizes many acorns, dogwood seeds, and beechnuts. The large flocks reported in 1600 suggest a population of at least 200 per 10 square miles, but the species was not uniformly distributed (Latham 1941, Bailey *et al.* 1951).

Large mammals enter all the seral stages and faciations.   The white-tailed deer was originally a dominant or major influent because of its abundance. The deer population varied greatly and sometimes cyclically between 100 and 840 individuals per 10 square miles with 400 the optimal number.   With their enemies and competitors removed, large populations have in many cases seriously damaged the forest reproduction and undergrowth (Ruff 1938). The wapiti or elk is similar to the deer in its impact on the forest vegetation but commonly grazes and browses more in open places.   It is less shy and more easily shot than the deer.   The elk was abundant in western Pennsylvania for a long time.   The bison was of some importance in Virginia and Pennsylvania.

In a "big hunt" or "circle drive" in 1760 that covered probably more than 250 square miles (648 km²) in central Pennsylvania and was participated in by 200 hunters, 98 deer, 111 bison, and 2 wapiti were secured (Seton 1929). The hunters started a half-mile apart so that there was opportunity for some herds to escape notice.   The area was one of very rough topography, containing two rather large branches of the Susquehanna River.   The great abundance of the enemies of the deer and wapiti is indicated by the killing of 109 wolves, 41 mountain lions, and 114 bobcats.

The puma or mountain lion, formerly present throughout the deciduous forest, ranged from two to three per 10 square miles.   The bobcat attacked only very young deer.   The gray wolf was a characteristic animal of the deciduous forest, hunting in packs (Young and Goldman 1944, Stenlund 1955).   Pristine numbers of one to three per 10 square miles are probable. Under recent Minnesota conditions, Olson (1938) estimates one wolf per 100 square miles (260 km²).

There is almost no data on the food of wolves within the deciduous forest. In second-growth forest in northern Minnesota, where conifers have been largely replaced by balsam poplar and birch and to a lesser extent by red maple, ash, sugar maple, basswood, hazel, and cherry, Stenlund (1955) found winter food of wolves in per cent of total volume to be: deer 95.5, hare 0.8, grouse 3.3, small mammals 0.3.   Olson (1938) states that they also eat mice, voles, marmots, fish, snakes, and insects. At the time of the big hunt in Pennsylvania small mammals were evidently numerous so that we may assume that only 40 per cent of the food of the wolf for about 20 weeks in winter was deer and wapiti. Probably 100 per year would be devoured by 5 predators (3 wolves and 2 mountain lions) on 10 square miles. The annual increase in deer and wapiti on 10 square miles of deciduous forest should have been about 20 per cent. If these were regularly destroyed by wolves and mountain lions, this would put a deer skeleton on each 64 acres (26 hectares) each year. In this connection it is worthy to note that, of all American mammals, the mountain lion has the largest home range in which it does its hunting, with a diameter of 20 to 60 miles (32 to 96 km) (Seton 1929). The wolf's home range commonly varies from 10 to 20 miles (16 to 32 km) in diameter, although Enos Mills knew of a wolf whose home range extended up to 40 miles (64 km).

There is also a good series of omnivores which eat everything from berries to snakes and from birds to grasshoppers. This group includes the red fox and gray fox. The gray fox is quite characteristic of deciduous forest and makes use of hollow trees for dens. Seton's estimates suggest 30 per 10 square miles or one for every 25,000 trees.

The black bear probably reached its greatest population density in the deciduous forest, averaging perhaps 5 bears per 10 square miles. It appears to have been most abundant near large streams. The 1760 big hunt in Pennsylvania, however, rounded up only 18 bears. Estimates of recent populations for Pennsylvania are 2 per 10 square miles. Deer and bear do not compete severely.

These and other omnivores reduce seed crops almost to a vanishing point. The feet of large animals may, however, press seeds of trees into the soil where germination is favored. Seeds may be increased or decreased in germination ability after passing through the alimentary tract of animals (Krefting and Roe 1949).

Dead bodies, shed hair and feathers, and excreta of animals fall to the forest floor or originate from soil organisms in large quantities every year. The aggregate of a season's fall of leaves and stems from trees, shrubs, and herbs is considerable. This litter decomposes rapidly (Melin 1930). The bones of large vertebrates last perhaps ten years. Each six acres should hold a deer or wapiti skeleton along with an occasional one of fox, wolf, mountain lion, and bear. Each life and death struggle of large animals tramples the herbs, shrubs, and seedlings of a large area. The soil and litter are also disturbed in these struggles and by the frequent burying of unconsumed meat by some carnivores. Where deer or wapiti bed down for the night, herbs and shrubs are crushed. When traveling, they cause much damage to seedlings, and in winter heavy browsing destroys many young trees. Several hundred mouse and bird bones are egested each year in owl pellets. The excrement of the large mammals and birds is an important fertilizer that is widely distributed.

White-tailed Deer as a Major Permeant Dominant

In the cutover and second-growth Pisgah National Game Preserve, there were in 1938 about 4600 deer on 98,408 acres (39,360 hectares). On the portion of the preserve given special study there were 38 individuals per square mile (15/km²). Bears and bobcats were common; turkeys appeared to be absent. Approximately 30 per cent of the area was in tulip–oak or cove hardwoods (mixed mesic) forest; 61 per cent in oak–chestnut, 2 per cent in northern hardwoods; 7 per cent in pine–hardwoods; and 0.5 per cent in spruce–balsam (Ruff 1938). The timber was largely in the 20-year age class, with an average diameter, breast high, of 7.5 inches (19 cm), and with 80 per cent

Table 2-1. Composition and Deer Use of Cove Hardwoods[a]

Species	Per Cent of Canopy	Palat- ability[b]	Per Cent of Understory	Sprout Palatability
Characteristic Dominants				
Tuliptree	31.0	40	5.5	55
White oak	5.8	10	1.0	20
Northern red oak	5.1	10	0.6	15
White basswood	2.7	45	0.2	60
Chestnut	2.6	20	1.4	30
Mountain silverbell	1.9	20	1.6	35
American beech	0.9	20	0.2	+
Sugar maple	0.2	..	...	..
Black cherry	0.1	20	...	..
Yellow buckeye	...	30	...	..
Local or Occasional Dominants				
Red maple	8.9	20	3.2	35
Sweet birch	5.5	30	1.8	35
Pignut hickory	...	10	1.3	30
Bitternut hickory	4.0	10	...	+
Eastern hemlock	3.9	10	0.6	+
Blackgum	0.9	40	0.2	75
Butternut	0.9	..	+	+
Yellow birch	0.8	30	+	+
White ash	0.2	35	+	55
Cucumbertree	0.2	..	...	..
Seral Species				
Scarlet oak	2.6	5	0.3	15
Black locust	2.3	30	0.7	50
Chestnut oak	1.9	10	0.5	15
Black oak	0.2	5	0.1	15
Sycamore	0.2	0	...	..
Eastern white pine	0.2	20	...	0
Pitch pine	0.1	20	...	0

[a] From Ruff 1938.
[b] The palatability ratings are for the summer, except for the three conifer species, which are used chiefly during the winter.

crown density. The relatively low crown density permitted a greater than usual development of forest herbs and shrubs.

In evaluating the role of animals on vegetative growth and reproduction, one must realize that competition occurs vigorously among plants, and seedling loss is startling even in areas protected from the browsing of animals. In Ruff's (1938) two-acre (0.8 hectare) fenced plots that were fully protected from large animals, white ash started with 27 trees, but at the end of five years there were only 7 small trees left. Only 56 per cent of the white oak seedlings survived the period, but hickories increased from 3 to 5 trees.

Ruff (1938) rates all the important plants as to their palatability to deer, based on observations of plants nipped or browsed and on stomach contents (Tables 2-1, 2-2). Palatability is based upon the per cent to which a plant is eaten under biotically balanced conditions in which both the plant and the animal can survive without retarding normal growth of the plant or causing the animal to be improperly nourished.

Browsing has a detrimental effect on trees, the seriousness of this depending on its intensity, and the palatability and resistance of the species. One of the trees most heavily damaged is the tuliptree. When young it is succulent and highly palatable. No seedlings become established under heavy deer use. Planted specimens less than five feet in height do not survive. Even during years when seeding is prolific, seedlings survive for little more than three months. The great abundance of tuliptrees in the area probably resulted from sprouting from stumps in 1916. The deer herd was very small about this time, thus giving the trees an opportunity to grow.

Black oak and white oak reproduce with difficulty, because the acorns are

Table 2-2. Per Cent to Which Plant Species May Be Browsed in the Several Important Vegetative Types

Species	Cove Hardwoods	Oak–Chestnut	Pine–Hardwoods
Red maple	20	20	20
Tuliptree	30	..	..
Blackberry	30	40	35
Flowering dogwood	40	35	35
Catbriar	50	50	50
Dog-hobble	10	..	..
Chestnut	10	20	20
Sassafras	40	40	45
Sourwood	35	30	40
Flame azalea	10	20	..
Scarlet oak	..	..	5
Pitch pine	..	..	10
Common sweetleaf	..	20	..
Pignut hickory	10	10	..
Chestnut oak	..	5	..

easily found by animals in autumn and nearly all are destroyed. Older seedlings, four and five feet high, (1.2 and 1.5 m) of both species will tolerate a medium amount of use. The black oak is more successful in reproduction than the white oak, largely because of its lower palatability.

Red maple is a common tree in the older age classes. It will survive after it has grown above the limit of the browse line, but before this age it barely maintains its own. Blackgum will not become established in areas used heavily by deer. Hickories cannot seed-in because, in addition to the factors affecting germination, the seed is palatable to deer, turkeys, and bears, and in some cases all seed is consumed. Seedlings are destroyed by deer soon after they germinate. Black locust does not suffer great mortality from wildlife use after it attains a height of three or four feet (1.0 or 1.2 m). White ash apparently cannot become established (Ruff 1938).

Hawthorn is found only in restricted areas, but it is an excellent food for game. Crab apple, under deer use, showed no establishment during five years. American hornbeam seeds heavily but is killed as soon as it attains a height of three inches (7.5 cm). Trees four and five feet (1.2 and 1.5 m) in height are damaged heavily due to their comparatively slow growth. Sassafras is a tree far too palatable to survive long in the "browse level." Regeneration must depend upon a rare seed production large enough to leave some after the game is satisfied. Sourwood sprouts heavily but produces comparatively slow height growth under shade. It may finally extend beyond the browse level. Flowering dogwood, a very palatable species, becomes established as sprout growth only and then merely for short periods of time. Trees that reach four and five feet (1.2 and 1.5 m) in height continue to survive, although generally they lose their lateral branches. Rhododendron and mountain-laurel (Fig. 2-5) make 9.5 and 6.3 per cent respectively of the undergrowth. They are seldom eaten during the summer, but during the winter rhododendron has a palatability of 40 and laurel of 15. The rhododendron is frequently destroyed. Grape has a high palatability rating of 65, but spicebush has only 5. The availability of these plants reduces the browsing pressure on the trees. Pitch pine, a seral tree, is an important winter food. Only one specimen on two acres attained a height of one foot in two years.

The climax dominant trees live perhaps 400 years. Few climax trees in the deciduous forest produce seed in less than 20 years after germination. Survival or perhaps even germination of seeds in the mesic portions of the forest can take place only where an adult tree has dropped out. Most of the seed crop of deciduous trees, such as oaks and beech, is regularly destroyed by deer, bears, gray foxes, and insects. Seeds of trees in mesic valleys ripen as the deer come down for the winter. Proportionate consumption by different animal groups is illustrated by the record obtained on the 1924 seed crop of 14,975 acorns from a 19-inch (47.5 cm) white oak, located southwest of Asheville, North Carolina, in the oak–chestnut area: destroyed by mammals, mice, deer, and other animals, 83 per cent; infested by weevils and moth larvae, 6; aborted embryos, 10; germinated but seedlings dead, 0.3; germinated but seedlings alive, 0.2; with normal embryos, 0.3. Chestnuts were destroyed by two chestnut weevils, *Curculio probiscideus* and *C. auriger*. Hickory nuts are destroyed by the hickory nut weevil (Korstian 1927).

The deer is not a year around dweller in the dense forest but frequents openings, edges, and burned areas. Large populations occur in closed climax stands only in winter. It is evident that deer do much to retard community development by removing seedlings of trees.

Fig. 2-5. Relative utilization by deer of rhododendron and mountain laurel; the rhododendron is entirely consumed (from the U.S. Forest Service, Southern Division, photo by F. J. Ruff, 1938).

In general, the deciduous forest is not the exclusively plant-controlled unit which the vegetational studies of 50 years ago assumed. The plants control certain conditions, but the animals control the plants of the next generation to a noteworthy extent. Thus the forest stands are what the deer and associated animals permit to grow from what they miss in their feeding on the seed crop.

Arthropod Influents and Dominants

About 60 species of insects do serious damage to climax trees. Bark beetles are especially important since they work in the cambium and kill trees by girdling. The larvae of cerambycid and buprestid beetles also girdle trees. Defoliations by insects are numerous in the deciduous forest. This weakens the plant but rarely causes the death of trees over large areas, as the bark beetles occasionally do in coniferous communities. The bark beetle *Scolytus quadrispinosus* does much damage to hickories but causes no extensive tree destruction. Insects destroy large numbers of nuts. Sucking insects also weaken and damage trees. In case of the birch, a treehopper, attended by ants, damages stems and twigs.

Aspection

It is difficult to assign time limits to the seasonal aspects of the deciduous forest (Table 2-3), since one aspect merges gradually into the next, especially with some groups such as the vertebrates. There are, however, characteristic fluctuations in the activities and abundance of both plants and animals in the various aspects that should be recorded (Weese 1924, Smith-Davidson and Shackleford 1928, Williams 1936).

Prevernal aspect. In March, the green leaves of wintergreen and trailing arbutus brighten, and flies appear. In April, the spikes of both red and white trillium emerge, and a dozen other plants may be recognized. Sapromyzid flies appear from pupae of the soil and litter. Flea beetles and leafhoppers come out of hibernation.

Vernal aspect. In May, the yellow violets bloom. Numerous leafhoppers and lace bugs return to the trees to deposit their eggs. The pilot blacksnake starts taking its food from the population of frogs, toads, and salamanders which are themselves snatching insects. Snails start moving about. The northward flight of birds reaches its peak. In June, seedlings of beech and maple are prominent. Spiders and beetles are numerous. Forest fruits are

Table 2-3. *The Seasonal Distribution of Prominent Herbs in Slope Woods near Nashville, Tennessee, in 1936*[a]

DATES	APRIL 11	APRIL 27	MAY 26	JULY 24	AUGUST 28
Total number of species.............	3	13/19	8	12	21
Everlasting........................	5				
Violet wood-sorrel................	3	1			
Shooting-star.....................		5			
Cinquefoil........................		3	2		
Mandrake.........................		5			
Toothwort........................		2			
Ipecac............................			5		
Flowering spurge.................			3	3	
Prairie-tea.......................				5	
False aloe........................				4	
Sunflower.........................					5
Horsemint........................					3

[a] The figures in the columns refer to relative abundance when in flower on a scale of 1-5, 5 being abundant and 1, rare. Sample plots were 10 m² on each date except April 27 with two plots, 6 m² and 8 m², respectively. After flowering, many of the plants died or disappeared (Frick 1939).

developing, and squirrels are already cutting down the unripe beechnuts, hickory nuts, and acorns.

Aestival period. As the fruits of the tupelo, hemlock, cucumbertree, and oaks develop, they are gathered, while still immature, by blue jays, gray squirrels, red squirrels, and chipmunks.

Serotinal aspect. Nesting territories have broken down, and birds tend to move about in small flocks. The premature gathering of nuts and fruits by squirrels continues.

Autumnal aspect. The fall of leaves, the death and disintegration of the herbage, and preparation for winter by the animal constituents are evident. Squirrels leave their summer nests to find shelter in hollow trees. There is a large movement of leafhoppers, lace bugs, and moth larvae down the tree trunks to the leaf litter.

The winter or hiemal aspect. Summer nesting birds are gone, and some northern birds are present for the winter. A few mammals are in hibernation, others, together with the birds, subsist on buds, nuts, and seeds. The arthropod population is concentrated in the litter and soil.

In addition to seasonal variations, the activities of organisms also vary with the time of day. The red-backed salamander passes the day beneath stones and logs but becomes active on the forest floor at night. Slugs and snails and many insects regularly become active at night with the lowering of temperature and increase in humidity (Park *et al.* 1931).

Variations in activities and especially in abundance also occur from year to year. The annual crop of plant seeds, including seeds of trees, varies sharply from year to year and thereby affects the abundance of animal species that depend on them for food.

TULIP–OAK FOREST

The tulip–oak forest (*tulip–deer–oak faciation*) includes Braun's (1950) mixed mesic and western mesic forests and what has been called the Cumberland fauna. The climax stage is dominated by tulip (tuliptree), white oak, beech, white basswood, yellow buckeye, sugar maple, American chestnut, red oak, chinkapin oak, and others. Less abundant but important trees are the cucumbertree, white ash, red maple, shagbark and bitternut hickories, blackgum, black cherry, certain species of birch, and hemlock. The number of dominant species varies from place to place between 2 and 20 but decreases irregularly toward the west (Braun 1950). Chinkapin oak occurs in about 60 per cent of the western mesic areas listed by Braun but in only two of her mixed mesic forest lists; it thus appears to increase in importance toward the west. The western boundary of the main forest of mixed mesic (Fig. 2-1) or tulip–oak in Tennessee corresponds roughly with the western base of the Cumberland Plateau, although considerable mixed mesic forest occurs on the Highland Rim immediately to the west.

Shaver (personal communication) indicates that more species of the mixed forest of eastern Tennessee occur in southwestern Tennessee and in greater abundance than was previously recognized. Lack of good relics has rendered study of the western mesic forest difficult. However, Adams (1941) listed young sugar maples, American elm, mockernut hickory, red mulberry, sassafras, chinkapin oak, white ash, tulip, black walnut, and beech near Nashville. Frick (1939) also found pignut hickory, chestnut oak, sweetgum, black tupelo, and winged elm. American chestnut is known to have been present not far from Nashville and near the Mississippi River along with white basswood. The understory trees, magnolias, sourwood, and holly, are noteworthy. Shrubs are spicebush, witch-hazel, pawpaw, wild hydrangea, and alternate-leaf dogwood. The herbs in the disturbed subclimax near Nashville, Tennessee, are shown in Table 2-3 which shows the seasonal distribution for the slopes of a hill in a forest that was thinned many years before. All but one of the 12 herb species, the false aloe, occur also outside the tulip–oak area.

The central portion of the Appalachian Plateau in Ohio, West Virginia, and Kentucky had one of the scantiest human populations in North America at the time of discovery by white man (Kroeber 1939).

Animal Constituents

The dominant and influent animals described in the general account above were present. The white-tailed deer was exceptionally abundant in Tennessee during the early white settlement. This was due to its tendency to occupy forest edges and the openings which became numerous. In the mixed mesic forest of Great Smoky Mountains National Park, mice and shrews are more abundant than in other adjacent forests. Among the large fauna of breeding birds, only the solitary vireo and black-throated blue warbler are somewhat restricted. The reptiles are those widely distributed in the deciduous forest. The ranges of several amphibians which appear to center in the moist mixed mesic forest region are not restricted to this type of forest (Fig. 2-6).

Rogers (1930) found 40 or more species of crane flies in two tulip–oak forest stands near Allardt in north-central Tennessee. Ten species are confined to tulip–oak, but three, or perhaps five, of these also occur near Ann Arbor, Michigan. Whittaker (1952) reported 18 species of crane flies in the Great Smoky Mountains. He noted one species, listed by Rogers, *Atarba picticornis*, which places its larvae in decayed wood.

Herb-frequenting insects reported by Whittaker were species of long-legged flies, including *Diaphorus leucostomus*, which are predatory on smaller flies and are, in turn, taken by birds. One sphere-headed predatory empidid fly *Leptopeza compta* was very frequent in cove forest collections. *Minettia lupulina*, a sapromyzid fly, was common. The larvae of all these flies live in the litter.

Of the beetles present, *Xanthonia decemnotata* feeds on oaks and beech, *Baliosus nervosus* on basswood, *Anoplitis inaequalis* on nettles, and *Xenochalepus dorsalis* on black locust. All four beetles as well as *Minettia* above, also occur in northern Illinois. The click beetle *Athous acanthus* is widely

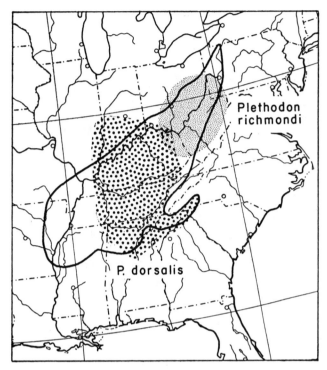

Fig. 2-6. Outline of the distributional limits of the mixed mesophytic (mesic) forest (Braun 1950). It is climax on the Mississippi floodplain but does not predominate in southern Indiana (Potzger *et al.* 1956) or in the western Appalachians. The salamander *Plethodon richmondi* occupies mixed mesic and large areas of maple–beech forests. *P. dorsalis* occurs in the maple–beech forest of Illinois and southern Indiana and mixed and western mesic forests in Alabama, Georgia, and east of the mountains (Grobman 1944).

distributed, while the sap beetle, *Meligethes rufimanus*, is associated with beech in southern Indiana.

Of the Homoptera, *Agalliopis novella* and *Cicadella flavoscuta* are widely distributed, while *C. vanduzeei* is perhaps limited to the mountains. *Typhlocyba rubriocellata*, which occurs here, has also been described from Illinois. The hemipteron *Neolygus belfragii* feeds on the widely distributed mountain maple and alternate-leaf dogwood. However, nine species of snails appear to be restricted to the moist forest of the mountains.

Community Development

Except for the cedar glades on Lebanon limestone, few or no studies of the xerosere have been reported, and hydroseral studies are limited to floodplains (Chap. 4). Cedar glade succession begins as a bare rock and herb stage (Freeman 1933, Meyer 1937). In the spring, certain ground beetles and the grasshopper *Dissosteira carolina* occupy limestone covered only with crustose lichens. The first flowering plants to appear are stone-

Table 2-4. *Succession of Earthworms and Snails in the Cedar Glade Succession*[a]

STAGE	HERB	SHRUB	CEDAR	OAK
ENVIRONMENTAL CONDITIONS (FREEMAN 1933)				
Relative plant density.....................	1	3	6	8
Soil pH at 15 centimeter depth..............	7.5	8.0	7.3	7.2
Wilting coefficient at 15 centimeters..........	13.61	13.91	17.58	17.52
Temperature, August 1, 1931, at 15 centimeters	34.4° C.	30.5° C.	26.6° C.	23.3° C.
Evaporation during summer of 1931..........	5732 cc	3835 cc	2951 cc	1952 cc
EARTHWORMS				
Eisenia roseus...........................	1	2		
Bimastos tenius..........................	1	2	1	
Allolobophora iowana......................	1	1	2	
Diplocardia communis.....................	11	11	4	
Diplocardia riparia.......................	25	44	44	
SNAILS				
Pupoides albilabris........................	6	1	1	
Gastrocopta armifera......................	1	4	1	
Ventridens...............................	1	54	17	
Stenotrema monodon aliciae................		5		
Polygyra trootsiana.......................			1	
Triodopsis albolabris......................				2
Allogona profunda........................				1

[a] The figures in the columns indicate the total number of specimens recorded from each habitat (mostly from Meyer 1937, the last two species from Green 1928).

crop, sometimes 400 per square meter. Pitcher's sandwort comes just a little later and numbers 100 or more per square meter. Both these species are in bloom in early May and both die down in midsummer with almost nothing taking their places. The first shrubs to appear are clumps of coralberry, wild privet, and fragrant sumac. Winged elm and eastern redcedar make a nearly continuous shrub stage. Bedstraw, mistflower, panic grass, and a few other herbs are present. The crane fly *Eroptera cana* and the red-legged grasshopper come in early and decline in numbers as the differential grasshopper makes its appearance, along with the long-necked bug, the forest-edge leafhopper, and such spiders as *Linyphia communis*.

When the cedars come to dominate the area, the herb stage disappears. Due to the reduction in herbs and shrubs, the animal inhabitants of these levels decline and grasshoppers are no longer important. Such leafhoppers as *Kolla bifida* make their appearance, along with the spider *Acacesia folifera* (Table 2-4). A broad-leaved tree stage begins with the sugarberry and is followed by oaks, hickories, elms, walnuts, and others.

OAK–CHESTNUT FOREST

The *oak–deer–chestnut faciation* has been destroyed through lumbering and

the death of the American chestnut by chestnut blight (Fig. 2-7). Introduced into the United States with Chinese chestnut trees about 1900, the
blight was diagnosed in the Zoological Park, New York City, in 1904. The
New York Legislature failed to provide funds to combat it and the disease
spread. It reached the Harlan, Kentucky, mesic forest in 1932 and had killed
nearly all the chestnuts by 1935 (Braun 1950). The largest areas of oak–
chestnut were on lower ridges of the Blue Ridge Mountains and the Piedmont
Plateau (Fig. 2-8). Oak–chestnut occurred on both sides of the Appalachians
from Massachusetts to northern Georgia and in most continuous stands at
an elevation of 1500 to 2000 feet (450 to 600 m). Scattered areas occurred
in the southern mountains up to 4900 feet (1485 m) and at a maximum
height of 3000 feet (900 m) a little farther north.

Fig. 2-7. A. Leaves, burs, and nuts of the American chestnut tree (from Charlotte Hilton
Green's *Trees of the South,* courtesy University of North Carolina Press). B. Dead chestnut tree with dead and living sprouts below and a chestnut oak at its left, Jefferson National Forest (author's 1943 photo).

Near the northern boundary of Tennessee and about midway between the
valleys of the Tennessee and Powell rivers, the ridges were originally covered
by open oak–chestnut woods (Rogers 1930). American chestnut was most
abundant on thin rocky soil on mountain ridges of Maryland (Shreve *et al.*
1910). In southwestern New York it occurred between 1600 and 2000 feet
(485 and 600 m) (Gordon 1940). In Cheat State Park at 1600 feet (485
m) near Morgantown, West Virginia, chestnut and chestnut oak originally
made up 80 to 90 per cent of the overstory trees, and were accompanied by
scattered sweetgum, red oak, red maple, black locust, and sassafras. The
subordinate plants included the mountain-laurel, rhododendron (both somewhat local), and trailing arbutus. Whorled loosestrife and bracken, which

were of greater abundance here than elsewhere, dominated the low layer. In June, 1885, near Mountain Lake, Virginia, Rives (1886) found *Trillium grandiflorum* (large rose-colored petals), deep purple *T. erectum* (a characteristic northern species), lady's slipper, small jack-in-the-pulpit, wood-betony, and others. The flame azalea was in bloom.

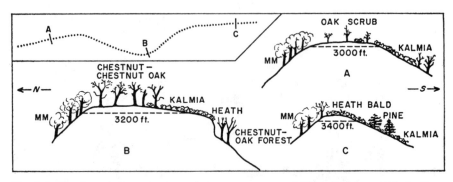

Fig. 2-8. The line of dots represents the east–west crest of River Ridge east of the Poor Fork of the Cumberland River in Kentucky. A, B, and C represent north–south transects across the ridge at the points designated: MM, mixed mesic or tulip–oak forest; the pine is *Pinus rigida* (after Braun 1940).

Since the removal of the American chestnut, the forest is becoming a red oak–chestnut oak–white oak forest on the Blue Ridge (Keever 1953). In the Smoky Mountains, chestnuts made 50 to 80 per cent of the canopy and were being replaced by dense stands of small red maple, sourwood, and similar trees; chestnut oak remains important (Whittaker 1956). Changes are less striking in a red oak–chestnut faciation in western North Carolina (Fauver 1949).

In the Appalachian Mountains adjacent to the oak–chestnut forest, there are several species of pine which grow in pure stands or are mixed with hardwood trees to form pine–hardwoods. The pines are Virginia pine at an altitude of 1400 to 2400 feet (425 to 730 m), pitch pine at 2400 to 3500 feet (730 to 1060 m), and table-mountain pine above 3500 feet (1060 m) (Figs. 2-3, 2-8, 2-9).

Animal Constituents

The deer and wapiti were formerly abundant, but the latter was gone before scientific study began. The wapiti was undoubtedly important in oak–chestnut south into Georgia until some time in the 1800's. Ruff's (1938) studies show the deer to be a dominant in the oak–chestnut forest.

The Allegheny wood rat is a resident. On the average 70 per cent of the materials in and around its nests and 85 per cent of its food are from trees and other plants occurring in the oak–chestnut forest (Newcombe 1930, Pool 1940). These items include fruits and materials from chestnut, rhododendron,

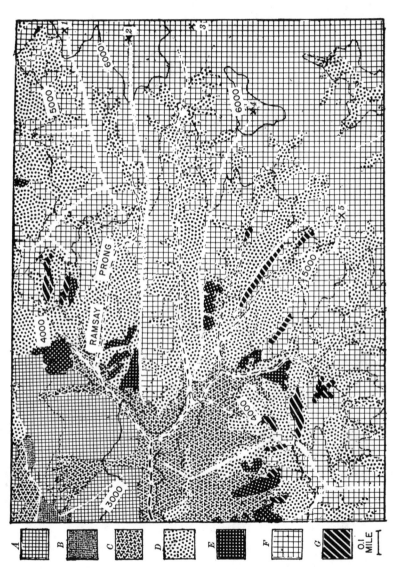

Fig. 2-9. The distribution of forests in the headwaters of the Little Pigeon River in Great Smoky Mountains Park; the top of the figure is west. Dash lines indicate streams; wide white lines, ridge crests. Important high points are: 1, Old Black; 2, Mount Guyot, 6621 feet (2000 m) ; 3, Tricorner Knob; 4, Mount Chapman; 5, Mount Sequoyah. A, oak–chestnut; B, shortleaf pine–hardwoods, also with other pines, American chestnut, and scattered oaks and hickories; C, tulip–oak (mixed mesic or cove hardwoods) ; D, maple–basswood or northern hardwoods; E, hemlocks; F, spruce; G, balds, grassy and shrubby. There is a tendency for tulip–oak to be in the deep valleys and on north-facing slopes; spruce is on south-facing slopes at high altitude. Oak–chestnut occurs on south-facing slopes of broad valleys with pitch pine on exposed headlands (drawn from F. H. Miller's 1941 map, courtesy U.S. National Park Service).

juneberry, and chestnut oak. The wood rat is in part dependent upon the presence of large boulders, cliffs, and ledges. Outside the oak–chestnut forest it occurs on cliffs along the Tennessee, Cumberland, Green, Kentucky, and Ohio rivers. Of mice, the white-footed mouse shows the largest population in oak–chestnut forest and the deer mouse in the hemlock–yellow birch (Wetzel 1939).

Before the chestnut blight appeared in the forest about Mountain Lake, Virginia, the common birds were wood thrush, least flycatcher, black-throated blue and Canada warblers, ovenbird, and eastern phoebe (Rives 1886). These species are no longer abundant in the modified forest (Fauver 1949).

The five-lined skink and the green anole have their upper distributional limits in oak–chestnut. The common forest snakes are present; the eastern milk snake is more conspicuous here than elsewhere. *Plethodon jordani metcalfi* is characteristic of the southern part of the oak–chestnut forest and *P. wehrlei* of the northern part.

In 1922 and 1924 before the blight arrived near Fentress, Tennessee, Rogers (1930) noted the oak-chestnut crane flies *Tipula umbrosa, T. flavoumbrosa, Nephrotoma ferruginea, Dicranoptycha megaphallus, D. sobrina, D. winnemana,* among others. The first two have a distinct preference for this forest under present conditions. In 1943, two hundred species of invertebrates taken at ten oak–chestnut stations between North Carolina and New York showed most species widely distributed in the cool moist deciduous forest. The fly *Pseudogriphoneura gracilipes* is very widely distributed, having been collected near Tapachula, Mexico. Two other flies, *Sapromyza incerta* and *S. philadelphica,* are also widely distributed in the deciduous forest. The ants are the widespread deciduous forest species. *Aphaenogaster fulva aquia picea* was exceptionally abundant in oak–chestnut in 1943 (Sparkman 1943). The click beetle *Althous acanthus* and the leaf beetle *Xenochalepus dorsalis* occurs here and in the mixed mesic; the cerambycid *Pidonia aurata* and the predatory ground beetle *Dicaelus ovipennis* are prominent. The hemipterons are best represented by the numerous mirids, of which *Hyaliodes harti* is very widely distributed and feeds especially on oak and elm throughout much of the deciduous forest, and *Dicyphus famelicus* which feeds upon purple-flowering raspberry and belongs to open spaces especially in rocky woods.

FOREST ZONES IN THE APPALACHIAN MOUNTAINS

Tulip–Oak and Oak–Chestnut Forests

In the Appalachian Mountains, the tulip–oak or mixed mesic forest lies below the oak–chestnut forest (Figs. 2-3, 2-9). Near Weston, West Virginia, the valley floors are at 1000 feet (300 m) altitude and the hills at about 1500 feet (450 m). Mixed mesic forest probably originally covered much of the surface eastward until the area reached 1500 to 2000 feet (450 to 600 m), when it became confined to the valleys. The oak–chestnut forest extended up to about 2700 feet (820 m). The marked variations in the altitude reached by the two communities is brought out by Braun's (1940, 1942) studies in the Cumberland Mountains of Kentucky. Figures 2-3 and 2-9 show tulip-oak forest on favorable north slopes up to 3200 feet (970 m). The oak-chestnut forest probably begins at a higher altitude in southern Kentucky than in West Virginia, which is 100 miles (160 km) farther north and has a little less rainfall.

Maple–Beech–Basswood Forest

In West Virginia, the maple–beech–basswood forest resembles the northern hardwood forest and is likewise climax in character. The forest is well developed over large portions of the Cheat Mountain slopes. It comes in just above the oak–chestnut at about 2500 feet (760 m) and extends to 3500 feet (1060 m), depending on topography, exposure, and edaphic factors. Beech, sugar maple, striped maple, mountain maple, red maple, basswood, white basswood, sweet birch, cucumbertree, black cherry, and white ash are the prominent trees in it (Brooks 1943). Brown (1941), working on Roan Mountain on the North Carolina-Tennessee boundary, found basswood absent but yellow birch and a little red spruce present.

In the Pisgah National Game Preserve, Ruff (1938) reports second-growth containing beech, northern red oak, yellow birch, sugar maple, buckeye, and red maple in the order of declining abundance. The most important shrubs are mountain-laurel, blackberry, blueberry, and white snakeroot. Several species of ferns are important herbs with none of them occurring in the oak-chestnut.

Ruff's data indicate that the deer was a dominant here also. It utilized seedlings of yellow and sweet birch, beech, red maple, basswood, and most of the subordinate plants. The breeding birds are veery, solitary vireo, black-throated blue warbler, blackburnian warbler, and rose-breasted grosbeak (Brooks 1943). The turkey probably breeds here. It finds its food most abundant in this zone during the autumn and has learned to fly downslope to the rhododendron thickets when pursued by hunters.

The eastern garter snake is common. This species goes up to 6400 feet (1940 m) in the Great Smoky Mountains (Necker 1934). The red-bellied snake, milk snake, timber rattlesnake, and eastern ringneck snake are present; the rattlesnake is present sparingly.

Among the invertebrates collected in 1943, snails were numerous including *Triodopsis fraudulenta, T. tridentata, T. notata, Mesodon sayanus, Mesomphix cupreus, M. inornatus, Haplotrema concavum,* and *Retinella indentata.* Nearly all were widely distributed. The ground myriapods included *Pseudopolydesmus serratus, Saiulus canadensis, Otocryptops sexspinosus,* and *Linotaenia bidens.* Phalangids, which appeared in numbers, were *Leiobunum longipes, L. politum,* and *L. vittatum.* Of nine species of insects, one or two had mountain preferences.

Birch–Hemlock–Spruce Forest

Between the maple–beech–basswood forest below and the spruce forest above, at approximately 3500 to 4200 feet (1060 to 1275 m) are birch, hemlock, and spruce.

In the West Virginia forest, yellow birch is by far the commonest deciduous tree, sometimes occurring in nearly pure stands. Hemlock grows to the tops of many of the ridges, also in stands that are often practically pure. Characteristically at the higher levels, yellow birch, hemlock, and red spruce are interspersed. Bigtooth aspen, Fraser's magnolia, American ash, serviceberry, and red, striped, and mountain maples are widely distributed and locally common (Brooks 1943). This community resembles the forest in northern New York, New England, and Quebec much more than the hemlock communities farther south, which are at lower altitudes and associated with cove hardwoods. There are gentle slopes, swampy areas, sphagnum bogs, jungle-like abundances of ferns, and thickets of such arborescent species as rhododendron, menziesia, mountain-laurel, and skunk-currant.

In addition to yellow-cheeked vole, cloudland deer mouse, and smoky shrew, the common deciduous forest vertebrates are present.

Brooks (1943) lists 35 species of small birds breeding in the birch–hemlock–spruce forest. About 20 are not generally distributed through the deciduous forest. The Cheat Mountain salamander and spring salamander are recorded. The arthropods are mostly those of the deciduous forest generally; a few also occur on the southern plain, and a few range northward.

The swamps lie at the level of the conifer–hardwood ecotone. Among these is Cranberry Glades at an altitude of 3100 feet (940 m). Conifers come down locally as low as 2500 feet (760 m). Rumsey (1926) describes a swamp near Cranesville, Maryland, at 2500 feet (760 m) which may be typical for the belt.

The spruce–fir forest that lies above the birch–hemlock–spruce forest is an extension south on the mountains of the coniferous forest biome and will be discussed in a later chapter.

Balds

Shrub balds (Fig. 2-8) occur on crests of the main spurs or finger ridges, especially where the topography is very rugged. The principal species are mountain-laurel, great and/or catawba rhododendron, highbush blueberry, smilax, red-twig leucothoe, occasional red spruce, southern balsam fir, yellow birch, and mountain ash (Cain 1930).

The seven species of plants present in all the grassy balds on Roan Mountain are *Danthonia compressa, Rumex acetosella, Agrostis hyemalis, Fragaria virginiana, Carex debilis rudgei, C. flaccosperma glaucodea,* and *Veronica officinalis.* The harvestman *Leiobunum calcar,* the spider *Lycosa rabida,* the millipede *Aporiaria deturkiana,* and the centipede *Linotaenia fulva* are usually present. These species of invertebrates also occur in northern Illinois.

Snails that are present include *Vitrinozonites latissimus* and *Mesomphix andrewsae*. The latter species occurs also in the Savannah River Refuge, Georgia.

MAPLE–BEECH AND MAPLE–BEECH–HEMLOCK FORESTS

The largest, relatively uniform forest is the maple–beech or, more completely, the *maple–wapiti–deer–beech faciation* (Fig. 2-10). It covers most of Indiana, Ohio, extreme southern Ontario, and the Lake Erie and Lake Ontario plains of New York. The conifers associated with it in the northern localities are confined to soils and topography quite unfavorable for other trees. Studies of maple–beech forest relict areas, that were earlier supposed to be virgin, have indicated that the stands have lost tulip, basswood, red

Fig. 2-10. The characteristic form of snail shells of deciduous forests. A. *Triodopsis fraudulenta*, Indiana. B. *T. notata*, West Tennessee. C. *T. caroliniensis*, Mississippi. The forest at the right is a moist deciduous forest in Indiana commonly called beech–maple, although a number of important trees have been taken out.

oak, cherry, and other valuable woods or about 10 per cent of the original timber by logging, while beech and much of the maple were allowed to remain. An example of such a relic is Warren Woods near Three Oaks, Michigan (Cain 1935). The witness trees blazed by surveyors of Indiana lands between 1799 and 1846 showed more than 80 per cent of the land not in prairie or swamp to be beech–maple–ash (Potzger *et al.* 1956).

The North Chagrin Reservation of the Cleveland Metropolitan Park System in Ohio is a mixed forest (Williams 1936). The location of hemlock and American chestnut in this forest is typical (Fig. 2-11). In Indiana and Ohio, hemlock and white pine are usually in ravines or on sand. Figure 2-10 shows the red oak associated mainly with chestnut and that a number of trees had been removed, probably less than 30 years before. The scattered tulip had 30 trees removed in recent years. Some large oaks and tulips were removed in 1871. The trees in the beech–maple portion of this area showed the following densities in per cent: beech, 53; sugar maple, 33; red maple, 6; tulip 3; miscellaneous, 5. The removal of trees together with the absence of six large mammals, some birds, and various other animals indicates the poor and unrepresentative condition of many relict forest areas. Another important unnatural feature of these areas is their small size.

The relative importance of small mammals may be partially evaluated by their numbers, but duration of their activity periods and their food habits must also be considered (Table 2-5). The most numerous and ceaselessly active animal in some moist forests is the short-tailed shrew (Williams 1936). A short-tailed shrew in one month may eat meadow mice or other mice, 8; adult insects (size of May beetles), 90; insect larvae (size of May beetle larvae), 78; earthworms 4 centimeters long, contracted, 53; snails, 18 (Shull 1907). The barred owl is a consistent hunter of the shrew and an important check on its numbers. In two or three local areas, shrews have been known almost to eliminate the mice. Since the large mammals are gone, this shrew appears to be the most influential mammal. From the standpoint of numbers and year around activity, the white-

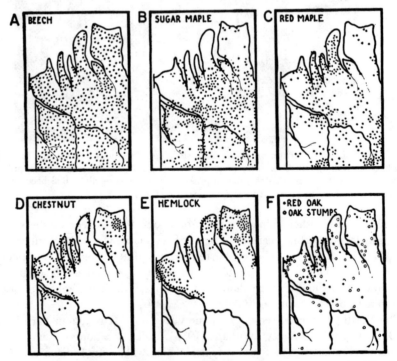

Fig. 2-11. The distribution of some important tree species in an area near Cleveland, Ohio. The top of the figure is toward the east and the valley of the Chagrin River (after Williams 1936). Note location of American chestnut and hemlock.

footed mouse probably ranks second in importance. Woodland white-footed mice are nocturnal and feed largely on tree seeds (Johnson 1926).

The permanent resident bird population in the Cleveland area in 1933 was tufted titmouse, 8 pairs; black-capped chickadee, 5; cardinal, 5; white-breasted nuthatch, 3; rufous-sided towhee, 1; downy woodpecker, 4; hairy

Table 2-5. *The Activitiy and Populations of Mammals in a Cleveland Park Area of 65 Acres (26 Hectares)* [a]

COMMON SPECIES	ACTIVITY[b]	POPULATION 1932
Short-tailed shrew....................................	ADN	...
Northern white-footed mouse..........................	AN	...
Eastern chipmunk.....................................	DW	650
Northern gray squirrel...............................	AD	2
Southern red squirrel................................	AD	20
Southern flying squirrel.............................	AN	24 (1935)
Cottontail...	ADN	30
Woodchuck..	DW	10
Eastern raccoon......................................	NW	12
Eastern striped skunk................................	NW	6

[a] Williams 1936.
[b] A, active through the year; D, active during the day; N, active at night; W, inactive

woodpecker, 3; barred owl, 1; and ruffed grouse, 1. There were also 141 breeding pairs of summer residents. The only reptile present in important numbers was the rat snake.

The red-backed salamander occurs in moist woods of all kinds, including floodplain woods, northern parts of the oak–pine forest, and southern sand areas. The population is small and scattered in the south, probably largest in the maple–beech forests of Indiana and Ohio, and scattered in the east and northward to the southern end of James Bay in Canada. It is not found in pure coniferous forests. The wood frog, commonly called the beech frog, has a lesser range than the red-backed salamander but is abundant in maple–beech forests.

Maple–Beech with Conifers

Maple, beech, basswood, yellow birch, hemlock, and white pine occur in several combinations in climax and late subclimax stages. At low altitudes the sugar maple appears to be the most important species near the northern limits of the deciduous forest from Minnesota to Quebec, with beech taking a secondary role. However, in the plateaus of western New York and Pennsylvania, the beech becomes more important and is commonly mixed with hemlock to form beech–hemlock or maple–beech–hemlock stands. The hemlocks sometimes occur separately in ravines or on north slopes. White pine is often present with the hemlock. These two conifers occur south to Georgia. There

Table 2-6. A Representative Distribution of Pairs of Breeding Birds per 100 Acres (40 Hectares) in the Edmund Niles Huyck Preserve near Albany, New York[a]

Species	Maple–Beech	Beech–Hemlock[b]	Pine–Hemlock–Hardwoods	White and Red Spruce
Great crested flycatcher	3	1	1	
White-breasted nuthatch	2	5	1	
Ruffed grouse	3	2	1	
Red-eyed vireo	25	1	10	
Ovenbird	28	32	9	1
Hermit thrush	5	6	10	2
Magnolia warbler	2	25	4	26
Black-throated green warbler	9	32	13	33
Blackburnian warbler	6	14	18	13
Slate-colored junco			10	23
Swainson's thrush		#		10
Golden-crowned kinglet		#		23
Myrtle warbler			2	23

[a] After Kendeigh 1946.
[b] # indicates absent but originally present.

were white pine, hemlock, and beech trees of notable proportions in the original forests of Connecticut (Nichols 1913). The Edmund Niles Huyck Preserve, near Albany, New York, presents a large variety of groupings of trees (Odum 1943). Birds evidently distinguish a difference between broad-leaved and needle-leaved coniferous trees (Table 2-6) (Kendeigh 1946).

MAPLE–BASSWOOD FOREST

The density of beech trees in forests declines in the northwestern corner of Indiana (Fig. 2-12) (Potzger and Keller 1952). Its place is taken by white oak, black oak, and red oak. Farther south there are maple–basswood with oak and oak–hickory forests with occasional beech trees extending about ten miles (16 km) inside Illinois. Across northern Illinois maple–basswood occurs generally in the valleys and other favorable spots and oak–hickory on the bluffs and in contact with the prairie. Myers and Wright (1948), working in western Illinois opposite the southern boundary of Iowa, found white oak, black oak, and shagbark hickory to be the principal trees of the upland. A little farther north and east in Brownfield Woods near Urbana, Illinois, Gleason (1912) and Vestal and Heermans (1945) note the absence of white oak and black oak in this more mesic forest. They stress the preponderance of sugar maple, red oak, basswood, and elm. At Lake Michigan near Waukegan, Illinois, oaks and hickories are seral trees (Gates 1912). Adams (1915) and DeForest (1921) suggest that oak–hickory occurs on the dry uplands and that maple–

basswood is the probable climax in the better mesic habitats.   In eastern Nebraska, Weaver, Hanson, and Aikman (1925) found red oak–basswood communities only near streams.   The basal area (listed first) in per cent of the dominant trees of 10 inches (25 cm) diameter or more and frequency (listed second) in per cent of all plots in which the species occurred in a relict stand near Minneapolis, Minnesota, are as follows: sugar maple, 47-44; American basswood, 28-28; American elm, 11-10; slippery elm, 7-12; and northern red oak, 6-5.   American elm appears to be a regular constituent of the climax (Daubenmire 1936).

Some of the subordinate plants in this same stand were elder with a frequency per cent of 48, climbing bittersweet, 16; moonseed, 12; frost-grape, 12; and prickly-ash, 4 (Daubenmire 1936).   The first species is rare, but the other species are also present in central Illinois (Jones 1941).   The absence of pawpaw, burning bush, and spicebush in the Minnesota forest is noteworthy.

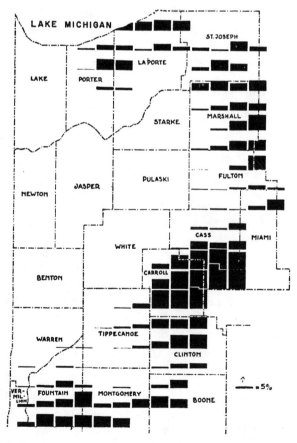

Fig. 2-12. The decline in abundance of beech as the prairie is approached in northwestern Indiana (after Potzger and Keller 1952).

The most abundant and representative herbaceous species were bellwort, sweet cicely, enchanter's nightshade, downy yellow violet, and bloodroot.

Eggler (1938) found elm nearly absent, a little white pine present, and red oak more important than basswood in a Wisconsin area, 50 miles (80 km) northeast of Minneapolis. The forest was growing on soils of *p*H 4.5 to 6.0.

An ecotone between maple–beech, oak–hickory, and maple–basswood is suggested along the Sangamon River southwest of Urbana, Illinois. Here the exposed uplands are covered with oak–hickory, and larger tracts in more mesic areas are maple–basswood (Dirks-Edmunds 1947). In a 13-acre (5.2 hectare) upland farm woodlot, the percentages of trees were white oak, 49; red oak, 12; black oak, 8; slippery elm, 9; American elm, 10; black walnut, 5; butternut, 2; and shagbark hickory, 1. Hickories had probably been removed for use in farm equipment. Goff (1931) reported the two hickories as constituting 16 per cent of another grove, with black cherry 4 per cent. This resembles Oosting's (1942) white oak faciation of the oak–hickory. The northern third of Illinois, a little of northern Indiana, and a little of southern Wisconsin are covered by an ecotone between three associations: maple–beech, post oak–hickory, and maple–basswood (Fig. 2-1). Differences in animal constituents of the associations are of a minor character.

Influence of Drought upon Vegetation

During the drought period of the early 1930's, the leaf litter in Trelease Woods, Urbana, Illinois (Rice 1946), continued nearly normal through the winter of 1933-34, giving a good winter ground cover. In 1934, the very low rainfall and high temperatures of the early part of the growing season caused herbs such as *Geranium, Hydrophyllum, Floerkea,* and *Asarum* to die rapidly and early so that by June the ground was almost bare. Some of these plants are normally 3.5 feet (1 m) high in early summer. Wood-nettle was scarce and restricted to low wet spots. The pawpaw, the buckeye, and many of the forest trees began to lose their leaves in July and were almost completely defoliated by August. Several trees died. The leaf fall during the autumn was small and in many places the ground was barren. The heavier rains and cooler weather of September caused the herbs and shrubs to make a second-growth, and at the first light freeze on October 28 many were in bloom and had an abundant growth of leaves. In 1935, the woods were thoroughly wet for the first time in three years on May 3. A rank growth of vegetation followed. A heavy leaf canopy was formed, and no additional trees died.

The canopy of the late subclimax maple–basswood in the ecotone of northern Illinois covers about 90 per cent of the forest floor, leaving 10 per cent open to the sun and sky. This open space is increased when leaf growth is reduced during droughts. Presumably more light enters the woods in dry years than in wet years.

COMMUNITY DEVELOPMENT IN MAPLE–BASSWOOD AND MAPLE–BEECH CLIMAX AREAS

Sand Plains, Ridges, and Dunes near Lake Michigan

Sand plains with depressions and ridges are often laid down under water. Sand ridges (Fig. 2-13) occur especially on depositing shores as the water level is gradually lowered. Such topography originally predominated on the southwest shore of Lake Michigan.

Shoreward and beyond the reach of ordinary waves, there is a nearly level wide beach where the sand is unstable and frequently blown onto adjacent ridges (Fig. 2-14). *Artemisia* in scattered stands and sea-rocket are the first plants to take possession of the level loose sand. Termites occur under driftwood, and white tiger beetles, both as flying adults and as larvae in vertical burrows, are characteristic animals. There are a few burrowing spiders and a few larval lepidopterons occurring on the sea-rocket.

COTTONWOOD–WHITE TIGER BEETLE FACIES

With the addition of more sand, *Artemisia* is followed by a grass or sand-reed, sand-cherry, and cottonwood. The three species, each of a different life form, become the principal dominants. White tiger beetles become abundant, and the openings of their cylindrical burrows over the ground surface are numerous. Termites continue to be present, and burrowing spiders increase in numbers. This is preeminently the stage of digger wasps. Here the holes of the *Microbembex monodonta* are numerous. They store flies,

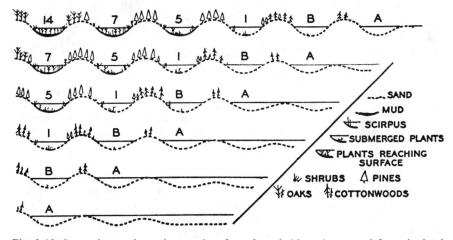

Fig. 2-13. Succession as shown by a series of ponds and ridges that were left as the level of Lake Michigan gradually fell. The upper horizontal series shows four representative ponds, the oldest (14) at the left and two earlier hypothetical stages (A, B) at the right. The vertical series at the left shows the same series of stages as a pond gradually grew older on one locality, the oldest stage being at the top (after Shelford 1911a, from Pearse 1939, copyright by McGraw-Hill Book Company, Inc).

Fig. 2-14. The middle beach and first ridge of Lake Michigan near East Chicago, Indiana.
A. View showing line of cottonwoods and scattered driftwood. B. Vertical burrow of the
spider *Geolycosa pikei* protected from cave-in by a web. C. *Artemisia caudata,* a common
plant on the inner side of the flat beach. D. First ridge. At right the low shrubs are wild
plum *Prunus pumila* and willow *Salix cordata*. The grass is chiefly *Calamovilfa longifolia*.
E. The adult white tiger beetle *Cicindela lepida,* twice natural size. F. The burrow of the
larva of the white tiger beetle (modified from Shelford 1913).

probably secured from the beach, in their larval nests. Velvet ants are present.
Two species of robber flies, *Erax* and *Promachus vertebratus,* are common;
their larvae live in the sand as parasites on other invertebrates. The bee fly
Exoprosopa lays eggs at the entrances of the burrows of *Microbembex*. The
white grasshopper is characteristic, and the long-horned locust occurs commonly.

The herb layer includes an occasional red-legged grasshopper, and various
sparrows are common in the autumn, feeding on grass and weed seeds. The
crab spider *Philodromus alascensis* is found on the young cottonwoods, often
with its appendages stretched out on the petiole or midrib of a leaf. In early

spring, the blossoms of willows that occur in the wet depressions are frequented by pollen-gathering insects (Andrenidae, Apidae, syrphus flies, etc.). The plum is attacked by aphids which attract lady beetles and syrphus flies.

In the tree layer, the cottonwood is attacked by many borers. The most characteristic is *Plectrodera scalator*. There are many leaf gall aphids. The larvae of the large tiger beetle *Cicindela formosa generosa* appear, along with their pits in the sand. They rarely invade the dense pine areas. Another grasshopper *Melanoplus bilituratus* is added. The burrowing spider *Geolycosa pikei* continues to occur in open places.

PINE–BRONZE TIGER BEETLE FACIES

The jack pine is the principal dominant, and the bronze tiger beetle is a characteristic insect. In the subterranean ground layer, the larva of the bronze tiger beetle with its straight, cylindrical burrow is common. Several digger wasps of the earlier stages continue to be present. The ant *Lasius niger americanus* nests beneath the surface and swarms in early September. The six-lined racerunner, the blue racer, and the pond-inhabiting midland painted turtle all bury their eggs beneath the sand. There is an occasional thirteen-lined ground squirrel. The grasshopper of the previous stage continues, and two other species are added, so that there are the long-horned locust, narrow-winged locust, lesser migratory locust, mottled sand locust, and sand locust. The ruffed grouse nested here originally.

In the herb layer, the little bluestem, a bunch-grass prairie climax species, appears along with the first seedlings of jack pine. *Arabis lyrata* is a common herb. Occasional monardas support the crab spider *Dictyna sublata* which resembles the blossoms closely in color. The flowers are also visited by bees and flies.

In the shrub layer are young pines, the juniper *Juniperus communis,* and bearberry. The evergreens support several spiders, *Philodromus alascensis, Theridion spirale,* and *Xysticus punctatus,* and with them sometimes the assassin bug *Zelus exsanguis.*

In the tree layer, the pine is attacked by borers and a few leaf-feeders. The bark beetle *Ips grandicollis* is common under the bark of dead and dying trees. Hairy and downy woodpeckers nest in the hollow trees. Their deserted holes may be used in later years by the black-capped chickadee. The pines prepare the way for the oaks by improving soil conditions and providing shade for seedlings. Oaks become more dense with time and finally crowd out the pines. There are places where the two grow together (Shelford 1913, Sanders and Shelford 1922).

BLACK OAK–ANT-LION FACIES

Among the black oaks there still remain open spots of relatively stable sand. These small areas possess some of the same species as the pine areas, as well as additional ones. The abundant bronze tiger beetle larvae are parasitized by the larva of a bee fly. In the subterranean ground layer several digger wasps, not found in the earlier stages, occur among the more closely placed herbs. Once, the digger wasp *Ammophila procera* was seen carrying the black oak caterpillar *Nadata gibbosa*. A megachilid or leaf cutter makes a nicely matched thimble-shaped cell, placed at the end of a burrow about 2 inches (5 cm) below the surface of the sand. Ant-lions are rare in this sere except among black oaks, where they make funnel pits in the sand at the bases of trees. The hognose snake is common. The six-lined racerunner occurs in open spaces.

The number of herbs is large and insects numerous. The shrub layer was attenuated in the area studied probably because of the second-growth character of the forest. The shrub layer includes scattered common chokecherry, young oaks, rose, and blueberry. The lace bugs occur in numbers on chokecherry.

In the tree layer, the black oak is attacked by *Nadata gibbosa* (see above) and several unidentified slug caterpillars. Feeding on the leaves are several species of leafhopper. The oak treehopper is a common leafsucker. Squirrels are occasional visitors as they come to feed upon acorns.

The black oak is invaded by white oak and the white oak–black oak stage by red oak. With the red oaks come large snails, earthworms, woodchucks, and squirrels. At this stage, the vegetation has become late seral deciduous forest (Shelford 1913).

Sand Dunes

This history of low and high sand dunes is closely similar to the ridge stages described above. The grasses *Ammophila* and *Calamovilfa longifolia* grow on them. The latter grass is the more effective in holding sand from blowing. *Prunus pumila* may follow and prepare the way for cottonwoods, jack pine, and oaks. Cottonwoods stand burial and re-exposure remarkably well.

Deep pits which are open on one side, often called blowouts, are frequently produced by wind. The bottoms of such depressions may be wet. Bare wet sand is soon occupied by *Juncus balticus,* which is followed by willows, usually *Salix glaucophylloides,* and cottonwoods. At this stage such grasshoppers as the locust *Melanoplus differentialis* and numerous small leafhoppers occur. When the surrounding dunes become stabilized, a basswood–maple–ash forest, sometimes with tulip, becomes established, and the trees grow to a large size. The south end of Lake Michigan is on the boundary between maple–beech and maple–basswood climaxes (Gates 1912, Shelford 1913).

Bog Lakes and Bogs Terminating as Deciduous Forest

Bog lakes and bogs are scattered over the glaciated portions of the deciduous forest area north of 42° N. Lat. from Illinois to New England. Welch (1935) divides bog lakes into two classes: those with a central area of open water and those without open water at the present time but with large quantities of water held in mats of vegetation and bog deposits. Floating mats of vegetation may be made up of mosses, yellow waterlilies, sphagnum, cat-tails, or bulrushes. These bogs belong properly to the coniferous forest. They were left behind when the coniferous forest retreated northward at the close of the ice age. Dansereau and Segadas-Vianna (1952) have depicted the cross-section of a bog (Fig. 5-14) where succession goes to deciduous forest and another which terminates in coniferous forest.

So much of the general setting and the plants and animals are the same as those in bogs described in Chapter 5 that a description of the bog stage in the deciduous forest would be repetitious. A few special facts regarding the animals in the bogs near the southern end of Lake Michigan are, however, worth noting. The floating mat supports aquatic insects. The water-holding leaves of pitcher-plants contain the mosquito *Wyeomyia smithii.* Frogs are numerous, and the swamp rattlesnake or massasauga is common in Indiana. In the tamarack forest which follows the floating mat and shrub stages, sphagnum occurs in small pools and supports two species of sphagnum crickets. On the higher ground occur numbers of leopard frogs, green frogs, chorus frogs, eastern wood frogs, and northern spring peepers. There are few herbs. The brindled grasshopper has been found on the low branches of the tamarack. It deposits eggs on the bark of trunks (Shelford 1913). The tamarack at the south end of Lake Michigan was infested by bark beetles in 1912, and many tamaracks were killed by *Dendroctonus simplex,* a seral arthropod dominant. *Polygraphus rufipennis* was also present as a secondary dominant. The predatory clerid beetle *Thanasimus dubius* supposedly feeds upon bark beetles. The tamaracks of the Volo Bog in northern Illinois were heavily damaged by bark beetles around 1942.

Chapter **3**

The Temperate Deciduous Forest Biome
(Southern and Lowland Regions)

The southern and lowland regions cover most of the southeastern United States (Fig. 2-1). The innermost of three concentric zones is the oak–hickory forest, next the pinelands (not mapped in Fig. 2-1 but prevalent throughout the ecotone and beyond), and on the outside, the magnolia–maritime forest. The oak–hickory forest would replace the pinelands were it not for the frequent fires. The magnolia forest includes a large percentage of coreaceous broad-leaved evergreen trees.

The annual rainfall varies from 40 to 60 inches (100 to 150 cm), greatest near the Gulf of Mexico and declining northward to Missouri and southern Illinois. Rainfall is greatest in the spring and summer (Fig. 2-2).

In 1600, the forests of the south Atlantic and Gulf states were probably less modified by man than those in New England. In eastern Virginia and North Carolina, the Indian population had a density of 8 to 20 per 10 square miles (3 to 8/10 km²), in Virginia 1.3 to 3.1 (0.5 to 1.2), and in Georgia and the Gulf states 3 to 8 (1 to 3). There was a lesser requirement for firewood, but village stockades were built of small trees (Kroeber 1939). The natives cleared the land by girdling and burning trees but did not remove the larger ones of five feet (150 cm) or more in diameter. The economy of the Indians was based on agriculture, hunting, and fishing. They burned the lowland and gentle slopes to "favor deer, wapiti, and bison," or more likely, to facilitate hunting. This burning evidently produced a savanna type of vegetation in the valleys. Early white explorers in the western portion of the area reported great herds of wapiti and deer on the valley flats, and bears were seen eating mast on the slopes. None of the expeditions encountered any bison, except for a few stragglers in Virginia. One explorer saw a doe with a wildcat (reported as a mountain lion) on her back fastened to her shoulders. Wolves were numerous. A rattlesnake, seven

56

and one-half feet (2.5 m) long, which had swallowed a squirrel, was killed (Lederer 1672, Alvord and Bidgood 1912).

OAK–HICKORY FOREST

The *post oak–turkey–hickory faciation* occurs on both the piedmont and the coastal plains south of the oak–chestnut region in New Jersey and Maryland. The oak–hickory forest is characterized by the presence of post oak, white oak, and black oak. Near the mouth of the Potomac River in Maryland a relict area of very old trees was composed of post oak, 47 per cent; southern red oak, 21; black oak, 9; white oak, 7; chestnut, 6; and hickory, 3 (Chrysler 1910). The oak–hickory forest of Oosting (1942) is largely included in the oak–pine of Shantz and Zon (1924). South of the James River in Virginia the oak–hickory is separated from the maritime and magnolia forest by a rather wide interdigitated ecotone of pineland (Figs. 2-1, 3-1). Pineland occurs from New Jersey, where it is called pine barrens, to Texas and Missouri.

Plant Dominants

The dominant trees in scattered relict areas of climax and late subclimax forest are the post, white, black, blackjack, and scarlet oaks, and shagbark, mockernut, and, in some areas, pignut hickories. Three faciations are widely distributed: (a) white oak–post oak–hickory, in the Duke Forest in Virginia (Oosting 1942), in St. Charles County, Missouri (Brown 1931), and elsewhere; (b) white oak–hickory, in the Duke Forest and in Arkansas; (c) post oak–blackjack oak–hickory, in xeric habitats in the Duke Forest and on small areas in Arkansas. Telford (1926) maps the latter faciation for the southern third of Illinois with a species composition roughly post oak, 74 per cent; blackjack oak, 12; hickory, 7; and miscellaneous, 7. Near the western edge of the forest, the black hickory replaces other hickories (Tharp 1926, Bruner 1931). A black oak–white oak–hickory forest also occurs in southern and central Illinois (Telford 1926) and in Maryland, with scarlet oak and two other species of hickories added in the latter state (Hotchkiss and Stewart 1947). Red oak occurs along with the white and black oaks in the best habitats, and there is a scattering of species from the tulip–oak forest (Shreve *et al.* 1910, Wells 1928, Brown 1931, Putnam and Bull 1932, Oosting 1942, Arend and Julander 1948).

Competition was observed among plants in small trenched plots in the Duke Forest in climax and second-growth mixed forests. At the end of three years, one white oak seedling had survived and grown while apparently all other seedlings had been replaced by about the same number of mainly different species (Korstian and Coile 1938). An immense mass of living roots was exposed in the trenching process. These roots are of importance, especially on their decay, in providing food for various early stages of insects, earthworms, and other small animals.

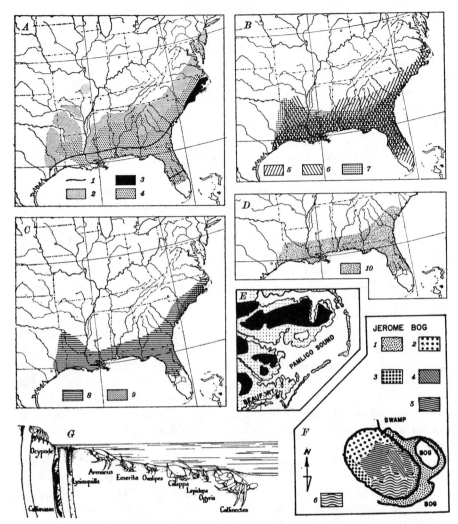

Fig. 3-1. Features of southeastern forests. A. Relations of magnolia and maritime forests to the oak–pine forest and longleaf pine fire climax: 1, northern and southern limits of climax magnolia forest (magnolia–laurel oak–redbay); 2, oak–pine forest of Shantz and Zon (1924); 3, maritime forest (live oak–yaupon–sweetbay); 4, distribution of longleaf pine which is dependent on fire (after Preston 1948). B. Distribution of three character-istic dominants of the magnolia forest: 5, laurel oak; 6, redbay; 7, magnolia (after Preston 1948, courtesy Iowa State College Press). C. Distribution of amphibians in magnolia–maritime area: 8, distribution of the dwarf salamander (after Conant 1958, Houghton Mifflin Company); 9, Brimley's chorus frog (after Wright and Wright 1932, courtesy Comstock Publishing Company, Inc.). D. 10, distribution of Florida free-tailed bat (after Burt 1952), Houghton Mifflin Company. E. The black areas, peat deposits near Pamlico Sound. F. Distribution of communities in Jerome Bog. Note oval shape of bog and northwest–southeast orientation: 1, sand area around bog depressions; 2, pine forest; 3, bay forest (*Magnolia virginiana* and *Acer rubrum*); 4, Atlantic white cedar for-est; 5, tall shrubs (*Ilex coriacea, Zenobia pulverulenta,* etc.); 6, low shrubs (Buell and Cain 1943). G. Crabs which work over marine deposits and prepare them for occupation by shore plants and animals (Pearse *et al.* 1942).

The loblolly pine and pitch pine, important subclimax trees, are mixed with the oaks in late seral stages. Some Virginia pine and shortleaf pine occur in the oak–hickory climax area. The longleaf pine belongs to more southerly communities (Fig. 3-1). The largest primitive areas of oak–hickory lacking pine were in southern Missouri and in northern Arkansas.

Animal Dominants and Influents

Among the larger animals, only the species common to the deciduous forest as a whole appear to have been present under pristine conditions, if a study done in Missouri is representative (Bennitt and Nagle 1937). The turkey may have had its largest populations in oak–hickory because of edibility of the acorns of post oak and blackjack oak. Large carnivorous animals were represented on the east-central portion of the Ozark Plateau in 1934 by one wolf per 10 square miles (26 km²). The U.S. Biological Survey reported 218 wolves killed on the Ozark Plateau in 1923, 1924, 1925, and 1933. In the same period 42 bobcats were killed.

The deer's impact on the forest has not been fully analyzed. The bear was credited with destroying acorns, but bear populations were small. A low population of the gray fox extended north nearly to the limit of the post oak as a dominant, but there are no records of gray fox in northern Oklahoma (Blair 1938) where the forest is poor. The gray squirrel population was estimated by Bennitt and Nagel at 22 per square mile (8.5/km²). Fox squirrels were more numerous in open woods. Raccoons, opossums, and striped skunks have a breeding preference for stream sides and forest edges but range through the forest for food. Their populations in Missouri were declining in 1934 and fluctuating around six to nine raccoons, two skunks, and three opossums per square mile (2 to 3.5, 1, and 1/km², respectively). Few mammals appear to have large populations in oak–hickory forests or in pinelands.

In Illinois, squirrel food comes from 57 species of deciduous trees. Only about 10 per cent of these species are not found in the climax. This food includes nuts, seeds, buds, and fruits (Brown and Yeager 1945).

There were about five turkeys per square mile (2/km²) in Missouri before settlement by white man. The fruits of all the common deciduous forest trees, shrubs, and vines were utilized by turkeys with the acorns of the post oak and blackjack oak being favored because of their small size. The flowering dogwood contributed much turkey food. The enemies of the turkey included all the predatory mammals and many of the larger birds. The predation of three species of hawks on the turkey was nearly twice that of owls, crows, or wolves. The ruffed grouse, another gallinaceous species, also formerly occurred (Bennitt and Nagel 1937, Dalke *et al.* 1942).

Those smaller animal constituents not generally distributed through the deciduous forest do not often follow the concentric or U-shaped distribution of the oak–hickory forest as shown in Figure 2-1. However, there are mammals, such as the golden mouse (Fig. 3-2) and cotton mouse, that are characteristic of the lowland and southern portion of the deciduous forest as a whole. The northern limits of these species usually fall near the lower Ohio and lower Potomac rivers. This is also the boundary line between the *borealis* and *virginianus* subspecies of the deer, *Odocoileus virginianus*.

Fig. 3-2. A night photograph of a golden mouse and nest on a grapevine several feet above the ground (photo by Maslowski and Goodpaster).

The forest snakes, reported by Clarke (1927, 1949), include the copperhead, which eats frogs, 40 per cent; mice, 27; and birds, 18; the rough green snake, a shrub climber which feeds on katydids, crickets, and other small invertebrates; the rat snake, which eats mice, 35 per cent; rats, 20; rabbits, 8; birds, 17; and squirrels, 10; the coachwhip, a tree climber which feeds on birds and mice; and the speckled kingsnake which prefers poisonous snakes. Three hundred and one stomachs of the speckled kingsnake contained 13 cottonmouth moccasins, 17 copperheads, 5 coral snakes, and 9 timber rattlesnakes. These poisonous snakes made up 13 per cent of its diet, the rest being eight more species of snakes along with birds and mammals. Similar food habits prevail throughout the deciduous forest (Surface 1906). However, reptile populations decrease toward the north. Lizards occur, and some appear in the food of snakes. The slimy salamander is the only salamander found regularly in the oak–hickory forest.

Invertebrates and the Fallen Tree Trunk

Invertebrates of the ground and fallen tree trunks are very widely distributed in the deciduous forest. The fallen tree presents a locus for a succession of plants and animals correlated with stages in its disintegration. The history of 125 oak logs in the Duke Forest outlines some features of the process. In the first year after falling, the phloem becomes bored into by the common cerambycid beetles *Xylotrechus colonus, Graphisurus fasciatus,* and *Romaleum atomarium.* Subcortical fauna invades where the bark is loosened. The diptera larvae *Megaselia* and *Lonchaea,* mites, and collembolons are present. The carabid *Tachyta nana* feeds on the mites and collembolons.

During the second year, wood-rotting fungi soften the sapwood, and their mycelia become thickly extended through the unburrowed inner bark. Fourteen species of *Polyporus* and three species of *Fomes* were recorded. The tenebrionid *Alobates pennsylvanica* burrows in wood attacked by the "white-rot" fungi. Adults of *Popilius disjunctus* form burrows in rotten patches of wood. Wood-eating termites and carpenter ants enter, the ants being most abundant during the third year. Fifty additional insect species eat fungi.

In logs over three years old, the heartwood is attacked by the "red-rot" fungi *Polyporus sulphureus* and *P. graveoiens* so that it can be crumbled in the hand. The rotten-wood caterpillar occurs while a small amount of sapwood still remains.

After the third year, the number of species and general population declines. However, larvae of the scarabaeid *Polymoecus brevipes* often occur in large numbers (150/m³) in logs in advanced stages of decomposition (Savely 1939, Wolf 1938). In northeastern Illinois, the red-rotted wood is the favorite hibernating site of white-faced hornet queens.

There are about 12 per cent fewer species in pine logs than in oak logs, and populations are smaller. The large larvae of the lucanid *Pseudolucanus capreolus* occur under nearly every log. Probably the most important predaceous species is the larvae of the eyed-elater *Alaus oculatus.* The millipedes *Pseudopolydesmus serratus* and *Apheloria (Fontaria) coriacea* are generally present as they are widely distributed. The most prevalent snails and slugs are *Mesodon thyroidus, Zonitoides arboreus,* and *Philomycus carolinensis* (Savely 1939).

Comparison of Eastern and Western Borders

In the cross timbers of eastern Texas, post oak, blackjack oak, and black hickory are the principal plant dominants, and there are a few southern red oaks and white oaks. The understory trees are *Vaccinium, Ilex vomitoria, Ilex* (probably *decidua*), shining sumac, a small dogwood, littlelip hawthorn, coralberry, and a few others. The invertebrates include the spiders *Schizocosa crassipes* and *Marpissa lineata,* which are common also in Georgia and North Carolina, and the centipede *Neolithobius suprenans.* The ants *Monomorium minimum* (usually in oak–pine), *Ponera trigona opacior,* and *Leptogenys elongata* were recorded only in oak–hickory by Sparkman (1943). The snail *Polygyra leporina* is largely confined to parts of the oak–hickory and tulip-oak and to pure pine. There is a click beetle, *Crigmus texanus,* in the oak-hickory which occurs also in Arizona.

The oak–hickory climax in the east is in contact with the Atlantic shore from near the southern boundary of Virginia northward (Fig. 3-1). In an inland area near Norfolk, Virginia, containing a substratum of moist sand and silt with five centimeters of black top soil, Kearney (1901) found that pines were largely replaced by hardwoods. This late subclimax stage, made up of red maple, large sweetgum, large tuliptree, American beech, willow oak, swamp chestnut oak, white oak, black oak, blackgum, and American hornbeam, suggests that the final climax may be much more of a mixed forest than is commonly recognized. Understory trees include sourwood, rhododendron, wild crab, and Hercules-club. Five common species of vines are present. In a second-growth forest on Cape Henry, Egler (1942) reports post oak, sand post oak, black oak, white oak, and bitternut and pignut hickory mixed with Virginia and loblolly pine.

Succession on Bare Earth

In central Missouri, land abandoned after a grain crop is first covered with various species of grasses and forbs in a successional series during the first four years, and woody shrubs, such as sumac, and trees appear in the fifth year. Post oak and shagbark hickory invade by the sixth year, black oak by the twelfth year, and white oak after 15 years (Drew 1942).

In North Carolina, the plants that appear during the first year include 60 per cent crab grass, a European species, and about 11 per cent horse-weed. The crab grass would not, of course, have occurred in the succession in primitive days. During the second year, *Aster pilosus* becomes characteristic and horse-weed becomes dwarfed. In the third year, 23 species disappear and one new species invades. Beard grass becomes important (Oosting 1942, Keever 1950).

In abandoned fields in Georgia and North Carolina, insects of cultivated fields occur during the first year in great numbers. The least shrew, oldfield mouse, and harvest mouse appear during the first year and attain maxima populations early in the second year. The bobwhite and eastern meadowlark also become conspicuous during the second year. Cottontail rabbits and cotton rats appear during the third year.

Fields abandoned for six years usually show a scattering of pine seedlings, *Pinus taeda, P. echinata,* or at higher altitudes and farther north, *P. virginiana.* When the pines attain two to ten feet (1 to 3 m) in height, beard grass becomes scattered. This produces a condition similar to a forest edge. The cotton rat usually then disappears, the cottontail population declines, and a different group of birds appears. Forest species begin to visit the area. By the time the pines are 20 to 25 years old, seedlings of broad-leaved trees such as *Liquidambar styraciflua* and *Nyssa sylvatica* invade. In stands 30 to 40 years old, climax oaks and hickories are present (Oosting 1942, Billings 1938, Odum, personal communication).

Succession in Coastal Marshes

Along the Atlantic Coast from Virginia to New Jersey, salt and brackish water marshes grade from the open shore to the oak–hickory climax. Tide flats are an intricate network of rivers, creeks, ponds, bays, and lagoons separated by stretches of salt meadow of varying exposure, bottom, and biotic composition.

Protected areas in the outer sandy flats support abundant water ditch grass. The salt meadows here include fewer species of sedges than are found farther south. As the tide goes out of the marshes, various marine invertebrates become exposed, and shore birds flock in to feed on them (Stone 1937). The greater scaup duck feeds on the water ditch grass during the winter months (Cottam 1939).

Chrysler (1910) gives a list of 38 plants arranged in the order of their occurrence in soils with decreasing salt content. The first four are glasswort, nearest the sea, sea-lavender, salt reed grass, and salt-water cord grass. The composites *Iva* and *Baccharis* and the grass *Panicum* occur near the center of the series and usually in a stage preceding cedars and loblolly pine. Cat-tails come in a little later. The last four of the 38 salt-tolerating plants are swamp rose, arrowhead, lizard's tail, and bur-marigold. Marsh nesting birds extend from the salt reed grass into the cat-tails.

On the strand or sea beach at Cape Henry, Virginia, cedar, persimmon, hackberry, loblolly pine, and various broad-leaved trees occur. Farther from the water on higher ground and somewhat protected from salt spray are forests of bluejack oak being invaded by live oak and loblolly pine (Egler 1942). An oak–pine stage apparently precedes the climax (Harshberger 1916).

Farther inland, fresh water swamps usually have Atlantic white-cedar growing around the margins of ponds which they gradually invade and fill. Pines follow the cedar and are in turn followed by oaks (Harshberger 1916).

The New Jersey pine barren area was originally a pitch pine forest with trees 75 feet (25 m) in height. In A.D. 1600, it was probably surrounded by deciduous forest. This pine forest served as a refuge for deer, wolves, and other large animals driven out of the deciduous forest under pressure from the early settlers. The beaver was extirpated first, followed by the bear and later the deer (Harshberger 1916). Wolves were put under a bounty in 1682.

South of Virginia the oak–hickory forest is separated from the Atlantic and Gulf shores by a narrow maritime forest and by the magnolia forest.

MAGNOLIA AND MARITIME FOREST

This broad-leaved, largely evergreen forest makes up the magnolia–deer–oak faciation (Figs. 2-1, 3-1, 3-3) and extends along the Gulf and Atlantic coasts from east Texas to central South Carolina, including most of the Florida Peninsula. Climax areas are small and scattered, lying in large areas of mixed hydric and xeric vegetation.   Pine forest is interspersed between the other communities and occupies a wide ecotone area between the maritime and magnolia forests and the oak–hickory forest to the north.

Coastal Plain

At the outer edge of the piedmont is a downward slope, called the "fall line."   The fall line represents the inland limit of a former marine submergence and extends roughly from northeastern Mississippi through central Alabama, Georgia, South Carolina, east-central North Carolina, and Virginia to northern Delaware.   In the west, submergence covered a Mississippi embayment area from Texas to Kentucky.   The withdrawal of the sea exposed a number of rock layers thinly overlaid by sea bottom deposits.   This is the "coastal plain."

A concentric series of different surface materials stretch from the middle of Mississippi around the southern end of the Appalachian highland into South Carolina (Harper 1913, 1914, 1930).   In the vicinity of the fall line

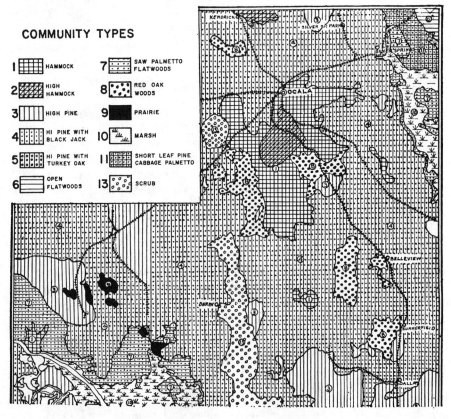

Fig. 3-3. Map of about 90 square miles (233 km²) of late subclimax high hammock near and south of Ocala, Florida, with seral vegetation in the magnolia climax area, showing 13 types of vegetation (after Sellards *et al.* 1915, modified by J. R. Watson).

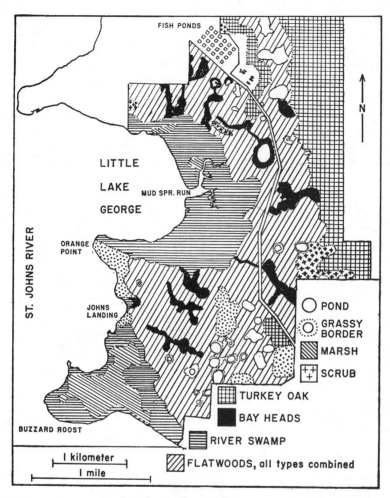

Fig. 3-4. Types of vegetation which appear to influence mammals (Moore 1946).

there is usually a line of sand hills. This dune area is poorly developed or absent in Alabama where there occur belts of red soil, black soil, and silty sand. A zone of red soil or red hills reaches from Tennessee to Virginia. It averages about 30 miles (48 km) wide across Alabama and Georgia and dips southward at one point to within 25 miles (40 km) of the Florida border. The soils range from sandy loam to clay loam. Ponds are common; streams are numerous. Next comes a limestone belt that extends from Mississippi southeasterly into western Florida and then northeasterly across Georgia into the Carolinas, maintaining a width of about 20 miles (32 km). The surface is undulating and contains many ponds. Finally, beginning in Louisiana and crossing Mississippi, Alabama, western Florida, and Georgia and extending into North Carolina and probably Virginia, there is a broken strip of rolling lands, varying in width from 15 to 100 miles (24 to 160 km), covered with oaks, pines, and wire grass (Wells and Shunk 1931).

The flatwoods substratum, occurring in flat, poorly drained sandy areas underlaid with a layer of impervious clay, cover large areas in Florida and scattered areas elsewhere. Dunes occur along the coast and for some distance inland in Florida and Alabama.

The entire pineland, magnolia, and maritime area is geologically, physiographically, and ecologically young, progressively so from north to south. Its physiographic youth is shown by the poor drainage in many parts, the numerous swamps and shallow lakes, and the sluggish, ill-defined, and crooked streams.

The continuity of the concentric zones is broken by the valley of the lower Mississippi River. Alluvial deposits may be separated into low terraces, second terraces, and high terraces (Putnam and Bull 1932). Loess banks occur adjacent to the Mississippi Delta. There is an abundance of oxbows and ponds on the Mississippi River floodplain and in the Delta area.

Climate

Rainfall from March through September varies from 42 inches (105 cm) along the Gulf of Mexico to 32 inches (80 cm) along the northern boundary of the magnolia forest area (Fig. 3-5). Annual rainfall ranges from 45 to 62 inches (112 to 155 cm) with spring being the driest season. There is usually

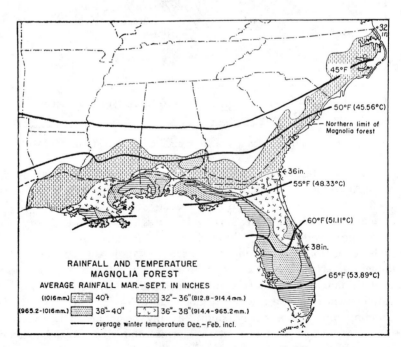

Fig. 3-5. Variations in rainfall and temperature in the magnolia climax area.

less than one inch (2.5 cm) of snow. Local variations in rainfall and the zonation of winter temperatures are shown in Figure 3-5.

General Characteristics of the Magnolia Forest

Much of the climax forest occurs in "hammocks." A hammock is a dense growth of mesophytic broad-leaved trees growing on a slightly raised substratum

Fig. 3-6. Magnolia (magnolia–holly–bay) late subclimax forest near Gainesville, Florida (1930 photos by J. R. Watson). A. Trees are chiefly magnolia, a few redbays, hollies, and a somewhat larger number of hophornbeams. The shrub–herb layer consists of French mulberry, two-eyed-berry and the grass *Oplismenus setarius*. In the more open places are also *Vernonia* and *Bidens bipinnata*. B. A large grapevine in the magnolia–sweetgum low hammock. C. A red oak–hickory hammock containing *Quercus falcata, Carya glabra,* and relict pine. Undergrowth consists of two-eyed-berry, French mulberry, tree sparkleberry and poison ivy. D. Bromeliads, *Tillandsia,* and orchids, *Epidendrum,* growing on a trunk of magnolia.

and not wet enough to be a swamp. A "low hammock" is somewhat moister. Important trees are southern magnolia, American holly, redbay, laurel oak, American beech, and live oak. Swamp chestnut oak, Carolina basswood, winged elm, sugarberry, pignut hickory, and white ash are also widespread (Fig. 3-6).

The understory includes the following trees: eastern hophornbeam, American hornbeam, redbud, hawthorns, and yaupon. Shrubs and vines are tree sparkleberry, common sweetleaf, yellow jessamine, cross-vine, and Virginia creeper. Characteristic herbs include two-eyed-berry, wild cucumber, and the Spanish moss that occurs on many trees.

The climax occurs in two subfaciations: magnolia–laurel oak–redbay (*Persea borbonia* of Watson 1926) and magnolia–laurel oak–beech. There is evidently no dominant beech east of a line from the mouth of the Apalachicola River in Florida to the mouth of the Savannah River. This serves as a rough boundary between the two faciations. The magnolia–laurel oak–redbay subfaciation (magnolia–bay–holly of Watson) is confined to Florida and the Atlantic coast of Georgia. It occurs on the best soil. The magnolia–laurel oak–beech subfaciation differs in (a) the presence of beech as a climax dominant, (b) an increase in the importance of laurel oak, and (c) an increase in abundance westward of Carolina laurelcherry. In most cases some pines are present. About half of the understory is evergreen.

Outlying Areas of Magnolia Forest

Bartram (1914) describes the vegetation of Moore's Island near Charleston, South Carolina, in 1765. He mentions a magnolia, probably *Magnolia grandiflora,* redbay, numerous evergreen oaks, American beech, and basswood. Goodrum, Baldwin, and Aldrich (1949) report a similar grouping of species on Bull Island northeast of Charleston. Black Beard Island, off the east coast of Georgia, has seral stages of the magnolia forest, while Okefinokee Swamp has both seral stages and the climax (Wright and Wright 1932). Goodrum (1940) found scattered small areas of magnolia forest on the higher floodplain bottoms of east Texas rivers.

Animal Influents

The three subspecies of deer in the magnolia forest exhibit no differences of ecological significance. Pristine abundance of deer is suggested by the many predators. In the Gulf Hammock, a large diversified subclimax and early seral area located on the west side of the Florida Peninsula, the fall and winter food of deer was found to be largely acorns of water oak, 16 per cent; live and laurel oaks, 15; deer's tongue, 15.6; and tassel-white, 9.6 (Strode 1954 and personal communication). Mushrooms constituted 9.6 per cent and the berries of saw palmetto and two gallberries totaled 8.8 per cent. Leaves of magnolia, laurel oak, and live oak were taken, and about 40 other plants were sampled. The summer food of a deer herd near Ocala was 25.5

per cent mushrooms. The flowers and fruits of palmettos constituted 12.9 per cent of the food; laurel, water, and five other oaks 7.5 per cent; and the water plants *Bacopa caroliniana* and *Brasenia schreberi* 32 per cent.

Black bears occurred through the magnolia forest and some survived for a long time in the Mobile Delta and Mississippi floodplain areas. Wolves are occasionally still seen in some areas. The mountain lions of Florida are the largest of the species. The Florida race of bobcat *Lynx rufus floridanus* is limited to the magnolia forest. The wood rat and the golden and cotton mice cover the climax area but extend their ranges farther north. The gray squirrel is probably more abundant in Louisiana than in Florida. In east Texas, it has a preference for the moist hammocks (Goodrum 1940). Other species are restricted to the faciation but occur in seral stages.

The turkey is a prominent bird. In hammocks along the Gulf, acorns of live oak make up 49 per cent of its food during the winter. Seeds of love grass and grasshoppers also enter into its diet (Strode, personal communication). The small climax areas afford nesting sites for a large number of birds.

Two lizards are especially noteworthy in Florida: one, *Eumeces egregius onocrepis,* inhabits dry late seral areas, and the other, a worm lizard, *Rhineura floridana,* occupies climax hammocks. The common snakes of the oak–hickory forest are present in Louisiana. The "chameleon" *Anolis carolinensis,* brown skink, and spadefoot toads are typically present (see also Fig. 3-1C).

Dozier (1920) found that most of the insects of the magnolia hammocks near Gainesville, Florida, are distributed throughout the southern states. In the soil litter and in decaying logs he found the rhinoceros beetle *Stratigus antaeus* that belongs to the Gulf states, the cerambycid *Stenodontes* with tropical relatives, and the darkling beetle *Alobates barbata* which is widely distributed in the moister portions of the deciduous forest. Darkling beetles, in general, are numerous. The butterfly larva *Papilio polydamus* feeds on *Aristolochia* and the "orange dog" larva *P. cresphontes* feeds on prickly-ash (Watson 1926). Both plants are hammock shrubs. The larvae of *Gelechia cercerisella* fold the leaves of the redbud. The magnolia foliage is only slightly damaged by the scale *Toumayella turgida* and the leaf miner *Phyllocnistis magnoliella.* The flowers are injured by *Thrips spinosus* and *Trichiotinus piger* (Dozier 1920).

Three or more species of the crane fly *Dicranoptycha* are characteristic of mesic hammocks (Rogers 1933). The spider *Conopistha globosa* occurs only in hammocks. The jumping spider *Lyssomanes viridis* was found in hammocks but also occurred in the tulip–oak faciation and elsewhere. The snails *Mesodon inflectus, Discus patulus,* and *Zonitoides arboreus,* and slug *Philomycus carolinianus* occur from Louisiana to Florida but records are few.

Mud-covered Shores of the Gulf of Mexico

One such shore near the mouth of the Mississippi coast west of Cat Island supports salt-water cord grass, with spike grass on slightly higher ground. The spike grass probably succeeds the cord grass (Penfound and O'Neill 1934).

On the mud in and near the cord grass are found fiddler crabs, the marine snail *Littorina irrorata*, tiger beetles, and two species of shore bugs. The large arthropod population on the grasses, about 80 per square meter, includes the spiders *Argiope trifasciata*, *Tetragnatha lacerta*, and immature *Phidippus*. Grasshoppers are represented by numerous immature *Conocephalus* and *Ophulella olivacea*. Among the flies are the picture-winged species *Chaetopsis aenea* and *Gymnopsoa texana*. The spotted lady beetles *Naemic seriata* are numerous.

The arthropod population is smaller in the spike grass (45/m²). The spiders are the tube weaver *Serigiolus variegatus* and *Tetragnatha pallescens*. Orthopterons include immature *Conocephalus* in much smaller numbers than in the cord grass. The picture-winged fly *Chaetopsis debilis* replaces *C. aenea* and is much more abundant. Tachinid flies are represented by numerous *Myiophasia metallica* and the sarcophagids by *Gymnopsoa texana*. *Polistes fuscatus* is the principal hymenopteron.

Salt and Brackish-Water Marshes

Marshes are extensive all along the coast. The coast of Louisiana has an area of brackish and fresh-water marshes, ponds, and lakes about 20 miles (32 km) wide and nearly 200 miles (320 km) long. Included along the shore in the western half of the area are sand ridges 2 to 5 miles (3 to 9 km) wide. The continuity of the marsh is also broken by small clay and shell ridges. The marshes are fed by an annual rainfall of nearly 60 inches (150 cm) and a network of rivers.

The marshes are largely covered by grasses, rushes, and sedges. Beginning with the most salt tolerant, the principal species, with the per cent of salt in the water that is tolerated, rank as follows (Penfound and Hathaway 1938): salt-water cord grass, 0.55 to 4.97; spike grass, 0.45 to 4.97; black rush, 0.9 to 0.89; saw-grass, 0.0 to 0.2; and maiden-cane, 0.0. There are many other species present, notably the water-millet which is confined to fresh water. On the Gulf coast of Florida, *Juncus roemerianus* is often the principal plant in the marshes, forming dense, almost impenetrable tangles four to six inches (1.5 to 2 m) high.

The Mississippi River Valley is one of the main flight routes of North American migratory waterfowl from the north and funnels a large number of birds into these coastal marshes for overwintering. The most important foods of these waterfowl are the cord grass, spike grass, and saw-grass (Fig. 3-7).

The Sabine National Wildlife Refuge of nearly 150,000 acres (6000 hectares) is largely a grass marsh somewhat disturbed by intersecting canals. In 12 inches (30 cm) of water in a stand of saw-grass, there were the top minnow *Gambusia affinis*, the leech *Placobdella*, and numerous amphipods *Hyalella azteca*. Along the canals, the vegetation is composed principally of *Phragmites*, *Spartina patens*, *Baccharis halimifolia*, and *Hibiscus lasiocarpus*. The shrimp

Palaemonetes paludosus, amphipods, top minnows, water beetles, water-boat-men, the sunfish *Lepomis symmetricus,* otters, and muskrats are present. The 1952 Christmas bird count (National Audubon Society) recorded 88 species of birds of which 10 or more individuals were seen. The largest populations were blue goose (estimated at 24,000) and ducks (Lowery *et al.* 1952).

The Aransas National Wildlife Refuge on the coast of Texas contains waterfowl, turkeys, deer, and peccaries. The wooded parts are characterized by live oak and redbay (bay brush). Both species are invading the grass-covered areas. Goodrum (personal communication) calls attention to coreaceous-leaved plants along the Texas coast. Among these is southern bayberry which invades the coastal prairies (Tharp 1926) and is widely distributed to the east and northeast in late seral stages of magnolia forest (Harper 1914, Sellards *et al.* 1915).

Fig. 3-7. Ducks, chiefly pintails, on a Savannah River marsh. In the background salt-water cord grass is fringed by *Salicornia virginica.* At the extreme right, eastern baccharis and seashore-mallow occur in front of yaupon and oak (U.S. Fish and Wildlife Service, photo by F. W. Quradnik).

The salt marsh grasshoppers around the Gulf and on the Atlantic Coast are the salt marsh locusts *Ophulella o. olivacea,* and *O. halophila,* the latter being restricted to the Gulf Coast, and meadow grasshoppers *Conocephalus spartinae, C. nigropleuroides, Neoconocephalus lyristes,* and *Orechelimum fidicinium.* Some of these also occur in freshwater marshes. On the Gulf coast of Florida, the thrip *Haplothrips leucanthemi* and the southern white butterfly *Ascia monuste* are characteristic. The salt marsh mosquitoes are abundant except during the winter months. The malarial mosquito *Anopheles* is comparatively scarce. The horse fly *Tabanus* and deer fly *Chrysops* are abundant in late spring. The mangrove water snake is present along with common marsh birds (Watson 1926). Raccoons and deer occur on the "islands" or cheniers in the coastal marshes.

The substratum of these islands grows by deposition, and transects from the marsh onto islands show a series of developmental stages leading to forest (Penfound and Hathaway 1938). Surrounding one such island there was a zone of *Spartina–Distichlis–Juncus* marsh, 10 feet (3 m) wide, with the salt reed grass *Spartina* reaching 12 feet (3.7 m) in height. With a rise in the substratum inward of 8 inches (20 cm) and a salinity of 0.2 to 0.7 per cent shrubs occurred. The marsh elder was most abundant, but eastern baccharis and dwarf palmetto together about equaled it in numbers. Of the herbaceous components, the versatile salt-meadow grass was predominant. Marsh bindweed, marsh morning-glory, switch grass, and seaside goldenrod characterized both the shrub and the salt reed grass communities. With an increase of 14 inches (35 cm) in the substratum and a further reduction of salinity, redbay, sweetgum, and various oaks appeared, followed by live oak on the highest part of the island.

In another case, there was a waterlily community in a pond with water of 0.4 per cent salt and a bottom 17 inches (42 cm) below the center of the island. Next came saw-grass, then pondcypress and swamp tupelo, and finally pine.

The time at which plant species enter into particular seres depends on the amount of salt in the substratum (Fig. 3-8). The salt tolerance of the important woody species of the area is as follows: eastern baccharis, 0.00 to 1.98 per cent; common buttonbush, 0.00 to 0.89; black mangrove, 3.68 to 4.97; marsh elder, 0.21 to 1.98; water tupelo, 0.00; pondcypress, 0.00 to 0.89; and

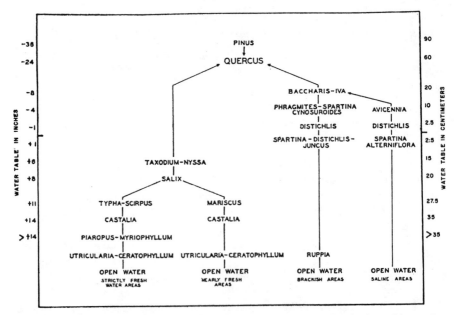

Fig. 3-8. Diagram of succession and convergence of communities beginning with varying degrees of salinity on the coast of Louisiana (Penfound and Hathaway 1938).

swamp tupelo, 0.21 to 0.53 (Penfound and Hathaway 1938).   Evidently live oak and redbay tolerate only minute amounts of salt as do the trees of the oak–hickory and maple–beech–white pine forests farther north.   Beech and magnolia both occur on small islands off the coast of Georgia and South Carolina, which indicates that magnolia forest may eventually enter into these marshes.

Freshwater Ponds and Marshes

Lake Mize, near Gainesville, Florida, is a typical, small, deep lake of this limestone area.   It contains semifloating islands with cutgrass, St. John's wort, submerged bog-moss, and spike-rush.   The emergent plants around the

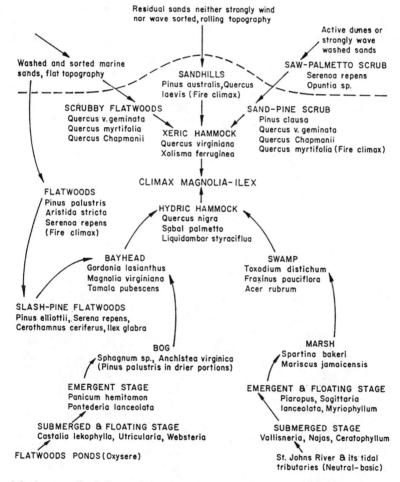

Fig. 3-9. A generalized diagram showing the principal stages of plant successional convergence to subclimax stages and then to the climax. The stages below the dashed line are absent from the Welaka Reservation and, except for the saw palmetto scrub, have not been described (Laessle 1942).

lake are sand-weed, the grass *Sacciolepis striata,* water-pennywort, cutgrass, St. John's wort, smartweed, umbrella-grass, and pipewort.

There are a number of both northern and southern species of vertebrates here. Probably having their largest populations in seral stages of the magnolia faciation are the Florida water snake, Florida snapping turtle, Florida cooter, Florida red-bellied turtle, chicken turtle, and the softshell turtle. Some 25 or more species of insects and large crustaceans (Fig. 3-1) also influence the succession to terrestrial communities (Harkness and Pierce 1941).

The shallow freshwater ponds, often called "prairies," in Okefinokee Swamp in Georgia are similar to those in the Welaka Reservation in Florida which were studied by Laessle (1942). Waterlilies and scattered bladderworts *Utricularia inflata* and *U. purpurea* are noteworthy in the Welaka Reservation at depths of two to five feet (0.6 to 1.5 m), and succession proceeds to a hydric hammock (Fig. 3-9). In Okefinokee Swamp, golden club and yellow-eyed grass tend to distinguish these ponds from those of the northern interior. In midsummer, a limpet-like snail, damselfly nymphs, squatting dragonfly nymphs, cricket frogs, shrimp, chironomids, and top minnows are found. There are alligator holes surrounded by vegetation differing from that noted above. Redroot, an important food of the sandhill crane, pickerelweed, and pitcher-plant are also present.

The maiden-cane is best developed in water which averages one to three feet (0.3 to 1 m) in depth. Common associates with it are the floating-heart and the bladderwort *Utricularia purpurea.* There are commonly alligator trails through such vegetation, and nests are made from the grass (Fig. 3-10). In July, meadow grasshoppers, slender grasshoppers, and the immature leaf-

Fig. 3-10. Alligator nest surrounded by cleared area in maiden-cane, a buttonbush is present; located at Grand Prairie near Little Cools Lake in Okefinokee Swamp, July 7, 1922 (photo by F. Harper for A. H. Wright's Cornell Expedition).

hopper *Draeculacephala* are numerous. Two species of ants, *Pseudomyrma pallida* and *Dolichoderus plagiatus pustulalus,* are present on the grass along with the spider *Clubiona tibialis.* Frogs are generally numerous in both species and individuals.

Shrubs and vines that invade around the edges of the ponds are buttonbush, swamp loosestrife, *Smilax laurifolia,* tetter-bush, cat tail grass, *Cyrilla racemiflora,* and others.

In the larger hammocks in Okefinokee Swamp, such as Floyd's Island, occur water oak, live oak, laurel oak, and southern magnolia (Wright and Wright 1932). Dwarf palmetto is present. Some shrubs are *Hypericum,* wild olive, and huckleberry. Due to shade and lack of herbaceous growth, invertebrates are generally few. The snail *Polygyra pustula* is common, and largely limited here and south through Florida. The numerous ants include *Odontomachus haematoda insularis, Ponera trigona opacior,* and *Crematogaster m. minutissima.* The spider *Nephila clavipes,* a showy representative of a tropical group, is conspicuous. Other animals present are mainly the familiar deciduous forest species. The Okefinokee Swamp includes various other types of aquatic communities, especially the cypress swamps.

Cypress and Other Wooded Swamps

Cypress swamps characteristically occur adjacent to large rivers, sluggish streams, lakes and ponds on the coastal plains, in the lower Mississippi Valley, and on the Florida Peninsula. The pondcypress occupies ponds and lakes. Baldcypress is commonly found in river swamps and floodplains and along lake shores, and is the more common species in the "Big Cypress Swamp" west of the Everglades (Laessle, personal communication). Numerous projections, "knees," extend up into the air from the roots of this tree and are frequently covered with mosses and ferns. "Knees" are not very frequent on pondcypress. The trunks of the trees frequntly bear orchids, and long festoons of Spanish moss droop from the branches. Commonly associated trees are redbay, red maple, and swamp and water tupelo.

Common animals of cypress swamps are swamp rabbits, cottontails, opossums, raccoons, skunks, bobcats, mountain lions, and bats (Watson 1926, Blair 1935). Animals are most numerous in the long narrow cypress ponds of southern Florida that are surrounded by thickets of cocoa-plum and other shrubs (Davis 1943). Throughout much of their range, these swamps were the regular home of the ivory-billed woodpecker and are still inhabited by pileated woodpeckers, red-bellied woodpeckers, red-shouldered hawks, short-tailed hawks, and many smaller birds, including the red-eyed and white-eyed vireos, parula and prothonotary warblers, Acadian flycatchers, cardinals, and Carolina wrens. Egrets, herons, and ibises place their nests in cypress trees. Cottonmouths are usually present and in the larger swamps, alligators. Insects, except mosquitoes, are not abundant in the dense shade of the cypress swamps, but certain crane flies, *Limonia (Geranomyia) vaduzeei* and *Polymera georgiae,* are characteristic (Watson 1926, Rogers 1933, Davis 1943).

Saw-Grass Pseudo-Savanna and Prairies

Large areas in the southern two-thirds of Florida have the appearance of savanna. Areas of tall grass, or most often the sedge saw-grass, are skirted with shrubs or interspersed with islands of shrubs, among which dense clumps of cabbage palmettos occur. The saw-grass areas are largely under shallow water during the summer and comparatively dry during the late winter and early spring. Where the water table had gradually become permanently lowered, succession apparently proceeds to flatwoods, particularly where there is an impervious substratum.

The marsh rice rat and marsh rabbit frequent these marshes as do most of the larger deciduous forest mammals (Watson 1926). The rice rat builds a nest supported by stiff sedges and grasses. Limpkins, Everglade kites, anhingas, ibises, and herons frequent such marshes in the upper St. Johns River. Large numbers of common and purple gallinules and sandhill cranes also breed in the community (Howell 1932). The reptiles are species commonly found in moist places. The mud snake burrows into the mud. Adults of various moths, butterflies, and skippers are frequent. The grasshoppers *Mermiria intertexta* and *Bucrates malivolans* (Friauf 1953), the earwig *Doru davisi*, a club-legged plant bug *Acanthocephalus femoratus*, and the leaf-footed plant bug *Leptoglossus phyllopus* (Watson 1926) are abundant.

"Prairies" appear to be a temporary stage between the shallow marshes and flatwoods, particularly in the Kissimmee and other areas of central and south-central Florida (Harper 1927). The important plants are wire grass, Indian grasses, saw palmetto, dwarf wax-myrtle, sand live oak, and inkberry. Small hammocks and clumps of cabbage palmetto are scattered over the prairies. The characteristic birds are the common nighthawk, loggerhead shrike, eastern meadowlark, burrowing owl, caracara, and red-tailed hawk (Howell 1932).

Flatwoods

These peculiar, poorly drained areas in Florida and southeastern Georgia are underlaid with an almost impervious hardpan. They contain ponds, sometimes with cypress, and appear to belong to the hydrosere. The vegetation consists of an open forest of pine, with longleaf pine the most common. Slash pine and pond pine are present locally. Saw palmetto is common, interspersed with gallberries, the blueberry *Vaccinium myrsinites*, pitcher-plant, sundew, and other species characteristic of boggy soil. Wire grass is generally prevalent. The longleaf pine extends into Louisiana and Texas, both on the flatwoods and in hilly regions (Brown 1945). In Alabama, there are also flatwoods composed of post oak. Northward in South Carolina, the longleaf pine grows in restricted flat areas underlaid by hardpan (Wells 1942), and in Georgia, some features of this community occur mainly seaward from a ridge about 40 miles (64 km) from the Atlantic (Watson 1926, Laessle 1942, Harper 1930).

In the Welaka area of Florida, the fox squirrel, cotton rat, cotton mouse, and short-tailed shrew are most abundant in flatwoods of various types (Fig.

3-4). The mouse population, *Peromyscus floridanus,* is high. All the common mammals of the region frequent flatwoods. The wood duck and various predatory birds use the pines for nesting. Characteristic smaller species are the bobwhite, common nighthawk, eastern meadowlark, Bachman's sparrow, rufous-sided towhee, pine warbler, yellow-throated warbler, brown-headed nuthatch, yellow-shafted flicker, red-cockaded woodpecker, and eastern bluebird. An occasional eastern diamondback rattlesnake and dusky pygmy rattlesnake occur along with the common deciduous forest snakes. Lizards are few (Moore 1946).

Friauf (1953) found 47 species of Orthoptera in the longleaf pine flatwoods, of which two-thirds were common. Especially noteworthy are *Gymnoscirtetes pusillus, Aptenopedes sphenarioides, A. aptera,* the katydid *Odontoxiphidium apterum,* and the crickets *Falcicula hebardi* and *Nemobius abitiosus.* Characteristic crane flies are *Tipula sayi, Limonia (Dicranomyia) liberta, Limonia (Rhiphidia) domestica,* and *Gonomyia sulphurella.* During the late spring, horse flies, deer flies, and mosquitoes are very abundant. Grass-feeding hemipterons are numerous, including the black thrip *Haplothrips graminis.* Gallberries are often covered by the sooty mold fungus growing in the honey dew given off by the Florida wax scale *Ceroplastes floridensis.* Seven species of butterflies and moths frequent the flatwoods, but none of them feeds on important plants (Watson 1926).

Live Oak Xeric Hammocks

Higher portions of the flatwoods succeed to live oak hammock through a shrubby stage of various dwarf oaks (Fig. 3-9). This is frequently the stage at which the xerosere and hydrosere converge. The dominant oak *Quercus virginiana* and subdominant trees of other species are commonly covered with drooping Spanish moss. These "live oak hammocks" are a characteristic feature of the landscape throughout the coastal area of the Gulf states, except in Texas. Bluejack oak, laurel oak, cabbage palmetto, and scattered relict longleaf and loblolly pines also occur in them. Shrubs are abundant, the most characteristic being myrtle oak, Chapman oak, tree sparkleberry, shining sumac, French mulberry, squaw huckleberry, and staggerbush. Common vines are the wild grapes *Vitis rufotomentosa* and *V. vulpina,* bullace-grape, Virginia creeper, and the "wild bamboo" *Smilax auriculata.* A dozen species of grass, restricted to the magnolia and maritime forest areas, include *Panicum villosissimum pseudopubescens, P. arenicoloides, P. aciculare,* the awn grass *Aristida tenuispica,* and some beard grasses of which *Andropogon elliottii* is very widely distributed. The pinweed *Lechea prismatica,* the golden aster *Pityopsis graminifolia, Crocanthemum corymbosum,* the goldenrod *Solidago chapmanii,* the tick-treefoil *Desmodium canescens,* the partridge-pea *Chamaecrista brachiata,* and the vervain *Verbena carnea* occur.

Moles and gray squirrels are abundant in these hammocks, while flying squirrels are found almost exclusively in them. The *Tillandsia*-inhabiting cockroach *Eurycotis floridana,* the peculiar slant-faced locust *Radinotatum c. carinatum,* the widely distributed *Amblytropidea occidentalis,* the crepitating grasshopper *Spharagemon crepitans,* and the wingless Florida locust *Aptenopedes aptera saturiba* are usually present, the last species being commonly found on staggerbush (Blatchley 1920, Watson 1926, Friauf 1953).

Pinelands

Xeroseral pinelands occur on the coastal plains with their inner boundary at the outer edge of the Piedmont Plateau. Four species of pines are important. Foremost among these is the longleaf pine which is very intolerant of shade; grows on dry sterile, sandy soils, as well as in flatwoods; and rarely mixes with broad-leaved trees. The species is absent from the alluvial valley of the Mississippi River (Fig. 3-1). In contrast, loblolly pine is tolerant of shade and occurs mixed with broad-leaved trees. It persists into the oak–hickory forest in North Carolina (Oosting 1942) and in early white oak–post oak stands in the alluvial valley of the Mississippi from northern Louisiana northward (Putnam and Bull 1932). The shortleaf pine is evidently less shade tolerant than the loblolly pine; it occurs as a relic from earlier stages in the more xeric post oak faciation of the oak–hickory (Oosting 1942) and goes farther inland than any of the others. It barely reaches Florida and covers only a little of the magnolia and maritime forest regions. The slash pine extends only a little way north of Florida but reaches westward into Louisiana. It is more tolerant of shade than the longleaf pine.

Other species of pines in the region are the sand pine, which grows on the most sterile sand and is limited to Florida; spruce pine, which grows on the coastal plain from South Carolina to Louisiana but mainly outside Florida, mixes with broad-leaved trees, and is quite shade-tolerant; Virginia pine, which is a short-lived, shade-intolerant tree; and the pond pine, which is limited to the Atlantic Coast.

Pine scrub occupies areas of deep white sand. A central Florida highland area known as the "Big Scrub" covers much of the Ocala National Forest and the Ocala Game Preserve. Sand pine and relatively tall evergreen shrubs peculiar to Florida and Alabama make up the plant constituents of the community. Herbaceous vegetation is sparse. The large shrubs are sand live oak, Chapman oak, myrtle oak, staggerbush, saw palmetto, semaphore cactus, and rosemary. Some of the smaller shrubs are scrub holly, *Polygonella, Vaccinium myrsinites,* and golden buttons. Other characteristic plants are the bluecurl *Trichostema dichotomum, Petalostemum feayi,* the umbrella-plant *Eriogonum longifolium,* and the dodder-like parasitic vine *Cassytha filiformis* (Watson 1926). This vegetation resembles that on coastal dunes (Fig. 3-11).

Animal populations are small and species few but the "Big Scrub" supports a deer herd (p. 68). Cottontails, ground doves, mourning doves, scrub jays, rufous-sided towhees, and blue-gray gnatcatchers are present (Watson 1926, Howell

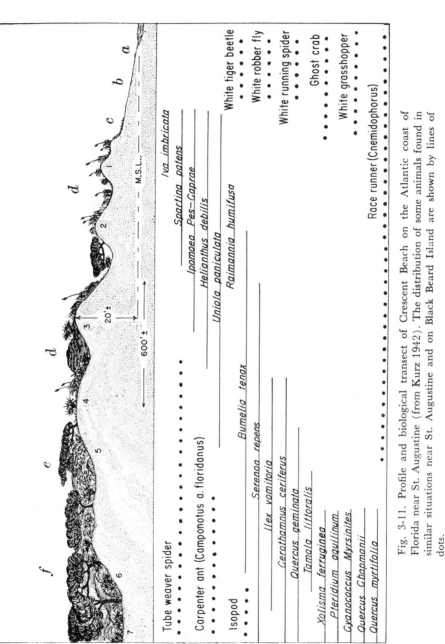

Fig. 3-11. Profile and biological transect of Crescent Beach on the Atlantic coast of Florida near St. Augustine (from Kurz 1942). The distribution of some animals found in similar situations near St. Augustine and on Black Beard Island are shown by lines of dots.

1932). The lizard *Sceloporus woodi* is indigenous. In the Ocala area, arthropods are scarce during both July and December. The coarse tough plants, however, support the locusts *Spharagemon collare, Scirtetica marmorata picta, Psinidia fenestralis,* and *Melanoplus k. keeleri.* The great southern white butterfly *Ascia monuste* is present (Watson 1926).

On Cat Island, Mississippi, new sand deposits are quickly covered with the sea-oats *Uniola paniculata* followed by *Andropogon maritimus, Cakile edentula,* and several other species. If there is protection from wind, this group is quickly replaced by the rock-rose *Helianthemum arenicola,* and the goldenrod *Chrysoma pauciflosculosa.* Slash pine, yaupon, and a few live oak appear in older areas. The low vegetation is chiefly grasses and forbs (Penfound and O'Neill 1934).

In plant communities similar to these but at Bayou Cady, Mississippi, the spider *Grammonota texanus,* the hemipterons *Creontiades debilis, Lygus apicalis, Orius insidiosus,* and *Peritrechus,* and the fulgorid *Pissonotus delicatus* were found.

Large dunes are not important along the Gulf Coast in the west and are of irregular occurrence along the west coast of Florida. Extensive areas of sand beach, backed by low sand dunes, are general, however, on the Atlantic Coast.

A series of belts or zones (Fig. 3-11) occur on the outer shore and low dunes of Black Beard Island National Wildlife Refuge, off the coast of Georgia (Kurz 1942).

a-b. The lowest zone extends from the moist beach up to the drift line. Marine worms occur in the sand near the water's edge and periwinkle eggs and an occasional horseshoe crab are washed ashore. Ghost crabs burrow into the sand over a wider belt. Marine turtle tracks are common. The white tiger beetle, the brown tiger beetle, and a white robber fly are conspicuous in early summer.

c. From the fluctuating drift line to the beginning of vegetation, numerous tracks of raccoons and sea turtles are found, and the white running spider is present.

d. Where vegetation begins to appear, salt-meadow grass, sea-oats, *Yucca,* and bayberry occur. Deer tracks are common, and the six-lined lizard is frequently found. Sea turtle nests of buried eggs are usually present in early summer. White grasshoppers, robber flies, and 18 species of shore birds were noted.

e. In the saw palmetto stage, grasses of the earlier stages drop out and some small trees make their appearance. The shapes assumed by these trees show the effects of wind. Insects and spiders on the vegetation number about 40 individuals per square meter, but on patches of open sand, they are reduced or wanting.

f. The live oak stage has in the soil the resident ants *Myrmothrix abdominalis floridanus,* the isopod *Porcellionides virgatus,* insect larvae, and various spiders. The trees have a small insect population.

The refuge has a good mixture of various types of seral vegetation. In 1943, the deer population was estimated at 40 per square mile ($15/km^2$), wood rats at less than 2 per acre (5/hectare), and raccoons at a little more than 1 per 2 acres (0.8/hectare). Alligators were common in the ponds.

Longleaf Pine–Turkey Oak–Wire Grass–Pocket Gopher Community

This is a very open forest dominated by longleaf pine and turkey oak. Blue-jack oak occurs locally, and live oak in lower and more fertile spots. Where fire has not been frequent, persimmon is present. Wire grass is a dominant in frequently burned over areas, although in Louisiana and farther west, *Andropogon tener* is more important than wire grass (C. A. Brown, personal communication).

This community covers a large area in southeastern North Carolina (Wells and Shunk 1931), but there is little mention of it in South Carolina. Harper (1930) estimates an area of 10,000 square miles (26,000 km²) in Georgia. These areas, together with a large section in western Florida and scattered spots west into Texas, add up to 20,000 square miles or more (52,000 km²).

Of special interest is the occurrence in the Gulf states of two burrowing species, the gopher tortoise and the southeastern pocket gopher or "salamander." The pocket gopher is abundant. Both species exert a considerable influence on the vegetation because the mounds of dirt which they shove up from their underground burrows afford a foothold for young plants in a place where competition with other plants is reduced. The gopher tortoise removes foliage of plants, but this does not appear as destructive as does the root-feeding habits of the pocket gopher. The gopher tortoise is particularly fond of both the leaves and the fruit of the gopher-apple. The seeds of the gopher-apple are not digested and hence are dispersed by the animal.

The Florida mouse is present in the longleaf pine–turkey oak community in small numbers. Cottontails are fond of the habitat. The bobwhite, Carolina wren, summer tanager, pine warbler, and yellow-throated warbler are the most characteristic birds (Watson 1926, Howell 1932). Common snakes in these open woods are crowned snake, eastern coachwhip, racer, rat snake, corn snake, and hognose snake.

Among the characteristic insects are the orthopterons *Amblytropidea occidentalis* and *Melanoplus rotundipennis;* the katydids *Arethaea phalangium, Orchelimum minor,* and *Conocephalus saltans;* the cricket *Falcicula hebardi;* and the yellow cockroach *Cariblatta lutea.* The dermapteron *Prolabia pulchella* occurs. Butterflies, which breed here, are *Precis lavinia coenia* and *Danaus gilippus berenice. Atlides halesus* is extremely abundant in the more southerly area about blossoms of chinkapin and *Eupatorium serotinum;* the larvae occur on oaks. The fiery skipper *Hylephila phyleus* feeds on grasses (Watson 1926, Friauf 1953).

Florida Red Oak–Hickory Forest

This forest is known only from small areas near Ocala and Gainesville, Florida. It contains as many hickories as oaks but little undergrowth. Watson (personal communication) indicates that it follows longleaf pine and turkey oak and precedes magnolia forest. It is sometimes followed by live oak forest. Sellards *et al.* (1915) lists the following overstory trees near Ocala: southern red oak, mockernut hickory, sweetgum, redbay, redbud, laurel oak, flowering dogwood, pignut hickory, and American holly. Understory trees and shrubs are also named. There are no animal studies in this community.

ATLANTIC MARITIME LIVE OAK FOREST

The climax forest on Smith Island (Cape Fear) in North Carolina (Wells 1939) is strongly dominated by live oak and contains some eastern redcedar and Hercules-club. The understory trees and shrubs are devilwood, tree sparkleberry, southern bayberry, yaupon, French mulberry, redbay, Carolina laurelcherry, and American holly. Several of these species, as well as red mulberry, laurel oak, and hornbeam or blue beech, also occur on Shackleford Bank, North Carolina. Woody vines include supple-jack, grape, and Virginia creeper. A broken area of maritime forest 300 miles (480 km) long and 2 to 20 or more miles (3 to 32 km) wide with its center near Cape Lookout, North Carolina, probably existed under pristine conditions, largely in seral stages (Fig. 3-1). Since a number of the species above listed also occur in the magnolia forest proper, this community may be designated the *live oak–yaupon–sweetbay–yellow-lipped snake subfaciation* of the magnolia forest.

Wells (1928) provides a diagram to show succession from salt marsh to forest, and Barnes (1953) provides further information. Trees growing along the shore where they are exposed to the salt spray blown in from the ocean have their canopies injured and deformed (Boyce 1954). Live oak may become abundant only 150 meters inland from mean tide. Although the live oak climax reaches inland only a short distance, it covers banks and islands a mile wide and sand deposits along shore, as well as filled-in swamps.

Animal Constituents

According to studies made near Beaufort, North Carolina, between 1869 and 1872 (Coues 1871, Coues and Yarrow 1878), black bear, wolf, white-tailed deer, and other common deciduous forest mammals and birds were present, ranging over both wet and dry seral stages.

The important snakes present are dusky pygmy rattlesnake, eastern diamondback rattlesnake, and yellow-lipped snake; the southern chorus frog and the southern toad *Bufo t. terrestris* are also present. These species are limited to magnolia and maritime forests and do not cross the Mississippi Delta. The dwarf salamander, squirrel tree frog, pig frog, and free-tailed bat cover the range of the magnolia–maritime forest. Brimley's chorus frog and the many-lined salamander are restricted to the maritime forest (Bishop 1943, Smith 1946, Wright and Wright 1949, Burt 1952, Conant 1958).

Succession

The area is immeasurably complicated as to physical conditions, but perhaps one may venture a biotic, rather than a vegetational, account of some of the seres.

A xerosere occurs on dunes and shifting sand. On the west end of Shackleford Bank, the lower or outer shore is covered by scattered sea-oats. The small beach beetles *Hymenorus* and chinch bugs are usually present in July. Barnes (1953) found the two spiders *Tibellus duttoni* and *Trochosa shenandoa* to be important. The former also occurs in forests. A little farther from shore, the bunch grass *Muhlenbergia capillaris* or *Andropogon scoparius littoralis* occurs along with the sea-oats. *Hymenorus* then becomes less abundant. The two spiders *Latrodectus mactans* and *Habronattus agilis,* appear on the grasses. The fulgorid *Megamelanus* occurs in the sea-oats (Metcalf and Osborn 1920). The spider *Aysha gracilis* is present here, in other coastal grass-covered areas, and in open oak–hickory forests of Texas, to suggest some of the peculiarities of spider distribution. The wasp *Polistes fuscatus,* the digger wasp *Microbembex monodonta,* and the ant *Dorymyrmex pyramicus flavis* are abundant. Two species of grasshoppers occur commonly: *Trimerotropis maritima* and *Schistocerca americana.* Nymphs of the small salt marsh cicada *Tibicen viridifasciatus* feed on the sea-oats and lay their eggs in the stems (Osborn and Metcalf 1920). The oldfield mouse also feeds on sea-oats. The ground-cherry *Physalis barbadensis glabra* and the croton *Croton puctatus* may appear along with the various grasses (Engles 1942). The marsh rabbit and the cottontail feed in such situations along with mourning doves, meadowlarks, mockingbirds, and song sparrows.

Yaupon in scattered clumps, American holly, the greenbrier *Smilax bona nox,* and northern bayberry appear among the grasses and forbs (Lewis 1917, Engels 1942). The soil now contains humus and will hold water. Here, *Meioneta beaufortensis* and *Schizocosa crassipes* are the principal ground spiders (Barnes 1953). The six-lined racerunner, mockingbird, song sparrow, and Fowler's toad are the important vertebrate inhabitants of the thickets (Engels 1942). The glass lizard is abundant and characteristic (Coues 1871).

In the woodland that follows, the soil is a light sandy loam, and the vegetation is very dense. The trees commonly occurring are redbay, water oak, devilwood, loblolly pine, and eastern redcedar. Of shrubs, the most striking are yaupon, bayberry, inkberry, and dwarf palmetto. Characteristic herbs are *Asplenium platyneuron, Lechea villosa, Elephantopus nudatus, Ascyrum hypericoides, Desmodium paniculatum,* and species of *Panicum.* Woody vines are very conspicuous: supple-jack, the greenbriers *Smilax bona nox* and *S. laurifolia,* the grape *Vitis rotundifolia,* and Virginia creeper (Lewis 1917).

Eastern chipmunks and striped skunks come in. Narrow-mouthed toads and squirrel tree frogs are common. Fish crows and boat-tailed grackles nest here. The next and final stage is the live oak–yaupon climax.

A salt-water sere occupies open unprotected shores and islands or banks. The sand is usually covered very early by a close stand of black rush which supports the seaside sparrow. With the building up of the substratum, succession proceeds to higher stages (Lewis 1917, Engels 1942).

Protected Tidal Flats

Most animals on the tidal flats are aquatic and either withdraw into the substratum when the tide goes out or follow the receding waters. The bacterial flora is abundant and important (Pearse *et al.* 1942).

In two series of four quadrats, each one square meter in size and arranged in a line extending upward from mean low tide line, Humm (personal communication) demonstrated the importance of water level and type of soil on the kind and size of the animal populations. On sand the invertebrate populations were as follows: *first quadrat,* 3 annelids exceeding 1 centimeter in length, 2 other unidentified invertebrates; *second,* 1 worm *Balanoglossus auranticus,* 8 annelids, 1 crab *Callianassa major; third,* 1 *Balanoglossus auranticus,* 5 annelids, 1 crustacean *Lysiosquilla excavatrix; fourth,* nothing; total, 22 specimens.

In another series of quadrats examined by Humm, the sand was black and slightly muddy, the slope was much more gentle and the tidal currents much slower than in the sandy area. The following collections were made: *first quadrat,* 15 annelids, 2 snails *Nassarius obsoleta,* 5 bivalves *Solemya velum,* and 3 other invertebrates; *second,* 7 annelids, 5 *Nassarius obsoleta; third,* 2 annelids, 51 *Nassarius obsoleta,* 1 worm *Terebra dislocata,* 1 *Balanoglossus auranticus; fourth,* 25 annelids, 2 worms *Arenicola marina,* 3 worms *Glycera americana,* 1 clam *Tagelus gibbus,* 1 clam *Solemya velum; total, 124 individuals.

A series of crab species occurs at various heights above water (Fig. 3-1G). These animals are subject to predation by enemies from both land and sea. When the tide is in they are set upon by killifishes, silversides, blue crabs, and other predators. When the tide is out, long-billed shore birds and raccoons dig them out (Pearse *et al.* 1942). The boat-tailed grackle also feeds on mollusks, aquatic insects, fiddler crabs, and small fry (Coues 1871). The grackle, red-winged blackbird, and green heron nest in the succeeding shrub stage.

With a slight building up of the tidal mud and sand flats, salt-water cord grass and glasswort appear. The glasswort is eaten by black ducks (Martin and Uhler 1939). The snail *Littorina irrorata* is often abundant. The fiddler crabs *Uca pugilator, U. minax,* and *U. pugnax* occur in immense numbers throughout the marshes and on the muddier banks (Coues 1871). The three species appear to associate indiscriminately at all seasons and to run through the scant herbage in troops and may gather about decaying plant and animal substances until the ground for several square meters is covered with them. The crabs form a considerable part of the food of herons, rails, gulls, fish crows, and grackles. An additional large number of crabs are destroyed by the larger crabs, such as the stone crab and the common edible species.

Salt-water cord grass becomes better established, and salt reed grass, knot grass, and seaside-spurge invade (Lewis 1917). The salt-water cord grass provides food for ducks and is inhabited by the grasshoppers *Ophulella olivacea, Neoconocephalus lyristes, Orchelimum fidicinium, Conocephalus spartinae,* and *Xiphidium nigropleuroides.* The leafhoppers *Laevicephalus littoralis* and *Deltocephalus marinus* and the lantern flies *Mydus enotatus, Megamelanus spartini,* and *Liburnia detecta* occur in abundance near low tide where they are associated with the troops of fiddler crabs. The abundant insects on the grasses are adjusted in varying degrees to survive the periods of submergence. Some of them have special adaptations for clinging to the grasses while submerged. The eggs of some are so attached that they can survive long submergence (Metcalf and Osborn 1920).

A little later in the sere or as a variation of it, grassy *Spartina–Distichlis* marshes may occur (Wells 1928, Barnes 1953). A series of forbs, including *Aster* and *Iva*, invade this community. Three marsh spiders present are *Eustala anastera, Grammonota trivittata,* and *Dictyna altimira* (Barnes 1953).

Eastern baccharis and seashore-mallow are the characteristic shrubs which make their appearance at the high tide line. As the area builds up higher, sea salt ceases to be a factor, and succession proceeds to a live oak–yaupon forest (Wells 1928).

Freshwater Sere

Ponds occur which have impervious bottoms and hold surface water. They are commonly surrounded by water-purslane, the marsh-fleabanes *Pluchea foetida* and *P. camphorata,* and the sedge *Cyperus haspan* (Brown 1912, Lewis 1917). The lakes are surrounded by trees and shrubs. Alligators, cottonmouths, mink, otter, and swamp rabbits were common when Coues (1871) was there.

Evergreen Shrub Bogs

Evergreen shrub bogs lie in interstream areas. For example, in the Holly Shelter Wildlife Management Area of North Carolina (Wells 1942, 1946), peat is 4 to 15 feet (1.2 to 4.6 m) deep (Fig. 3-1F) between small islands of sand. *Cyrilla racemiflora* and *Zenobia pulverulenta* make up 60 per cent of the cover and *Lyonia lucida, Smilax laurifolia,* and *Woodwardia virginica,* another 17 per cent. If these *Cyrilla* bogs are partially drained, as may happen naturally, or if the substratum is built up by the accumulation of debris, they give way to the "low fire frequency" bay community of *Clethra, Magnolia, Gordonia,* and *Ilex coriacea* or the "high fire frequency" *Arundinaria.* Fire may give the bogs a savanna-like appearance. In addition to these shrub bogs, there are many small and some large Atlantic white cedar bogs, and some of these have been invaded by pines.

On river slopes in the Holly Shelter Wildlife Management Area, sweetbay, redbay, turkey oak, and loblolly-bay make up the forest. Most species found in shrub bogs or bay lands are characterized by coriaceous leaves. Some species are found from Texas to Massachusetts.

Deer are being encouraged in the Holly Shelter Area. Other large mammals have been extirpated. Marsh rabbits are generally distributed in and about the swamps. Otters and muskrats are abundant, and several widely distributed amphibians occur (Robertson and Tyson 1950).

OAK–HICKORY–MAGNOLIA MARITIME FOREST ECOTONE

Xerosere

In attempting to reconstruct the condition of these forests about A.D. 1600, one must rely on relics and early accounts. Interest centers in the Carolinas, Georgia, eastern Alabama, and Florida. The only unburned, uncut area of the pine forest under consideration is a large park known as Hitchcock's Woods in the southwestern corner of the city of Aiken, South Carolina, on the inner edge of the coastal plain. Longleaf pine, with diameters up to three feet (1 m), is the principal dominant (Fig. 3-12). Tree sparkleberry and inkberry are the most abundant understory trees. The low shrubs and vines are *Smilax glauca, S. laurifolia,* and yellow jessamine; the principal herb is *Pteridium aquilinium.* Less common canopy species are shortleaf pine, loblolly pine, water oak, southern red oak, sweetgum, and mockernut hickory, and among the understory trees and seedlings are post oak, white oak, sweetbay, flowering dogwood, northern bayberry, *Viburnum,* and tetter-bush. In addition there may be a scattering of *Magnolia grandiflora,* blackgum, Carolina laurelcherry, hawthorn, sassafras, and a host of other shrubs and herbs (W. P. Kelly, personal communication). The vegetation appears to be approaching a deciduous forest climax, and from the location of the area, an oak–hickory climax is the most probable. There was undoubtedly much of this kind of forest in former days. In North Carolina, there are many early records of extensive hardwood forests in the lower coastal plain (Wells 1942).

Fig. 3-12. Longleaf pine with some shortleaf and loblolly pine in an unburned area near Aiken, South Carolina. The undergrowth in the area includes *Vaccinium arboreum, Ilex glabra, Cornus florida, Magnolia virginiana, Myrica pensylvanica, Viburnum,* and seedlings of white and post oak (R. M. Kelly, personal communication).

In this general region scrub oaks replace pines on coarse sand, while on drained medium and fine sand, southern red oak, flowering dogwood, and tupelo take possession. On flatwoods, laurel oak, water oak, sweetgum, and red-bay take over. Longleaf pine stands may be accompanied by turkey oak and blackjack oak along with dwarf huckleberry, wire grass, and a low herb, ipecac-spurge (Harper 1913, Wells 1942).

Hydrosere

Streams are numerous, and their wide floodplains contain a great variety of deciduous trees. Seral stages appear to lead to a mesic oak–hickory climax on the old terraces and, in some cases, for some distance onto the adjacent upland. Pines precede oaks in freshwater estuaries (Chrysler 1910). In a hammock in Texas, 70 miles (112 km) from the main body of magnolia forest, the trees were magnolia, holly, and various oaks (Goodrum 1940).

Scattered over the coastal plain and well toward the piedmont fall line, swamps of coriaceous-leaved plants occur, especially in North Carolina where the shallow elliptical Carolina bays are present (Fig. 3-1F). Peat deposits here are estimated to be 40,000 to 100,000 years old (Frey 1951). During periods of drought, marshes are readily invaded by ligneous plants, which remain for a long time and contribute to the final oak–hickory forest (Buell 1946). The Atlantic white-cedar grows near the edges of small coastal plain ponds, and white-cedar stands are invaded first by pines, later by oaks.

Fire Subclimax

Before A.D. 1600 fires were probably of low frequency, and large unburned areas were undoubtedly present. However, the natives used fire, and the highly inflammable wire grass and pine needles could frequently have been set afire by lightning. In areas of frequent fire, the pines are tall and slender, and the forest floor is covered with grass and green forbs. This represents the fire sub-climax in which longleaf pine is especially important but other species of pine may also occur. The fire subclimax occurs throughout the magnolia and maritime forests and possibly also into the oak–hickory forest. Shrubs and young deciduous trees are readily destroyed by ground fires, but the seedlings and saplings of longleaf pine are very resistant, more so than other species of pine. Although such pine stands doubtless constituted one of the pristine forest types, very little is known of their frequency. Fires appear to have become more common after white man appeared and began to log the area.

There should also have been a good population of animals some 350 years ago (Howell 1921). Recorded as occurring in pine forests are timber rattle-snake, white-tailed deer, gray fox, fox squirrel, eastern cottontail, gray wolf, and mountain lion. Absent from pine but common in oak–pine were bobcat, eastern chipmunk, gray squirrel, raccoon, white-footed mouse, opossum, and black bear.

Ungrazed fire areas are very favorable for bobwhite. Seeds of all species of pines furnish them food but the seeds of longleaf pine are the best. This species has a large seed crop about every seven years. When fire is eliminated, the quail population diminishes. This decline is caused by the accumulation of "pine straw," wire grass, broom-sedge, other grasses, and debris which hinders the birds in finding their food. The wild turkey can live under a wider set of conditions than the bobwhite (Stoddard 1939).

Cotton rats influence the quail population, not especially by destroying their eggs, but mainly by feeding heavily on quail foods. In the absence of fire and with the development of low bushes and tangles of vines, cotton rats are replaced by golden mice and cotton mice (Komarek 1939).

The southeastern pocket gophers and gopher tortoise in pinelands dig through the sand, injure plant roots, and throw up mounds of sand or soil. The southern pine beetle *Dendroctonus frontalis* and the broad-headed borer *Buprestis apricans* attack the pines but are not large-scale tree killers. The red-headed sawfly larva *Neodiprion lecontei* feeds on loblolly, slash, and long-leaf pines, and the tip moth *Rhyacionia frustrana* attacks loblolly pine.

Succession in the ecotone area was probably making considerable progress in 1600 toward a broad-leaved forest climax. Magnolia forest and maritime forest were extending up the stream valleys to meet the oak–hickory forest in the upper portions of these valleys. Oak–hickory forest was also invading the broad-leaved bogs. In the upland between the streams, elements of the post oak faciation were probably present, with blackjack oak important. Where fire occurred, there were, of course, fire subclimaxes, but had the pristine condition continued, climax forest would probably have been attained over a considerable part of the area of the magnolia forest faciation.

Chapter 4

Floodplain Forest Biotic Communities
in the Deciduous Forest and Grassland Biomes

The constantly shifting channels, islands, and bars in the floodplains of rivers (Fig. 4-1) provide continuous new ground for the initiation of succession. The resulting seres often stop short of the climatic climax of the region due to the regular flooding that occurs. This makes the large river floodplains as important as sand areas for the study of community development patterns. The early stages of seral development are similar throughout most of North America north of 28° N. Lat. The important sandbar willow, black willow, and eastern cottonwood are widely distributed, even along the Colorado River in the desert.

MISSISSIPPI RIVER FLOODPLAIN COMMUNITIES

The Mississippi River floodplain in northwestern Tennessee has been given special study. Here the light-colored cottonwoods along the river, partially submerged in the spring, are an outstanding feature of the floodplain forest. Where the forest has been cleared off, the processes of erosion and deposition are accelerated. Cultivated land is alternately washed away and built back again rapidly, sometimes in a period of 70 years. Between 1880 and 1900, nearly half of the inhabitants of Tiptonville, Tennessee, had their homesites washed away, and it appeared for a time that the entire village might be destroyed and Reelfoot Lake drained (Donaldson 1947). However, the river then changed its course (Fig. 4-1).

The area of special study (Fig. 4-2) centers around Island No. 10 of the Mississippi River (Shelford 1954a). Its eastern boundary is marked by the loess bluffs which rise 150 to 180 feet (46 to 55 m). The southern boundary is the southern tip of Reelfoot Lake. The northern boundary is at New Madrid

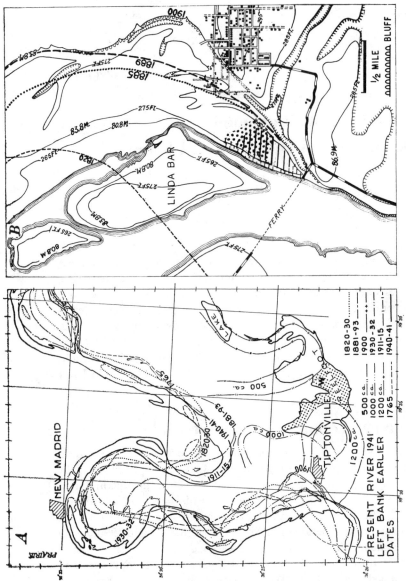

Fig. 4-1. A. Showing the channels of the Mississippi River from A.D. 500 to 1941. Approximate dates before 1765 from Fisk (1944). The dotted area was probably floodplain forest at the time of the earthquake. The character of the prairie west of New Madrid is unknown. The W near the lower right hand corner is the area from which furniture

walnut was removed in the 1880's (information by R. C. Donaldson). B. The encroachment of the river onto the village of Tiptonville, 1885 through 1900 (after Donaldson 1947). For the meaning of the legend north of the ferry see Figure 4-2 (courtesy Ecological Monographs).

and the western boundary near Marston, Missouri. This region was shaken by a terrific earthquake in the winter of 1811-12. Some land was uplifted, but oxbow ponds, cypress sloughs, and the floodplain forest sank 25 feet (8 m) at the south end and 5 to 10 feet (1.5 to 3m) near the center and well toward the north. Reelfoot Lake was formed at this time.

The original biotic communities have been obliterated from at least 98 per cent of the cultivable land. The remaining 2 per cent is second-growth forest in and about Reelfoot Lake Park. Also there are perhaps a score of pre-earthquake trees about farm buildings. The primeval communities within the area west of the Mississippi River are even more thoroughly destroyed than near Tiptonville.

The forested land too wet to cultivate has been thinned in the cottonwood areas and repeatedly thinned selectively on the older floodplain about Reel-foot Lake. Outside the area of special study east and northeast of New Madrid the obliteration of original communities has been less complete. The areas about the Obion and Forked Deer rivers in the southeastern one-fourth of the area (Fig. 4-2C) have not been investigated. The large influent animals of the region have been extirpated or their populations reduced to biotic impotency for at least a century.

Figures 4-2A, B, are attempts to represent the distribution of the plant communities before they were disturbed by white man. In preparing the map and in arriving at the conclusions in the discussion, the age of the deposits on which the trees are growing and the height of the deposits above mean low water were given serious consideration. The dating in Figure 33 of Fisk (1944) and all available maps of Mississippi River Commission and the U.S. Engineers were utilized. From the southwestern to the northeastern corner of the area, the mean level of low water in the Mississippi River varies from approximately 248 feet (75 m) to 266 feet (80 m) above the mean level of the Gulf of Mexico. The area east of Tiptonville is 305 feet (92 m) above the mean Gulf level and has had no channel through it for a very long period.

The climate of the area is favorable for the rapid growth of trees on the higher terraces of the floodplain. The annual rainfall is 45 to 50 inches (112 to 125 cm) and the mean annual temperature is 60° F. (16° C.) The lower portion of the floodplain is influenced more by upstream rainfall than by local precipitation.

About 1600, the native Indian population in the Mississippi and lower Missouri and Ohio floodplains was small, ranging from 0.5 to 1.3 per 10 square miles (0.2 to 0.5/10 km²), except in the Mississippi Valley south of the Ohio River where it ranged from 3 to 8 per 10 square miles (1.2 to 3.0/10 km²). The Indians were mound builders and of necessity inhabited only the higher parts of the floodplain. The remains of a snake mound, estimated to be about 2000 years old, are still discernible near Reelfoot Lake at Gray's Landing. Such structures disturbed the forest only in a relatively insignificant manner, but by 1950, white man had cleared all the cultivable land of timber, except for a few parks and reservations.

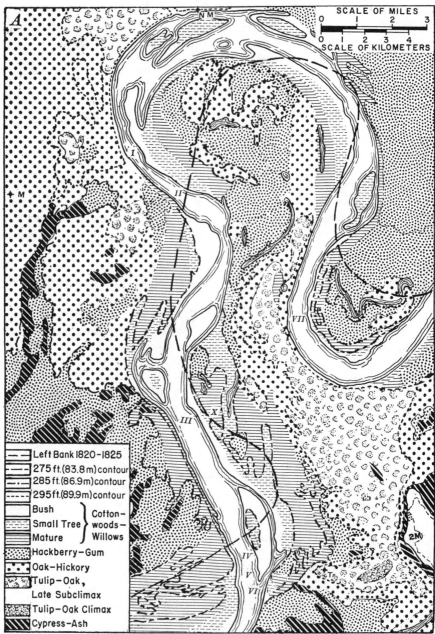

Fig. 4-2. The reconstructed vegetation of the Reelfoot Lake–Marston, Missouri, area. A. I–VII in the river channel are to locate features on land to the right or left. Contour lines are only partially shown. The X is the Donaldson virgin cottonwood near III; the relict tulip–oak is right and a little above X. The small circles with crosses and capital letters locate villages as follows: NM, New Madrid, Missouri; M, Marston, Missouri; T, Tiptonville, Tennessee.

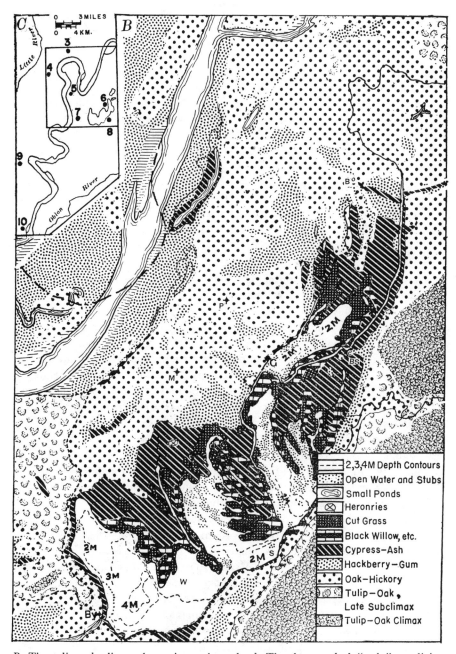

– – –	2,3,4M Depth Contours
	Open Water and Stubs
	Small Ponds
⊗	Heronries
	Cut Grass
	Black Willow, etc.
	Cypress–Ash
	Hackberry–Gum
	Oak–Hickory
	Tulip–Oak
	Late Subclimax
	Tulip–Oak Climax

B. The tulip–oak climax shown is on the upland. The dots marked "stubs" are living cypress trees sunk at the time of the earthquake. Curved line of dots in front of By is the line of cypress stubs suggesting the A.D. 1200 channel. B. Villages are: P, Phillipy, Tennessee; M, Markham, Tennessee; G, Gray's Landing; By, Boyette's Landing; S, Samburg. C. Insert map of study areas: 3, New Madrid; 4, Marston; 5, Darnell Point; 6, Gray's Landing; 7, Tiptonville; 8, Samburg; 9, Hayti; 10, Tyler (courtesy Ecology).

Floodplain Habitats

There are two types of floodplain habitats (Fisk 1944) : predominantly ter-restrial, which are dry at low water, and aquatic, which are covered with water the year round or nearly so. Two subtypes of terrestrial habitats may be distinguished: one, which lies close to the river channel, characterized by short annual submergence and molded by the gradual shifting of the river from side to side in the area known as the meander belt, and the other, a long submergence type, which occurs on low ground where floodwater stands for very much longer periods than in the meander belt. Habitats at 25 feet (8 m) and more above mean low water level and with short annual submergence are generally cultivated at the present time. The long submergence habitats are not ordinarily used.

With the exception of the extreme western part, most of the special study area lies in the meander belt. Nearly every year, two or more early stages of floodplain forest are inundated. The length of submergence varies from a week to two and one-half months. Usually the flooding comes early in spring but sometimes as late as May or June. Usually there is more than one period of submergence. Only a little attention has been given to the kind of material deposited by the flood waters. While the river from New Madrid to Tipton-ville Ferry was shifting its channel mainly westward, the cottonwood area near the Tiptonville Ferry received a layer of fine silt each year, probably amounting to several inches. The height of the flood and the duration of submergence determined the size of particles and amount of fine silt. On the other hand, as the river at Darnell Point (Fig. 4-2A–III) began to shift eastward in 1949 and to cut away a considerable portion of the cottonwood–willow forest (Fig. 4-2A–X), two or more feet (0.7 m) of sand were deposited. The forest had been thinned by a 1947 cutting, and seedlings of six or more species of trees were numerous in 1952. If the river should shift its channel again to the west, the sand will become covered with silt. Alternating layers of fluvial deposits of different sizes are thus to be expected in the meander belt.

Big Island, in the White River near St. Charles, Arkansas, 125 miles (200 km) south of Tiptonville, is of the long submergence habitat type. The White River National Wildlife Refuge is located here. The island is sometimes under 30 feet (9 m) of water for several weeks. Across the Mississippi River from Tiptonville and beyond a meander belt ridge near the west bank of the river, there is a vast, nearly level plain extending west for 50 miles (80 km) to the St. Francis River and Crowley's Ridge. In times of flood, before the present levees were installed, one could cross this entire area in a rowboat. Likewise, in the lower parts of the Little River drainage, water stands for long periods so that the habitat is of the long submergence type.

Sandbar Willow

This small tree grows in masses from seed and forms thickets by means of stolons. It must have a portion of the crown out of water in the summer of the second year in order to survive. Army engineer maps show willows down to five or six feet (1.5 to 2 m) above mean low water. In its second and third years, the sandbar willow tends to cause more sand to be deposited and to hold that already in place. The full grown sandbar willow is a small tree which produces roots along the entire length of its trunk when buried in sand. It is attacked by a large array of insects, including *Chrysomela knabi* and sawfly larvae, and is occasionally used as food by deer.

There is little or no evidence that sandbar willow prepares the habitat for later trees. Black willow and cottonwood come in whether preceded by sandbar willow or not (J. A. Putnam, personal communication). A dense stand of sandbar willow may prevent the establishment of other trees, but they are short lived.

Small Cottonwoods and Willows

Cottonwood seedlings survive in thick stands on well-drained ground, and willows survive in low moist spots at levels from 15 to 30 feet (4.5 to 9 m) above mean low water. The seeds of both willow and cottonwood are dispersed in spring, the willow a little earlier than the cottonwood. They germinate very quickly, and the seedlings are of notable size by midsummer. They are difficult to remove from agricultural lands after the second season. In dense stands, self-thinning is rapid and extensive.

In spring, willows and cottonwoods, 5 or 6 feet (1.5 to 2 m) high and about 15 feet (4.5 m) above low water, are usually accompanied by red earthworms and dipterous larvae. Often slugs are present on the soil surface. By late summer, cocklebur, grasses, and other pioneer herbs grow upon the bare areas. This is especially true where the substratum is silt. These plants, by the beginning of September, are occupied by a large series of native insects belonging to species common also in cultivated fields. One such area which was under water during most of April, 1943, supported 38 individual insects and spiders per square meter on September 7. These belonged to 19 different species. There were unidentified leafhoppers, lepidopterous larvae, the cucumber beetle *Diabrotica twelve-punctata howardi,* tarnished plant bug, the cocklebur mirid *Ilnacora stalii,* the stilt bug *Jalysus wickhami,* and the tree cricket *Oecanthus nigricornus.* There were 13 specimens of the cocklebur fly *Euaresta aequalis,* which feeds on cocklebur seeds, and the seed-storing ant *Pheidole vinelandica* was found. This aggregation was accompanied by a series of predators, including the red lady beetle *Coleomegilla maculata,* the predatory mirid *Deraecoris poecilus,* the crab spider *Misumenops celer,* the long-bodied spider *Tetragnatha elongata,* the spiny abdomen spider *Acanthepeira stellata,* and the crab spider *Phildromus.* It appears that these species must have withstood the long spring submergence, moved into the area from higher ground, or come down from the tops of the shrubs after the flood waters receded.

Small Tree-sized Willows and Cottonwoods

In 1937 small trees of willow and cottonwood occurred on an area 20 feet

(6 m) above mean low water, and the substratum had been built up considerably by silt deposition. The stand consisted of 90 per cent cottonwood and 10 per cent willow (Fig. 4-2A–VI). Willow grows on lower ground than cottonwoods, but by the time they reach small tree size, silt deposition may make them appear on the same level. Small-sized individuals of trumpet-vine occur in some numbers, and there is an occasional American sycamore seedling and young grape vine.

In early spring, there are earthworms in the soil at this stage along with a considerable number of collembolons and, on the surface, a few ground spiders. Also, there are usually a few scarabaeid larvae, red mites, staphylinid beetles, and spiders. The surface of the ground shows swamp rabbit pellets and occasional slugs. The willows and cottonwoods support only a small number of insects, particularly leafhoppers. The herbaceous vegetation is often entirely wanting. About 20 species of birds occur quite commonly in this, the preceding, and the next stages (list on p. 97).

By late summer, because of the rather dense shade cast by the thick stand of trees, there are only scattered herbaceous plants, usually the same species as those growing in the open. These support only a few animals. However, the woody plants show a rather large group of insects and spiders. On September 7, 1943, there were approximately 35 individuals per square meter. These included tree cricket nymphs, a considerable number of willow sawfly larvae, a few lepidopterous larvae, a few leafhoppers, and eight spiders belonging to seven different species. Those spiders old enough to be identified were *Theridion frondeum* and *Tetragnatha limnocharis,* species which occur also in magnolia forest. The stink bug *Podisus maculiventris,* which feeds upon other insects as do the spiders, was present.

Mature Cottonwood–Willow Forest

Areas with mature cottonwoods and willows are submerged for shorter periods than earlier stages. The two areas studied were 28 to 30 feet (8.5 to 9 m) above low water (Fig. 4-2A–V, X). The trees were more than 18 inches (45 cm) in diameter with about 60 per cent being cottonwood and 40 per cent willow. There were a few sycamores in the older area.

The trumpet-vine comes in with the cottonwood and willow on the ridges and persists at least up to the sugarberry stage. Poison ivy is frequently more abundant in the willows of the flats than elsewhere. It appears in the succession before the grape. Grape becomes abundant on the ridges. In some areas, pepper-vine takes the place of trumpet-vine. The trumpet-vine, poison ivy, grape, pepper-vine, honeyvine, sometimes buckwheat vine, and morning-glory make a tangled mass so dense and binding as to make passage very difficult except along trails (Fig. 4-3). The density and luxuriance of the vines vary from year to year. Their growth retards floodwaters in spring and thus increases the deposition of silt.

The principal shrub is the American elder, and there are occasional rough-leaf dogwoods. The herb layer is usually poorly developed. There are a few cockleburs and ironweeds in spaces not covered with vines or the thorny bramble.

Compared with earlier stages, mammals are numerous. The swamp rabbit continues to occur, the opossum and raccoon are common, and the white-footed mouse increases in numbers. Gray squirrels are present. Early travelers saw deer, wapiti, and, locally, bison, coming to the river to drink. These mammal species all continue into the climax forest.

Fig. 4-3. Floodplain stages with willow, cottonwood, hackberry and elm (author's photos 1947, 1948). A. Vine-covered cottonwoods 18 inches (45 cm) to 30 inches (75 cm) diameter at X in Fig. 4-2. B. Old willows, diameter 20 to 24 inches (50 to 60 cm) with a thin covering of trumpet-vine, poison ivy, crossvine, grape, and bur cucumber, Tipton-ville Ferry. C. Hackberry–elm forest at the Missouri Big Oak Tree Park near Bayouville. The large trees are elms; the one at the right is four feet (1.2 m) in diameter. Small trees are hickories and hackberries.

Resident and migrant birds are common throughout the cottonwood–willow stage from mid-March to early April. Their numbers, recorded in terms of individuals per 100 acres (40 hectares), have been estimated over several years by S. C. Kendeigh as follows: Carolina wren, 20; white-throated sparrow, 16; blue-gray gnatcatcher, 14; rufous-sided towhee, 14; cardinal, 13; myrtle warbler, 12; Carolina chickadee, 9; common crow, 9; downy woodpecker, 3; tufted titmouse, 3; brown thrasher, 3; song sparrow, 2; yellow-shafted flicker, 2; hairy woodpecker, 2; barred owl, 2; robin, 2; red-bellied woodpecker, 1; yellow-bellied sapsucker, 1; brown creeper, 1; catbird, 1; and red-eyed vireo, 1. Most of these species, except the owl, are partly or wholly insectivorous.

The invertebrates attain larger populations here than in the earlier stages. The following widely distributed species occur: the stink bug *Podisus maculi-*

ventris, the stilt bug *Emesaya brevipennis,* the plant bug *Phytocoris tibialis,* and the wedge beetle *Anoplitis inaequalis.* The cottonwoods are attacked by the larvae of *Plectrodera scalator;* to what extent this contributes to their natural thinning is unknown. The red maple leaves are damaged by small bladder galls caused by the micromite *Vasates quadripes,* which is only 0.2 millimeters long. Slugs, red annelids, ground beetles, the millipede *Pseudopolydesmus serratus,* adult insects, fly larvae, and beetle larvae are usually numerous. The snail *Succinea* and occasionally centipedes occur. Collembolons are numerous. Fallen logs with loose bark have a rich fauna of pyrochroid beetle larvae, carpenter ants, centipedes, millipedes, both white and red earthworms, the snail *Zonitoides,* and the slug *Deroceras.* The low vegetation is inhabited in mid-April by crane flies, syrphus flies, green midges, muscid flies, and mosquitoes, all of which are numerous. Leafhoppers, snout beetles, and minute spiders are numerous in the trees. During the autumn, the nursery web-weaving spider *Dapanus mira* and the harvestman *Leiobunum vittatum* are present, both here and throughout much of the deciduous forest.

In the spring when these areas are not under water, ground collections of animals are large. This means that the animals must survive flooding. Late summer collections, made from the lower vegetation, gave an estimate of 125 insects and spiders per square meter in 1943. This was following an April flooding. In 1947 following a June flooding, collections from a similar vegetation area yielded about one-third as many individuals.

Old Cottonwood–Willow Forest

The Donaldson area north of Tiptonville (Fig. 4-2A–X) was in the center of the river channel in 1912. In 1937 it was about 28 feet (8.5 m) above mean low water. Subsequent flooding raised it to 29 or 30 feet (9 m) by 1947. At this time, the cottonwoods were 20 to 24 inches (50 to 60 cm) in diameter at breast height and showed 20 annual growth rings. Willows of similar size occurred on the lower ground (Fig. 4-3).

Up to about 20 years of age, cottonwoods and willows are of approximately equal vigor and growth rate. Beginning about this time, the willow declines in vigor while the cottonwood continues to grow and remains in vigorous condition for another 10 to 20 years. In the Reelfoot Lake area, cottonwoods grow in diameter at the rate of 0.85 inches (2.1 cm) or more per year, but in Iowa, farther north, their growth has been measured at only 0.41 inches (1.0 cm) per year (Williamson 1913). The difference in vigor and growth of the two dominants after 20 years, together with the increase in shade and invasion of other species, brings a change in community composition.

Some idea as to the sequence and rapidity of invasion of new species may be obtained from a count of seedlings and small trees present in old cottonwood stands and from the measurement of their sizes (Table 4-1). The cottonwood stand near the Tiptonville Ferry was estimated about 40 years old, counting from the time the cottonwoods were seedlings. The area was a sandbar island in the river in 1896 or 52 years earlier.

Table 4-1. *Relative Number and Size of Seedlings and Small Trees in an Old Cottonwood Stand in a Disturbed Area near the Tiptonville Ferry in 1948*

SPECIES	NUMBER COUNTED	DIAMETER INCHES	DIAMETER CENTIMETERS
Boxelder	40	5–10	12.5–25.0
Red maple	35	4–7	10.0–17.5
American sycamore	10	9–13	22.5–32.5
Pecan	6	1–3	2.5– 7.5
American elm	3	1–8	2.5–20.0
Sugarberry	2	1–3	2.5– 7.5
Black locust	2	1	2.5
Sweetgum	+	..	....

According to Baker (1949), cottonwood and black willow have the lowest shade tolerance and red maple and boxelder the highest of any species listed in Table 4-1. This enables the boxelder and maple to enter under the shade of the cottonwoods and willows as soon as the soil and litter conditions are suitable. Stands of red maple and boxelder overtopped by old cottonwoods occur occasionally.

The seedlings appear after the accumulation of organic debris from leaves, small plants, twigs, animal droppings, and carcasses has modified the soil for some time. Meanwhile, bacteria, earthworms, and burrowing arthropods have worked over the material, converting it to humus. Flooding more often buries the accumulation of debris than removes it. It is probable that the presence of pecan, elm, mulberry, black locust, sweetgum, and sugarberry is governed by the improved soil conditions, as they are less tolerant of shade than boxelder and red maple. The preponderance of boxelder and red maple, evident here among the invading species, does not usually persist, so that the mature community that replaces the willow–cottonwood is more often characterized by the dominance of sugarberry, elm, and sweetgum.

Sugarberry–Elm–Sweetgum Forest

Sugarberry *Celtis laevigata* is often called "hackberry" since it has a similar appearance and replaces in the south the true hackberry *C. occidentalis* of the north. The designation "hackberry–gum" taken from old maps has met with objection from local people and from some foresters. Putnam (personal communication) thinks "hackberry–gum" and "hackberry–elm" result from timber cutting. The 1912 river maps show an area near Tiptonville, which was water in 1830, marked "hackberry" and another nearby area marked "gum." Eighty-two years is a short period for these trees to reach prominence unless an unusual flood had brought the substratum up to the necessary level. Nor is their prominence likely due to the harvesting of cottonwood in the 1890's, since better timber was then available.

The composition of the sugarberry–elm–sweetgum forest is subject to much variation, Putnam and Bull (1932) give a list of trees occurring in it. Sugarberry is predominant in an area one-half mile south of Gray's Landing in Reelfoot Lake Park, where there was lumbering in 1885. Nearly all stumps had rotted away by 1937. Black walnut stumps last for about 40 years and those of most other trees for perhaps half that time. An area of about 7.5 acres (3 hectares) contained 284 trees of 6 inches (15 cm) or more in diameter. The litter, herbaceous layer, shrubs, and understory trees were characteristic of floodplain forests. However, the American hornbeam or blue beech trees were larger and more numerous than in a pristine forest. The percentage composition of the various species in the forest in 1949 was: sugarberry, 32; American hornbeam, 17; elm, 16; sweetgum, 12; persimmon, 6; red mulberry, 6; shellbark hickory, 3; sycamore, 2; and boxelder, cherrybark oak, green ash, honeylocust, and bitternut hickory about 1 per cent each.

The tree composition of the City Park of Hayti, Missouri, by percentage was sweetgum, 71; sycamore, 17; elm, 1; and sugarberry, 1. Big Oak Tree Park (Fig. 4-3) near Bayouville, Missouri, contained sugarberry, 21; elm, 17; mockernut hickory, 21; three other hickories, 23; and a long list of miscellaneous species the remainder (Shelford 1954a).

In Reelfoot Lake Park, the understory trees and shrubs are hophornbeam, spicebush, and pawpaw. The foliage of the trumpet-vine and grape is principally high in the trees, that of poison ivy and Virginia creeper is both high and low, and that of pepper-vine and catbriar is low. The principal midsummer herbs are Virginia snakeroot, Virginia knotweed, smooth Ruellia, honewort, and elephant's-foot. There is a distinct vernal herbage, usually well developed by mid-April, composed of fleabane, rough bedstraw, jewelweed, and waterleaf.

No population data are available for the mammals. A few opossums, some raccoon signs, white-footed mice, golden mice, and short-tailed shrews have been seen or trapped. They are partly or wholly insectivorous in their food habits (Calhoun 1941). The gray squirrel, fox squirrel, and opossum appear to be present throughout the sere, from the cottonwood stage to the climax. Seton describes a migration of gray squirrels in 1914, estimated at 500,000 individuals, that developed on Island No. 20 opposite Tyler, Missouri, probably in cottonwood and sugarberry (Bennitt and Nagel 1937). None of the large mammals, deer, wapiti, black bear, mountain lion, bobcat, and wolf, now gone from the area except for the deer, appears to have shunned the floodplain. There is good evidence that bear and wapiti had a preference for areas adjacent to large rivers (Crockett 1834, Rhoades 1896, Wailes 1854).

Amphibians and reptiles, which appear at this stage, include the slimy salamander, small-mouthed salamander, gray tree frog, American toad, and copperhead.

The following invertebrates are not regularly found in earlier stages but appear in the sugarberry–elm–sweetgum forest and continue into the later stages of the sere: the beetles *Abacidus permundus* and *Notoxus bicolor;* the ants *Camponotus castaneus* and *Aphaenogaster fulva;* the snails *Mesodon thyroidus, Haplotrema concavum, Stenotrema hirsutum,* and *Mesomphix perlaevis* (largely limited to the area of the tuliptree–oak forest) ; the millipedes *Aniulus venustus, Pseudopolydesmus penetorum,* and *Cleidogona caesioannulata;* and the spider *Hahnia flaviceps.* In addition, the harvestman *Leiobunum cretatum* and adults of the orb-weaving spider *Mangora maculata* are abundant in early July but nearly absent in April and September. The jumping spider *Thiodina sylvana* and the nursery web-weaving spider *Pisaurina mira* have been present only in early September collections, while the water margin spider *Dolomedes sexpunctatus* is found in the woods only in spring.

A noteworthy spring phenomenon is the emergence from hibernation in April of six species of the leafhopper *Erythroneura.* On April 15, 1944, shrubs yielded 143 *E. vitis* per square meter, 210 *E. vitis corona* mixed with some *E. tacita,* and 32 *E. infuscata.* On the herbs, there were 152 *E. vitifex,* 83 *E.*

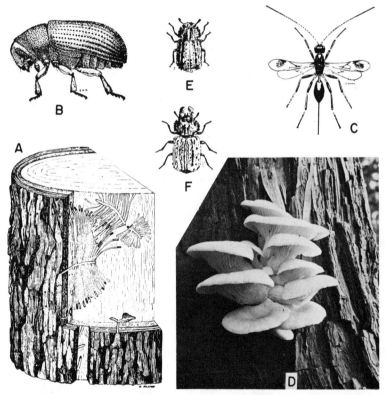

Fig. 4-4. Invertebrates and fungi which feed on floodplain trees, notably red maple and elm. A. Section of elm trunk showing the galleries of the native elm bark beetle. B. Adult beetle *Hylurgopinus rufipes.* C. Adult *Spathius canadensis,* the larvae of which are parasitic on the larvae of the beetle (A, B and C after Kaston 1939). D. Fruiting of the fungus *Pleurotus ulmarius,* most abundant on elm, causing hollow trees (courtesy L. Shanor). E and F. Male and female *Bilitotherus bifurcatus* which feed on fungus.

calycula mixed with some *E. tricinta,* 10 *E. aclys,* and 11 *E. vitis corona.*

The elm is attacked by elm scurfy scale and the heart rot fungus (Fig. 4-4). In some floodplain oak–hickory forests, half of the elm trees have hollow trunks. Apparently seral trees are more often attacked than climax trees.

Large cane *Arundinaria gigantea* is small and scattered along paths in the sugarberry forest and indicates disturbance. No cane brakes have been found in the Reelfoot Lake area, although Audubon describes one near Louisville, Kentucky. Normally it attains a height of 12 to 30 feet (3.6 to 9.0 m) and a diameter of from 1 to 2 inches (2.5 to 5.0 cm). Patches of plants occur with stems of all sizes and frequently are so close and tangled together that they make an almost impenetrable thicket. Such cane brakes may grow beneath gigantic trees and be interspersed with vines and other plants of many species.

Floodplain Oak–Hickory Forest

The oak–hickory forest on floodplains near Tiptonville requires one to two centuries for complete development (Table 4-2). The invasion of oaks is in the following order: *Quercus falcata pagodaefolia, Q. palustris* and *Q. lyrata, Q. macrocarpa, Q. falcata, Q. michauxii, Q. shumardii, Q. velutina, Q. alba,*

Table 4-2. Schedule of Community Development

	YEARS	
DESCRIPTION OF AREAS AND PROCESSES	ADDED	TOTAL
Building of sand bar to 5 feet (1.5 m) above mean low water; sandbar willow	2–4	2–4
Deposition of 10 to 20 feet (3 to 6 m) of sand, growth of sandbar willow, seeding of cottonwood and black willow	11–13	13–17
Dominance of cottonwood–willow[a]	19–21	32–38
Elimination of black willow dominance, development of red maple and boxelder under cottonwood, entrance of sugarberry and sweetgum[b]	28–30	60–68
Elimination of cottonwood dominance, beginning of sugarberry and sweetgum dominance[c]	16–18	76–86
Entrance of elms, early oaks, and hickories, subordination of sugarberry and sweetgum, entrance of intermediate oaks and hickories[d]	175–180	250–265
Elimination of early oaks and hickories, dominance of intermediate oaks and hickories[e]	80–100	330–365
Entrance of tulip and climax oak species[e]	60–100	390–465
Entrance of remaining tulip–oak forest species: American beech, probably cucumbertree, and others[e]	150–170	540–635

[a] Area in channel of river in 1912, with trees 20 to 21 years old in 1947; period of 35 years involved.
[b] Area was a sandbar island in 1896, probably with sandbar willow; approaching the sugarberry-sweetgum stage in 1947; 51 minus 21 gives 30 years for duration of stage.
[c] Area in river in 1830, surveyed as sugarberry-sweetgum in 1912; period of 82 years minus 65 gives 17 years for stage.
[d] Estimate of time involved based on location of stage in relation to the river channel in 1765.
[e] Time intervals largely a judgment based on known order of invasion, growth rates, shade tolerance, etc., of various species involved.

and *Q. muehlenbergii*. Likewise, the hickories appear in order: *Carya laciniosa* (not a primary dominant), *C. illinoensis* (scattered trees only), *C. cordiformis*, and *C. glabra*.

A small, well-drained area in Reelfoot Lake Park south of Gray's Landing in 1949 contained an early facies of this stage, consisting of sweetgum, 18 per cent; sugarberry, 23; swamp chestnut oak, 13; southern red oak, 13; pignut hickory, 10; elm, 8; and American hornbeam, 13. The undergrowth included redbud, climbing bittersweet, herbaceous mandrake, goosegrass, and Miami-mist. The redbud was not found in the nearby sugarberry–elm–gum forest.

An older facies near Samburg on the south side of Reelfoot Lake had cherry-bark oaks about 50 years old in 1948, making up two-thirds of the oaks and the pin oak about one-sixth. The hickories were shellbark and bitternut, the latter somewhat less numerous. The oaks composed more than 50 per cent of the stand and the hickories between 15 and 20 per cent; a scattering of trees from the preceding stage made up the remainder.

Many animals from earlier stages continue into this stage: The slimy sala-mander becomes more abundant, and the six-lined racerunner makes its ap-pearance here. Snails of the genera *Euconulus, Vertigo,* and *Bifidaria (Gas-trocopta)*, in addition to *Haplotrema concavum,* become relatively abundant.

Tulip–Oak Forest

The tulip–oak forest probably occurred on all areas 40 to 45 feet (12 to 14 m) above low water level which have not been disturbed by the river for several hundred years. It has been removed everywhere except for a few trees in a farm woodlot and around two houses. These trees are about 42 feet (13 m) above mean low water of the river. The area contained the fol-lowing number of trees, with the diameter at breast height and approximate age of some of the larger typical ones indicated: tulip, 2 (47 inches, 150 years; 38 inches); white basswood, 3 (48 inches); chinkapin oak, 9 (49, 30 inches); shumard oak, 13 (71 inches, 240 years; 40 inches); cherrybark oak, 1 (49 inches); silver maple, 1 (51 inches); winged elm, 10 (20 inches); pin oak, 1 (38 inches); sugarberry, 5 (20 inches); and elm, 7 (50 inches). American hornbeam was also present. This stand may be indicative of the climax of the area.

The complete development of the sere up to this stage probably required around 450 years. Attempt is made in Table 4-2 to give a time schedule for each stage. Probably 600 years, at least, would be required for the complete development of the climax equivalent to the western mesic forest of Braun (1950).

Long Submergence Habitats and Communities

On Big Island in the White River National Wildlife Refuge, Arkansas, the Mississippi River reaches or passes 6 feet (1.8 m) above flood stage two or more times during the average summer. Flood stage is regarded as 25 feet

(7.6 m) above low water level. In 1943, the spring flood stood at 35.4 feet (10.7 m). Big Island has a ridge around much of its circumference, and the central low portion is frequently flooded. The timber in the low interior is of the overcup oak–water hickory type of Putnam and Bull (1932). It includes overcup oak, 42 per cent; water hickory and bitter pecan, 26; red oak (probably cherrybark), 11; green ash, 11; sugarberry, hackberry, elm, Nuttall oak, and others, 10 (W. E. Houser, personal communication). Many trees were hollow due to fungus attacks.

In early July, 1943, the center of the island was just coming out of the water after two months submergence. There were few or no shrubs. Clearweed was growing on logs that had floated during the flood conditions. In the soil the crayfish *Procambarus blandingii acutus* was numerous. On the ground the carabid beetle *Loxandrus erraticus* was common. In and on the floating logs the ants found were *Camponotus herculeanus pennsylvanicus, Crematogaster lineolata,* and *Aphaenogaster texana* (var.) In the clearweed growing on the logs were a surprising variety of arthropods.

On the west side of the mainland adjacent to the Big Island Chute, water was draining out of the area. Shrubs and tree seedlings were without leaves on the lower two feet of their stems except on higher ground. The recognizable shrubs included supple-jack and Missouri gooseberry. Trumpet-vine and muscadine grape were present but not abundant. The forest is second-growth and contained four species of oak, sugarberry, and sweetgum. The sugarberry was most abundant. Raccoons are abundant; twenty thousand skins were taken adjacent to the refuge in the 1942-43 trapping season. The refuge manager (B. S. Webster) states that raccoons have a liking for frequently submerged communities and withstand flooding well. On the other hand, skunks and opossums cannot live in flooded areas and move in and out only to a minor degree. Deer and bears move back and forth rather freely with changes in water level. Both are nearly extirpated from the southern floodplain of the Mississippi River.

Big Island is in an area of the river with an immense flow of water. Its forest community cannot control the habitat. The forest will probably remain in its present condition for an indefinitely long time as an edaphic climax, with the species now present continuing to replace themselves. Whenever less hydric trees should invade during periods of low rainfall, they will likely be drowned in succeeding periods of high rainfall. Important changes in community composition will require intervals of 500 to 1000 years rather than of centuries.

The Little River drainage west of the meander belt opposite Tiptonville and Darnell Point is lower than the area near the Mississippi River. Putnam (personal communication) states that the original forest farther downstream was principally sweetgum, associated with water oak, Nuttall oak, willow oak, elm, sugarberry, and green ash. This forest occurred commonly on the high clay flats and low clay or loam ridges throughout the bottoms. The soils are usually silty or loamy where a clay top soil occurs and are underlain by a sandy or loamy subsoil.

Aquatic Sere in Small Ponds

Small ponds scattered over the floodplain are often the result of flood erosion where water whirls around masses of fallen trees, or they may have developed as oxbows from the river. They vary in size and, immediately after a flood, may have a depth up to 80 feet (24 m). The ponds may be connected with the river at high water and are more or less permanent.

A representative young, artificial pond of about 25 acres (10 hectares), two miles southwest of Markham, was available for study. The bottom had been scraped bare in 1934. Black willows of noticeable size appeared first in 1937, followed by swamp cottonwood and waterlocust in 1944. The trees were 15 to 20 feet (4.5 to 6 m) tall in 1949. As the pond increased in age, more and more animal species invaded (Table 4-3).

Table 4-3. Showing Development of Animal Life in a Newly Formed Floodplain Pond[a]

SPECIES	'37	'41	'43	'44	'47
Water-boatman	+	225	2	10	65
Crayfish, *Procambarus blandingii acutus* (Gib.)	+	2	2	0	35
Midge fly larvae	+	10	0	0	5
Dragonfly naiad	+	2	0	0	0
Damselfly naiad			2	0	0
Snail, *Physa*			5	0	0
Amphipods			1	0	15
Shrimp, *Palaemonetes paludosus* (Gib.)				5	0
Slider turtle, *Pseudemys*				1	0
Mayfly nymph, *Hexagenia*				1	0
Isopod, *Asellus*					5

[a] The figures are numbers of individuals per square meter; + indicates the species was present but populations were not measured.

A middle-aged pond near the north end of Reelfoot Lake east of Phillipy (Fig. 4-2B) had a depth at low water of about 3 or 4 feet (1 m), and sloping sides. The open water in 1949 contained considerable *Ceratophyllum* during the summer but no water-millet, probably due to the shade of surrounding trees, and few or no waterlilies. There was a buttonbush thicket around the margin about 20 feet (6 m) wide, and a number of young cypress trees from 4 to 10 feet (1.2 to 3 m) tall farther back from the water. *Hexagenia* naiads, found in the young pond (Table 4-3), had disappeared, but the shrimp and crayfish were still present. Small amphipods and isopods had increased in abundance. The snail *Gyraulus parvus* and bivalve *Musculium* had appeared.

With further development, ponds such as these become surrounded with cypress. The seeds of cypress will sprout and grow only when not submerged (Demaree 1932). The drought of the early 1930's brought exceptionally low water levels in Reelfoot Lake and permitted small seedlings to develop around the margins. By 1936, the seedlings were 1 to 3 meters high.

Aquatic Sere in Large Bodies of Water

Many Mississippi oxbows may be 1000 years old. If the river had cut through at point VII (Fig. 4-2) upstream from New Madrid, Missouri, in 1937, as it started to do, a lake would have formed similar to the present Horseshoe Lake in southern Illinois (Fig. 4-5).

Reelfoot Lake is large since it includes two or more oxbow depressions, one of them 1300 years old (Figs. 4-1, 4-2). The series of earthquakes which began in December, 1811, caused subsidence and flooding of these old cypress-filled oxbows to a depth that caused many cypress trees to die. Other cypress trees, submerged less than 10 feet (3 m), survived but grew very little afterwards.

A special feature is of interest. Fisk (1944) shows an A.D. 1200 channel (Figs. 4-1A, B) which curves around the southwest shore of the deep portion of the lake. It is probable that a narrow water-filled oxbow of the 1200 channel existed at the time of the 1811-12 earthquake. This former oxbow can now be traced near the present shore at Boyette's boat landing (Fig. 4-2B). There is a curved row of sunken but living trees in about seven feet (2 m) of water about 200 feet (60 m) offshore with no evidence of any trees between the row and the shore. This was probably an open space at the time of the earthquake.

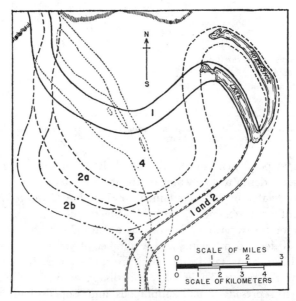

Fig. 4-5. Formation of a large oxbow—Horseshoe Lake, near Cairo, Illinois. The numbers indicate probable successive channels.

The row of living cypress was about ten inches (25 cm) in diameter at breast height when submerged. They have grown very little since but have developed large thickenings just below water level, with suggestions of roots. Beyond the row of trees is a large area of cypress stumps, still two or more feet (0.7 m) in diameter and standing in 15 feet (4.5 m) of water. This oxbow and cypress swamp must have developed during the 600 years from the time the channel was present to the time of the earthquake.

Since water flows through oxbows at times of flood for a long time after the main stream has changed its course, the sluggish portions of the Mississippi River are similar to young oxbows, although they do not permit the maintenance of a rich bottom or plankton community. Under these conditions, the bottom fauna near the shore of the river is limited to burrowing mayfly naiads, stonefly naiads, red annelids, and river shrimp. In the driftwood that accumulates against tree roots exposed by river cutting, crayfishes, isopods, and amphipods occur. During floods, bottom materials may be removed to a depth of 50 feet (15 m) or more, and then later replaced. The organisms are swept away, but some survive in eddies and connecting sloughs. Various fishes escape by heading into eddies and the mouths of smaller streams. River plankton is highly diluted and swept downstream. During early March it consists of some eighteen species of algae, protozoa, rotifers, and crustaceans. Only six of these species are also found in the open waters of Reelfoot Lake (Eddy, personal communication).

Since this river community is apparently not subject to seral development, it has been called a river climax (Clements and Shelford 1939). Succession occurs, however, in the oxbows, and the stages to be described for Reelfoot Lake are representative for large bodies of water. A century of silting from erosion of the surrounding upland has greatly reduced the depth and area of the lake, but the deeper waters are relatively free of wave action.

COMMUNITIES OF OPEN WATER

When an oxbow first forms, stillwater species invade over a period of 10 or more years (Eddy 1934). The plankton consists of diatoms and desmids and protozoans, small crustaceans, and rotifers. The glass minnow, white bass, yellow bass, and spoonbill cat frequent the open water and eat plankton or small fishes. The spoonbill cat has an unusual device for catching plankton in its flat expanded beak (Eddy and Shimer 1929).

BOTTOM COMMUNITIES

Bottom detritus is composed of plankton remains and fragments of higher plants and animals. Diatoms give a brown color to the layer just below the fresh deposit. *Hexagenia* mayfly naiads, such as are found in ponds, are absent, but the large red midge fly larvae *Tendipes plumosus* is abundant, and larvae of three other small flies are numerous. The small bivalved mollusk *Musculium transversum* and the carinated snail *Valvata bicarinata* are regularly present and sometimes abundant. The stomach contents of bottom-feeding fish, such as the drum, regularly contain these bottom organisms (Shelford and Boesel 1942, Shelford 1954a).

FLOATING AND SUBMERGED VEGETATION

Many protected coves in Reelfoot Lake are almost covered in summer with a mat of floating plants (Fig. 4-6). *Ceratophyllum* is important and the duckweeds *Spirodela* and *Lemna* occur in open areas. Submerged plants are from four to seven feet (1.2 to 2.1 m) tall. The pondweeds *Zannichellia palustris, Potamogeton filiformis, P. pusillus,* and *P. zosteriformis* occur in relative abundance in the order named (Davis 1937). In shallower water from two to six feet (0.6 to 2 m) in depth, the plant species listed in Figure 4-6 form dense beds but commonly die down in winter. Animals spend about one-third of the year without them. When present, the hornwort *Ceratophyllum demersum* supports a host of crustaceans and aquatic insects and is the resting and feeding place of the sunfish, bass, and crappies.

When waterlilies appear, the lily pads support a few animals on both their upper and lower surfaces. Cricket frogs and water striders are common. Leeches and the limpet snail *Ferrissia* are common on the stems. Below the floating and submerged plants, one to two feet of mud accumulates rapidly. This consists of silt and decaying organic matter and makes possible the invasion of the next stage (Shaver 1933, Davis 1937).

Bird species regularly present between mid-March and mid-April in the southwest corner of the upper Blue Basin, off Gray's Landing, are the American coot, double-crested cormorant, pied-billed grebe, lesser scaup duck, and blue-winged teal (S. C. Kendeigh, personal communication). The following were present in more than one-third of the counts over ten years: gadwall, baldpate, and shoveler ducks, great blue heron, and ring-billed gull. Coots were often estimated at 2000 on the open water. The double-crested cormorant and pied-billed grebe are fish-eaters. The lesser scaup and American coot feed on large amounts of animal material in winter when plants are absent. The bird composition in late summer is different. Whittemore (1937) recorded hooded merganser, mallard, pintail duck, and the common gallinule.

In the spring of 1937, nearly every stump within 1200 feet (360 m) of shore had a common egret perched on it. Most of the stubs were gone in 1952. The least tern nests on sand bars in the Mississippi River and comes to the lake area in late summer in family groups. The immature birds sit on the stubs and give a squeaking call which keeps the parents busy catching small fishes for them. Crook (1938) shows that young bluegills and other game fish of the vegetation areas make up the greater part of the food of the common egrets, double-crested cormorants, and great blue herons, all of which nest in a nearby heronry (Fig. 4-2B).

EMERGENT VEGETATION

As deposition of mud lessens the depth of water in the lily pad zone, watermillet comes in. This vegetation is generally rich in animals. The snails *Gyraulus parvus,* isopods, the amphipod *Hyallela azteca,* the shrimp *Palaemontes*

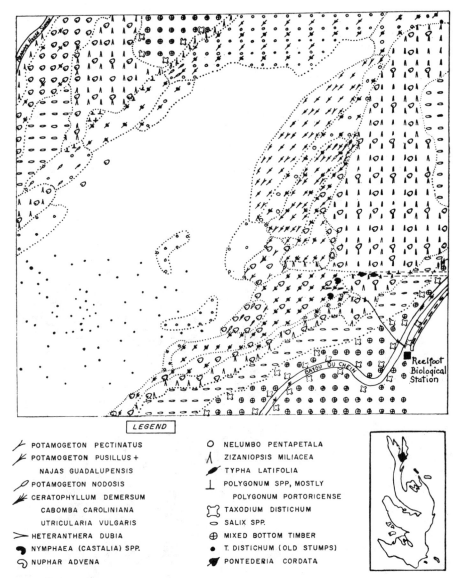

LEGEND

⊬ POTAMOGETON PECTINATUS	O NELUMBO PENTAPETALA
⊮ POTAMOGETON PUSILLUS +	⋏ ZIZANIOPSIS MILIACEA
NAJAS GUADALUPENSIS	⊁ TYPHA LATIFOLIA
⌀ POTAMOGETON NODOSIS	⊥ POLYGONUM SPP, MOSTLY
⋓ CERATOPHYLLUM DEMERSUM	POLYGONUM PORTORICENSE
CABOMBA CAROLINIANA	⊠ TAXODIUM DISTICHUM
UTRICULARIA VULGARIS	⊝ SALIX SPP.
⋗ HETERANTHERA DUBIA	⊕ MIXED BOTTOM TIMBER
⬤ NYMPHAEA (CASTALIA) SPP.	• T. DISTICHUM (OLD STUMPS)
⌒ NUPHAR ADVENA	⊁ PONTEDERIA CORDATA

Fig. 4-6. The submerged and emergent aquatic vegetation in an area at the north end of Reelfoot Lake in the summer of 1948 (prepared by John H. Steenis, U.S. Fish and Wildlife Service).

paludosus, and the small backswimmer *Plea* are usually numerous. The bowfin and the black bullhead make their nests in the water-millet. A cottonmouth was taken while swallowing a bullhead. The pirate perch, pygmy sunfish, and certain minnows are characteristic. An area of water-millet of 250 acres (110 hectares) in May, 1932, had 25 nests of the common gallinule, 75 of the American coot, 15 of the least bittern (Ganier 1933).

Flooding of aquatic vegetation is detrimental. Most flooding of Reelfoot Lake by small streams from the surrounding upland comes in March and April, occasionally in May or June. A June, 1945, flood raised the water 30 inches (9 m) above the normal lake level. A reduction of hornwort resulted, but the narrow-leaf pondweed grew more extensively later in the season. A whole series of other disturbances in growth and seeding occurred (Steenis 1947), showing that succession in oxbows is not a continuous smooth course forward.

SHRUBS AND YOUNG TREES

The first invader of the water-millet is usually buttonbush, followed soon by black willow (Figs. 4-6, 4-7), and later by cypress. As soon as the tall plants shade the water-millet, it disappears.

Fig. 4-7. The condition of the lake after the earthquake. A. Standing dead trees (photo by Ganier 1920). B. Young cypress submerged less than a meter by the earthquake, Gray's Landing (author's photo, April 1948). C. Invasion of the submerged aquatic plant area by water-millet or cutgrass and the displacement of water-millet by willows with young cypress behind. D. A regular inhabitant, the opossum (Illinois State Natural History Survey, Illinois River photo).

The small aquatic animals in the shrub and small tree stage are similar to those of the water-millet but decline in numbers as the shade increases. In spring, these areas often teem with top minnows. The buttonbush, willow, and young cypress afford nesting sites for the little blue heron and green heron, both of which are fish-eaters, and are visited by a long list of song birds and woodpeckers.

CYPRESS SWAMPS

It probably requires 100 to 150 years for water-millet to take possession of marginal areas of a large oxbow and form a proper bed for the seeding in of cypress. The cypress lives to be 300 to 400 years old, and almost pure stands of cypress occur locally. The usual understory consists of red maple, planertree, green ash, boxelder, and sometimes waterlocust. Starting at the edge of the water-millet and going toward shore, Gersbacher and Norton (1939) found the woody plants in the following order: buttonbush, black willow, green ash, cypress, planertree, fox-grape, and boxelder.

In cypress swamps, invertebrates are not numerous, probably due largely to the absence of submerged and emergent aquatic vegetation and low water or dry ground in summer. In spring when the water is high, gars are commonly seen. There are also a few pulmonate snails such as *Physa gyrina, Menetus exacuous,* and *Lymnaea columnella,* and the operculate snail *Viviparus intertextus.* The ground beetles *Poecilus louisinis* and *Oodes amaroides,* crane fly and lepidopterous larvae, harvestmen, and the ant *Aphaenogaster fulva* occur on floating logs.

The full-grown cypress trees are attractive to fish-eating birds, and a well-known heronry called "cranetown" (Figs. 4-8, 4-9) is located two miles south

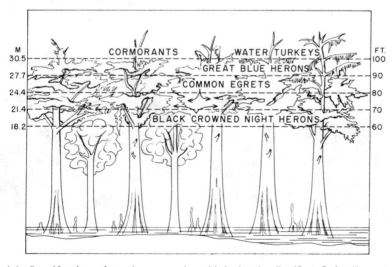

Fig. 4-8. Stratification of nesting gregarious birds in the Reelfoot Lake "cranetown" which occupied an oblong area 400 by 700 feet (120 by 212 m) when studied by Gersbacher in 1938.

Fig. 4-9. The "cranetown" in June. A. Typical base of a cypress tree. B. Cormorants and nests. C. Great blue herons. D. Egrets. E. An anhinga nest at a 90 foot (27 m) height (all photos by Albert F. Ganier).

of Markham in the center of a large area of cypress–maple–ash forest (Fig. 4-2B). Cranetown is a nesting colony of birds that at one time covered three acres (5 hectares) of cypress trees. In rainy periods trees are in one to three feet (0.3 to 1 m) of water, but during the summer the substratum usually becomes dry.

The cypress trees range from 80 to 130 feet (24 to 40 m) in height and 3 to 6 feet (1 to 2 m) in diameter. Nests are placed near the trunk of the trees or on the forks of strong branches, at heights from 60 feet (24 m) to near the top of the tallest trees. The ground and trees in the rookery are bespattered with the birds' excreta, mingled with feathers, broken egg shells, old nests, carcasses of dead birds, and disgorged fish. The number of nests at cranetown during 1938 (Gersbacher 1939) was as follows: common egret, 655; double-crested cormorant, 185; great blue heron, 85; black-crowned night heron, 45; anhinga or water turkey, 40; totaling 1010.

In addition to the birds in the heronry, Crook (1935) notes 500 double-crested cormorants in another area in 1934 and estimates the late summer population of common egrets to be 2500. He has a maximum record of 200 wood ibis, 25 green herons, and 14 belted kingfishers seen in one day. Other fish-eating birds in the area are least tern, least bittern, yellow-crowned night heron, osprey, and pied-billed grebe.

The gardener and forester limit the number of cultivated plants per unit of space to the number that can grow to maturity or to a desired life-history stage. In fisheries legislation, exactly the reverse has been the plan; provisions have been made to increase the number of fish indefinitely and levy penalties for taking game fish below a certain size.

Thompson (1941), Bennett (1947), and others maintain that fish-eating birds are important in preventing overcrowding and consequent dwarfing of individuals. Fish-eating birds are credited with maintaining the high sustained yield of fish at Reelfoot Lake, both for the hook and line fisherman and the commercial fisheries. The fertility and productivity of the lake is high. A high rate of annual cropping enables the fish population that remains to grow at a continuously fast rate. Fishing is much better with medium-sized populations of large individual fish than with large populations of very small fish.

Filling of the cypress swamp progresses slowly. The cypress and associated trees do not reproduce in their own shade, but new trees may develop in openings within the forest. Finally, the sugarberry succeeds the cypress along with sweetgum or elm. The last cypress trees persist for a long time, making the duration of the stage probably more than 800 years (Shelford 1954a). Only the deeper depressions in the swamp contain water the year around. These gradually fill to a point that they contain water only in spring. Thus the large oxbow lake becomes a small pond.

According to Fuller (1912), the shallower portions of Reelfoot Lake in 1905 were occupied by stubs of black walnut, ash, oak, elm, and mulberry. This suggests an oak–elm–ash forest. Part of this area lies in an old A.D. 800

channel (Fisk 1944). With 1000 years to develop, this forest could have succeeded one of cypress.

Convergence of Seres

The succession of biotic communities is generally from lower to higher levels and toward less and less contact with water. All the habitats, including the earliest oxbows and sandbar willow ridges, contain seres that converge toward a forest in which sugarberry is the most important dominant, along with elm or sweetgum. Seres initiated by the sandbar willow may under favorable conditions reach a sugarberry–gum forest in about 125 years. On the contrary, seres beginning in ponds, oxbows, and long submergence habitats may reasonably be expected to reach the sugarberry–gum stage in not less than 1000 years. The sugarberry–gum community would be essentially the same, however, regardless of whether it developed from a ridge or an aquatic habitat.

FLOODPLAIN COMMUNITIES IN THE MAPLE–BASSWOOD– OAK–HICKORY ECOTONE

The size of streams here under consideration is large enough for canoe travel during all except drought years. One of the outstanding features of small streams is the fact that over a large part of their course they flow in a ravine of trees 60 to 75 feet (18 to 23 m) deep and often narrower than the stream channel.

The Sangamon River in Illinois is skirted by wooded bluffs and tall streambank timber. Six stages of succession are recognized by Goff (1931, 1952). Goff censused animals in (0.1 m²) ground samples to a depth of 5 centimeters and took sweep net samples from the herbaceous vegetation and shrubs.

Heterocerus Facies

This first stage is a narrow area of silt without vegetation next to the stream that is repeatedly flooded. Shore bugs and toad bugs run about. The mud beetle *Heterocerus tristis* lives and breeds in galleries in the mud (Claycomb 1919). Four other beetles occur, but where they retreat to in times of flood is unknown. Tracks of mink, muskrat, raccoon, and occasionally cottontail are frequent (Goff 1931, 1952, Koelz 1936).

Bidens–Allolobophora–Rumex Facies

The first plant invaders may be willow or giant ragweed but are most typically the beggar tick *Bidens comosa*. The water-dock *Rumex orbiculatus* and an aster are sometimes present. The earthworm *Allolobophora iowana* makes its appearance at this stage and continues throughout the sere (Figs. 4-10, 4-11). Larvae of the fly *Parydra quadrituberculata* are less abundant than the earthworm. The fly *Leptocera fontinalis* and the ground beetle *Bembidiom affine* are active on the surface of the ground.

Maple–*Bimastos*–Willow Facies

Silver maple and black willow are the only young trees. The shade tolerance of silver maple gives it a distinct advantage over the willow. The more abundant herbs are an aster and false dragonhead. Several species of earthworms occupy the soil, but *Bimastos tumidus* is the most characteristic of the facies.

Flies are well represented by *Oscinella coxendix* and *Homoneura philadelphica* and beetles by *Chalcoides helxines*. The leafhopper *Erythroneura nigerrima* is abundant, and a few bugs are present. Ants are represented by *Ponera coarctata pennsylvanica*. The many young woody plants attract cottontails that eat the bark and cambium during the winter months. Seventeen muskrats, three opossums, and twenty-one cottontails were trapped during two months of early winter from about four miles (6.4 km) of the river margin (Koelz 1936, Harper 1938).

Maple–*Octolasum*–Elm Facies

On a sample area of approximately 1000 square meters, there were 12 dominant silver maples of from one to two feet (30 to 60 cm) in diameter, 2 American sycamores of about the same size, and 17 elms *Ulmus americana* and *U. fulva* of from two to six inches (5 to 15 cm) in diameter. Seedlings of hackberry and ash made a shrub layer. The abundant herb *Aster simplex* covered part of the ground.

The same earthworms as in the previous stage continue to occur here (Figs. 4-10, 4-11). Common beetles are *Telephanus velox, Arpedium cribratum,* and *Melanophthalma cavicollis.* Flies are represented by *Homoneura seticauda* in early summer and *H. philadelphica* in late summer. The most important leafhopper is *Empoasca fabae.*

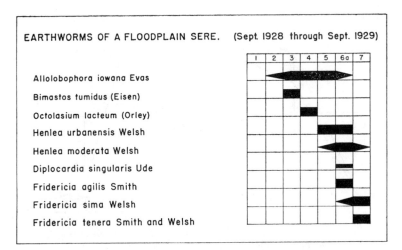

Fig. 4-10. Succession of earthworms in the Sangamon River floodplain (after Goff 1952); the stages are indicated by numerals and explained in the text and facies species lists.

Elm–*Henlea*–Bur Oak Associes

The ground level here is from four to five feet (1.2 to 1.5 m) higher than in the preceding stage, and was flooded only three times during the year of study. A tree count in a fairly typical sample area gave the following percentages of species: American elm, 32; pignut hickory, 13; hackberry, 12; bur oak, 10; hawthorn, 9; white ash, 6; and slippery elm, 6. A scattering of other trees made up 12 per cent. The shrubs and vines are catbriar, redbud, *Crataegus,* young hackberry, poison ivy, and grape. Young elms are very abundant. During the spring and early summer, buttercups, violets, wild onions, trilliums, nettles, sweet william, and a grass are the principal herbs.

Common animals in this stage are the leafhopper *Erythroneura campora,* the fly *Homoneura philadelphica,* and the snail *Stenotrema hirsutum.* Two species of white earthworms, *Henlea urbanensis* and *H. moderata,* and the snails *Succinea ovalis* and *S. avara* occur in or on the ground.

During the late summer and fall, *Ranunculus septentrionalis,* wood-nettles, and grasses are the outstanding plants. Associated animals are the beetles *Antherophagus suturalis* and *Melanophthalma distinguenda,* and the plant bug *Lygus oblineatus.* The leafhoppers are *Paraphlepsius irroratus* and *Erythroneura tricincta.* One bug, *Nabis ferus* is present. White-footed mice are numerous and the tracks of skunk, weasel, and muskrat are frequently seen.

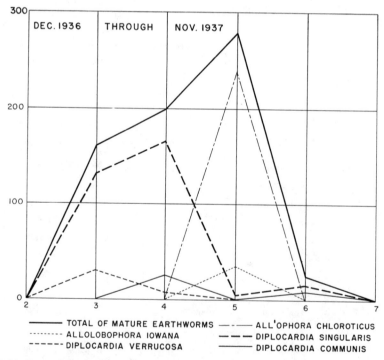

Fig. 4-11. Succession of red earthworms on the Sangamon River floodplain (after Harper 1938); numerals refer to the same stages as in Figure 4-10.

Elm–*Fridericia*–Shingle Oak Facies (Subclimax)

The sample of this community studied was on an old floodplain, two or three feet (0.6 to 1.0 m) higher than the preceding habitat. The percentages of different trees were American elm, 35; shingle oak, 19; pignut and shagbark hickories, 16; black walnut, 5; redbud, 5; red oak, 5; and slippery elm, 5. A scattering of other trees made up about 9 per cent. This list shows an invasion of typical upland trees.

Flying squirrels tend to be common in this stage. The earthworm *Diplocardia singularis* is usually present but few in number in soil collections down to five centimeters (Fig. 4-10). White annelids *Fridericia* are obtained only at this and the next or climax stage. The climax stage of white oak–black oak–hickory, occurring on dry upland, has few annelids.

The red earthworms of the Sangamon floodplain were also studied from December, 1936, through November, 1937, by Harper (1938), based upon digging 0.25 square meters to a depth of 0.6 meter every one or two weeks during the period (Fig. 4-11). They were found to occur both on well-drained ridges and in the shallow depressions but were more abundant in the former. In winter they were in the two inches (5 cm) of soil just below the frost line. None was present where the water table was nine inches (22.5 cm) from the surface, and over the period of study the size of populations collected varied sharply with the height of the water table. The elm–bur oak stage had the greatest number of worms. In laboratory experiments, *Allolobophora iowana* and *Diplocardia communis* were found to avoid air of 57 per cent relative humidity or less (Heimburger 1924). White earthworms selected 19° C. and soil instead of leaves or wood. They tended to keep near to the soil surface but to avoid daylight (Tucker 1943).

COMMUNITIES ON FLOODPLAINS IN OTHER REGIONS

Tulip–Oak Forest Region

Seven seral stages occur on the Mill Creek floodplain near Nashville, Tennessee (Shaver and Dennison 1928): the pioneer water-willow *Justica americana*, the spike-rush *Eleocharis*, the rice-cutgrass *Leersia oryzoides*, the annual–perennial herb, the willow–sycamore pioneer, silver maple–boxelder, and the oak, ash, walnut or mature floodplain forest (subclimax).

An island or bar with water-willow in the 20 foot (6 m) stream channel receives depositions of sand above the low water stage. Spike-rush appears among the water-willow, as the deposition increases. Water-willow is rapidly replaced by the spike-rush so that the island appears with a center of spike-rush and an outer zone of water-willow. Rice-cutgrass follows the spike-rush on the higher ground. The sedges *Cyperus lancastriensis* and *Carex frankii* then invade, the latter species becoming the next dominant. Perennials occurring with the sedges are knot grass, late-flowering thoroughwort, water smartweed, and others. Seedlings of willows, cottonwoods, and sycamores appear and a willow–sycamore forest becomes established. This is followed by silver maple, boxelder, and sugarberry. When the ground becomes high enough rarely to be flooded, black walnut, white ash, redbud, bur oak, and southern red oak invade. The most typical plants growing beneath these trees are the cane *Arundinaria gigantea* and the ground ivy *Glechoma hederacea*.

Maple–Beech–Hemlock Forest Region

On the floodplain of the Galien River at Warren Woods, near Sawyer, Michigan, an almost never flooded terrace between five and ten feet (1.5 and 3.0 m) above the river contains black walnut, basswood, butternut, elm, hackberry, and a large relict sycamore. Underneath the trees are spicebush and an occasional dogwood. There is no beech here, but on a higher terrace nearby beech is the principal tree, as it is on the surrounding upland.

In Turkey Run State Park near Marshall, Indiana, the Sugar Creek floodplain forest is composed of sycamore, elm, walnut, butternut, and a few other species. On the adjacent bluffs and upland the forest is largely beech with scattered oaks and sugar maple. The animal populations in the floodplain and upland forests have been estimated annually for 20 years as shown in Table 4-4. The salamander populations were at their maximum during the period of study. More insects were present on the floodplain than on the upland because of the greater variety and density of herbs and shrubs that were present.

Table 4-4. A Typical Estimate of the Numbers of Organisms per Hectare, Made on May 4, 1935, at Turkey Run State Park, Indiana

	CLIMAX MAPLE–BEECH	FLOODPLAIN SYCAMORE–ELM
Slimy salamanders	10,000	500
Red-backed salamanders	10,000	2,500
Birds	33	155
Invertebrates, largely insects	3,868,004	5,107,384
Fungi counts, August, 1936, by C. L. Porter	82	34

Oak–Hickory Forest Region

In the Piedmont region of North Carolina the earliest woody vegetation on floodplains is willow and alder, followed by sycamore, elm, red maple, and sweetgum (Table 4-5). The alder *Alnus rugosus* disappears when the willows become more than 20 years old (Oosting 1942).

Grassland Regions

Oak–hickory forest occurs on river bluffs, old terraces, and in ravines for two-thirds of the distance across the midcontinent grasslands. The Canadian River in Oklahoma is a turbid, sand-laden stream characteristic of this general area. The shallow bed and the adjacent low bars are composed of silt and

Table 4-5. Ages of Trees of Different Species Compared with Willows in Different Stands on Floodplains in North Carolina

WILLOW	SYCAMORE	ELM	RED MAPLE	SWEETGUM	FLOWERING DOGWOOD
10 years		seedlings			
10–20	2–10	seedlings			
46–48	12–15	?–16			
50–52	45–55	27–60	25–42	38	seedlings

sand, saturated with water except during hot dry portions of the year. Willow and cottonwood are the important early trees in the seral development. The principal willows are the sandbar, black, and peachleaf. These species are succeeded by elm and later by oak. Hefley (1937) recognizes the elm–bur oak forest near Norman, Oklahoma, as an edaphic climax. Hackberry forms local facies of this forest and ash is also present. The undergrowth includes *Smilax, Vitis,* and *Symphoricarpos.* Elsewhere through the midwest the netleaf hackberry, little walnut, and green ash occur (Bruner 1931).

Vertebrates present in this floodplain forest are almost entirely deciduous forest species. Invertebrates are also of wide distribution. The westward extension of these species has also been aided by the planting of trees around farm buildings in Nebraska, the Dakotas, and Canada. Birds and mammals are principally forest-edge species.

Chapter 5

The Boreal Coniferous Forest

Coniferous forest extends over 60 degrees of latitude from the high mountains of Costa Rica to the mouth of the Mackenzie River in Canada and the Brooks Range in Alaska. The boreal–montane coniferous forest biome extends from Labrador to western Alaska, and south through the Rocky Mountains and Sierras. In Mexico, between latitudes 27° and 23° N., it makes an ecotone with the mountain coniferous forest of the tropics. *Pinus ayacahuite,* for instance, extends from 27° N. Lat. south into Guatemala, while Douglas-fir, ponderosa pine, black bear, and certain tree squirrels extend from 23° N. Lat. northward into Canada. The lowland coniferous forest of southern United States, Cuba, Yucatan, and elsewhere are treated in Chapters 3 and 19. The hemlock–cedar forest of the Pacific Coast constitutes a different biome and is considered in Chapter 8.

The climate in this biome varies from cool to cold. There is precipitation in all seasons, but most of it comes in the summer over much of the area. Near the Pacific Ocean and in the mountains of Idaho and eastern California, the greater part of the precipitation falls in the winter. Precipitation is least in central Canada north of the Great Plains.

The evergreen climax trees have long needle leaves, as in pine, or short thick leaves, such as characterize spruce, hemlock, and fir. Dead leaves persist for a time before falling so that crown fires are frequent. There is a thick layer of needles and dead twigs on the ground. Stream sides are occupied by tamarack, willow, birch, alder, and poplars. Most of the animals, especially those of the climax, are shy and retiring. They utilize the dense growths of evergreens as cover for breeding and for protection against the weather in winter. They are capable of tolerating cold winters with much snow. Some species burrow into the snow for the night; several are migratory. The herbivorous

120

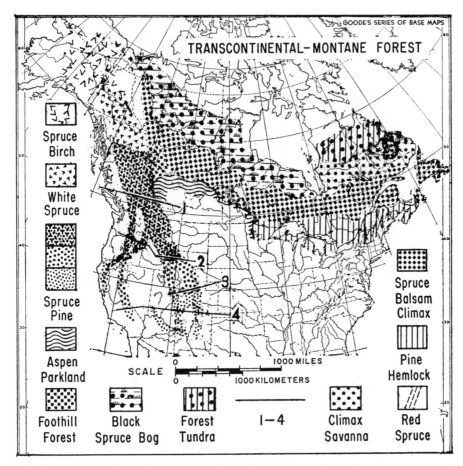

Fig. 5-1. Transcontinental–montane coniferous forest (according to Dansereau, personal communication, Halliday 1937, Hare 1950, and Marr 1948, in Canada; Shantz and Zon 1924, Preston 1948, courtesy Iowa State College Press, in the United States; University of Chicago base maps). The numbered lines indicate the fadeout of the biome toward the south. The southern limits of important species are indicated: near 1, black spruce, white spruce, balsam fir, tamarack, and spruce grouse; near 2, paper birch, moose, and caribou; near 3, balsam poplar, lynx, and snowshoe rabbit; near 4, wolverine; and near end of the stippling, the Colorado blue spruce.

animals are largely browsers. Insects often destroy large areas of forest.

Found throughout nearly the entire biome are paper birch, balsam poplar, quaking aspen, certain species of *Ribes,* wolverine, lynx, red squirrel, and certain races of snowshoe rabbit. The larger animals progressively drop out southward on the mountains (Fig. 5-1). The spruce budworm (Fig. 5-2B) and the tent caterpillar (Fig. 5-10A) are dominants or influents everywhere. Various birds, small mammals, insects, herbs, and shrubs of subordinate character have a general distribution. Outside of those species above listed, the plant elements

Fig. 5-2. A. Open spot in white spruce climax in Prince Albert National Park, Saskatche-wan, Canada (photo by J. A. Macnab). B. Life history stages of the spruce budworm (after Swaine 1931-32, F. Hennessey).

show two major divisions: the montane forest discussed in Chapter 6, and the transcontinental boreal forest described here.

In 1600, the native population in the Arctic drainage area in Canada was 0.2 to 0.4 persons per 10 square miles (0.1 to 0.2/10 km²). Their cultural development was low, and they subsisted mainly by hunting and fishing. In the area draining into the Great Lakes and upper St. Lawrence, the population of Algonquin, Ottawa, and Ojibwa tribes was 1.3 to 3.2 persons per 10 square miles (0.5 to 1.2/km²). The woodland caribou was important in their economy. The wild rice marshes in the mixed coniferous and deciduous forest area of Wisconsin supported a large population of 3.2 to 8.0 per 10 square miles (1.2 to 3.1/10 km²). There were many Indian tribes in the area with some agriculture, but the general cultural level was low (Kroeber 1939). Since 1600, there has been widespread movement of natives out of the area, much lumbering and disastrous fires especially since 1870, and depletion of many large mammals. Northern Ontario and Quebec have been least disturbed because of the low commercial value of the timber.

Regions of the Boreal–Montane Coniferous Forest

Three large natural regions are recognized (Fig. 5-1) :

1.   The boreal forest east of the Rocky Mountains. It includes several forest types. (a) The white spruce–balsam fir forest is climax from Newfoundland to the foothills of the Rocky Mountains, with balsam fir relatively unimportant west of Lake Winnipeg. Halliday (1937) mapped 23 faciations and facies of this forest. (b) Red pine–white pine–hemlock forest of the Great Lakes region and the lower St. Lawrence and New Brunswick lowlands. Halliday mapped 12 faciations and facies of this forest in Canada. (c) Red spruce–balsam fir forest beginning in the New Brunswick highlands where Halliday recognized two faciations and extending south across the highlands to northern Georgia. Balsam fir is replaced by Fraser fir in the Appalachian Mountains. (d) Forest–tundra or savanna with about equal ground coverage of trees and lichen tundra or where a thin forest occurs in the valleys and lichen tundra on the upland between streams (Hare 1950). (e) The black spruce bog forest west of Hudson Bay which appears to be early seral and subclimax. (f) The aspen parkland which appears to represent the first stage of invasion of the grassland by the coniferous forest (discussed in Chap. 12).

2.   The forest of the valleys and lower mountain slopes of the northern Rocky Mountains which is transitional between the boreal forest and the mountain forests in the west. This region includes three types of forest. (a) The white spruce–lodgepole pine forest of northern British Columbia and southern Yukon. Halliday mapped two faciations or facies in this type. (b) The white spruce–black spruce–birch forest of Yukon and Alaska. Porsild (1951) found four faciations and facies along the Canol Road, and Dice (1920) reported almost the same series in Alaska.

3.   The montane forests of the Rocky Mountains and Sierra Nevadas. In this chapter the first two regions will be described, leaving for the next chapter the description of the forest communities in the Rocky Mountains and Sierras. The boreal forest may be called the white spruce–woodland caribou association.

Topography and Climate

Elevation ranges from sea level to about 1200 feet (360 m) in the boreal forest, except for the highlands of the north end of the Appalachian Plateau where the forest extends above 2500 feet (758 m). Mountains are few and of low elevation. There is permafrost beneath the northern part of the forest. The annual rainfall ranges from 10 to 20 inches (25 to 50 cm) in the western half and from 20 to 50 inches (50 to 125 cm) in the eastern half. Snowfall also increases eastward, reaching 200 inches (500 cm) in parts of Quebec. East of Lake Superior the forest lies chiefly between the mean January isotherms of 14° and –4° F. (–10° and –20° C.), and west of Lake Superior between –4° and –20° F. (–20° and –29° C.). Mean monthly temperatures may fall as low as –30° F. (–34° C.) and go as high as 73° F. (23° C.). Minimum daily temperatures may attain –35° to –60° F. (–37° to –61° C.), and the daily maximum, 100° F. (38° C.) or higher.

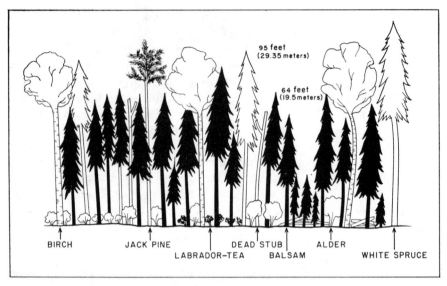

95 feet
(29.35 meters)

64 feet
(19.5 meters)

BIRCH   JACK PINE   DEAD STUB   ALDER
          LABRADOR-TEA   BALSAM   WHITE SPRUCE

Fig. 5-3. The structure of the vegetation near Lake Nipigon, Ontario (after Kendeigh 1947).

Dominant and Influent Plants and Animals

The dominant climax trees of the boreal forest are white spruce (Figs. 5-3, 5-4), balsam fir, black spruce, red spruce, aspen, and an occasional paper birch. White spruce and aspen are the principal dominants from Manitoba westward. Black spruce is more important in seral than climax stages (Fig. 5-5). The principal shrubs of climax and late subclimax communities are speckled alder, Labrador-tea, bunchberry, sour-top-bilberry, and mountain-ash. Prominent herbs are wild sarsaparilla, twinflower, sweet-scented bedstraw, and star-flower.

Mammals and insects are important animal constituents of the forest. In a study of the white spruce climax of Prince Albert Park, Saskatchewan (Fig. 5-2), the midsummer arthropod population averaged 100 per square meter, almost as large as in the deciduous forest at the same season. The following vertebrate signs were found on eight square meters: moose, 2; grouse, 9; snow-shoe rabbit, 14; and red squirrel, 4. Cruising one hectare revealed 42 squirrel cone piles, 11 squirrels seen, 30 rabbit forms, and 8 moose signs.

Permeant Dominants

These animals enter all faciations and facies of the forest and destroy many other organisms. The woodland caribou at one time was very abundant in the central area of the transcontinental forest and probably occurred throughout. The species migrated irregularly with the seasons; in Newfoundland, it left the forest for the open ridges in October (Fig. 5-6). The food of the caribou included mosses, lichens such as *Cladonia*, paper birch, aspen, cherry, speckled alder, mountain maple, striped maple, bearberry, hazel, black crow-

Fig. 5-4. A. Undisturbed forest south of The Pas, Manitoba (1939); trees near the foreground are black spruce and jack pine (photo by J. A. Macnab). B. Jack pine near The Pas (photo by J. A. Macnab). C. Red pine in the Superior–Quetico (Minnesota–Ontario) area. D. Egg. E. Larvae. F. Adults of the jack pine sawfly (Department of Agriculture, Science Service, Canada, by Dunn 1931).

berry, low birch, pussy willow, and red-osier dogwood. Grasses and various, mostly hydric, herbaceous plants, formed an important part of their diet during the warm period. During the winter, when the ground was deeply covered with snow, they fed on lichens such as the common *Usnea barbata,* which hangs from the lower branches of black spruce and larch, and on *Sticla pulmonaria,* which also grows on trees (Seton 1929, Shelford and Olson 1935).

The moose (Fig. 5-7) inhabits climax stands of white spruce and balsam fir and subclimax stands of jack pine, black spruce, aspen, birch, and cherry, especially during the winter. The last three species are used for food as well as for shelter. Most of its relations in the spring, summer, and early autumn are with early seral stages. At this time much of its food is white and yellow water-lilies, sedges and grasses, pondweeds, the twigs of willow, alder, birch, and aspen, and some *Equisetum* (Fig. 5-10). Moose and caribou destroy subclimax species, undergrowth, and early seral species. This results in a tendency to maintain a seral or subclimax condition which is probably their optimum. Where the moose become overly abundant, as they have on Isle Royale in

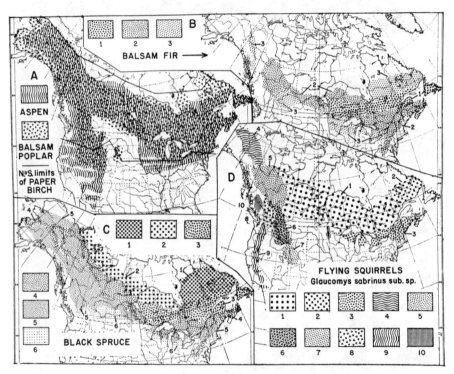

Fig. 5-5. A. Distribution of quaking aspen, interrupted to suggest a scattered distribution, balsam poplar, and the northern and southern limits of paper birch. B. Suggested function of balsam fir; 1, a climax dominant in the eastern transcontinental forest in association with white spruce and black spruce; 2, scattered climax dominant; 3, scattered trees. C. Suggested function of black spruce: 1, climax dominant in savanna and forest–tundra in Quebec; 2, a seral dominant in extensive swamp or muskeg forest; 3, a climax dominant in the eastern transcontinental forest; 4, a scattered climax dominant in Yukon and Alaska; 5, a dominant in scattered bogs; 6, isolated individuals or small groups of trees (after Shantz and Zon 1924, Halliday 1937, Preston 1948, courtesy Iowa State College Press; University of Chicago base maps). D. Distribution of subspecies of flying squirrels: 1. *Glaucomys s. sabrinus;* 2, *G. s. makkovikensis;* 3, *G. s. macrotis;* 4, *G. s. yukonensis;* 5, *G. s. alpinus;* 6, *G. s. fuliginosus;* 7, *G. s. bangsi;* 8, *G. s. columbiensis;* 9, five additional Sierran subspecies; 10, *G. s. reductus* (after Howell 1918 and Anderson 1946).

Fig. 5-6. A. Woodland caribou on Mount Albert, Quebec (3775 feet, 1145 m), in subalpine meadow during the October rutting season. The meadow consists of sedge with the grasses *Agrostis* and *Agropyron;* the shrub is not identified. B. General view; X indicates the location of the herd. The forest is chiefly balsam fir with white and black spruce, alder, and paper birch also present (courtesy Quebec Department Fish and Game and G. Moisan; photo by E. L. Desilets). C. Distribution of some subspecies of *Rangifer caribou* (1, 2, 3) and *R. arcticus* (4, 5, 6).

Lake Superior since 1905, they do great damage to their food supply, even reducing bogs and lakes to mud holes.

The white-tailed deer is not a coniferous forest species (Stenlund 1955) but is able to flourish on the variety of trees and shrubs (Swift 1946) which spring up following lumbering and fires. It has extended its range about 500 miles (800 km) north from central Wisconsin. Westward, the white-tailed deer has displaced the mule deer and perhaps the wapiti from the aspen parkland (Bird 1930).

Arthropod Dominants of the Climax

Approximately 50 species of insects are important in the spruce–fir forest, not counting those that infest black spruce, jack pine, and seral broad-leaved species. Several of these species may be considered dominants, especially in the areas of heavy infestation that shift from place to place through the forest.

Spruce budworm (Fig. 5-2) is always present in destructive numbers in one or more areas of the forest. In recent outbreaks, balsam fir and red spruce of the Maritime Provinces of Canada have been severely injured or extensively killed, white spruce injured less severely, and black spruce checked in growth but rarely killed. The moths appear in early July and lay their eggs on the needle leaves. The caterpillars, on hatching, spin silken cases in crevices of the twigs near the buds, where they pass the winter. In spring, as the buds are opening, the caterpillars emerge and feed for three to five weeks on the buds and growing shoots at the top of the tree, causing defoliation. Two or more defoliations result in the death of the tree. A heavily infested forest has a

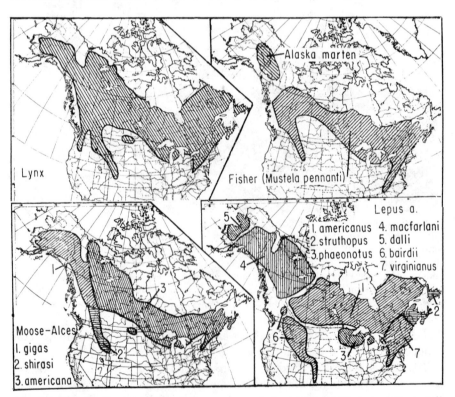

Fig. 5-7. The distribution of four influent mammals in the forest. Under pristine conditions they were absent from the wet coast forest. The range of *Alces alces americana* has recently been divided near Lake Winnipeg with the western half becoming *A. a. andersoni*.

scorched appearance and later becomes grayish in color (Swaine *et al.* 1931).

The eastern spruce beetle (Fig. 5-8) is also a dominant. This species is nearly always present and very destructive of white and red spruce. The beetles make galleries in intricate patterns just under the bark (Swaine *et al.* 1931). Many trees have the bark partly removed by woodpeckers which feed upon the larvae and pupae.

The black-headed budworm occurs in very moist regions and attacks balsam fir, spruce, and associated trees chiefly where balsam fir and hemlock are nu-

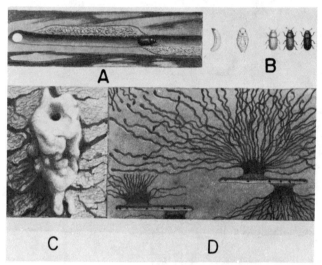

Fig. 5-8. Eastern spruce beetle *Dendroctonus piceperda*. A. Inner surface of bark with a tunnel, a beetle, and eggs (the white spot). B. Life history stages. C. Pitch tube which indicates beetles present. D. Pattern of the bark tunnels (after Swaine 1931).

merous. Both large and small trees may be completely defoliated in one season. The moths lay their eggs on the under side of the needles in autumn, and the eggs hatch about the first of June. The caterpillars crawl into the opening buds, make a web, and feed on the new tissue. When the new needles are consumed, larger caterpillars move down on the stem and feed on the older needles, especially of balsam fir. White, red, and black spruce are only lightly attacked (Swaine *et al.* 1931).

Major Permeant Influents

Wolves feed largely upon small mammals and do not attack the larger ungulates unless forced to do so by the failure of their normal food supply. Moose, caribou, and deer are generally attacked by packs rather than individuals. It is doubtful if they can control the size of the deer population (Stenlund 1955). Although the wolf ranges widely over both climax and subclimax in pursuit of food, the bulk of its food supply is found in the subclimax (Fig. 5-9) (Olson 1938).

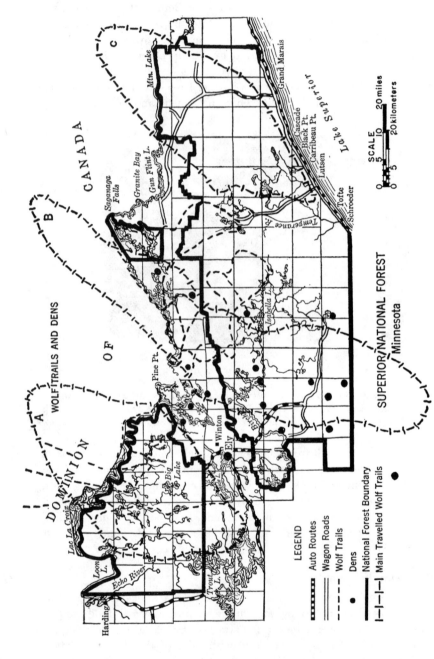

Fig. 5-9. Wolf pack routes in the Superior-Quetico (Minnesota-Ontario) area. The long axis of ellipse A is approximately 55 miles (88 km), B is 95 miles (152 km), and C is 60 miles (96 km) (after Olson 1938, courtesy John Wiley & Sons, Inc.).

LEGEND

Auto Routes

Wagon Roads

Wolf Trails

Dens

National Forest Boundary

Main Travelled Wolf Trails

SCALE

The lynx (Fig. 5-7) prefers the densely wooded climax but travels widely. Rabbits are their chief food, but they also eat grouse, squirrels, and bog lemmings. The wolverine or glutton eats only flesh, everything from fish to woodchucks. The black bear and grizzly bear are omnivorous, consuming leaves, buds, fruits, nuts, berries, fish, and occasionally the flesh of other mammals. In getting fruits they frequently destroy the shrubs and trees that produce them. The porcupine (Fig. 5-10D) is largely a bark-eater and often girdles or seriously damages trees, chiefly conifers. Seton (1909) estimates the population level of the species to average one per 13 acres (5 hectares), but they may have been more numerous than this. The great horned owl is an important large bird predator (Shelford and Olson 1935).

Minor Permeant Influents

Included among the minor permeant influents are the grouse, two hawks, woodchucks, weasels, woodpeckers, jays, and rabbits (Shelford and Olson 1935).

The snowshoe rabbit (Fig. 5-7) girdles poplars about two and a half inches (6 cm) in diameter by removing the bark for an average of about two feet (0.6 m) above the snow. Twelve hares made up the population of an area in western Ontario in which during one winter 168 quaking and bigtooth aspens, 11 small willows, 15 speckled alders, 2 paper birch saplings, 1 jack pine, and some serviceberry were used (MacLulich 1937). Forty per cent of the snowshoe rabbit population occurred in willow–alder swamps, 25 per cent in poplar–birch and cut-over lands, 14 per cent in upland spruce forest, and 12 per cent in jack pine. By their feeding, they tend to prevent invasion of conifers and to maintain the vegetation suitable to their own needs. The population is subject to cyclic fluctuations and varies greatly from place to place. The rabbits are subject to a number of diseases, and the lynx, fox, and wolf all feed on them.

Woodchucks occur in the climax community only in the mountains of the northwest where they are called thick woods badgers.

Spruce grouse has a preference for the white spruce–balsam fir forest (Fig. 5-11). It feeds on spruce and other conifer needles in winter, doing damage to seedlings and eats a variety of green plant material at other seasons. It also requires low shrubs and seedlings for nesting. Populations are generally low, one breeding pair of this species and two pairs of ruffed grouse being recorded on 100 acres (40 hectares) (Kendeigh 1947).

Other Climax Species

Northern flying squirrels (Fig. 5-5) have a preference for the food and shelter found in the climax forest. The arboreal marten rarely ranges outside of dense climax coniferous forests. They feed largely upon the smaller herbivorous squirrels, mice, and voles but also take insects, grubs, small reptiles, carrion, and even the berries of the mountain-ash. Red squirrels gather great quantities of cones of pine and spruce as well as the fruits of hazel, rose, cherries, and plums found sparsely throughout the climax or around its edges. These they store in hollow trees, other cavities, or bury in the ground.

Fig. 5-10. Coactions. A. Defoliation of aspen by the forest tent caterpillar south of The Pas, Manitoba, in 1939 (photo by I. A. Means). B. Large aspen cut by beaver in the winter of 1938-39, in Riding Mountain Park, Manitoba (photo by J. A. Macnab). C. Aspens topped by moose in the Superior–Quetico area (photo by A. R. Cahn). D. White spruce nearly girdled by a porcupine near Castle Mountain, Alberta (photo by J. A. Macnab). E. Waterlilies with leaves removed by moose in the Superior–Quetico area (photo by A. R. Cahn).

Small mammals are inhabitants of ground, fallen tree trunks, and litter in the climax and late subclimax forests. The following were trapped on two acres (0.8 hectare) in the Lake Nipigon area, Ontario, by Kendeigh (1947): deer mouse, 3; rock vole, 3; Gapper's red-backed mouse, 5; woodland jumping mouse, 1; masked shrew, 8; least chipmunk, 2. The chipmunks go up into bushes and, when abundant locally, exert a marked influence by feeding on fruits, seeds, green foliage, and insects. Reptiles and amphibians play a very minor role in the forest.

Breeding birds in the same area, with more than two pairs per 100 acres (40 hectares) in 1945, were yellow-shafted flicker, 4 pairs; red-breasted nuthatch, 3; brown creeper, 9; winter wren, 5; Swainson's thrush, 4; golden-crowned kinglet, 8; red-eyed vireo, 7; Tennessee warbler, 59; Nashville warbler, 8; magnolia warbler, 6; Cape May warbler, 28; myrtle warbler, 3; black-throated green warbler, 6; blackburnian warbler, 6; bay-breasted warbler, 92; ovenbird, 10; slate-colored junco, 3; and white-throated sparrow, 18. The food relations of the warblers determined from the stomach contents of specimens secured several years earlier were as follows: Phalangidea, 0.2 per cent; Araneida, 15.2; Orthoptera, 1.2; Neuroptera, 0.2; Ephemerida, 6.1; Hemiptera, 1.5; Homoptera, 11.5; Coleoptera, 19.5; Trichoptera, 1.5; Lepidoptera, 22.3; Diptera, 9.3; Hymenoptera, 7.0; and undetermined insects, 3.9 (Kendeigh 1947).

The soil in the spruce–balsam forest is a damp, closely packed leaf mold, varying in depth, and composed of decaying balsam and spruce needles, birch leaves, and sticks interwoven with fungus mycelia. Open areas may have a ground cover of *Aster macrophyllus* and the wintergreen *Pyrola virens*. Few or no shrubs are present (Gleason 1909).

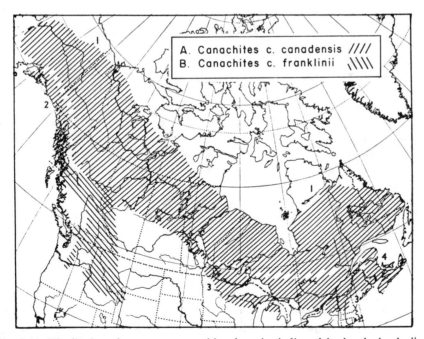

Fig. 5-11. Distribution of spruce grouse, with subspecies indicated by breaks in shading (after Pitelka 1941).

A fairly thorough study of a balsam fir–paper birch–white spruce forest has been made on Isle Royale in Lake Superior (Adams 1909). In the leaf mold were found a few species of spiders, the collembolon *Tomocerus flavescens,* two species of myriapods, a few enchytraeid earthworms, and the snails *Discus cronkhitei-catskillensis, Zonitoides arboreus,* and *Retinella binneyana.* A large carpenter ant, *Camponotus herculeanus,* and a black carabid, *Calathus gregarius,* foraged over the ground surface.

The trees shelter a more varied population, including buprestids and cerambycids. Dead trees of balsam and spruce are bored into by larvae of the longhorned beetles *Anoplodera canadensis, Monochamus scutellatus,* and others. Among the bark beetles are *Dendroctonus valens,* which works at the bases of trees, and *Ips perturbatus,* which is confined to white spruce. Under the loose bark of trees which have partly decayed, the spider *Amaurobius bennetti* frequently builds its web and the beetle *Pristodactyla advena* forages. A nest of the ant *Formica sanguinea* was found in the rotten wood of a fallen tree. In prostrate decaying trunks the fauna is similar to that in the leaf mold. The staphylinids *Gyrophaena* and *Bolitobius cincticollis* and the erotylid *Triplax macra* feed on the mushroom *Pleuronotus ostreatus.*

Only a few species of flying insects occur, especially mosquitoes. When the fresh mold is turned over, a few of the moisture-loving flies *Hydrophorus philombrius* appear and rest on the most exposed surface. No butterflies or moths were in the climax studied by Gleason (1909). Several other species found on Isle Royale extend their ranges eastward and are mentioned in the lists below.

VARIATIONS IN THE CLIMAX EASTWARD

Quebec

Black spruce is a climax dominant in Quebec and Ontario where the rainfall is 1.3 to 2.0 times the evaporation, and boggy conditions occur. Black spruce is climax in Laurentides Park, and shrubs and herbs are almost absent. At Passe Dangereuse, black spruce is the chief dominant, and there is an undergrowth of speckled alder, mountain-ash, bunchberry, corn-lily, willow, and interrupted fern. The number of species and population of birds are small, due to lack of forest-edge conditions (Godfrey 1949).

In climax communities the following are usually present in the ground layer: the slug *Deroceras laeve;* the snails *Retinella electrina, Zoögenetes harpa, Euconulus fulvus,* and *Discus cronkhitei-catskillensis;* the millipede *Underwoodia;* the spiders *Amaurobius bennetti* and *Bathyphantes nigrinus;* the carpenter ant *Camponotus herculeanus whymperi;* the ground beetle *Agonum sinuatum;* and larvae of Tenebrionidae and Pyrochroidae.

Inhabitants of the lower vegetation are commonly the spiders *Estrandia nearctica, Ceraticelus atriceps, Linyphia marginata, Theridion montanum;* the black fly *Simulium venustum;* the helomyzid *Suillia bicolor;* the leaf beetle

Chrysomela falsa; and the aphid *Macrosiphum ambrosiae.*

Yellow birch, balsam fir, and mountain maple occur in an area along the road to Passe Dangereuse near 50° N. Lat. Shield-fern, wild sarsaparilla, common wood-sorrel, and the moss *Hylocomium splendens* were present in 1944. Spiders in the vegetation were *Araneus trifolium* and *Helophora insignis.* On the ground were the carabid beetle *Calathus ingratus* and some rove beetles. Diptera included *Allophyla laevis.* In fallen tree trunks were found pyrochroid beetle larvae, the boreal spider *Amaurobius bennetti,* and the red-backed salamander. The northern limits of this deciduous forest salamander and of the yellow birch are not far apart.

Mount Katahdin

The forest on Mount Katahdin is representative of the climax south of the St. Lawrence River. Black spruce and balsam fir occur at an altitude of 2500 feet (760 m) (Harvey 1903, Blake 1926). Glacial boulders and partly buried rocks project from the decomposing humus on the ground. There is an almost continuous forest floor covering of the mosses *Hypnum crista-castrensis* and *Hylocomium splendens* (Harvey 1902). Blake (1926) notes the shrubs *Kalmia angustifolia, Viburnum cassinoides, Amelanchier bartramiana,* and *Clintonia borealis.* The following mammals are present (Blake 1926): moose, woodland caribou, lynx, marten, black bear, long-tailed weasel, white-footed mouse, bog lemming, red-backed mouse, short-tailed shrew, and masked shrew. All these species occur in central Quebec. The birds also are essentially the same as those already mentioned. There are no terrestrial amphibians, and the common garter snake alone represents the reptiles.

Invertebrates are abundant and varied, including phalangids and four species of ground-dwelling mites. There are 11 species of spiders of which *Serropalpus barbatus, Amaurobius sylvestris,* and *Neoscona arabesca,* together with the staphylinid *Anthobium pothos,* contribute most to the large ground populations. The most characteristic beetle is the borer *Pogonocherus pennicillatus.* The two leafhoppers *Laevicephalus sylvestris* and *Graphocephala coccinea* are found feeding on many plants of the lower forest layers.

The muscid *Fannia canicularis* and three species of crane flies are common. The mosquitoes are represented by *Aedes fitchii* and *Culiseta impatiens;* the biting flies by two species of *Chrysops* and one of *Tabanas;* the black flies by *Prosimulium hirtipes.* The syrphids *Syrphus torvus* and *Xylota curvipes* and the mycetophilid *Zelmira subterminalis* are abundant in favorable local spots. There are large numbers of butterflies and moths.

Cape Breton Island

On North Cape Breton Island, the climax forest above 700 feet (210 m) is balsam fir, 75 to 85 per cent; white spruce, 8 to 12; black spruce, 7 to 10; and mountain-ash, 6 to 7, with a scattering of aspen, yellow birch, and red maple (Nichols 1918). The common shrubs are *Taxus canadensis, Ribes*

glandulosum, Nemopanthus mucronata, Acer spicatum, A. pennsylvanicum,
Lonicera *canadensis,* and *Viburnum cassinoides.*

RED SPRUCE–BALSAM FIR–FRASER FIR FACIATION

In Nova Scotia, New Brunswick, extreme southern Quebec, New England, and the higher areas of New York, the red spruce is a dominant in addition to the black and white spruce and balsam fir. On Mount Marcy in the Adirondack Mountains at an elevation of 4250 to 4890 feet (1288 to 1480 m), the forest is 85 per cent balsam fir, 10 per cent paper birch, and 5 per cent red spruce. At lower altitudes, red spruce occurs mixed with conifers and hardwoods in bogs. The black spruce extends south into West Virginia, and the red spruce into northern Georgia. Important areas of spruce occur between southern Pennsylvania and extreme northern Georgia (Oosting and Billings 1951). Balsam fir of the north is displaced by Fraser fir in the southern Appalachians.

The largest existing forested area lies in the Great Smoky Mountains National Park where all sites above about 4500 feet (1400 m) which are not occupied by beech or heath are occupied by forests of red spruce and Fraser fir. The proportion of fir increases at higher elevations and more mesic sites, that of spruce at lower elevations and drier areas. Yellow birch enters the forest at lower elevations and mountain-ash at higher ones. On north slopes and flats the undergrowth is a complex, five-story structure with a ground layer of moss, chiefly *Hylocomium splendens;* a low herb layer of common wood-sorrel and corn-lily; a high herb layer dominated by wood-fern; a low shrub layer of mountain-cranberry; and a high shrub layer of hobblebush (Whittaker 1952).

A little farther north on Roan Mountain, the characteristic tree species, in order of frequency, are red spruce, Fraser fir, yellow birch, mountain maple, and mountain-ash. The first two are overstory species, the last two are understory ones, and birch may occur in either layer. The important shrubs are the catawba rhododendron, mountain scinberry, hobblebush, alternate-leaf dogwood, and gooseberry. The ground surface is dominated by the fern *Dryopteris dilatata* in the upper stratum, and by common wood sorrel nearest the ground. Important associated herb species in order of frequency are foamflower, two species of aster *Aster divaricatus* and *A. acuminatus,* corn-lily, the goldenrod *Solidago glomerata,* the shining club moss *Lycopodium lucidulum,* and the fern *Polypodium virginianum* (Brown 1941).

The original spruce forest in West Virginia consisted of dense stands of large trees (Hopkins 1899). A typical relict stand of red spruce and yellow birch occurs on Spruce Mountain at near 4800 feet (1450 m) altitude (Core 1929). Several of the plant species listed above also occur here.

Permeant Animal Influents and Arthropod Dominants

The woodland caribou was never more than a straggler in New York and

New England except in northern Maine. The moose and wolverine, however, came into the highlands up to timberline as far south as northeastern Pennsylvania (Cory 1912, Seton 1929). Most of the other large animals influencing the spruce forest, such as the wapiti, deer, black bear, mountain lion, wolf, and bobcat, came from the surrounding deciduous forest.

The red squirrel occurs in dense spruce forest, the fox squirrel enters open stands of spruce, and the West Virginia flying squirrel probably also penetrates into spruce forests, although only hemlock and birch forests are mentioned by Hamilton (1943). The porcupine was formerly present but has now almost disappeared. The snowshoe rabbit was formerly present throughout the Appalachian highlands and has a preference for dense forest but utilizes hemlock and white pine as well as spruce. Certain predators, the ruffed grouse and the wild turkey, are the only large birds regularly reported in the mountains.

The southern pine beetle *Dendroctonus frontalis* may be considered a dominant. It is confined to southeastern United States (Hopkins 1909) and has repeatedly been responsible for the destruction of large areas of spruce forest. At low altitudes it also attacks other conifers.

Vertebrate influents include the red-backed mouse, subspecies of which occur throughout the transcontinental boreal–montane forest. The deer mouse is widely distributed in the coniferous forest. The rock vole and the woodland jumping mouse occur in the coniferous forest of the Great Smoky Mountains (Wetzel 1949).

The most characteristic breeding birds are red-breasted nuthatch, brown creeper, winter wren, golden-crowned kinglet, solitary vireo, black-throated green warbler, black-throated blue warbler, red crossbill, and slate-colored junco (Fauver 1949). All species also breed in the transcontinental coniferous forest and the golden-crowned kinglet in the Rocky Mountains. Salamanders are fairly numerous, including *Desmognathus wrighti* and *Plethodon jordani*. Other vertebrates which enter the mountain coniferous forest have been discussed in Chapter 2.

Invertebrates observed in 1942 at Collins Gap in the Great Smoky Mountains National Park inhabited the ground duff, moss, and fallen tree trunks and included the snails *Stenotrema altispira, Vitrinozonites latissimus,* and *Mesodon andrewsae.* Common millipedes were *Aporiaria deturkiana* and the common centipedes *Cryptops hyalinus, Otocryptops sexspinosus,* and *Sogona minima.* The most abundant ants were *Aphenogaster fulva aquia picea.* The mosquito *Aedes vexans* was present. Dipterous larvae, including crane flies, were common.

A great number of adult crane flies occur in the Smoky Mountain communities. Among them are *Prolimnophila areolata, Limonia indigena,* and *Elephantomyia westwoodi.* Among other flies, the empidids are most abundant, especially *Leptopeza ruficollis.* The drosophilid fly *Scaptomyza terminalis* and the dolichopodid flies *Dolichopus dorycerus, Hercostomus frequens,* and *Gymnopternus chalcochrus* are common.

The small cantharid *Cantharis fraxini,* which is widely distributed in the coniferous forest, is by far the most numerous beetle. The chrysomelid beetle *Syneta ferruginea* which occurs as far north as Newfoundland is common. The homopterons are represented by numbers of the leafhopper *Cicadella flavoscuta,* which occur also in deciduous forests, and *Oncopsis variabilis,* which is widely distributed in coniferous forests (Whittaker 1952).

COMMUNITY DEVELOPMENT AFTER FIRE

In an inflammable community like the coniferous forest, fires are frequent and often cover large areas. A very hot fire produces essentially a bare habitat, hence the sere that follows may be called a primary one; where humus, roots, and seeds are left undamaged, the resulting sere is secondary (Nichols 1918). The predominant plants which appear in the next growing season after the fire depend upon the degree of destruction of the humus.

An area near Passe Dangereuse, Quebec, where a fire had destroyed a pure stand of conifers, had developed a community of low shrubs by 1944. Two species of blueberries present were *Vaccinium myrtilloides* and *V. angustifolium;* other low shrubs were *Ledum groenlandicum, Kalmia angustifolia,* and *Amelanchier bartramiana.* There was also the grass *Deschampsia.* The fruit of the blueberry had ripened in midsummer and had attracted birds and fruit-eating mammals. No tree seedlings were present. Such open spaces provided places for the October rutting activities of caribou (Fig. 5-6).

Spiders, including *Araneus trifolium, Misumena vatia, Tibellus oblongus, Paraphidippus marginatus,* and *Ceratinopsis interpres,* were numerous in the Passe Dangereuse area in late August. All of these species had previously been found in Georgia. Diptera were next in number, including *Hydrotaea militaris, Allophyla laevis, Minettia lupulina,* and one species each of fungivorid and anthomyiid. The Hemiptera were represented by about 40 *Ischnorhynchus resedae* per square meter and by *Nabis ferus.* Ants of the genus *Myrmica* were numerous on the vegetation.

In Maine, some 300 miles (480 km) farther south, cultivated areas of blueberry have largely different insect constituents, since none of the ten insect species and genera taken at Passe Dangereuse are found there. Phipps (1930) lists 19 species of flies and 20 species of bees and wasps that were collected during two seasons while pollinating blueberry blossoms. The larvae of the trypetid fly *Rhagoletis pomonella* destroys many blueberries.

Under natural conditions the invasion of the shrub community by alder and birch starts it toward forest, and the blueberries are suppressed in a few years unless another fire occurs. Following a light fire, blueberries come up from roots the following season while the tree seedlings are killed.

The Lake Nipigon area in Ontario was burned over about A.D. 1770. Previous to that time white pine was fairly common. After the fire, birch, aspen, and jack pine invaded and were succeeded first by white spruce and then by balsam fir. The composition of the late subclimax forest 175 years after the fire was balsam fir, 59 per cent; black spruce, 2; white spruce, 10; jack pine,

4; paper birch, 17; and quaking aspen, 8. There were two tree strata, the upper one averaging 101 feet (30 m) high. The lower tree stratum, chiefly invading balsam fir, averaged 65 feet (20 m) high, and the age of the larger trees was between 90 and 100 years. One jack pine in the upper stratum was 140 years old. Birch and aspen were represented mostly by overmature and dying trees. Black spruce occurred in low moist areas. Speckled alder formed a high shrub layer and Labrador-tea a low one. There was a thick moss carpet over most of the surface. The prevalence of shrubs in this forest is in contrast to their lack in the Isle Royale white spruce–balsam fir climax (Kendeigh 1947).

XEROSERAL DEVELOPMENT ON ISLE ROYALE AND IN QUEBEC

Community Development on Sand

On the shifting sand dunes in central Alberta, two grasses *Bromus ciliatus* and *Agropyron dasystachum* are grown in an open arrangement with *Crispermum marginale*. *Astragalus gracilis*, *Artemisia caudata*, and *Solidago elongata* are present. When the shifting of the sand becomes arrested, the most important early invader is *Cornus stolonifera* which soon forms a thick growth. Other species that quickly appear are the shrubs *Amelanchier alnifolia*, *Elaeagnus commutata*, *Juniperus horizontalis*, *Symphoricarpos pauciflorus*, *Rosa acicularis*, *Salix interior*, *Artemisia caudata*, and *Solidago nemoralis;* the grasses *Calamagrostis*, *Bromus ciliatus*, and *Agropyron dasystachum;* and the blanket flower *Gaillardia aristata*. Under the shrubs occur *Anemone patens*, *Comandra livida*, *Smilacina stellata*, and a dense growth of *Arctostaphylos uva-ursi*. The shrub stage is usually followed by jack pine (Dowding 1929).

On a sand area near the St. Lawrence River at Tadoussac, Quebec, the principal pioneer plant in 1944 was the grass *Festuca rubra* which was very scattered. Of animals, there were a few tiger beetle larvae but no adults. A second stage in the sand sere was dominated by the grasses *Scirpus* and *Elymus arenarius*. Spiders, gnats, parasitic hymenoptera, and the braconid *Meteorus vulgaris* were abundant. Present also were Mexican grasshoppers; common clear-winged grasshoppers; the leafhoppers *Ophiola cornicula* and *Polyamia inimica;* the hemipterons *Nabis ferus*, *Trigonotylus ruficornus*, and a green mirid; and the spider *Philodromus aureolus*.

In a third stage, locally called the pasture stage, the ground cover consisted of *Festuca rubra*, *Potentilla tridentata*, *Trifolium repens*, and *Empetrum nigrum*. Small and immature leafhoppers were numerous, mainly *Polyamia inimica* and *Cloanthus acutus*. There were numerous ichneumonids, including two species of *Syrphoctonus* and the fly *Egle parva*. The two grasshoppers of the second stage were still present. The grasshoppers have a northern range and are widely distributed in grassy areas (Parker 1930). It is of interest that evaporation in this area is lower and rainfall at least twice that of most areas in which the grasshoppers are commonly found. There were no signs of resident mammals.

Sand and gravel ridges are frequent throughout the forest, but especially so near Lake Winnipeg. Jack pine in pure stands or mixed with white spruce commonly covers these ridges. Trees 60 to 80 feet in height (18 to 24 m) are reported. The undergrowth in the older stands contains many of the climax herbs and small shrubs. Permeant and climax animals utilize these areas quite generally. From scat counts made near Mafeking, Manitoba, and elsewhere in 1944, this forest appeared to be the favorite location for snowshoe rabbits and porcupines. The jack pine sawfly and spruce budworms are serious defoliators. They attack older trees especially (Graham 1925).

In Quebec there are areas of jack pine mixed with black spruce and containing *Salix humilis, Kalmia angustifolia, Vaccinium canadense, Potentilla tridentata,* and *Cladonia rangiferina* as subordinate species. Invertebrates here include the common ones noted above, and also the hemipteron *Nabis ferus,* a willow sawfly *Cimbex americana,* and the two spiders *Lycosa frondicola* and *Tibellus oblongus.*

Community Development on Rock

Animals precede plants on the rocky beaches of Lake Superior. Some of the spiders *Pardosa groenlandica* wander upon the rock from gravel beaches. The ant *Formica rufa obscuriventris* is a common visitor seeking dead caddisflies for food. The fly *Hydrophorus philombrius* is sometimes very common on rocks that are constantly washed by waves but nearly absent from rocks that are dry (Gleason 1909).

Crustose lichens and the moss *Grimmia unicolor* are the first plants to appear on rocks along the lake shore. A brilliant orange band of the lichen *Placodium elegans* is a prominent feature of vertical cliffs two meters above the water. Foliose lichens invade rock surfaces after the crustose lichens, and these are followed by the mosses *Hedwigia albicans* and *Orthotrichum anomalum.* Finally fruiticose forms appear, prominent among which are *Cladonia rangiferina, C. sylvatica* and *C. alpestris.* These lichens sometimes form dense aggregations 50 feet (15 m) across. Cooper recorded changes in the vegetation on rock areas taking place within 17 years (Gleason 1909, Cooper 1928).

Numerous plants next make their appearance in the narrow crevices of the rocky surface that contain a little soil. These plants include harebell, three-toothed cinquefoil, swamp saxifrage, and wormwood. The insectivorous butterwort occurs in rocky pools. The most common crevice shrubs are *Juniperus procumbens, J. communis saxatilis,* bearberry, soapberry, and ninebark.

Scattered among the low clumps of *Cladonia* are *J. communis saxatilis, J. procumbens, Arctostaphylos uva-ursi,* and *Vaccinium angustifolium.* Open areas of considerable size on the ridges are called *Cladonia* heaths. These open areas are invaded by junipers and bush-honeysuckle.

Ants make their nests in the soil under dense *Cladonia.* A wasp *Eutypus americanus* was found backing over the ground dragging the spider *Alopecosa kochii.* The grasshoppers *Melanoplus huroni* and *Circotettix verruculatus* are abundant and rarely leave the sunny open ridge. The sprinkled locust, which deposits eggs in decaying wood, was collected from pin cherry, bush-honey-suckle, and canker-root. The grasshopper *Melanoplus bruneri* was taken on cudweed and grass. This species is also abundant on barren rock ridges. Immature specimens of the sprinkled locust hop over the lichens. In thickets of *Juniperus communis saxatilis,* grasshoppers are abundant, including nymphs of *Melanoplus fasciatus* and the fiddling grasshopper. A small carabid beetle *Carabus serratus* forages over and through the *Cladonia.* Bush-honeysuckle is visited by the bumblebee *Bombus terricola,* and the bee *Coelioxys moestra* visits associated plants.

The white admiral butterfly, which feeds on birch, aspen, and balsam poplar, is numerous and characteristic of these openings. Black flies are abundant and are preyed upon by dragonflies at considerable distance from water. Numerous horntails *Urocerus flavicornis* and *U. californicus* were observed flying 2 to 3 feet (0.7 m) above the ground, presumably searching for balsam trees in which to deposit their eggs. Two species of leaf cutter bees, *Xanthosarus melanophaea* and *X. latimanus,* were taken on the flowers of the harebell. Insects were more numerous on the flowers of *Physocarpus,* including the butterfly *Speyeria atlantis,* the fly *Tubifera dimidiata,* the blow fly *Protophormia terraenovae,* and the anthomyiid *Phaonia serva.* Reptiles, except for garter snakes (Dansereau, personal communication), amphibians, and birds appear lacking in the *Cladonia* heaths (Adams 1909).

HYDROSERAL STAGES IN THE SPRUCE–BALSAM AREA

Bogs

Shallow water (Fig. 5-12) occupies immense areas. Aquatic vegetation declines in luxuriance toward the northern edge of the forest. The early seral stage of *Nuphar* commonly also contains *Nymphaea odorata, Potamogeton natans,* and *Sparganium americanum.* In Saskatchewan these species range about halfway across the forest region, but none of them is recorded for the muskeg black spruce forest (Fig. 5-5) (Frasier and Russell 1954).

The mammals that frequent these old bog areas are the moose, muskrat, mink, otter, and beaver. The floating mat, especially the shrub-covered part, is visited by the snowshoe rabbit and its predators.

Few or no birds breed on the floating mat. With the invasion of alder, conditions are favorable for American redstart, white-throated sparrow, and yellow-bellied flycatcher. When trees appear, golden-crowned kinglet, cedar waxwing, black-capped chickadee, gray jay, and white-winged crossbill enter. Later, as the bog conifer forest becomes continuous and compact, waxwing, redstart, and white-throated sparrow disappear (Adams 1909).

An old small bog at Passe Dangereuse in central Quebec was surrounded by forest and had a large number of invertebrates present in August, 1944. The plants were black spruce, tamarack, water avens, *Carex tenuiflora, Galium palustre,* Labrador-tea, the currant *Ribes lacustre,* and *Equisetum sylvaticum.*

Fig. 5-12. A. A stream and margin in the region of the headwater of the Severn River, west-central Ontario. The trees and shrubs are black spruce, speckled alder, sweet gale, and willow. B. Aerial photograph of muskeg country about 50 miles (80 km) inland from James Bay between the Albany and Kapiskan rivers (both photos by Harold C. Hanson). C. Map of Knife Lake area on the boundary between Minnesota and Ontario; black areas are water bodies.

The spiders were *Linyphia marginata, Hypselistes florens,* and *Estrandia nearctica.* Flies were not numerous, but tipulids and the fungivorid *Dynatosoma* were present. A few aphids included *Macrosiphum pisi.* Leafhoppers were *Laevicephalus canadensis* and *Elymana acuma,* hemiperons were *Nabis limbatus* and *Homaemus aeneifrons,* and the hymenopterons were all parasitic species.

Floodplains

The earliest stages on a small island in the floodplain of the Saskatchewan River near The Pas, Manitoba, consist of scattered willows among herbs. Spiders present are *Tetragnatha extensa, Neoscona arabesca, Crustulina alterna.* Flies are numerous, consisting of *Coenosia nigrescens, Ectecephala similis, Microchrysa polita,* and *Oscinella coxendix.* The sawfly *Pachynematus aurantiacus,* two or three species of short-horned grasshoppers, and numerous leafhoppers were found. One meadow vole was taken.

In the central and northern parts of the forest, the willows are followed by balsam poplar. Many mature trees on higher floodplain levels are 2 feet (0.6 m) in diameter and 100 feet (30 m) high. Spruce is infrequently a later seral stage, and tamarack occurs commonly on the floodplain near the northern edge of the forest.

On floodplain deposits of increasing height in the Yukon Delta, occurred the willow stage consisting in progression of seedlings, thickets, and forest, with a rust infesting the willows throughout, and old forest with the trees containing heart-rot fungi; a balsam poplar stage with four disease-producing fungi; and a white spruce stage with two damaging fungi (Baxter and Wadsworth 1939).

SOUTHERN ECOTONE BETWEEN SPRUCE–BALSAM AND THE DECIDUOUS FOREST

Red Pine–White Pine Faciation

The status of this faciation is controversial since it was largely destroyed before studies could be made of it. It has been called lake forest because of its occurrence in the Great Lakes region (Fig. 5-1). In northern Minnesota, Wisconsin, and Michigan it has been mapped (Shantz and Zon 1924) as patches of pine and hemlock in an area mainly of maple–beech–hemlock forest in which there were also areas of spruce–balsam fir. In 1660 before deforestation had occurred, it was probably more extensive and extended eastward to the Atlantic Ocean. The dominant trees are white pine, hemlock, and red pine. The red pine, however, does not persist into the climax, while hemlock and white pine do (Nichols 1935, Potzger 1946). The pines go farther north than hemlock. There appears to have been a white pine forest near Sturgeon Lake west of Lake Nipigon, Ontario, before the fire of about 1770 which is now replaced by spruce–balsam (Kendeigh 1947). Similar replacements are reported in eastern Ontario.

This forest community may best be called the *red pine–moose faciation,* rather than an association, since it has the same seral dominants as the spruce–balsam forest, it contains the same permeant influents as the transcontinental boreal forest, and there are few species of animals characteristic of the region. The bark beetle *Pityokteines sparsus* attacks both balsam fir and pines; another species, *Ips pini,* attacks white spruce and white pine. The birch leaf skeletonizer is an influent in both the pine–hemlock and spruce–balsam forest. White pine and hemlock may occur mixed with the deciduous trees, sugar maple and beech. A forest owned by the University of Maine contains a heavy growth of large white pine, red pine, and hemlock, although individual trees have been removed (Blake 1926). There is a thick carpet of coniferous needles, twigs, and other organic debris. The shrub stratum is poorly developed with beaked hazelnut, white ash saplings, fly-honeysuckle, and currant. The most prominent herbs are canker-root, false lily-of-the-valley, whorled loosestrife, corn-lily, and wild sarsaparilla. During a short study by Blake (1926), the ground stratum contained larvae of Diptera, Lepidoptera, and Hymenoptera with the springtail *Tomocerus flavescens separatus* as the only resident species in the litter. Herb animals included an abundant mirid, *Dicyphus famelicus.* In the alder shrubs and in the herbs, the spittle insect *Clastoptera obtusa* was abundant. The leafhopper *Graphocephala coccinea,* the aphid *Macrosiphum coryli,* and the mirid *Diaphnidia pellucida* were also present.

Near Albany, New York, there is little evidence that birds recognize this forest as different from pine–hemlock hardwoods or maple–beech–hemlock (Kendeigh 1946) (Table 2-6).

In the Harvard University pine forest at Petersham, Massachusetts, species occur which are not common in spruce–balsam forest, including masked shrew, short-tailed shrew, raccoon, woodchuck, eastern chipmunk, gray squirrel, flying squirrels (*Glaucomys volans* as well as *G. sabrinus*), white-footed mouse, snowshoe rabbit, the cottontail *Sylvilagus transitionalis,* and white-tailed deer (Hatt 1930).

Bogs in the Lake Superior Region

A considerable number of bogs occur in the pine–hemlock faciation in the Lake Superior region.

The major influents and dominants in them are the same as in the spruce–balsam association. Olson (1938) has traced the routes taken by wolf packs in their winter hunting (Fig. 5-9). The most important arthropod influents attack the white pine. The white pine aphid sucks the sap from the base of the needles while the white pine weevil larva bores into the center of the terminal twigs. The red pine cones are attacked by the pine cone beetle. The hemlock spanworm attacks hemlock, pines, and balsam. It causes the death of some trees and retards growth in others.

Most of the bogs are of considerable depth and support waterlilies (Fig. 5-14B). A cedar bog in Anoka County, Minnesota, covers about a square mile, but only a very small area of water still exists. The rooted aquatic plants have nearly disappeared and at low level the water is a soupy flat of floating plants (Lindeman 1941). In a bog near Rhinelander, Wisconsin, the first seral stage supports more aquatic vegetation (Jackson 1914). In the bogs described for the area, there is usually a zone of sedge, 20 meters in width. Water-willow is a mat-builder at the water's edge; cat-tails and various grasses occur scatteringly. Near Rhinelander, the sora, marsh hawk, short-eared owl, and long-billed marsh wren frequent this associes and are not found elsewhere (Fig. 5-13). The muskrat, American toad, leopard frog, garter snake, redwinged blackbird, and meadow vole are numerous, and the mink frog occurs at the edge near the water. The swamp sparrow, Minnesota mink, masked shrew, and short-tailed shrew are generally present.

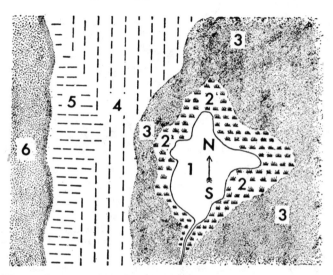

Fig. 5-13. The southwest portion of Ridgeway Bog, northern Wisconsin: 1, water; 2, sedge; 3, leatherleaf; 4, tamarack–spruce; 5, cedar–balsam–hemlock; 6, sandy ridge (after Jackson 1914).

Leatherleaf (Fig. 5-13), Labrador-tea, lambkill, and speckled alder come onto the sedge mat later. Sphagnum grows everywhere in dense spongy masses often a foot or more deep and forms a damp layer about the roots of the shrubs. Meadow-sweet, pitcher-plant, and marsh five-finger become abundant; small cranberry creeps extensively over the surface of the sphagnum (Figs. 5-13, 5-14). The swamp sparrow is an influent species here. The mink dwells in chambers in the sphagnum. The leopard frog, American toad, garter snake, redwinged blackbird, song sparrow, meadow vole, muskrat, masked shrew, and short-tailed shrew occur regularly.

With an increase in the density of shrubs noted above, tamarack and black spruce invade; paper birch does not occur here as it does in Quebec, northern New York, and northern New England (Fig. 5-14). The deer mouse, red-

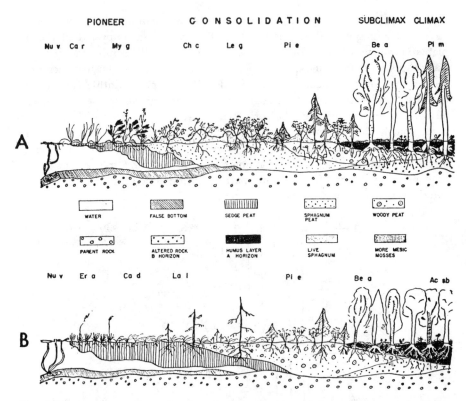

Fig. 5-14. A. Cross-section of a bog leading to spruce–balsam forest — early stage: Nu v, *Nuphar variegatum;* Ca r, *Carex rostrata;* My g, *Myrica gale;* Ch c, *Chamaedaphne calyculata;* Le g, *Ledum groenlandicum;* Pi e, *Picea mariana* and ericaceous shrubs *Chamaedaphne, Ledum,* and *Kalmia;* Be a, *Betula papyrifera* and *Abies balsamea;* Pi m, *Picea mariana.* B. Cross-section of a bog leading to deciduous forest: Nu v, *Nuphar variegatum;* Er a, *Eriophorum angustifolium;* Ca d, *Carex disperma;* La l, *Larix laricina;* Pi e, *Picea mariana;* Be a, *Betula papyrifera* and *Abies balsamea;* Ac ab, *Acer saccharum* (after Dansereau and Segadas-Vianna 1952).

backed mouse, snowshoe rabbit, and the shrews *Sorex a. arcticus* and *Sorex palustris* are visitors and occasional residents. Birds increase in numbers. The spruce grouse, olive-sided flycatcher, winter wren, and golden-crowned king-let are restricted to this part of the bog.

As the forest matures, the masked shrew and the Minnesota red squirrel become common, and the hairy woodpecker, yellow-bellied flycatcher, cedar waxwing, hermit thrush, red-breasted nuthatch, white-throated sparrow, song sparrow, and slate-colored junco appear.

Buell and Buell (1941) suggest that the Anoka bog may go to a deciduous forest climax, while Jackson (1914) indicates that the Rhinelander bog is headed toward a coniferous forest climax. *Pinus banksiana, P. resinosa,* and *P. strobus* occur together near the Rhinelander bog on a sandy ridge. The subordinate plants are *Epigaea repens, Gaultheria procumbens,* and *Mitchella repens.* The least flycatcher, common crow, chipping sparrow, red-eyed vireo, ovenbird, wood thrush, and the eastern striped skunk are present (Jackson 1914).

NORTHERN ECOTONE BETWEEN SPRUCE–BALSAM AND TUNDRA

West of Hudson Bay

The northern portion of the transcontinental coniferous forest east of Yukon Territory is made up of white spruce, black spruce, and tamarack, which become dwarfed in size northward. How much of the dwarfed forest is climax is unknown. There is much wet land, and the many ponds (Fig. 5-12) contain little aquatic vegetation. Sand and gravel ridges, especially old Hudson Bay beaches, are occupied by jack pine. Shrubs in this forest are mainly species that invade from adjacent communities. Among those which seem to reach their greatest perfection in the Athabaska–Mackenzie area are *Empetrum nigrum, Ledum palustre, Vaccinium uliginosum* var. *alpinum, V. vitis-idaea, V. oxycoccos,* and *Betula michauxii* (Preble 1908).

The area is commonly mapped as the Hudsonian Life Zone. East of James and Hudson bays the boundary drawn for the Hudsonian Life Zone (Fig. 1-9) agrees fairly well with the northern limit of white-cedar and the southern boundary of open boreal woodland (Hare 1950). West of James Bay, Halliday (1937) maps a "Northern Transition Section" of the boreal forest which agrees with the Hudsonian Life Zone except that it extends farther north almost to the Arctic Ocean. The width of the stunted tree area averages about 300 miles (480 km) in Quebec (Hare 1950) but becomes narrower westward and northward.

Among birds, the great grey owl, hawk-owl, pine grosbeak, and tree sparrow breed principally within this area. The lower Mackenzie region has no strictly characteristic mammals, although the range of a subspecies of red-backed mouse *Clethrionomys rutilis dawsoni* is partially confined within its limits. Most of the boreal forest mammals have their northern limit within this zone. The Richardson caribou may be limited to the central part of this area. The woodland caribou, moose, red squirrel, flying squirrel, red-backed mouse, porcupine, lynx, and snowshoe rabbit are present in relatively small populations (Preble 1908).

A study was made in 1931 of a stand of full-grown jack pine with a few young white spruce and an undergrowth of Labrador-tea, hazelnut, and *Rosa* on a ridge near the Nelson River at Gillam, Manitoba. There was evidence of an abundance of snowshoe rabbits. In the conifers there were 13 species of flies, four of which are common on the tundra. Included were the mosquito *Aedes,* the phorid fly *Bicellaria uvens,* and the flesh fly *Sarcophaga peniculata.* Hemiptera were few or absent. A few beetles were present, including *Corticaria ferruginea, Podabrus,* and *Cantharis.* The melydrid beetle *Dasytes hudsonicus,* which also occurs in the mountains of Colorado, was found. Spiders were numerous; *Xysticus triangulosus* has been taken farther north but not to the south; *Dictyna volucripes* occurs also in Georgia; *Tetragnatha extensa* is found in the United States and Europe.

A wooded area south of Churchill, Manitoba, some 180 miles (288 km) farther north, is characterized by much smaller spruce than at Gillam. The banks of a small stream that flows through the area are lined with tamarack. Subdominant trees and shrubs are dwarf birch, arctic willow, Labrador-tea, black crowberry, and sour-top-bilberry. Sphagnum moss and *Cladonia* lichen are abundant.

There was about one pair of red squirrels per 24 hectares. Mammals reported by trappers in this and similar areas are flying squirrels, porcupines, and red foxes; most of them are uncommon or rare. There is one or more records of garter snakes. The leopard frog and gray tree frog were taken near the stream.

The nesting birds on 24 hectares adjacent to and immediately north of the above area include the permanent resident gray jay, willow ptarmigan, goshawk, and great horned owl. The tree sparrow, Harris' sparrow, white-crowned sparrow, and pigeon hawk are among the first arrivals in May and early June. The sparrows appear in small flocks and at once scatter over their breeding territory. The hawks appear in pairs. The arrival of fox sparrow, gray-cheeked thrush, robin, blackpoll warbler, and Lincoln's sparrow, in groups of two or three, occurs during the second week of June. Nesting activities are well underway on June 10. Flocking, especially of sparrows, begins with the completion of nesting the latter half of August and continues until the birds leave on southward migration the latter part of September. Food is still plentiful at this time (Twomey 1937).

East of Hudson Bay

The forest–tundra or spruce savanna in Quebec occurs between 56° and 59° N. Lat. (Hare 1950) and occasionally as far south as Passe Dangereuse. Trees are several meters apart with the intervening ground covered by lichens. The distribution of trees in the ravines resembles some parkland types in the tropics (Marr 1948). Near Passe Dangereuse, black spruce occurs with lamb-kill, mountain-holly, low sweet blueberry, Labrador-tea, bracken, mountain-ash, and a mass of lichens including *Cladonia alpestris*. The fulgorid *Cixius misellus* and the leafhopper *Idiodonus subcupraeus* are present in considerable numbers. The hemipteron *Lygus* and the widely distributed *Nabis ferus* are associated with flies, drosophilids, *Leucophenga*, and the sapromyzid *Minettia lupulina* whose larvae thrive on decaying vegetation. The ant *Formica sanguinea subnuda* is abundant. The spiders are *Thanatus lycosoides* and *Misumena vatia*. A single snail *Retinella indenata* was taken.

West of the Rocky Mountain Divide and North of 58° N. Lat.

White spruce mixed with lodgepole pine or paper birch covers the west portion of the transcontinental forest from British Columbia to Alaska. This

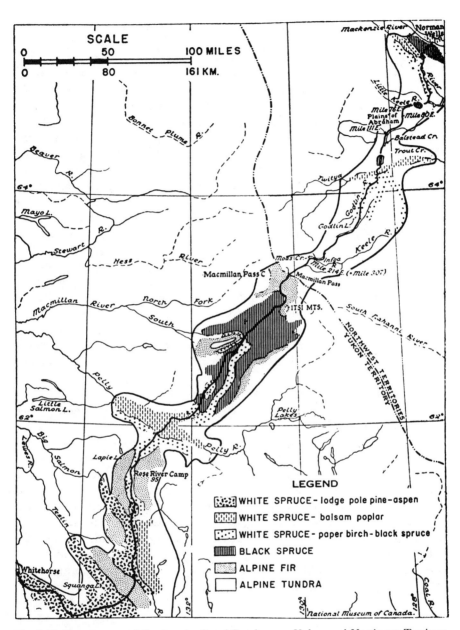

Fig. 5-15. Transect map following the Canol Road across Yukon and Northwest Territory showing the various forest types and the mountain tundra; the line showing the limits of the observations (after Porsild 1951).

may be called the *spruce–giant moose–lodgepole pine–paper birch faciation.* The white spruce–lodgepole pine forest is largely limited to the mountains of British Columbia and Yukon between 61° 30′ N. Lat. and the Stikine River and to branches of the Liard River (Fig. 5-15). Although lodgepole pine is subclimax over most of its range (Halliday 1937), doubtless some of it is climax near the northern limit of its distribution. The climax in most of central Alaska is white spruce–paper birch with the forest that fringes the streams extending up over the sunny slopes of the hills (Dice 1920, Preston 1928, Hanson 1953). There are occasional stands of paper birch alone; stands of spruce alone are less common. Trees are 6 to 12 inches (15 to 30 cm) in diameter, and some are 65 feet (20 m) in height. Other trees found in this forest are willows, balsam poplar, red birch (perhaps a color phase of *Betula papyrifera humilis*), and black spruce. A considerable amount of alder underbrush occurs. Where the shade is not too heavy, there are grasses and a few shrubs of squashberry, common juniper, and bearberry. The forests reach their best development all along the Nisultlin River, along the south slopes bordering Quiet Lake, and in the lower, well-drained slopes bordering the Roos River Valley (Rand 1945).

Dice (1920) lists 44 vertebrates in this forest, with some subspecies restricted to the region. The influent and dominant mammals are moose, wolverine, porcupine, Osborn's and Stone's caribou, black bear, northern gray wolf, and lynx. The moose ranges from the climax down through the seral stages. Stone's caribou is confined to black spruce. The grizzly bear enters the forest only in autumn. Most of its activities are above treeline (Osgood 1909). The lodgepole pine beetle *Dendroctonus murrayanae* is important as an arthropod dominant.

Influents of minor character are flying squirrel, red fox, marten, least weasel, mink, red squirrel, spruce grouse, ruffed grouse, bald eagle, varied thrush, pileated woodpecker, gray jay, and great horned owl. The red squirrel and least weasel are binding influents occurring in many subdivisions of the transcontinental and montane forest (Fig. 5-5). Another group of species occurring in the forest is western in distribution: dusky shrew, coyote, least chipmunk, bushy-tailed wood rat, long-tailed vole, and western jumping mouse.

Community developmental stages are suggested by descriptions of floodplain vegetation. *Equisetum* in practically pure growth is often extensive on wet mud bars along the rivers, perhaps preceding willows and alders, which are followed in turn by poplar thickets. Frequent fringes of poplars occur between the willow–alder thickets and white spruce–paper birch forest. Sedge and grass are present locally on river banks, and extensive areas covered with tussocks of tough grass occur in parts of valleys where drainage is poor (Dice 1920).

A black spruce open forest covers the low hills of the greater part of interior Alaska. Some of it is doubtlessly seral, but an area in Yukon (Fig. 5-15) is probably climax. The trees are mostly stunted, 2 to 9 centimeters in diameter. One tree, 6.5 centimeters in diameter, had 113 rings. The tamarack occurs sparingly with the black spruce in damp habitats. The ground is covered with wet sphagnum up to a depth of a few meters. The principal shrub species are dwarf birch, blueberry, dwarf willow, dwarf alder, and Labrador-tea. Dice lists 28 species of vertebrates here; 17 of the 30 species of birds in the spruce–birch are absent, including the ruffed grouse, all the woodpeckers, the pine grosbeak, and others. The lynx and porcupine are absent, but Stone's caribou is present (Dice 1920, Porsild 1951). Timberline at Tanana is about 2000 feet (600 m). Above this are numerous patches of blueberry and dwarf birch, with occasional growths of scrub willow or alder. These vegetation types and others described by Hanson (1953) are better discussed in connection with tundra.

Chapter 6

Montane Coniferous Forest and Alpine Communities

South of the upper Stikine River (58° N. Lat.) in British Columbia, the coniferous forest increases rapidly in extent for the first 200 miles (320 km) (Figs. 5-1, 6-1) and then is uniform and continuous southward through the Rocky Mountains for 2000 miles (3200 km) to its ecotone with tropical forests in southern Mexico. It may be designated the Engelmann spruce–Douglas-fir–Shiras moose association.

The western boundary of the montane area is the upper eastern slope of the coast mountains of British Columbia, the Cascade Mountains of Washington and Oregon, and the coast range of northern California, and includes the Sierra Nevadas. The eastern boundary of the forest is with the boreal forest in the north and the grassland of the Great Plains in the south. Alpine meadow occurs above the forest at high elevations, and woodlands of various types are below the forest at low elevations.

Separation of the mountainside forest into separate montane forest and subalpine forest associations (Weaver and Clements 1929) is not here recognized, since dominant and influent animals pass freely back and forth. The two plant associations will be referred to as the lower montane and upper montane faciations of the montane forest association (Rasmussen 1941, Hayward 1945). The montane forest has close relations with the transcontinental coniferous forest described in the last chapter. Hayward (1942) lists 82 plants, 35 mammals, and 32 invertebrates of the same or closely related species as occurring in both forests.

NORTHERN ROCKY MOUNTAINS

The southern boundary of this subregion is near the Idaho–Utah border in the main chain of the Rocky Mountains, the south edge of the Blue Moun-

152

tains of eastern Oregon, and Bachelor Butt, southwest of Bend, Oregon, on the east slope of the Cascades.

Rainfall is highest in northern Idaho and southeastern British Columbia, averaging 30 to 40 inches (75 to 100 cm) annually. In northern Montana and in the lower mountains of British Columbia, it ranges from 20 to 30 inches (50 to 75 cm). South-central British Columbia has far less rainfall than either the Pacific slope or the Rocky Mountains to the east; the January rainfall of the Pacific slope is more than 8 inches (20 cm), of the interior only 2 inches (5 cm), and of the Rocky Mountain area about 4 inches (10 cm). The Pacific slope has been called the wet belt and the interior of British Columbia the dry belt. The region is divided into three areas for consideration: the northern transition, the dry belt, and various national parks.

Northern Transition to Boreal Forest

It is convenient to begin the discussion with the alpine meadow because important animals of the forest move into and out of the alpine meadow at different seasons of the year.

The *alpine grass–pika–sedge alpine meadow* in northern British Columbia is of vast dimensions with forest confined to the valleys. Very little is known about the alpine vegetation of this area. Deer, black bear, mountain goats, rock ptarmigans, hoary marmots, and Townsend's solitaire birds have been reported by the Stanwell-Fletchers (1940), and they show a photograph of a freshly killed *Rangifer arcticus montanus* in the Amineca Mountains.

A one-day study of the upper montane forest was made in early September, 1926, in the Aleza Lake Provincial Forest in northern British Columbia, and additional information was supplied by Percy M. Barr (personal communication). The lower limit of the upper montane forest is at 1600 feet (485 m). The principal plant dominant is Engelmann spruce. Numerous small subalpine firs are present, and there is an occasional western paper birch and Douglas-fir (var. *glauca*). The shrubs are the rose *Rosa woodsii*, black twinberry, bunchberry, viburnum, and Queen's cup.

Fresh moose tracks in abundance and a red squirrel have been seen. The Canada woodchuck, skunk, badger, and wolverine are common. Beaver and mink occur near streams. Litter, soil, and decaying wood support the harvestman *Leiobunum;* the spiders *Xysticus, Lycosa pratensis,* and *Lepthyphantes zebra* (6/m²); mites (14); fly and beetle larvae (9); the engraver bark beetle *Cyrpturgus borealis* (15), found in a dead Douglas-fir; the engraver *Dryocoetes betulae* (6), taken from a dead birch; the staphylinid *Quedius briviceps;* the clerid *Thanasimus;* the silphid *Nicrophorus investigator;* the Oregon fir sawyer *Monochamus oregonensis;* and a single snail *Goniodiscus anthonyi.*

The herbs, shrubs, and low conifers support six spiders per square meter; the flies *Pegomya anorufa* and *Suilla;* mycetophilids and drosophilids; the hemipteron *Lygaeus elisus;* the leafhoppers *Macrosteles divisia, Colladonus montanus,* and *Empoasca;* and the hymenopteron *Bathythrix triangularis.*

On the floodplains in this provincial forest, willow appears to be followed in series by poplar, aspen, and conifers. Much of the area was lumbered and burned years ago and has grown up to aspen.

Dry Belt

The interior of British Columbia is a basin crossed in various directions by ranges high enough to be covered with lower montane forest. Some small areas are in grassland. In the northern two-thirds of the basin, the forest is chiefly Douglas-fir (var. *glauca*); in the southern one-third, the altitude is lower and the forest is chiefly ponderosa pine. Near Lac la Hache, the ridges are covered with rather open Douglas-fir with a little aspen. The higher shrubs here are soapberry, the rose *Rosa sayi,* and the willow *Salix bebbiana.* A low shrub is bearberry. There is a thin and scattered ground covering of the forest grass *Oryzopsis asperifolia,* the strawberry *Fragaria glauca,* and mats of the everlasting *Antennaria parvifolia* and other plants.

The chief major permeant influents of the forest are moose and mule deer.

Fig. 6-1. Upper montane forest at Lake Louise, Alberta, with Engelmann spruce and subalpine fir (photo by E. Cieslak 1939).

The wolverine, lynx, black bear, mountain lion, and wolf occur. The coyote is reported as numerous. The minor influents are the snowshoe rabbit, red squirrel, flying squirrel, yellow-pine chipmunk, and yellow-bellied marmot. Birds have not been recorded; reptiles are represented by two species of garter snake. Insects commonly present include the mountain swallowtail butterfly and the northern silverspot. A tent caterpillar *Malacosoma disstria erosa* had defoliated all aspens and taken some leaves from rose, willow, and birch, and, along the lake margin, alder and dogwood. Six genera of long-horned borer beetles feed on conifers, notably the Oregon fir sawyer, along with six genera of flat-headed borers, including the flat-headed fir borer.

The ponderosa pine community of the dry belt has apparently not been studied biotically but has a striking resemblance to similar forests on the eastern slope of the Rocky Mountains bordering on grassland.

National Park Areas

A number of important studies have been made in the National Parks of Canada and the northern United States. These parks cover an area of about 60,000 square miles (150,000 km²) of mountainous terrain. There are also forest and game studies outside the parks in the general area.

On Mount Tekarra in the Jasper Park area of Alberta, with an altitude of 8818 feet (267 m), the alpine meadow is a *Cassiope–Phyllodoce* community with dry habitats containing *Vaccinium scoparium*. Sedges and some alpine grasses are abundant and herbs are conspicuous. Birds present are horned lark, water pipit, and the less common willow ptarmigan (Cowan, personal communication).

In Glacier National Park, Montana, *Poa alpina, Carex tolmiei*, and *Luzula parviflora* are the most prominent of the grasses and sedges at 7500 feet (227 m). There are local colonies of *Phyllodoce empetriformis* and *Salix petrophila* (Waterman 1925). In this region Elrod (1902) has described how the sulphur butterfly *Colias ochraeus meadii*, the silver spot *Speyeria eurynome*, the checker spot *Melitaea nycteis*, the alpine butterfly *Boloria helena*, the copper *Agriades glandon rustica* and other species sometimes hover in very large numbers over a flower carpet of *Rydbergia, Silene, Castilleja, Polygonum, Sedum*, and *Potentilla* (Fig. 6-2).

In Mount Revelstoke Park are found the white-tailed ptarmigan, which originally occurred throughout the Rocky Mountains, and pipits. Dense colonies of Columbian ground squirrel are characteristic, and populations of the western jumping mouse are larger here than elsewhere. Pikas and colonies of voles are present. Signs of grizzly bear activity are usually numerous (Cowan and Munro 1944-46).

Upper montane forest occurs in Idaho at 5500 to 6500 feet (1667 to 1970 m) (Larsen 1930). It consists of subalpine fir, Engelmann spruce, and whitebark pine. In Mount Revelstoke Park Engelmann spruce is dominant over a wide area at 3000 to 5000 feet (909 to 1515 m) (Cowan and Munro 1944-46). The forest floor contains much fallen timber and little undergrowth, although

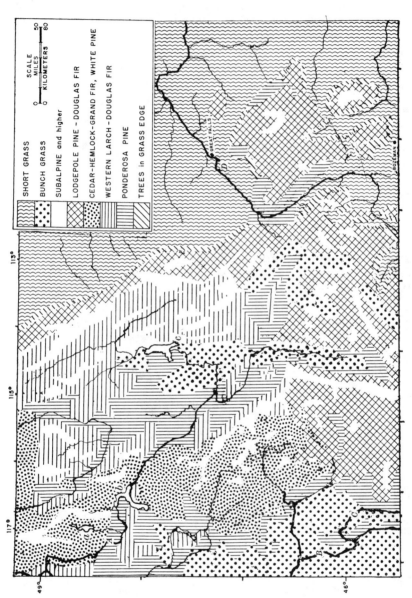

Fig. 6-2. A diagram of the distribution of the northern Rocky Mountain forest types in Idaho, Montana, and southern Canada based primarily on one by Larsen (1930) but modified by the map of Shantz and Zon (1924) and a map of commercial forest types in Montana supplied by the U.S. Forest Service which added considerable ponderosa pine. Halliday's map (1937) of the Canadian part is much more generalized than the parts in the United States. The trees at the forest edge are probably Douglas-fir where ponderosa pine is not present. Attention is called to the cedar–hemlock. a. Priest Lake; b, Cour de Alene Lake; c. Flathead Lake; A, Clark Fork of the Columbia River; B, Snake River; C, Salmon River; D, Missouri River.

the prostrate berry-producing teaberries *Gaultheria myrsinites* and *G. ovatifolia* are abundant and bunchberry is usually present. Spruce grouse, boreal chickadee, golden-crowned kinglet, pine grosbeak, and white-winged crossbill are characteristic birds. The marten and red-backed mouse are at their greatest abundance here, and the region is the chief winter range for caribou.

On the western slopes of the ranges west of the Bitterroot (Selkirk) Mountain divide in Idaho, the lower montane forest consists of two belts, hemlock–cedar–white pine (Larsen 1930) between 2000 and 5500 feet (600 and 1667 m) and ponderosa pine with western larch in patches at 1000 to 4000 feet (300 to 1200 m). The two belts together appear to constitute the *Larix–Pinus* association of Clements (1920). East of the Bitterroot (Selkirk) range, the western larch–Douglas-fir forest replaces the hemlock–cedar–white pine.

In Mount Revelstoke Park near the center of the cedar–hemlock area, western white pine, western hemlock, Douglas-fir (var. *glauca*), and western red-cedar predominate on western slopes at 2000 to 3000 (600 to 900 m) (Cowan and Munro 1944-46). Salmon-berry and devil's club are important in the undergrowth which locally forms dense thickets. Birds with centers of abundance in this habitat are ruffed grouse, chestnut-backed chickadee, red-breasted nuthatch, and Swainson's thrush. Mammals taken here are flying squirrel, snowshoe rabbit, and long-tailed weasel; and red squirrels are more abundant here than elsewhere. There appear to have been no studies of the ponderosa pine community in the park.

The lodgepole pine is the most extensive forest type in Wyoming, where it occurs from 7000 to 8500 feet (212 to 260 m). There is a considerable amount of spruce–fir and alpine meadow from 8500 to 10,000 feet (260 to 300 m), depending on the slope and latitude. Most of Yellowstone Park is at the level of the upper montane forest, but vast areas are in mountain meadow.

The upper montane forest of Yellowstone Park is dominated by Engelmann spruce and subalpine fir with little shrubby undergrowth and the following herbaceous plants: *Shepherdia canadensis, Aquilegia chrysantha, Astragalus alpinus, Potentilla gracilis* and *Geranium richardsonii*. Spiders present in July, 1945, were *Pityohyphantes cristatus, Clubiona kulczynokii, Dictyna minuta,* and *Pardosa uncata*. The crane fly *Tipula banffiana* was also taken. Ants found in rotting wood and the soil were *Camponotus herculeanus modoc* and *Formica fusca neorufibarbis*.

The lodgepole pine forest is commonly seral. The herbaceous growth includes *Carex, Vaccinium scoparium, Gilia aggregata,* and *Haplopappus*. Common spiders include *Cyclosa conica, Theridion zelotypum, Gnaphosa brumalis,* and *Coriarachne brunneipes;* common ants, *Myrmica lobicornis scabronodis fracticornis, M. mutica, Camponotus herculeanus whymperi,* and *Formica fusca neorufibarbis*.

Major Permeant Dominants and Influents

Trails cross ranges at frequent intervals wherever a pristine population occurs, and these "trails used by elk are game trails shared by moose, mule deer, and Rocky Mountain sheep" (Green 1946). Mountain caribou, white-tailed deer, black bear, grizzly bear, wolverine, coyote, lynx, bobcat, mountain lion, and wolf also utilize these trails when it is convenient. The trails follow streams, traverse mountain passes, and approach favored watering places from all directions. Much used trails are from four to six inches (10 to 15 cm) deep, forming trunk highways from which side paths diverge, but when the snow is deep, trails are seldom followed.

The deer, wapiti, moose, and caribou normally occur in large aggregations. The wapiti is abundant especially in the eastern foothills of Canada and southern foothills of western Wyoming. There were from 25,000 to 40,000 wapiti in Yellowstone National Park at the beginning of the century. In 1925 the northern herd in the park was estimated at 17,242 and the southern herd at 19,443. At the first snowfall in the autumn, the northern herd moves down the rivers and out of the park for the winter, and the southern herd goes into Jackson Hole. The summer food of the wapiti includes grasses, wild strawberry, geranium, thistles, clover, lupine, and wild sunflowers. The winter food from October to March includes wild sunflower, larkspur, low huckleberry, mushrooms, grasses, sedges, aspen, bearberry, shrubby wild rose, gooseberry, currant, willows, rabbitbrush, sagebrush, and other plants. The wapiti also eats foliage and bark of aspen, cottonwood, western birch, redcedar, Douglas-fir, and limber pine, thereby exercising considerable control over the composition of various types of vegetation. The wapiti appears to have increased greatly with the destruction of the wolf about 1916 and in many areas has caused extensive damage to the vegetation. According to Wright, Dixon, and Thompson (1933), the wapiti requires shrubs and mountain meadow plants near the upper forest edge for summer range, the higher forested slopes for bedding down at night, and the midslopes and valleys for wintering.

In the mountains of Idaho, the white-tailed deer and the mule deer are the common species, making up 40 and 60 per cent respectively of the population. The mule deer starts down slope in the autumn when the snow becomes two feet (0.6 m) deep but can travel in three, four, or even five feet (1 to 1.5 m) of dry snow. The white-tailed deer avoids heavy snow and is usually at the lower edge of the mule deer range (Mass 1938). The winter food of deer and wapiti differs sharply. Deer eat shrubs, 52.5 per cent, and grass, 2.3 per cent; wapiti eat shrubs, 15.5 per cent, and grass 56.0 per cent. The white-tailed deer prefer heavy cover on gentle slopes while the mule deer prefer a rough, rocky terrain at higher elevations where food and scattered cover is available.

Moose have a preference for willow, but permeate through all the communities of the park (Fig. 6-3). The mountain caribou was originally in Idaho and in the extreme northern part of the montane forest, but the size of populations is unknown.

Mountain sheep were more abundant than deer in 1870. In Idaho their summer range is from 8000 to 9000 feet (2400 to 2700 m) (Godden and Gutzman 1938). The mountain goat occurs commonly in steep rocky areas.

Their population was estimated at 4000 in the Bitterroot Mountains and adjacent ranges in Idaho.

The wolf is a normal and common inhabitant of montane forests. Four or five young, sometimes more, are generally born in the latter part of March or early in April. Dens are usually situated in caves or hollows among rocks or sometimes in large burrows on steep hillsides (Bailey 1930, Cowan 1947). The annual diet of wolves in Jasper and Banff national parks has consisted of 80 per cent big game, with wapiti alone contributing 47 per cent and mule

Fig. 6-3. Moose in deep snow along small stream in willows near Jenny Lake, Grand Teton Park, Jackson Hole (photo by A. W. Gabbey).

deer 15 per cent. Although mountain sheep are in great abundance, they are seldom hunted by the wolves. The wolves of the Rocky Mountain parks take wapiti in preference to sheep even when both are equally available and both are in excess of the carrying capacity of the ranges. Eighteen per cent of the annual diet consists of lagomorphs and rodents, of which snowshoe rabbit and beaver are most important. There are records of wolves chasing deer in the dry belt. The coyote, which feeds on mice, insects, and small birds, is common throughout the open country of the mountains but according to Bailey (1930) is absent from densely timbered areas. It is less gregarious than the wolf but has similar habits.

The black and grizzly bears are common in montane forest throughout the region. They pick the bones of dead animals and rarely kill game animals. They dig mice and ground squirrels from burrows and eat the young. In summer they eat the tops of the cow parsnip, thistles, other coarse but tender and succulent plants, and berries. Spruces and pines are peeled on one side from the ground up three or four feet (1+ m) and are well marked with the incisor teeth of the bear. The droppings of the bears often are full of skeletons of ants and beetles. They pass the winter in cavities of various sorts. Among other large predators, mountain lions and bobcats are generally present.

Resident Birds and Insect Dominants

In Banff Park, blue, spruce, and ruffed grouse and white-tailed ptarmigan occur. Sharp-tailed grouse sometimes come into the valleys. As bud eaters these birds do damage to seedlings.

The mountain pine beetle has caused the death of large areas of lodgepole pine and damaged Engelmann spruce and whitebark pine. The lodgepole pine needle miner has done serious damage in the Canadian national parks and elsewhere from time to time. The western hemlock suffers from the western hemlock looper *Lambdina fiscellaria lugubrosa*. The spruce budworm is a serious pest much of the time.

CENTRAL ROCKY MOUNTAINS

The communities of the central Rocky Mountains in Colorado, their extensions into New Mexico, and the Wasatch and Uinta mountains in Utah will be described first. The communities on the eastern slope of the Cascades in Oregon, in the Coast Range of California, and in the Sierra Nevadas constitute another natural group.

At high altitudes in the mountains occurs a bleak, windswept region of rocky peaks and cliffs with excessive snowfall and arctic temperatures in winter and frequent squalls of rain, sleet, or snow in the short summer (Fig. 6-4A). On all ranges, snow fills the gulches and partly covers cold slopes even in the warmest months, and on the high massive ranges there are extensive snow fields and even ice fields or glaciers in protected alpine valleys (Cary 1917).

Alpine Meadow (Fig. 6-4B)

In general appearance the alpine meadow is one of low vegetation with scattered succulent herbs and taller grasses forming a higher stratum. It occurs on all peaks and spurs of sufficient altitude but is broken by valleys and gulches that fall below its lower limits. The chief climax dominants on Mount Emmons in the Uinta Mountains are the sedges *Carex pseudo-scirpoidea* and *Carex albonigra* and the grasses *Festuca ovina brachyphilla*, *Poa rupicola*, *Agropyron trachycaulum*, *Calamagrostis purpurascens*. Among the forbs, *Geum turbinatum* may persist in the climax. Elsewhere the grass *Trisetum spicatum* is commonly present along with the rush *Luzula spicata* (Hayward 1945, 1952). Cox (1933) found *Elyna* the principal species in the climax on James Peak in Colorado. The seasonal period of plant growth and animal activity is from June to August, varying in exact time with latitude, altitude, and slope exposure.

In the Rocky Mountain National Park in July, 1936, a herd of mountain sheep was observed to graze bunches of *Trifolium dasyphyllum*, *Polygonum viviparum*, *Poa*, *Carex filifolia*, and other species of *Carex*. Wapiti feed in the alpine and mountain meadows in the daytime and bed down at night in the forest. A herd that was under observation left evidence of having cropped two unidentified species of *Carex*, *Poa wheeleri*, *P. interior*, *P. epilis*, *Trisetum subspictum, Agropyron scribneri,* and *Danthonia intermedia,* which were the commonest plants of the area.

The coyote and western red fox find good mouse and pocket gopher hunting and abundant insects and small birds in these areas, as well as an occasional snowshoe rabbit (Cary 1911). Mountain lions hunt in the area. These in-

Fig. 6-4. A. General character of the high portion of a mountain (Mount Timpanogos, photo by C. L. Hayward). B. Alpine meadow on Pikes Peak (12,500 to 13,000 feet, 3790 to 3940 m). Climax vegetation in the background; in the foreground an area disturbed by pocket gopher digging (photo 1930 by Whitfield).

fluents do not, however, remain in the alpine meadows throughout the year.

Regular residents include the marmot which on Mount Timpanogos in the Wasatch Mountains of Utah feeds mainly on climax and late subclimax vegetation and is active from May to October. The Uinta ground squirrel is abundant and feeds on forbs. Northern pocket gophers attain populations of 18 per hectare; in winter they pack soil into snow burrows which appear as casts when the snow melts (Fig. 6-6) (Hayward 1945, 1952). On Pikes Peak in Colorado at about 12,500 feet (3788 m), there were three active mounds and six abandoned ones found on two hectares (Fig. 6-4). The pocket gopher destroys the roots of *Campanula rotundifolia, Polemonium viscosum, Mertensia alpina,* and *Geum turbinata.* With the destruction of these species, *Polygonum bistorta* takes possession of the area. The long-tailed vole occupies old gopher mounds (Whitfield 1933). The deer mouse on Mount Timpanogos has populations of 32 to 46 per hectare (Hayward 1945, 1952).

The migrating water pipit is the most characteristic bird of the alpine meadows of the Wasatch and Uinta mountains, and on Pikes Peak, but breeds also in mountain meadows at lower elevations. Large flocks move down the mountains into the valleys in August or September. In late summer and autumn, flocks of small birds are occasionally seen feeding in the alpine meadows, and mountain bluebirds are frequent. The white-tailed ptarmigan remains in the meadow over winter.

Ground-dwelling invertebrates on Mount Timpanogos live under the cover of rocks and piles of droppings left by the large animals. One species of spider, eight of beetles, one burrowing hemipteron, three species of ants, and several species of mites comprise 40 per cent of th ground populations in open places. Ants comprise about 14 per cent, beetles about 23 per cent, and the bug *Geocoris bullatus* about 20 per cent, at least in August. *G. uliginosus* replaces *G. bullatus* in the climax on Pikes Peak. The snout beetle *Sitona* is numerous. The ground beetle *Harpalus carbonatus* and the click beetle *Hypolithus* are common.

The herbs support five species of Orthoptera, seven of beetles, eleven of flies, and three species of parasitic Hymenoptera. The leafhopper *Dikraneura carneola* is 10 to 15 times more numerous than any of the other groups. The hemipterons *Nysius californicus* and *N. grandis* are present on both Mount Timpanogos and on Pikes Peak along with the grasshopper *Melanoplus.* Ten species of grasshoppers reside throughout the year in alpine meadows of Colorado, but none of them is limited to this habitat (Alexander 1951). On Mount Timpanogos the grasshoppers are eaten in large numbers by mountain bluebirds. Bees of the genus *Halictus* are generally present. Butterflies are numerous, including the alpine butterfly *Parnassius clodius.* Leafhoppers comprise 84 per cent of the average herb population during the summer, Hemiptera 5 per cent, Diptera 4 per cent, and grasshoppers 4 per cent (Hayward 1952).

Succession occurs on a variety of primary bare areas: fell fields (smooth barren rock areas), boulder fields, talus slopes, snow flushes (soil saturated with water by late melting snow), and stream and lake margins. Regardless of how initiated, there is convergence of the vegetation to the climax (Cox 1933). Wet meadows around ponds are dominated by sedges and grasses and have small populations of all the invertebrates above mentioned except flies. These wet meadows afford food to the pipit, pika, marmot, and visiting ungulates. Dry meadow is dominated by *Trisetum spicatum* and *Arenaria* and has the largest number of invertebrates. These include nearly 20 species of flies, the ant *Formica fusca pruniosa*, and numbers of hemipterons including *Lygaeus kalmii* and *Stenodema vicens* (Hayward 1952).

Rock slides and rock fragments support a scrub composed of *Salix petrophila*, *S. saximontana*, and other species which rarely exceed two to three inches (5 to 7.5 cm) in height (Cox 1933). The Wilson's warbler and white-crowned sparrow breed regularly here and through the elfinwood zone (Cary 1911). Dry areas in the Uinta Mountains hold the pika, both in the alpine region where it is most abundant and in loose rock piles at lower elevations. It stores the plants *Mertensia leonardi*, *Frasera speciosa*, *Ranunculus adoneus*, *Zygadenus elegans* and *Ligusticum filicinum* for winter food in large piles called "haystacks." Material is added so gradually that it becomes completely dry. The golden-mantled ground squirrel also utilizes this habitat. Black rosy finches are confined in summer to early xeroseral stages in the vicinity of permanent snowbanks; they feed to a large extent on insects found frozen there. In winter they move down to the foothills. Rock wrens are common in loose rock piles in alpine areas as they are also at lower elevations.

Elfinwood or Wind Timber Belt

The elfinwood or wind timber belt (Fig. 6-5) in Colorado varies in width from 600 to 1200 feet (180 to 360 m). There are no mammals restricted to this area, but most of the major permeant visitants of the alpine meadow seek shelter here, especially in winter. The pine grosbeak, brown creeper, and golden-crowned kinglet breed here. The distribution of wind timber is determined to a large extent by the amount of snow and its deposition by the wind. The sprawling form of the trees results from the weight of the snow and from the branches that project above the snow being shorn off (Cox 1933).

Well-protected elfinwood is a climax of spruce–fir, Engelmann spruce, and subalpine fir. Bristlecone or limber pine forms an edaphic climax on dry rocky slopes and windswept ridges. A species of *Vaccinium* forms a layer of much importance to both plants and animals. Dominant in drier locations are *Polemonium pulcherrimum*, *Epilobium angustifolium*, and *Arnica cordifolia*. Very exposed communities contain species of the fell fields above the tree limit or stands of *Dryas octopetala* (Cox 1933). Hayward (1952) found the least chipmunk, a deer mouse, two species of meadow voles, a lemming mouse, and a red-backed mouse taking advantage of the shelter of the elfinwood. Mountain sheep, wapiti, deer, bears, coyotes, and wolves probably spend much more time in the wind timber than in the alpine meadow.

Fig. 6-5. Northeast slope of Kingston Peak, Colorado, showing the effect of wind and snow upon the distribution of alpine scrub (photo July, 1928, by C. F. Cox 1933).

Montane Forest

Below the wind timber zone, the upper montane forest is dominated by Engelmann spruce, subalpine fir, bristlecone pine, limber pine, and lodgepole pine. The principal shrubs are *Vaccinium caespitosum, Linnaea borealis, Lonicera involucrata, Ribes lacustre, Shepherdia canadensis,* and *Salix nuttallii,* some of which occur in the wind timber (Clements 1920). The lower montane area is dominated by Douglas-fir, white fir, blue spruce, ponderosa pine, and limber pine.

The upper montane climax is variable in structure. At its contact with the lower montane forest, the trees are often 100 feet (30 m) high, the canopy is closed, and a typical undergrowth is present. In the ecotone between the two forests, the respective dominants meet on nearly equal terms to form an apparently homogenous forest. At higher altitudes the forest mass becomes more and more open or fragmented and near timberline is broken into isolated groves and clumps. In altitude, the community ranges from 8000 to 12,000 feet (2400 to 3600 m). Burned areas are dominated by lodgepole pine or aspen or by the two in varying mixtures.

In the Medicine Bow Mountains of southeastern Wyoming, the climax at 9800 feet (3000 m) is composed of Engelmann spruce with a little subalpine fir. Soapberry is reported in the most exposed parts. The low-growing blueberry *Vaccinium scoparium* is the most widely distributed ground plant and glacier lily the most conspicuous herb. Most of the invertebrates found here (Fichter 1939, Blake 1945) belong to the same genera as occur in the Wasatch Mountains, but only a few species are the same (Hayward 1952).

Ponderosa pine, Douglas-fir, and white fir are the major dominants of the lower montane forest. Lodgepole pine is typically the subclimax dominant of burned areas in both the lower and upper montane forests. Limber pine has a wide altitudinal range and hence is found in both the montane and sub-alpine faciations. Ponderosa pine is the most xeric of the dominants. It often grows in open stands with shrubs of the Rocky Mountain oak bush or may be found scattered in grassland with few or no associated shrubs. Subordinate plants include Rocky Mountain maple, ninebark, dogwood, mountain-ash, bear-berry, and squashberry.

The absence of the caribou, moose, mountain goat, spruce grouse, and mountain pine beetle among the animal influents and dominants is noteworthy. The wapiti and mule deer have received more attention in this forest than farther north. These species are not inhabitants of dense forests. When fire destroys the forest, it is followed by a growth of herbs, shrubs, and tree seedlings, especially aspen. Usually then the deer and wapiti populations increase greatly. By the time that a new coniferous forest becomes established, they have usually declined in numbers.

The Engelmann spruce bark beetle has caused extensive destruction of mature trees in parts of the Rocky Mountains, most notably in Colorado, where it approaches the status of a dominant. Associated with it are several other species of bark beetles which attack weakened trees.

Animals designated as minor influents in one area may become of major importance in another, depending on their abundance. Of the larger rodents in the montane forests of the Wasatch and Uinta mountains, the yellow-haired porcupine is present in populations of about one per hectare. In winter, it feeds preferably on the bark of the white and subalpine fir, Engelmann spruce, and lodgepole pine. Aspen is untouched and Douglas-fir rarely eaten, but the common shrubs of the area are utilized. At lower elevations on Mount Timpanogos, the porcupine plays an important role, along with deer and snowshoe rabbit, in retarding the replacement of aspen by white fir. Hayward found that 78 per cent of 41 seedlings of fir were so damaged in one year's time that they failed to reach maturity. In the lower montane climax, young white fir is nearly always killed before it reaches a diameter of six inches (15 cm).

Snowshoe rabbits are common throughout the montane forest: lower montane climax, 10 per square mile (4/km²); aspen subclimax, 2 (0.8); upper montane climax, 16 (6); upper montane subclimax, 16 (6). Other influent mammals are the long-tailed weasel and the striped skunk which occur in scattered numbers throughout the montane forest and are mainly nocturnal. The Uinta ground squirrel is numerous locally and the red squirrel widely distributed. The flying squirrel, pocket gopher, the chipmunks *Eutamias quadrivittatus* and *E. minimus,* deer mice, red-backed mice, long-tailed voles, vagrant shrews, and western jumping mice inhabit the entire montane forest (Hayward 1952).

Of about 57 species of birds whose breeding distribution is established through extensive field work, about 61 per cent are known to occur in both upper and lower montane forests, 14 per cent seem to be confined to the lower montane, and 10 per cent to the upper montane.

The small leafhopper *Dikraneura carneola* is exceptionally abundant and flies up in great swarms from the herbs at all altitudes; it reaches its greatest abundance at the higher elevations (Hayward 1952).

Abert's squirrel, least chipmunk, and rock mouse are in the lower part of the montane forest, especially in ponderosa pine forests. Hayward (1952) lists eight species of birds largely in lower montane forest, including ruffed grouse, purple martin, red-breasted nuthatch, and western tanager.

Fig. 6-6. A. Summer earth piles of the pocket gopher *Thomomys talpoides uinta*. B. Winter earth cores made under snow by pocket gopher (photo by C. L. Hayward).

Seral vegetation on rocky slopes, containing scattered aspens and clumps of squaw currant and the elderberry *Sambucus microbotrys*, support abundant spiders and ants. The widely distributed wolf spider *Pardosa groenlandica* is commonly present. The land snail *Oreohelix strigosa* lives among debris in crevices among the rocks and in aspen groves. The centipede *Nadabius pullus* burrows into the shallow soil under rocks. Numerous flying insects frequent the rocks, include the clear-winged locust (Blake 1945).

In the Medicine Bow Mountains, aspen and lodgepole pine appear to alternate on burned areas. A study was made of mature lodgepole forest of several acres on a shallow, sandy loam soil, poor in humus. Water content was good through June from retention of melted snow but was low by mid-July. Pine needles covered the ground thickly. Many logs from earlier fires were present in an advanced stage of decay. An even-aged stand of mature lodgepole pines (var. *latifolia*), with an average diameter of 8 inches (20 cm) and a height of 30 feet (10 m), dominated the forest. Both shrub and herb layers were almost absent. A few coniferous saplings were growing at normal shrub level; too little light was present for the lodgepole seedlings to be healthy, but spruce saplings were thrifty. Where the slope was to the northwest, common juniper grew. Soapberry and the widely distributed blueberry *Vaccinium scoparium* were present. Of ants, *Tapinoma sessile* was abundant among the dry needles, the red slavemaker *Formica fusca subaenescens* foraged for insects into the lower pine branches, and the widely distributed carpenter ant *Camponotus herculeanus modoc* was found everywhere among the forest floor debris. The fungus gnat *Sciara* was abundant. The golden-mantled ground squirrel and least chipmunk were present, but the most common mammal was the deer mouse (Blake 1945).

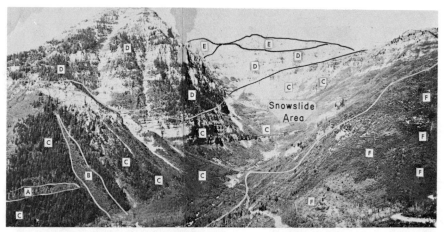

Fig. 6-7. Portion of east face of Mount Timpanogos (courtesy *Great Basin Naturalist*): A, remnants of aspens being replaced by conifers; B, snowslide paths; C, lower montane; D, upper montane subclimax forest; E, mostly barren cliffs; F, chaparral ecotone on south-facing slopes (photo 1938 and labels by C. L. Hayward).

The aspen forest on Mount Timpanogos represents a typical late seral stage (Figs. 6-7, 6-8). It resulted from a fire in 1860. Fallen logs, stumps, and branches are much in evidence, covering 10 to 20 per cent of the ground surface. Herbs and aspen leaves, accumulating over many years, have built up a rich loam several feet deep. Soil-turning activities of animals are here more important than in the conifer forest. A maximum of 40 gopher diggings and 60 ground squirrel burrows per hectare is not uncommon.

The ground layer of the aspen subclimax contains many of the invertebrates of the coniferous forest, but in greater numbers. Eight species of ground beetles, including *Pterostichus protractus,* are present, and beetles make up nearly 50 per cent of the total population.

In early spring, immediately following the snow, there is a vegetational aspect composed of two early blooming species, the spring-beauty *Claytonia perfoliata* and the dog-toothed lily *Erythronium parviflorum,* together with *Stipa pinetorum* and a common sedge *Carex festivella.* The high population of invertebrates includes 8 species of dipterons, 5 of hemipterons, 9 of leafhoppers, 4 of psyllids and aphids, 3 of ants, and 14 of parasitic hymenopterons.

In the shrub layer, leafhoppers are the most abundant invertebrates, forming about 34 per cent of the total population. Diptera, including many species of Anthomyiidae, form about 20 per cent of the total summer population. The leafhoppers *Idiocerus formosus, I. suturalis,* and *I. lachrymalis* are by far the most conspicuous animals in the canopy of the aspen forest, making up about 87 per cent of the total population. Leafhoppers are eaten by a number of birds. The warbling vireo and Audubon's warbler are important in the canopy and the house wren on the ground.

Fig. 6-8. A. Claw marks of bear on aspen, Mount Timpanogos. B. Effect of the feeding of snowshoe rabbit and deer on the white fir, Mount Timpanogos (photos by C. L. Hayward).

California and Oregon

The inner ridges of the Coast Range in the northern third of California, the Sierra Nevadas, and the eastern slopes of the Cascade Mountains in Oregon have certain features in common. The eastern slopes of the Cascades are dry, and the communities here are more similar to those of the forested parts of the Interior Dry Belt of British Columbia, the Blue Mountains, and Rocky Mountains than to the hemlock–cedar forests of the moist western slopes.

Crater Lake in Oregon is near the northern limit of three characteristic trees, California red fir, Jeffrey pine, and sugar pine, which are included in the upper montane forest. Lodgepole pine, mountain hemlock, California red fir, and species of *Abies* are the principal trees in this forest. There is little undergrowth, chiefly *Ribes* and *Carex*. The common invertebrates in July, 1945, were the ants *Camponotus herculeanus modoc* and *Formica fusca subaenescens,* the beetles *Leptoferonia inanis* and *Pterostichus protractus,* many collembolons and the centipede *Ethopolys integer*. There were tracks of bear and deer and many mounds made by pocket gophers.

The lower montane forest is principally ponderosa pine and Douglas-fir with some Pacific silver fir and sugar pine. South of Crater Lake Park, there is a great profusion of undergrowth in contrast to the very open condition of the ponderosa pine forest of the Rocky Mountains. The principal shrubs are sticky laurel, antelope brush, currant, and waxberry, and two tall grasses are *Trisetum canescens* and *Festuca rubra.*

The following invertebrates are common: the spiders *Misumenops celer* and *Pardosa uncata,* the ant *Lasius niger sitkaensis,* the beetle *Centronopus parallelus,* the camel cricket *Pristoceuthophilus,* ichneumonids, the centipede *Ethopolys integer,* and the termite *Zootermopsis.*

On Mount Shasta in northern California, Merriam (1899) describes avalanche paths and "rock slopes" which were largely pumice sand strewn and mixed with fragments of gray volcanic rock among which individual plants were scattered. Deer mice were feeding on seeds of *Polygonum* and other timberline plants. Mountain pocket gophers had mounds at various elevations up to 9000 feet (2700 m). The pika, a ground squirrel, and a chipmunk were present but no marmots were found. Mountain sheep and mule deer were once present.

In the upper and lower montane forest of the Mount Lassen area in the upper Sierras, the principal trees are ponderosa pine, lodgepole pine, white fir, red fir, and western hemlock. In the lower montane, the important ponderosa pine grows in a characteristically open stand. The lower limit of the forest agrees fairly closely with the average lower limit of the snow cover in winter. Upper montane forests, however, are dense at altitudes between 6000 and 8000 feet (1800 and 2400 m). Shrubby growths of chaparral plants occur in the forests up to about 7000 feet (2100 m). The most conspicuous groups of birds in the montane forest, the woodpeckers, nuthatches, chickadees, and creepers, are closely dependent upon the trees. Most of them remain in the forest throughout the year. The mammals of the mature lower montane forest are limited; only the red-backed mouse was noted (Grinnell *et al.* 1930).

The distribution of trees and various mammals, reptiles, and amphibians in the southern Sierra Nevadas is shown by transects made over the mountains near 38° N. Lat. (Figs. 6-9, 6-10, 6-11).

The mule deer in recent years goes much higher in the Sierras in summer than formerly, due to the extensive growth of food plants that has come into cutover areas. The forest is dense from 2500 feet (760 m) in some areas to about 5500 feet (1670 m). In the rougher topography at higher elevations, the forest is interspersed with bush fields, such as talus chaparral (Fig. 6-10), and the pristine deer food is composed of greenleaf manzanita, huckleberry oak, mountain whitethorn, and many other species.

Mountain lions, coyotes, and black bears permeate through the montane forests (Fig. 6-9) as scattered individuals. The more stationary constituents of the forest are the bushy-tailed wood rat and flying squirrel. Of the birds, the mountain quail, common nighthawk, Lincoln's sparrow, Audubon's warbler, pine siskin, and Oregon junco occur throughout. The bark beetles *Ips emarginatus* and *I. integer* infest ponderosa and lodgepole pines.

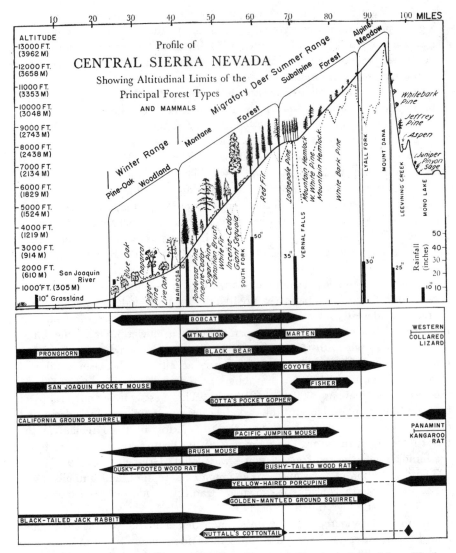

Fig. 6-9. Transect of the Sierra Nevada Mountains in California (modified from Hughes and Dunning 1949), showing the principal climaxes and tree species. Rainfall is indicated by heavy vertical lines with figures in inches at the top. Horizontal distance at sea level is shown by the scale at the top. The distribution of several mammals and birds (after Grinnell 1933 and Grinnell and Storer 1924) is shown by elongated ellipses below. The width of these is varied to suit habitat variations and broken lines connect ellipses which represent the distribution of species on both sides of the mountains.

The lower montane forest consists of ponderosa pine with a light scattering of incense-cedar and more scattered sugar pine and Douglas-fir. The undergrowth consists of species of *Ceanothus* and *Arctostaphylos* (Klyver 1931). The long-eared chipmunk and Trowbridge's shrew occur here. The summer birds are pigmy owl, band-tailed pigeon, purple finch, solitary vireo, Nashville warbler, black-throated gray warbler, Macgillvray's warbler, and winter wren. The California mountain kingsnake occurs (Grinnell and Storer 1924).

The spruce budworm and the Douglas-fir beetle are important in the Sierra forests. The mountain pine beetle has killed large numbers of western white pine, sugar pine, and ponderosa pine. The Jeffrey pine beetle destroys Jeffrey

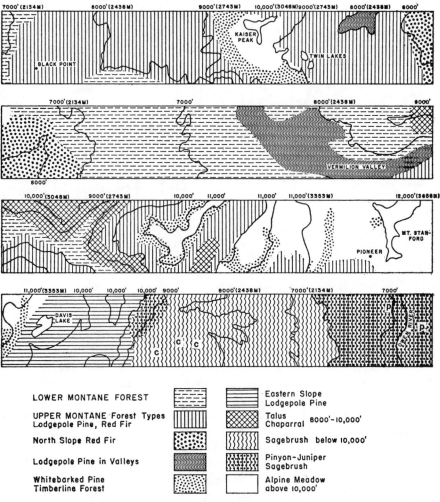

LOWER MONTANE FOREST				Eastern Slope Lodgepole Pine
UPPER MONTANE Forest Types Lodgepole Pine, Red Fir				Talus Chaparral 8000'–10,000'
North Slope Red Fir				Sagebrush below 10,000'
Lodgepole Pine in Valleys				Pinyon–Juniper Sagebrush
Whitebarked Pine Timberline Forest				Alpine Meadow above 10,000'

Fig. 6-10. A portion of Klyver's (1931) surface transect of the Sierras at about 7000 feet (2120 m) and above. The transect is about 1.5 miles (2.4 km) wide and a little less than 45 miles (72 km) long, cut into four pieces. The right end of each strip matches the left end of the one below. Among the high altitude sagebrush, c indicates mountain mahogany; in sagebrush, p indicates jeffrey pine. The characteristic animals of the different forest types are noted in the text. The lower montane forest is chiefly ponderosa pine with incense-cedar and sugar pine at the higher altitudes. Douglas-fir was present at only one point. The eastern end of the transect is discussed on page 278.

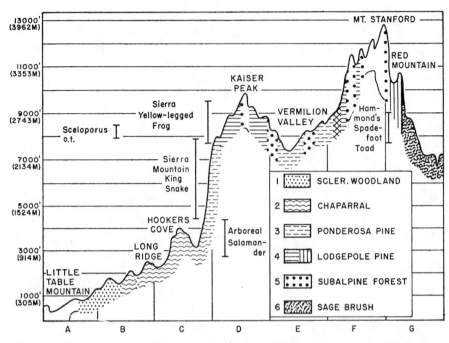

Fig. 6-11. Vertical transect of the mountains shown in Figure 6-9, vegetation from sources cited in Figures 6-9 and 6-10. Some reptile and amphibian populations are indicated (after Grinnell and Storer 1924, Klyver 1931).

pine, especially those trees weakened by defoliation by pandora moth larvae and other insects. Insects that attack trees are subject in turn to predators and disease. The pandora moth larvae are affected by wilt disease which kills many at times, ground squirrels and chipmunks which dig and eat quantities of pupae from the soil, birds which destroy some larvae, the Mono Indians who use them for food, and four species of parasitic insects which destroy larvae. The mountain pine bark beetle is devoured by the clerid beetle *Euoclerus* and is parasitized by the fly *Medeterus* and the parasitic wasp *Coeloides.* The sugar pine is attacked by the sugar pine tortrix *Archips lambertiana,* which may kill trees which are weakened by mountain pine beetles, and the western six-spined engraver *Ips confusus* (Keen 1938).

The red fox, fisher, and yellow-haired porcupine are probably most abundant in the upper montane forest. The Pacific jumping mouse and mountain beaver are known to occur. The blue grouse, great gray owl, calliope hummingbird, Williamson's sapsucker, mountain chickadee, hermit thrush, ruby-crowned kinglet, and Audubon's warbler are relatively numerous. In pure red fir (Fig. 6-10) Hammond's flycatcher is usually present. In the valley lodgepole pine forest occurs the Williamson sapsucker (Grinnell and Storer 1924). On the eastern slope of the Sierras the lodgepole pine is attacked by *Ips integer.* The fox sparrow is the characteristic bird. The mountain pine bark beetles attack lodgepole pine and the Sierra fir borer penetrates into the sapwood. These insects destroy many trees.

The timberline forest includes whitebark pine and mountain hemlock. The pika, heather vole, and pine grosbeak appear to be characteristic of this forest. The hemlock looper attacks the hemlock.

SOUTHERN ROCKY MOUNTAINS

In this subregion, the San Francisco Mountains and the Kaibab Plateau have received special study. Both areas are in northern Arizona. The zonation of vegetation and animal life on San Francisco Mountain has been described by Merriam (1890). Its elevation is 12,565 feet (2807 m). The mountain is covered with volcanic material and the top is largely unweathered rock. Merriam lists 18 species of plants of far northern distribution, some northern butterflies, a weasel, and the mountain sheep (Fig. 6-12).

The Kaibab Plateau rises from about 5000 feet (1500 m) to above 9100 feet (2760 m) (Fig. 6-13) and lies north of the Grand Canyon, about 60 miles (18 km) north of the San Francisco Mountain. Its ecology has been studied by Rasmussen (1941).

Upper Montane Forest

The *Picea–Abies–Tamiasciurus hudsonicus fremontii* faciation occurs on the Kaibab Plateau generally above 8200 feet (2485 m) and on cool slopes at lower elevations (Figs. 6-13, 6-14). The dominant trees are Engelmann and blue spruce, Douglas-fir, and white fir. Quaking aspen is a seral tree throughout

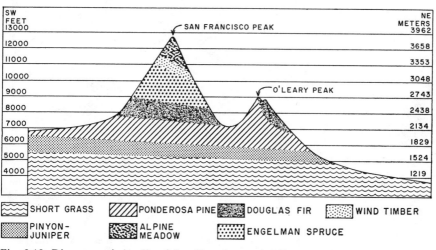

Fig. 6-12. Diagrammatic profile of San Francisco and O'Leary peaks from southwest to northeast, showing the several vegetation zones and effects of slope exposure (after Merriam 1891).

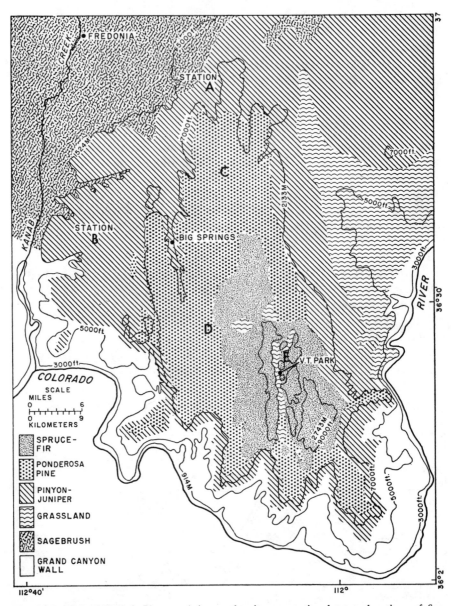

Fig. 6-13. Map of Kaibab Plateau, Arizona, showing vegetational types, locations of five study stations (A, B, C, D, E), and 1000 feet (304.8 m) contours (after Rasmussen 1941, courtesy Ecological Monographs).

the area. Blue spruce shows a preference for the edge of meadows, and ponderosa pine is on exposed hillsides. According to Pearson (1931) deficient surface moisture, caused by high insolation and attendant evaporation stress, is likely to be the critical factor in limiting the down slope occurrence of

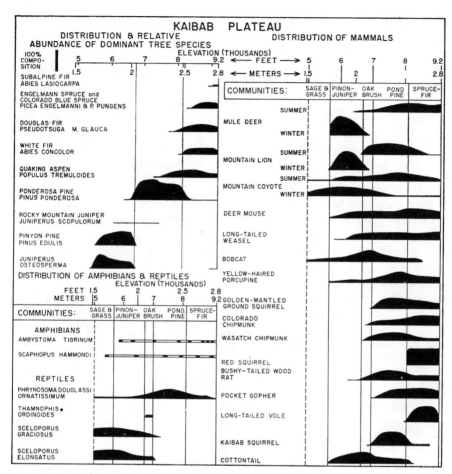

Fig. 6-14. Distribution and relative abundance of dominant tree species, reptiles, amphibians, and permeant mammals on Kaibab Plateau, Arizona. Total width of all shaded areas at any designated elevation equals 100 per cent, i.e., the total number of trees present (after Rasmussen 1941, courtesy Ecological Monographs).

Douglas-fir and white fir, while with ponderosa pine deficient heat limits its appearance at higher elevations.

Juniperus communis is especially abundant at the higher elevations, and is the only shrub of any importance. Snowberry was formerly quite a conspicuous plant, but deer have now greatly reduced it to small compact bushes. There are a number of old dead clumps of willow that have been killed by deer browsing. The creeping barberry, elderberry, *Ceanothus fendleri,* and a species of rabbitbrush are sparsely represented.

A great amount of lichens, liverworts, and mosses exist in the densest forest. Common plants occurring where there' is more sunshine are the strawberry *Fragaria platypetala,* pussy's toes, *Geranium fremontii,* brome grass, and twinflower. In still more sunlight, many of the meadow herbs appear: *Potentilla, Silene douglasii,* and *Erigeron.*

There are large areas of mountain meadow interspersed through the upper montane forest. The deer ordinarily concentrate in these meadows where 90 per cent of the vegetation consists of plants with net-vein leaves. Two species of mountain clover are important. *Agoseris, Leontodon, Erigeron, Achillea lanulosa,* and *Ranunculus subsagittatus* are less numerous. Knotweeds occur on newer denuded areas. The grasses are mainly *Phleum alpinum* and species of *Poa, Deschampsia,* and *Carex.* In the higher and more xeric portions of the plateau, grass makes up nearly 50 per cent of the vegetation.

MAJOR PERMEANT INFLUENTS

The mule deer is the most important major permeant animal in the community because it prevents the growth of several species of woody plants through its selective browsing. Its trampling is important locally. Deer eat a great amount of quaking aspen and white fir, but an insignificant amount of other conifers (Fig. 6-15). Certain lupines are heavily utilized in the late summer, and mushrooms, including *Amanita muscaria,* often compose 50 per cent or more of their food in July and August when rainfall is heavy. Very

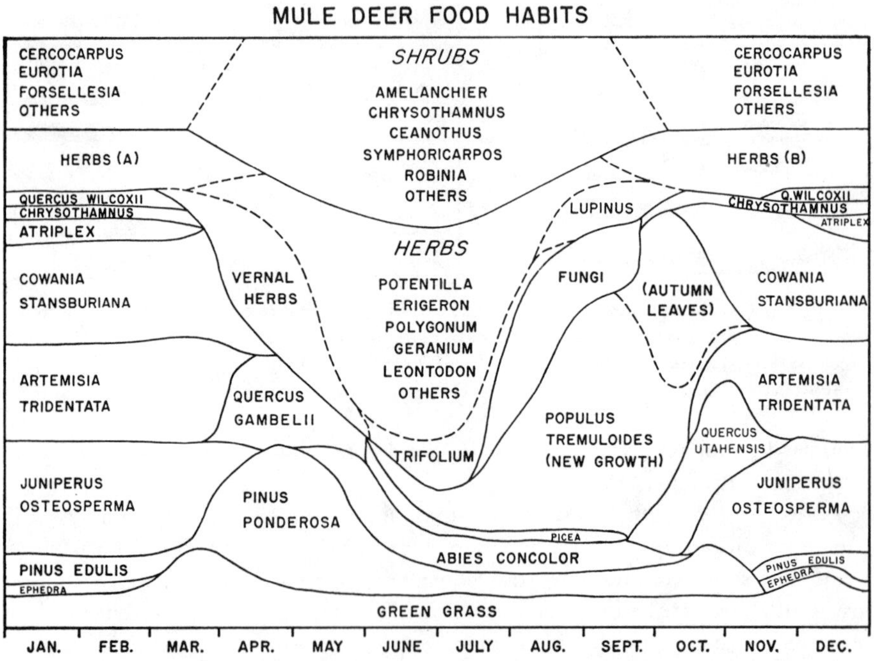

MULE DEER FOOD HABITS

Fig. 6-15. Generalized year-long food habits of mule deer, Kaibab Plateau, Arizona, 1929-31 inclusive. Based on analysis of approximately 60 stomachs and year-long field observations. The herbs omitted are: (A) species of *Eriogonium, Sphaeralcea, Lotus,* and *Castelleja;* (B) *Erodium, Ranunculus, Saxifraga,* and *Oreobroma* (courtesy Ecological Monographs).

little grass is eaten. Deer concentrate about the mountain meadows in the later hours of the day to feed on mountain clover. Through their browsing of seedling aspen and conifers they tend to maintain the meadows in a grassy condition (Fig. 6-16).

Deer exert their greatest influence in the pinyon–juniper woodlands of the lowlands where they spend the winter. The major part of the deer herd enters the lower montane forest in middle and late April and reach the mountain meadows (a local opening in the forest) in May. During June, July, and August they are throughout the mountain coniferous forests. In September and October, depending on weather conditions, the ponderosa pine of the lower

Fig. 6-16. The margin of one of the mountain meadows showing a fenced (exclusion) plot at the left and a staked plot at the right, ten years after they were set aside. The plots were 27 by 21 feet (8.2 by 6.4 m) (courtesy H. G. Johnson, personal communication, U.S. Forest Service, Kanab, Utah; author's photos).

montane forest is again the area of their maximum concentration. Dense timber provides little deer food, but great numbers of fungi in late summer cause the deer to leave the meadows and become fairly uniformly distributed through the forest. The deer begin to enter the pinyon–juniper woodlands in late September, following well-defined routes in their migration through the lower montane forest. The mountain lion and coyote follow the deer, but their movements are due, in part, to similar responses to seasonal changes and storms.

The mountain lion is dependent upon rough country for denning. The maximum summer concentration of the lion is in the ponderosa pine forest; their population in the upper montane forest is relatively small.

Coyotes are present on the plateau throughout the year, but a number move to lower altitudes in the presence of deep snows. Their distribution is fairly uniform through the forest with a summer population of 36 to 40 per township. The coyote's winter diet while in this community is primarily rabbits and small rodents. They kill some fawns in the upper montane forest in summer. Some fruits and insects are taken (Rasmussen 1941).

The gray wolf occurs as a small band of less than a dozen individuals. Of the four bears formerly recorded on the area, two were black, one "brown," and one a grizzly.

MINOR INFLUENTS

Of the minor influents in the forest, the porcupine is a wide-ranging resident in the coniferous forest and pinyon–juniper communities, active the year round. It shows a preference for ponderosa pine, and many areas show 15 to 25 per cent of these trees with signs of porcupine damage. Small trees are sometimes girdled, and death results. Stomach analyses show the summer food to be a preponderance of net-veined leaf material. The paucity of shrubs over the plateau, due to their exploitation by deer, has augmented the amount of damage done by porcupines to the trees.

The red squirrel occurs throughout the upper montane community. During late August, while they gather the cones of blue spruce, there may be several animals per acre, but in early summer, they average one squirrel per 15 acres (1/6 hectares). The pocket gopher is fairly abundant throughout the meadows where the soil is suitable for burrows and not subject to flooding. They obtain most of their food from roots but occasionally take other food from near the mouth of the burrows. The long-tailed vole is found in mesic habitats with thick grass. The golden-mantled ground squirrel and the Colorado and least chipmunks exercise considerable influence on plant growth. Deer mice, long-tailed weasels, and bushy-tailed wood rats also occur.

There are no birds restricted to this community, but the mountain bluebird and the robin eat an enormous number of the abundant grasshoppers. During the summer, flocks of common ravens also feed on the grasshoppers and on crickets in these high meadows.

The most important invertebrate group is the grasshoppers. The most abundant species are *Melanoplus m. mexicanus, Camnula pellucida,* and *Trimerotropis suffusa.* The large Mormon cricket also destroys a large amount of plant material.

Lower Montane Forest

The *Pinus ponderosa–Sciurus kaibabensis* faciation has ponderosa pine generally dominant between 6800 and 8200 feet (2000 and 2500 m) extending downward on cold slopes and upward on exposed slopes. The forest is open, with the large mature trees growing in groups or so widely spaced that sunlight reaches the ground in almost all parts of the forest. The shrub *Ceanothus fendleri* is characteristic and quite uniformly distributed. Mexican locust is in the lower portions of the forest and snowberry in the upper; creeping barberry is small and inconspicuous while manzanita occurs only locally. Rabbitbrush, elderberry, the currant *Ribes inebrians,* and ocean spray are present.

The grass *Muhlenbergia* is the predominant herbaceous element. A yellow sedge grows among the needles under the trees throughout the community. A number of lupines are present, particularly *Lupinus barbiger*. Other herbs are *Potentilla, Erigeron divergens, Solidago, Castilleja, Lotus wrightii, Artemisia caudata calvens, A. ludoviciana,* and *Astragalus.* Mistletoe is parasitic on 10 to 20 per cent of the older trees and in some cases causes death (Rasmussen 1941).

ARTHROPOD DOMINANTS

There is considerable evidence of former engraver beetle infestations, and there are large areas covered with dead trees (Fig. 6-17), some very old, others more recent. The insect responsible is probably the Black Hills beetle *Dendroctonus ponderosa,* although search failed to show the presence of a single live specimen during the summers of 1929, 1930, and 1931. An estimated 15 million feet of standing ponderosa pine was destroyed between 1920 and 1925.

Fig. 6-17. A. Damage to ponderosa pine by Black Hills bark beetle, 10 to 15 years after the trees were attacked. B. Badly riddled ponderosa pine bark still showing beetle burrows at the top (photos in 1936 by H. L. Andrews). C. Black Hills beetle (x 8, courtesy of Hopkins).

In trees that are freshly killed or weakened by earlier attacks of *D. pon-derosa* are great numbers of another bark beetle, *Ips integer*. In trees dead for several years, the shot borer *Orthotomicus ornatus* penetrates the wood. The carpenter ant *Camponotus herculeanus modoc* is under the bark of all dead trees, and the smaller *Camponotus sansabeanus vicinus nitidiventris* is also frequently found. Late in the summer, the large-winged form of *Camponotus* is extremely abundant throughout the forest (Rasmussen 1941).

MAJOR PERMEANT INFLUENTS

Major permeant influents are the same as in the upper montane forest. The coyote occurs in the same numbers as in the upper montane forest during the growing season; very few remain all winter. The summer population of mule deer in the more concentrated areas amounts to 20 to 30 per square mile (8 to 11.5/km²), but for the entire ponderosa pine forest the average is near 15 to 20 per square mile (6 to 8/km²). The home range of a doe with fawns seldom exceeds two miles (3.2 km) in radius; with abundant available water it would probably be much less. During the summer the deer feed mostly on raspberry, snowberry, and *Ceanothus*. When deer are numerous, these shrubs as well as young aspen trees may be greatly reduced in numbers or completely eliminated for periods of 10 or 15 years.

MINOR PERMEANT INFLUENTS

The Kaibab squirrel is a minor influent and is limited to the ponderosa pine forest of the Kaibab Plateau. It is active throughout the year. In 1931, their numbers were estimated at six to eight per square mile (2.3 to 3/km²). This squirrel feeds principally on the bark of year-old terminal shoots of ponderosa pine. During summer it also eats numbers of green pine cones, a small amount of herbage, and occasionally some fungi.

The two species of chipmunks forage in the trees and bushes as well as on the forest floor. The least chipmunk is more numerous than the larger Colo-rado chipmunk, usually in populations of 3 to 5 per acre (7.5 to 12.5/hectare). The most abundant mammal is the tawny deer mouse. The bobcat and the rock squirrel are present in small numbers in rocky habitats.

The pigmy nuthatch, white-breasted nuthatch, Mexican junco, and Steller's jay are year-long residents of the forest. The abundance of the red-shafted flicker and the Rocky Mountain sapsucker is influenced by the presence of aspens. The sapsucker usually nests in aspen but feeds to a considerable extent on ponderosa pine trees during the summer. The goshawk feeds on insects and other birds. The summer resident red-tailed hawk feeds in part on the Kaibab squirrel. The only generally distributed reptile that is represented in any numbers is the short-horned lizard.

At the time of study, jumping plant lice, plant lice, and leafhoppers were abundant on both small and large ponderosa pines. The abundant spiders were immature stages of *Araneus displicatus*. The weevil *Tricolepis inornata* and anthomyiid flies were also abundant, but perhaps the most important invertebrate group was the ants. They constituted 18 per cent of all invertebrates taken and were found from the ground to high in the trees. Species included *Lasius niger sitkaensis, Formica fusca subaenescens,* and *Formica neogagates lasioides vetula.* The sawfly *Xyele,* a geometrid larva, and the syrphus fly *Syrphus opinator* were also constituents of the shrub and tree layer.

The herb and grass layers cover only 2 to 10 per cent of the ground surface. The most conspicuous small animals in this layer are the banded-winged locusts, many of them immature, that normally rest on bare ground and feed on the herbaceous plants. The most abundant species are *Trimerotropis suffusa, T. p. pallidipennis,* and *Circotettix coconino.* Also abundant on the herbs are the two aphids *Phis* and *Macrosiphum,* leafhoppers, the spider *Philodromus pernix,* and the chermid *Paratrioza cockerelli.*

On the ground, the most numerous species are the same ants that were noted for the tree strata and *Myrmica scabrinodis* and *Liometopum apiculatum luctuosum.* The carabid beetle *Dyschirius globulosus,* the tenebrionid *Eleodes,* and the centipede *Scolopendra polymorpha* are present. Following the summer rains, myriads of collembolons appear in the soil; these are not in evidence earlier in the season.

Chapter **7**

The Tundra Biome

The term tundra has been applied to all types of vegetation in the treeless Arctic, both in Europe and North America. In the United States, "barren ground" sometimes has been used in a similar sense. The greater part of North America, north of 57° N. Lat., is tundra (Fig. 7-1). Tundra often appears as a gray-green plain or rolling country with innumerable lakes and ponds of all sizes. The high mountain tundra may have the same gray color. Small areas of the Aleutian–Bering Sea tundra have a different appearance because of the prevalence of shrubs.

Precipitation on the tundra is low, ranging from 12 to 20 inches (25 to 50 cm) annually, decreasing northward. Much of it falls as snow. At its southern boundary the mean temperature for January varies from –20° to –26° F. (–29° to –32° C.).

Large areas of climax tundra vegetation consist of heath which has plants growing in two or more inches of unconsolidated peatlike material full of roots and stems which make up a mat (Fig. 7-2) (Trapnell 1933, Shelford and Twomey 1941). The heath plants are principally *Vaccinium, Ledum, Cassiope,* and *Rhododendron,* known as bilberries and dwarf huckleberries. The crowberry family is represented by *Empetrum.* Among these shrubs is a growth of reindeer moss *Cladonia* and *Cetraria.* In other areas sedges appear to make up the climax. The dominance of sedges and grasses is, however, particularly characteristic of the numerous marshes and poorly drained areas. The tundra community may be called the *bilberry–caribou*–Cladonia *biome.*

In 1600, the tundra communities had been little modified by Europeans. At the present time, they are subject to considerable damage from fire which burns off the *Cladonia*-heaths. The caribou, once abundant throughout, have been greatly reduced along the coasts of Bering Sea and Arctic Ocean and in

Keewatin, due to the excessive hunting of them for skins, food, fox bait, and dog food (Harper 1955). In Ungava, their numbers have become reduced due to the loss of winter food when fires burn into the northern fringe of the coniferous forest where they migrate for the winter. Introduction of the Asiatic reindeer into the North American tundra may result in serious consequences for the caribou. Populations of all tundra animals of value to trappers have doubtlessly decreased considerably.

The continental and northeastern tundra constitutes the main central body of the tundra and covers an immense area (Fig. 7-1), extending approximately 2000 miles (3200 km) from Seward Peninsula in Alaska to the east end of Pearyland in north Greenland, 1100 miles (1760 km) from treeline on the west shore of Hudson Bay to the northernmost end of the Canadian Arctic Islands, and 1700 miles (2720 km) from the same point to the northwestern Alaska coast.

The high altitude tundra on Baffin Island and Greenland has been described in much detail by Trapnell (1933). The high coast of Labrador, Newfoundland, Gaspé Peninsula, Mount Katahdin, Mount Washington, and Mount Marcy is a series in which high tundra is replaced by alpine meadow. The southern limit of the mat described above, which distinguishes high tundra from alpine meadow, is somewhere between 55° and 50° N. Lat.

The western Alaska and Aleutian tundra is an area similar to continental tundra and extends along the west coast of Alaska from the Seward Peninsula

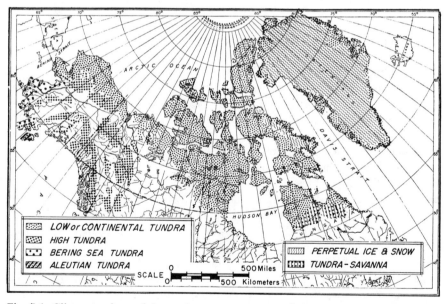

Fig. 7-1. Climax tundra and its seral stages cover most of the eastern 1,250,000 square miles (3,250,000 km²) of North America. The scale applies only to the southern part (Canada after Halliday 1937; Greenland, American Commission International Wildlife Protection Publication 5, 1934; Alaska, after Brooks 1906).

south to the Aleutian Islands. Hanson (1953) has described five types of shrub-covered areas, some species of which are common and widely distributed elsewhere over the tundra. Atypical tundra occurs on Kodiak Island and in the Aleutian Islands where the annual rainfall amounts to 50 to 60 inches (125 to 150 cm).

The high altitude tundra of the northern Rocky Mountains is an important area in Alaska, Yukon, and British Columbia south to about 55° N. Lat.

Fig. 7-2. A section of a hummock showing the mat as a dark area four inches (10 cm) thick with *Cladonia, Dryas integrifolia, Arctostaphylos rubra,* and *Carex misandra.*

CONTINENTAL AND NORTHEASTERN TUNDRA

Topography and Climate

The main area of tundra ranges from sea level to 1000 feet (300 m), except near Greenland and northwest of Great Bear Lake. In and near Greenland occurs a mountainous plateau country with small areas of mat-forming tundra at the mouths of streams and along the coast. The rougher topography contains mountains, glaciers, ice caps, and local snow areas.

The climate in the far east is modulated by the Atlantic Ocean. The influence of the Arctic Ocean has not been investigated. In eastern Greenland and on Resolution Island, off the north tip of Labrador, the mean annual temperature is 5.2° F. (−14.9° C.). In the extreme northeast it is 19° F. (−7.2° C.) (Manniche 1912). Temperatures in western Greenland vary from +0.5° F. (−17.5° C.) in the extreme south to −35° F. (−37° C.) along the northern coast. At Pond Inlet on the north end of Baffin Island, the mean annual temperature is −29° F. (−34° C.). At Craig Harbor on Ellesmere Island it is −25.5° F. (−32° C.) (Ekblaw 1926). Snowfall ranges from 70 to 117 inches (175 to 292 cm) at the south end of Baffin Island. Total annual precipitation was 6 inches (25 cm) in northeast Greenland in the period 1906 to 1908, and 9.3 to 21 inches (23 to 52 cm) on Baffin Island.

The mean annual temperature at Fort Reliance on Great Slave Lake, 490 miles (784 km) west of Hudson Bay and 350 miles (560 km) south of Coronation Gulf, is 19° F. (–7.2° C.), the rainfall 3.86 inches (9.6 cm), and snowfall 51.7 inches (129 cm). The annual temperature and precipitation is the same at Churchill, Manitoba, on the Hudson Bay. In the Thelon Game Preserve, northeast of Great Slave Lake, there were only ten readings above 0° F. (17.8° C.) during one winter (Clarke 1940). For the area as a whole, mean annual temperature ranges from –15° to –27° F. (–26.1° to –32.8° C.), with extreme daily minima of –42° to –57° F. (–41.1° to –49.4° C.) and extreme daily maxima of 64° to 96° F. (17.8° to 35.6° C.). The snowfall ranges from 43 to 68 inches (108 to 170 cm). Total precipitation ranges from 11 to 17 inches. The winters are characterized by complete cover of snow and ice. Ice usually breaks up in July on the lakes and in June on the rivers. It freezes again in September or October. The months of July and August may pass without frost in the southern parts, but the growing season is too short for any north temperate agricultural plants.

All soil material above the underlying bedrock that remains continually frozen throughout the year is called permafrost. These materials include loose talus rock, boulders, gravel, sand, clay, peat, sphagnum, and water. Permafrost is characteristic of tundra but extends down into the transcontinental boreal forest and into other forested and bush-covered areas. At the Nelson River along the Hudson Bay Railroad in Manitoba, the permafrost is 23 feet (7 m) thick, but near Churchill it is only about 8 feet (2.4 m). Associated with permafrost and the result of intermittent thawing of surface layers are the slow flow of soil down slopes, called solifluction, and the formation of

Fig. 7-3. *Dryas octopetala* growing on edges of boulder clay polygons (after Summerhayes and Elton 1928).

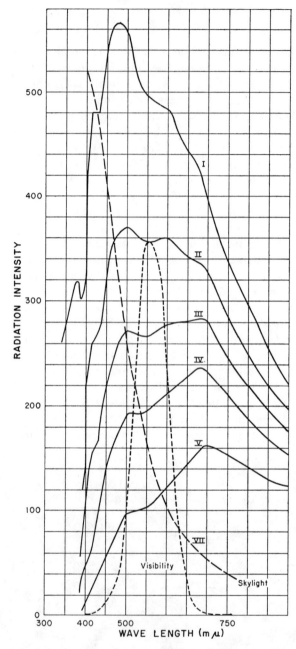

Fig. 7-4. Curves of relative intensity of solar radiation at Washington, D.C. (after Kimball 1924): I, outside the earth's atmosphere; II, zenith distance of 25° (1.1 air masses); III, zenith distance 60° (2 air masses); IV, zenith distance 70.7° (3 air masses); V, zenith distance 78.7° (5 air masses). Five air masses approached the condition at Chesterfield Inlet at noon about November 4, at Point Barrow, Alaska, about October 13, and northeastern Greenland, September 10. Note the sharp reduction in intensity of violet rays and the shifting of the maximum from 500 mu (green) to 700 mu (infra red) as the length of the air path increases.

polygon configurations in the soil (Fig. 7-3). The polygons are noteworthy throughout the northern part of the tundra, especially on the eastern Canadian islands and on the coastal plain of Alaska. Polygons are frequently 30 to 50 feet (9 to 15 m) in diameter but may be much smaller (Life 1954). They become covered by vegetation, the plants growing on them belonging to species widely distributed through the tundra (Summerhayes and Elton 1928, Polunin 1934-35, Wiggins and Loren 1953). The vegetation associated with various permafrost phenomena is usually that of early seral stages, due to the disturbance involved (Hopkins and Sigafoos 1951).

About two-thirds of the tundra is above the Arctic Circle and is characterized by nearly continuous sunlight about half of the year. There are 46 weeks with some sunlight and 6 weeks of complete darkness at 70° N. Lat. At 80° N. Lat. there are 34 weeks with some sunshine and 18 weeks of complete darkness. Due to the angle of the sun's rays, they must at noon on June 20 pass through two air masses at Churchill in northern Manitoba, through three air masses at Point Barrow in northern Alaska, and through five air masses in northern Ellesmere Island. As the air masses increase, the maximum intensity of radiation shifts from wave length 470 mμ to 700 mμ (Fig. 7-4). Short wave lengths are absorbed rapidly, but nevertheless humans may sunburn on the tundra.

Communities on the Arctic Slope of Alaska, Canada, and Greenland

In climax and late subclimax stages, characteristic and widely distributed heath plants are black crowberry, alpine bilberry, white bell heather, and dwarf birches (*Betula michauxii* north and *B. glandulosa* south of 63° N. Lat.). Bearberry (*Arctostaphylos alpina* or *A. rubra,* the latter more southerly), Labrador-tea, and foxberry or squawberry are common. Lapland rosebay and salmon-berry or cloudberry are infrequent or absent north of 65° N. Lat. (Fig. 7-5). The lichens *Cladonia* and *Cetraria* accompany these shrubs (Holttum 1922, Trapnell 1933, Shelford and Twomey 1941, Oosting 1948). South of about 62° N. Lat. in Greenland, birch–willow scrub occupies considerable areas and may be considered climax. The thick scrub, five and one-half feet (1.6 m) high, consists of *Betula minor, Salix glauca, Sorbus americana,* and *Alnus.* Pure willow scrub occurs in moist, partially shaded areas (Holttum 1922, Longstaff 1932, Trapnell 1933).

MAJOR PERMEANT INFLUENTS

Influents of wide distribution are barren ground caribou (Fig. 7-6), which feeds largely in the climax; muskox (Fig. 7-7), which feeds largely in seral stages; tundra wolf, which penetrates all stages; and polar bear, which is a common visitant, especially for hibernation. Bears occur locally.

Using the caribou density in the Thelon Game Preserve as a basis, Clarke (1940) estimates the population of caribou in the central area of tundra at 3,000,000, with an annual natural increase of 750,000 and an annual loss of 400,000 to predators, 200,000 to hunters and the remainder to infant mortality and other factors. Caribou herds are small when feeding, sometimes four to

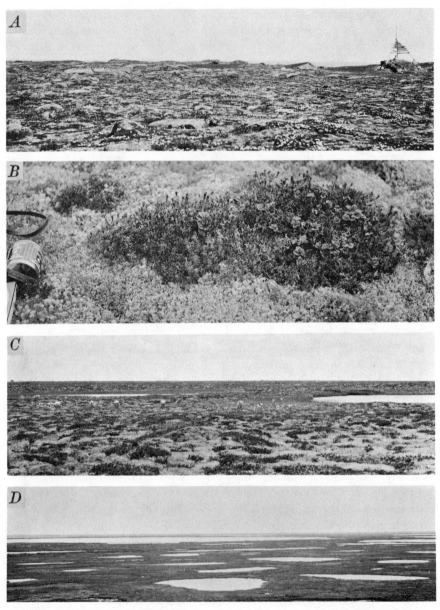

Fig. 7-5. Tundra vegetation. A. Climax tundra on rock (Churchill, Manitoba), *Ledum* in bloom, nesting site of horned lark and Lapland longspur, breeding place of the arctic hare, several foliaceous lichen-covered rocks protrude through the climax covering. B. Close view of one of the dark patches shown in C, the most abundant plants are *Ledum palustre decumbens* (in bloom), *Rubus chamaemorus,* and *Empetrum nigrum.* C. Climax tundra on 7 or 8 feet (2.2 m) of sphagnum, *Ledum* is past blooming, two ponds are marked by the white flowers of the arctic cottongrass. D. Ponds in the Little Barren, mile 474 Hudson Bay Railroad. Photos B and C are between the nearest two ponds.

six individuals or less. When migrating, they commonly contain 100 to 2,000 individuals.

Clarke describes one migration, however, of 20,000 animals. A band of caribou was seen near the Hanbury River in the game preserve; beyond them was another band, then another and another to the limits of vision. A herd would mill around in a compact mass; then one individual would head in the direction of migration and break into a trot. The others would follow until the entire herd was an extended line. In a mile or more, the leaders would start to graze, and as the others caught up, they also would graze or stand

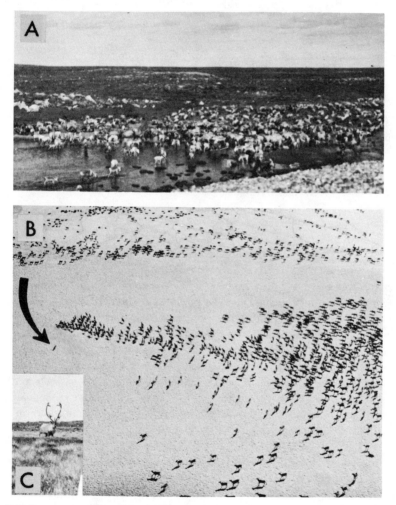

Fig. 7-6. Barren ground caribou in Northwest Territory. A. An undisturbed herd at a pond (courtesy National Museum of Canada, photo by J. Carrol). B. Wolf (indicated by arrow) approaching a herd. C. Bull approaching Eskimo hunter (photos B and C by A. W. F. Banfield in Ghost Lake District, 1949, Canadian Wildlife Service).

quietly about. Then they would mill around again, and again a leader would strike out. Where the caribou had come trooping up from the crossing, there was nothing but mud and dust, and hardly a foot of ground was without hoof prints. Only a few remnants of willow and drawf birch remained. Tree sparrows, Lapland longspurs, and Harris' sparrows were chirping or singing, but it was plainly seen that nests had been trampled and territories altered almost beyond recognition. A pair of ptarmigan was flushed with only a single two-day-old chick, the rest of the brood having succumbed before the avalanche.

Much trampling in summer, when the plants are dry and brittle, may entirely kill out the cover, but usually the devastation around river crossings is

Fig. 7-7. A muskox herd in a defensive ring against wolves (drawn from a photo by D. B. MacMillan, International Wildlife Protection).

limited in extent, and many roots send up shoots again the next year. The only persistent evidence of the passing of herds are the trails. From the air, these trails are seen as irregular lines all over the country, converging on river narrows, points of land, and at the bottom of bays (Clarke 1940) (Fig. 7-8).

Caribou possessing a comparatively southern summer range winter in the transition belt of the boreal forest. The herds that summer in the Thelon area winter near Fort Smith in northern Alberta; those that summer in southern Keewatin pass back and forth near Churchill on their way to or from their wintering area in northern Manitoba. Their southward migration continues until almost "freeze-up." On September 18, northwest of Lac de Gras, in the District of Mackenzie, L. E. Wray observed a long line of caribou pushing rapidly south, six or seven abreast, led by a buck a hundred or so yards in advance of the column. He estimated that there were several thousand animals in the herd. The rutting season is late October, and the young are born in June.

The species of caribou over the continent and in the northeastern islands, including Baffin Island, is *Rangifer arcticus,* often called the barren ground caribou (Manning 1943). The large herds that formerly occupied northern Ungava and Labrador have probably been reduced below a recovery point. The other species of caribou, *R. pearyi,* is found in the northern part of the

Fig. 7-8. Narrows of Lac de Gras showing caribou trails such as cover the barrens but rarely show in photographs (from Clark 1940, photo by Canadian Royal Air Force).

Arctic Islands, especially Ellesmere and Alex Heiburg islands. This species is believed to pass the summer in the highlands. They have little opportunity to migrate southward.

The chief enemy of the caribou is the wolf (Fig. 7-6B). According to H. N. Hamar, a trapper and wolf hunter of wide experience, one den per 100 square miles (259 km²) with a total of six wolves represents the average population on the tundra. Clarke (1940) estimates 36,000 wolves on 600,000 square miles (1,500,000 km²). About the first of April the wolves begin to establish themselves in pairs on their breeding grounds. Thereafter they stay near the den whether or not there are caribou around. Freuchen (1915) reports that when lemmings are abundant, wolf stomachs and excrement are full of their remains.

Other enemies of the caribou include warble flies (Murie 1935). Of two dried skins from southeastern Northwest Territory presented to the author, one rare albino skin contained 336 larvae holes, the normal-colored skin 164.

Muskoxen are relatively free of predators and insect pests (Fig. 7-7) (Clarke 1940). They prefer sedge-grass marshes with their favorite foods of sedges, grasses, and willows. They will eat upland vegetation, however, when it is more convenient, which reverses the preferences of the caribou. In 1700, muskoxen probably occurred over all the tundra of North America, except possibly southwestern Alaska, and up to 32,000 square miles (83,000 km²) in east Greenland. They may possibly have been as numerous as the caribou. In northeast Greenland, the arctic wolf has difficulty finding food during the winter and becomes very emaciated in spring. It supposedly stays near the muskox herds but feeds on any other animals it can secure, including lemmings. The wolf is not an important enemy of the muskox and is even less important than the grizzly bear farther south.

The polar bear is likely to be present anywhere near the sea coast. They rarely go more than 20 miles (32 km) from shore but eat any animal food, chiefly seals, that they can get. They feed on eider duck eggs and in summer may eat grass and roots. The polar bear may hibernate by merely lying down in a protected spot and letting the snow cover its body. In Greenland, Manniche (1910) states that he observed no "hibernation proper," but for a short period before the birth of her one to three young, the female remained quietly in her den. Male bears remain in their dens for "some days" during bad weather (Sutton and Hamilton 1932). On Southampton Island, they quite generally prepare dens for winter shelter.

MINOR PERMEANT INFLUENTS

The arctic fox is an important animal. They are not particular as to the kind of food they eat but prefer lemmings. They secure food with difficulty in winter because of snow, especially when lemmings are scarce. They do not often eat the arctic hare in Greenland. They mate in March in east Greenland and the 8 to 15 cubs are half-grown by August.

The arctic hare was seen near Churchill only on the rock ridges, which were largely covered with climax vegetation. This is the only community in which an examination of 30 square meters of ground surface by 18 observers in 1939 showed hare droppings. Trappers state that hares rear their young in the climax.

The brown lemming is found over most of the tundra north of Cape Eskimo on Hudson Bay and west of Hudson Straits. The varying or collared lemming occurs southward and in Labrador and is the only mouse that turns white in winter. The varying lemming belongs to the higher well-drained part of the tundra and occupies climax and late subclimax communities when they are on ground into which it can dig, such as on sand, gravel, or peat ridges (Sutton and Hamilton 1932). Its population varies cyclically (Fig. 7-9).

Parry's ground squirrel occurs only on the continental part of the tundra. It is often abundant but does not occur south of 60° 30′ N. Lat. on the west side of Hudson Bay. It burrows in alluvium and in sandy and gravelly ridges. The ground squirrel is the chief food of the wolverine in summer (Preble 1908).

Weasels make burrows among rocks and follow the runways of varying lemmings. They are also important destroyers of meadow voles that occur in the tundra.

The willow ptarmigan is usually present the year around, although there is some southward movement of individuals in the autumn and northward movement in the spring. The species feeds on any plant material it can find, especially willow buds. Its enemies are foxes, wolves, weasels, and snowy owls.

In east Greenland, the rock ptarmigan is wanting during the period of full darkness in midwinter but returns in February and March (Manniche 1910). Its food consists of buds of *Salix arctica* and leaves of *Dryas* and *Saxifraga*. At night during cold weather it roosts eight inches (20 cm) deep in the snow, which conserves body heat. Egg laying begins the first half of June.

Snowy owls are important permeant predators. Two nests were located in 1936 on long gentle slopes covered with *Cladonia* vegetation (Twomey 1937). The layer of lichens and mosses were scooped out to the bare gravel or sand, forming a cuplike depression 18 to 24 inches (45 to 60 cm) in diameter. Pellets about the nests indicated that a few Lapland longspurs and horned larks had been eaten, but lemmings formed the largest volume of the food.

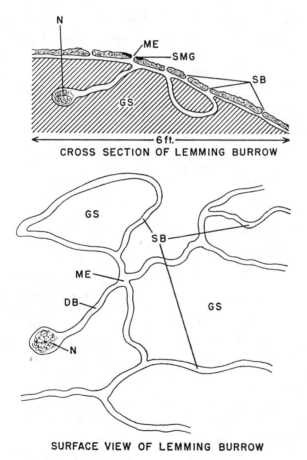

CROSS SECTION OF LEMMING BURROW

SURFACE VIEW OF LEMMING BURROW

Fig. 7-9. Surface and sectional views of a varying lemming burrow: N, nest with young; ME, main entrance; SMG, mound of gravel thrown out in digging; SB, surface burrow under the vegetation mat; GS, gravel and sand; DB, deep burrow (excavated and sketched by A. C. Twomey).

Their movement and local abundance are related to the lemming population (Shelford and Twomey 1941).

A long list of birds in their food relations contacts both seral and climax stages. At Churchill in the autumn, bearberries, bilberries, and squawberries are usually abundant on the larger climax areas and are eaten by the whimbrel, American golden plover, white-crowned sparrow, tree sparrow, and herring gull. The hoary redpoll, as well as the eastern states web spider, occurs in the birch–willow scrub of Greenland.

Invertebrates taken in July appeared to penetrate various habitats. An outstanding feature of the tundra community is the great number of dipterons. Shelford and Twomey (1949) found that the insects in three square meters of soil and mat consisted of 59 per cent dipterons, 18 per cent ants, 12 per cent beetles, 10 per cent ground beetles. Some of the flies taken in the climax were *Coenosia nigrescens, C. fraterna, Simulium vittatum, S. venustum,* and *Macrorchis majuscula.* The percentage of flies in the total collections of insects were 92 in wet seral stages, 86 in three stages of the sand sere, and 100 in the climax tundra. Many species are of wide distribution (Nielson 1907). Ants, grasshoppers, crickets, earwigs, cockroaches, and dragonflies were rare or wanting.

Coactions

Longstaff (1932) censused 728 pairs of nesting birds on about eight square miles (20.7 km²) of tundra in Greenland. Some 20.5 per cent were in willow scrub, 4.8 per cent about pools, and 71 per cent in the heath climax.

There were 321 pairs of the Lapland longspur. In the 15 stomachs examined, Diptera occurred in 11, including Mycetophilinae, Chironomidae, Anthomyiidae, Muscidae, and larvae of Syrphidae. Beetles occurred in 10 stomachs; of these, 6 contained 11 specimens of *Otiorrhynchus nodosus* and 4 contained *Byrrhus fasciatus.* In 3 stomachs there were 2 species of Curculionidae and in 1, *Coccinella transverso-guttata.* Moths, spiders, and parasitic wasps were identified in 4 stomachs, and the harvestman *Mitopus morio,* bugs, and the leafhopper *Psylla* occurred twice. The lacewing *Kimminsia betulina* was found once. A snail was found in a specimen killed in a fjord. Lapland longspurs were frequently seen picking small flies from the heath, and once they were seen to capture the crane fly *Tipula arctica.* In nest material were found fragments of dipterons including *Platycheirus albimanus, P. hyperboreus,* and *Limnophora arctica,* the spider *Lycosa furcifera,* a red mite, and various seeds. Several of these insects and spiders and the seeds occur in the climax.

Eighty-four pairs of snow buntings occurred, and 15 stomachs were examined. Of the animal food, single stomachs contained the remains of snails, moths, spiders, the harvestman *Mitopus morio,* mites, and two *Gamasina.* Bugs occurred in 4 stomachs, chiefly *Nysius groenlandicus,* with as many as 13 individuals in 1 stomach. Parasitic wasps occurred in 5 stomachs, represented by Ichneumonidae, Ophioninae, and Chalcidoidea. Beetles were identified in 9 stomachs, including *Otiorrhynchus nodosus* in 3, Curculionidae in 2, and *Byrrhus fasciatus* in 1. Flies occurred in 10 stomachs, including Chironomidae, Tipulidae, and *Sciara*. The crane fly *Tipula arctica* and the spiders *Dictyna major* and *Lycosa furcifera* were fed to the young. The ichneumonid *Phygadeuon* and the flies *Limnophora arctica* and *Hylemya profuga* were found in the bills of birds shot. Longstaff (1932) reports a great mixture of seeds in stomachs of the snow bunting with the commonest seed *Empetrum nigrum hermaphroditum*. The snow bunting was the most omnivorous in its food habits of the four common passerines of West Greenland.

Population of some other birds on the area were wheatear, 61 pairs; purple sandpiper, 20 pairs; ptarmigan, 28 pairs; parasitic jaeger, 6 pairs; a straggler snowy owl, and a straggler gyrfalcon.

Herbivorous animals on the Greenland heath are the caribou, arctic hare, and ptarmigan. Some dipterons, particularly crane flies, bumblebees, butterflies, moths, aphids, coccids, and the scarce beetles feed on plants.

Carnivorous animals include the arctic fox, various insectivorous birds, predatory birds, spiders, certain dipterons, mosquitoes, biting midges, larvae of hoverflies, parasitic hymenopterons, and ladybird beetles. The carnivorous dipterons are eaten by spiders and birds, and the spiders by birds. Large numbers of lycosid spiders hunt over the heath, notably *Lycosa furcifera*. A count of eight web entanglements of the spider *Dictyna major* yielded 34 insects belonging to five species.

Fluctuations in the populations of the arctic fox are supposedly controlled by the abundance of the lemming, its principal food. The lemming breeds in winter under the snow (Sutton and Hamilton 1932). Increases in its population are therefore favored by average or more than average snowfall well distributed throughout the winter. Temperature and rainfall above average in July and August probably also favor population increases. Two favorable breeding seasons in succession are usually necessary to build up a peak population. Declines in numbers may occur either suddenly or gradually, depending on predation by foxes and birds (Pitelka *et al.* 1955). Lemmings may also freeze to death in numbers during very cold winters when there is little snow cover (Shelford 1943).

There is a similar correlation between fluctuations in the populations of lemmings and of snowy owls. The snowy owl goes south into New England, New York, and other northern states during winters that follow seasons in which there has been a sharp decline in the lemming population (Shelford 1945). The Bureau of Animal Population at Oxford University compiled information on the fluctuations in populations of foxes, lemmings, and ptarmigans, as well as snowy owls, from 1933 to 1949 (Chitty 1950).

Community Development

POND SERE

The Little Barren area lies in northern Manitoba 40 miles (64 km) south of Churchill and extends the same distance east to the Hudson Bay. It is about 30 miles (48 km) from north to south. Permafrost lies seven to eight feet (2 to 2.5 m) below the surface, and the ground is covered largely with sphagnum, the rush *Scirpus cespitosus callosus,* and sedges, with the heath *Andromeda polifolia,* the lousewort *Pedicularis labradorica,* and cottongrass somewhat less common. On hummocks, resulting from frost heaving, occur sedges and climax plants, such as the lichen *Cladonia* and heath (Fig. 7-5). Numerous ponds occur and about 40 per cent of the low ground is filled-in bog. The ponds are shallow, have a muck bottom, and appear to have no aquatic plants. They come to be filled in, however, by sphagnum and sedges, invading from the margins. Plant communities appear similar on both the filled-in depressions and on wet clay substratum (Shelford and Twomey 1941).

Among the animal inhabitants of ponds, the stickleback and northern wood frog both survive being frozen solid in winter, but after thawing out, they die if frozen again (Hearne 1895). Some small rock pools are inhabited by the phyllopods *Branchinecta paludosa* and *Chirocephalopsis bundyi* and in July resemble ponds at southern latitudes in March. The entomostracans *Diaptomus tyrrelli, Cypris pellucida,* and *Daphnia pulex* occur consistently, and the caddisfly larvae Limnephilidae are regular inhabitants. Larvae of the mosquito *Aedes nigromaculis* and also the snail *Stagnicola vahli* were taken on a visit in 1939.

Nearly every hummock contained a burrow of a drummond meadow vole or a Richardson's varying lemming in July, 1939 (Fig. 7-9). Hummock patches in the climax were the nesting sites of the whimbrel, Smith's longspur, and stilt sandpiper. Willow ptarmigan were numerous.

On the low wet flats, collembolons made up 80 per cent of the invertebrate population, snails 12 per cent, and running spiders 5 per cent. On slightly higher ground with mountain willow and *Equisetum,* collembolons constituted 37 per cent, snails 24 per cent, spiders 17 per cent, and larvae and pupae of mosquitoes 16 per cent. Mosquitoes are often more abundant here than in the pools on higher ground. *Aedes nearcticus* is the common species. Several flies, *Limnophora subrostrata, Phytobia morosa, Hylemya repleta,* and *Hydrophorus signiferus,* were not taken elsewhere. Some homopterons occur; *Cantharis* is a common soldier beetle. A ground beetle *Lyperopherus punctatissimus* occurs rarely, and the carrion beetle *Silpha lapponica* is found on dead rodents. *Tetragnatha extensa* and *Pardosa modica* are common spiders. The slug *Derocerus* and the northern wood frog are seen occasionally.

In Greenland, there are few fresh water plants except *Potamogeton filiformis borealis,* and this species is not common. Pioneer mosses are *Funaria hygrometrica, Tortula,* and *Didymodon recurvirostris.* In older ponds *Pleuropogon* occurs and around their wet margins *Drepanocladus aduncus,* in which the buttercup *Ranunculus hyperboreus* grows prolifically and the sedge *Carex saxatilis* is present. *Aulacomnium palustre* invades later, followed by *A. turgidum.* Mossy hummocks are formed when mosses grow over the bases of higher plants that are clumped together. The willow *Salix herbacea* often helps to

bind these hummocks together. *Vaccinium* and less frequently *Cassiope* dominate in ericaceous heath in the highest and oldest bogs (Oosting 1948).

In pools within cottongrass bogs, Longstaff (1932) found caddisfly larvae,
three kinds of water beetles, numerous crustaceans, and a single species of
snail. Northern phalaropes were in the pools and catching insects in the surrounding grassy vegetation. Mallards had their nests nearby. Adults of the
caddisflies, the butterfly *Boloria chariclea arcticus,* other insects, and spiders
were in the bog vegetation, and the snail *Lymnaea* on the ground. The insectivorous Lapland longspur was also here.

In *Cassiope* heaths containing moss in east Greenland, Jorgensen (1934)
secured 104 collembolons and mites in 0.01 square meter of ground material.
The most abundant collembolon was *Onychiurus sibericus.* A dwarf willow *Salix
glauca chloroclados* is often covered with small galls caused by a mite. The
fly *Sciara* was obtained from the peat in a hummock. The conspicuous red
flowers of the willow-herb *Epilobium* were visited mostly by bumblebees and
syrphid flies. The flies *Calliphora terraenovae* and *Piophila vulgaris* came to
bird carcasses carried among the hummocks by arctic foxes, while *Scopeuma
squalidum* was seen visiting bird droppings. The purple sandpiper nests among
these hummocks and the Lapland longspur on the sides of hummocks clothed
with *Empetrum* near water.

ROCK SERE

Starting in small crevices in rock domes, fruticose lichens and xerophytic
mosses begin to form cushions which spread over the surface. The cushions
are tangled mats of *Cetraria islandica, Cladonia sylvatica, Alectoria jubata,*
and other lichens of similar habit. Species of *Xanthoria* and *Gyrophora* are
usually present. When these mats are well established, higher plants appear in
them, particularly *Lycopodium selago, Pyrola grandiflora,* and *Vaccinium
uliginosum* (Holttum 1922). Rock domes project above the heath surface.

Polunin (1948) found rock ridges at Chesterfield in Keewatin supporting
the lichen *Lecanora.* Crevices and slight depressions hold moss mats of
Rhacomitrium lanuginosum. At Churchill, Manitoba, the crowberry *Empetrum*
grows outward over adjacent rock, catching debris, and making a habitat for
Cladonia.

The climax on rock ridges is nearly the same as that on frozen sphagnum
or on the bog hummocks. *Cladonia* and similar lichens cover a large percentage of the surface. More of the shrub *Andromeda* appears to occur on
rocks. Some *Rhododendron lapponicum* is also found along with occasional
Arctogrostis latifolia. Sedges, grasses, and coarse forbs occur only occasionally
(Johansen 1924).

The young of the small bug *Nysius groenlandicus* crawl over bare rocks.
The dark speckled spider *Pardosa groenlandica* resembles rock spotted with
pale lichen. Many animals, such as ptarmigan, skua, Lapland longspur, snow
bunting, and wheatear, visit rock domes, which serve as lookout posts. The
arctic fox and a few arctic hares occur (Longstaff 1932).

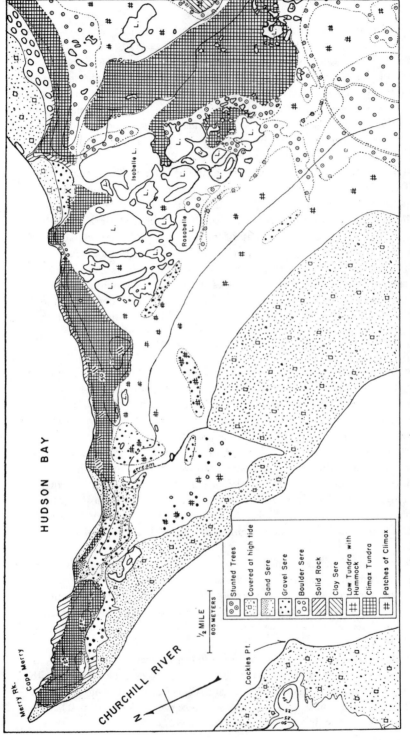

Fig. 7-10. Diagrammatic map reconstructing the biotic communities about Churchill, Manitoba. Lakes and ponds are marked with an L. Climax tundra near Hudson Bay is on solid rock; that south of Lake Isabelle is on deep frozen sphagnum or on 12 or more centimeters of mat over a clay surface. X (upper right) marks the location of a fox den.

GRAVEL SERE

Construction work near Churchill has provided large areas of bare gravel on which succession occurs (Figs. 7-10, 7-11). During the first four years, little or no invasion by plants takes place. In the fifth year, the small crucifer *Draba glabella* and a slender *Equisetum* form scattered but uniform stands. The crests of old gravel ridges near the shore of Hudson Bay still remain bare under natural conditions, but *Dryas integrifolia* and the bearberry *Arctostaphylos rubra* appear to be encroaching on the sides. The wintergreen *Pyrola grandiflora* plays a minor role in making up the ground cover. Similar vegetation occurs in east Greenland (Hartz and Kruuse 1911).

Fig. 7-11. A. Partially bare gravel ridge near Churchill. The vegetation around the bare spots is very largely *Dryas integrifolia* and *Arctostaphylos rubra*. B. View from a late subclimax ridge across a gentle slope to water, showing the light-colored grasses on the higher ground and the bright green of the wet portions, covered by tide at times (Hubbart Point 60° N. Lat.).

Birds arrange their nest sites with reference to seral plant stages at Churchill: on *Dryas* (Fig. 7-11A), semipalmated sandpiper, Lapland longspur, horned lark, and least sandpiper; on reindeer lichen, Lapland longspur and horned lark; in or near the climax where there are sedges, semipalmated plover, snowy owl, Lapland longspur, and Smith's longspur (Shelford and Twomey 1941).

Fig. 7-12. A. Lyme grass or wild rye, *Elymus arenarius*, inward from shore behind the pioneer *Arenaria peploides* which apparently may precede *Elymus* and continue after *Elymus* has dropped out. B. Old stand of lyme grass on gravel. C. Late seral stage on sand showing the grass *Poa glauca* with low herbs *Arenaria peploides*.

SAND SERE

The pioneer plants on newly deposited sand are the sandwort *Arenaria peploides* and the sea lyme grass *Elymus arenarius. Saxifraga tricuspidata* invades a little later (Fig. 7-12). The straw litter from these plants, especially the lyme grass, is sometimes present in sufficient quantity to form cover for the mouse *Microtus* and the lemming *Dicrostonyx*. The soil and mat in the lyme grass stage commonly support only one or two red mites and two to four running spiders, chiefly *Pardosa groenlandica*, per square meter. Occasional Microlepidoptera larvae and adults occur. Of the dipterons, *Coenosia fraterna, C. nigrescens, Fucellia ariciiformis*, and *Schoenomyza dorsalis* have been taken.

As the sand becomes stabilized *Poa rigens* and *P. glauca* invade and form "bunch" grasses. Some Greenland dandelion *Taraxacum lacerum*, mosses, and the chickweed *Stellaria longipes* also come in. At Godhavn, Greenland, Holttum (1922) reports the grasses *Festuca rubra arenaria, Poa alpina*, and *Poa glauca* appearing among the *Elymus*. The willow *Salix glauca* comes in later and clothes the bases of sand hillocks that are still crowned with *Elymus*.

Empetrum spreads over glacial sand near the shore. In East Greenland *Elymus arenarius* is replaced by *Lesquerella arctica* (Oosting 1948). *Empetrum* and *Arctostaphylos* follow these pioneers and crowd them out.

During July and August at Churchill, the homopterons *Lipurnia, Cicadula,* and *Deltocephalus,* and the hemipteron *Teratocoris herbaticus* are found in this *Poa* community. Most of these species also occur elsewhere in sedges. The grasshopper *Melanoplus borealis* is not common but reaches its maximum here. Animals of the mat and soil have increased about tenfold compared with the lyme grass stage. Dipterons present are mainly those found generally in the region: *Aedes nearcticus, A. spencerii, Fucellia ariciiformis,* and several species of *Coenosia* and *Tendipes.*

The arctic fox commonly makes its burrows in sand banks along streams, lakes, or on salt water beaches out of the reach of spray. The fox burrow marked on the map in Figure 7-10 was in a steep sand bank that was overgrown by herbs and decumbent shrubs along with scattered bunch grass. It was a typical den site.

When the populations of the varying lemming are small, they survive on sandy and gravelly ridges where they can burrow (Fig. 7-9). Lemmings are active at Churchill during the winter. They have the curious habit of placing their winter nest balls, made of grass or sedge, in the center of a snowdrift, in one case three feet (1 m) from the top surface of the drift and one foot (0.3 m) from the ground. Sometimes they are found in the same burrows with *Lemmus* (Sutton and Hamilton 1932).

CONVERGENCE

Whether the original bare area was aquatic, rock, or sand, community development tends in the main to converge to a heath climax (Fig. 7-13) (Shelford and Twomey 1941). There are only minor differences between the climaxes that develop on rock and on sphagnum. The *Cassiope* heath mat is probably postclimax, limited to regions with a generally wet and humid climate. Generally associated with the climax is the occurrence of a ground mat. This is composed of dead undecomposed plant remains interlaced with living roots and stems. On sand, gravel, and rock substrata, climax plants do not come in until the mat approaches six centimeters in depth, and the full climax develops only with seven to ten centimeters of mat. J. C. Tedrow (personal communication) states that in the north the mat reaches a maximum thickness of 7.5 centimeters on brown soil but thins out where bedrock is near the surface. In wet tundra it may reach 10 to 12 centimeters. *Dryas* appears to require close contact with mineral soil. Trapnell's (1933) best representative vegetation was on mats 5 to 10 centimeters deep.

Invertebrate animals found on areas which are approaching a heath climax, chiefly of *Dryas integrifolia, Cassiope tetragona, Salix, Saxifraga, Polygonum viviparum, Carex, Vaccinium uliginosum microphyllum,* and a number of mosses and lichens, are the spiders *Lycosa fuscula* and *Erigone arctophylacies,* the mite *Trombidium sucidum,* the butterfly *Boloria chariclea articus,* the

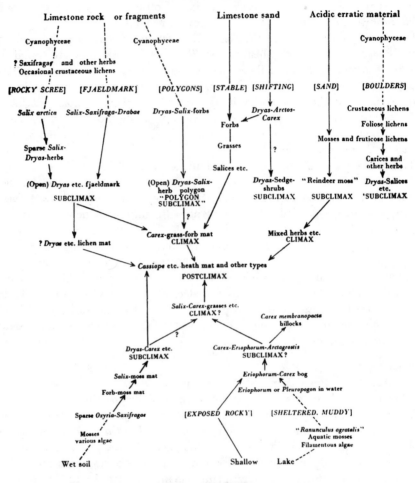

Fig. 7-13. Succession diagram (modified from Polunin 1934-35).

geometrid moth *Entephria polata brullei,* the weevil *Lepyrus labradorensis,* the hoverfly *Platycheirus,* the carrion fly *Acronesia popoffana,* the dung fly *Scopeuma furcatum,* the helomyzid fly *Scoliocentra thoracia,* and the springtail *Entomobrya erratica* (Davis 1936).

Aspection

Seasonal variation in the activity of invertebrates and in the flowering of plants was observed along the arctic coast of northeastern Alaska and north-western Canada between August, 1914, and July, 1915 (Johansen 1921, 1924).

CAMDEN BAY, ALASKA, SEPTEMBER TO APRIL

Insects went into hibernation in the latter part of September and remained in that state over winter. Some mortality occurred at 0° F. (−18° C.), but most hibernating insects withstood temperatures down to −50° F. (−46° C.). Aquatic insects and larvae either burrow into the mud or perish. Some species pass the winter in the egg stage.

CAMDEN BAY TO DEMARCATION POINT, MAY 1-21

The shrubs *Empetrum nigrum, Cassiope tetragona,* and *Ledum palustre* western Canada between August, 1914, and July, 1915 (Johansen 1921, 1924). About the middle of May, *Dryas integrifolia* and *Vaccinium caespitosum* showed new leaves, and the leaf buds of birch were opening.

Tipulid fly larvae remained in the ground until it thawed. Hibernating carabid beetles *Pterostichus mandibularis, P. agonus,* and *Nebria,* lepidopterous larvae, collembolons, flies, spiders, and mites were still found during May where they were in September in plant tufts, under stones, and under driftwood. They become mobile when exposed to the sun. The first active flies of the year were seen at Demarcation Point on May 13 when the weather was clear and at 35° F. (1.7° C.). One species, *Protophormia terraenovae,* was on the south side of a house where the temperature was 40° F. (4.4° C.) at two o'clock in the afternoon. Snow-free moss pillows held the muscid fly larvae *Rhamphomyia* and the sawfly pupae *Amauronematus cogitatus.* Beetle larvae and pupae and small staphylinid and carabid adults were present. Various tundra plants supported the moth *Byrdia rossi* in various life history stages. Some cocoons contained the parasitic tachinid fly *Murdockiana gelida.* There were spiders and leafhoppers in the grass.

When the snow melted, the collembolons *Podura aquatica* and *Isotomurus palustris* became noticeable in the ponds. The small dytiscid beetles *Agabus moestus* and *Hydroporus tartaricus* and red water-mites were active predators of the brown midge fly larvae *Tanypus,* probably their most important food.

MAY 22 TO JUNE 30

Salix pulchra came into bloom at the beginning of this period, and new leaves were on most of the plants. By June, *Anemone parviflora, Pyrola grandiflora,* and *Rubus chamaemorus* had new leaves. *Dryas integrifolia, Cassiope tetragona,* and *Luzula nivalis* came into bloom late in June.

Few insects were seen on the wing before the first of June, but then the flies *Cynomya cadaverina* and *Scaeva pyrastri* appeared. A black and white striped species, *Epistrophe sodalis,* was typical of high dry places on the tundra. When individuals were approached, they rose and hovered for a moment before flying away. The ichneumonid *Ophidnus nivarius* was seen. The first sawflies, *Amauronematus,* appeared in a short flight much like that of ants. Minute brown beetles were in flight on calm, sunny days. The carabid betles *Asaphidion* and *Amara brunnipes,* an occasional curculionid, the chrysomelid beetle *Chrysolina subsulvata,* small hemipterons, and the spider *Alopecosa pictilis* were found on the tundra.

Toward the end of June appeared a number of flying insects characteristic of places with rich vegetation, including the tipulid fly *Prioncera parryi,* the mosquito *Aedes,* the tineoid moth *Eucosma,* and the hemipteron *Euscelis hyperboreus.* Found in the tundra moss were beetles and beetle larvae, the fly *Tipula arctica,* other flies both adults and pupae, spiders, and mites. Larvae of the small orange-colored gall-gnat *Cecidomyia* were half hidden in wet sphagnum. The bumblebees *Bombus sylvicola* and *B. baltcatus* came into flight and began feeding on the male catkins of various species of willow. Queens of *Bombus hyperboreus,* various flies, and sawflies were also seen. The hemipterons *Euscelis hyperboreus* and *Calacanthia trybomi* and the spider *Lycosa* were on the tundra plants.

In the now ice-free tundra ponds thousands of mosquito larvae were hatching from overwintering eggs and *Daphnia pulex,* copepods, nauplii of the phyllopod *Branchinecta paludosa,* and the mite *Piona reighardi* were swimming in the water. Large red midge fly larvae were found resting in their mud tubes on submerged logs.

JULY 1 TO SEPTEMBER 30

The first butterflies *Boloria alaskensis, B. improba,* and *Colias hecla glacialis* appeared in early July.

Toward the end of July, the plants *Papaver nudicaule, Cochlearia officinalis, Oxytropis, Saxifraga oppositifolia,* and *Potentilla* had finished flowering or nearly so, but *Elymus arenarius mollis* and other grasses and sedges were in bloom. At about this time the arctiid moth *Platyprepia alpina* and the lymantriid moth *Byrdia rossi* appeared.

In late August insect life rapidly declined, especially among the less hardy neuropterons, lepidopterons, mosquitoes, wasps and sawflies. Flies, coleopterons, bees, and hemipterons were still numerous. The first signs of winter became apparent on September 3. Then, near a large lagoon, only the spider *Typhochraestus spitsbergensis,* colonies of collembolons, a few oligochaete worms, and fly larvae were found. Winter set in for earnest at Camden Bay in the middle of September.

McClure (1943a) studied aspection at Churchill, some 11° latitude farther south, but detailed comparisons are not practicable. It is of interest that mosquitoes appeared there at nearly the same time as at Camden Bay.

HIGH ALTITUDE TUNDRA OF GREENLAND, LABRADOR, AND MOUNTAINS TO THE SOUTH

The distance from the outermost islands and peninsular shores to the main glacier reaches a maximum of 100 miles (160 km) in west Greenland and 200 miles (320 km) in east Greenland. Trapnell (1933) divides the upland of west Greenland into a lower montane zone from 350 to 1000 feet (106 to 303 m) altitude, a middle montane zone from 1000 to 1700 feet (303 to 515 m), and an upper montane zone above 1700 feet, each differing from the others in physical conditions and in vegetation. The climate of the upper montane zone, that chiefly concerns us here, is more or less continuously cloudy in some parts and sunny in others.

The vegetation of the cloudy upper montane zone is open and diminutive and becomes extremely sparse or absent on the windy summit flats. *Luzula confusa, Poa glauca, Saxifraga rivularis,* and *S. stellaris comosa* occur. The common lichen is *Solorina crocea.* The vegetation of sunny areas extends from 2620 down to 1970 feet (800 to 600 m). The dwarf *Cassiope tetragona* is the chief shrub, and crustaceous lichens are common. Below 1970 feet down to 1640 feet (500 m), there is much open *Cassiope* mat with scattered *Dryas integrifolia* and *Rhododendron lapponicum.* The herbs are typically xerophytic, especially on talus knolls. Common mosses are *Rhacomitrium lanuginosum* and *Polytrichum alpinum,* but these are less conspicuous than the numerous lichens, especially *Alectoria ochroleuca* and *Sphaerophorus melanocarpus.* The muskox and caribou are known to enter the highlands of north Greenland.

In Spitsbergen, Summerhayes and Elton (1928) have pointed out that the ivory gull, great bearded seal, and polar bear are regular residents of ice floes. The bears feed on seals and the gulls on carcasses left by the bears. Red-colored snow results from the growth of the alga *Sphaerella nivalis* over the snow or ice surface. The springtail *Agrenia bidenticulata* and the oligochaete worm *Lumbricillus aegialites* occur in snow and ice.

In Labrador and Newfoundland there are relatively large areas above 2000 feet (600 m) and some peaks rise to between 5000 and 6000 feet (1500 and 1800 m). Ekblaw (1926) described typical heath–lichen tundra with their typical animals on most of these uplands. Mount Albert in the Gaspé Peninsula (Fig. 5-6), lacks the mat of roots and twigs which characterize the tundra climax, but some tundra species occur (Dansereau 1948). Near the top of Mount Katahdin, bare rock prevails, but where there is adequate soil, *Vaccinium uliginosum, V. vitis-idaea minus, Empetrum nigrum,* and *Ledum groenlandicum* are abundant. An equal number of nonarctic species are present. The animals are largely characteristic of more southerly areas. There are probably no resident large mammals, but the woodland caribou enters high-lands north of 45° N. Lat. in certain seasons (Blake 1926). On Mount Marcy, the alpine vegetation is the dwarf birch scrub *Betula glandulosa siberica,* along with the nonarctic grasses and sedges *Deschampsia flexuosa* and *Carex oligosperma* (Adams *et al.* 1920).

WESTERN ALASKA AND ALEUTIAN TUNDRA

Some of the tundra near Kotzebu Sound and northeast of Nome is similar to that of the Arctic Coast, but between Norton Sound and Bristol Bay and for some distance inland, the tundra and the climate are different. The annual precipitation varies from 30 inches (75 cm) in the south to 15 inches (3.75 cm) in the north, and mean annual temperature from 33° to 25° F. (0.6° to –3.9° C.) Both precipitation and temperature are higher in this Bering Sea tundra than in the continental tundra.

Willow and birch shrub covers much of the area, with the willow being 5 to 15 feet (1.5 to 4.5 m) tall. Hummocks between the willows support tall grasses and a good series of widely distributed arctic heath plants. Important constituents of the birch shrub, mostly 2 to 12 inches (5 to 30 cm) high, are *Betula glandulosa siberica, Salix pulchra,* and *Carex cryptocarpa* above a lichen layer of *Cladonia, Arctostaphylos alpina,* and *Ledum palustre decumbens.* Faciations of the shrub vegetation merge into white spruce forest (Hanson 1950, 1951, 1953).

Snails, slugs, and centipedes, which are almost absent from continental tundra, occur here. Bumblebees are plentiful, and flies which resemble them are numerous about flowering *Epilobium spicatum, Iris, Aconitum, Delphinium,* Leguminosae, *Campanula* and *Pedicularis,* but willow catkins are preferred. The large flowers of *Heracleum maximum* attract tipulid flies and sawflies. Barren ground caribou, the grizzly bear *Ursus kidderi,* and wolves occupy this area (Johansen 1921).

The vegetation of Popof in the Shumagin Island group of the Aleutians appears to be scrub and grass. Black crowberry, a circumboreal species, is almost the only tundra plant (Kincaid *et al.* 1910). The giant grizzly bear *Ursus gyas* occupies Unimak Island and the Alaska Peninsula. The western end of Kodiak Island is dominated by grasses and sedges (Hanson 1953). The giant Kodiak bear *Ursus middendorffi* is confined to this and adjacent islands.

ARCTIC SLOPE OF THE BROOKS RANGE AND ANAKTUVUK PASS

Forests are mostly absent from the arctic slope of the Brooks Range (Brooks 1906). Some snow fields persist all summer. Below 2000 feet (600 m) and down to the coastal plain the area is covered with niggerhead meadow. A "niggerhead" is a tussock of tough grass, in this case the cottongrass *Euphorium vaginatum pissum,* that elongates each growing season until it may be several feet above ground. This type of growth is related to the frequent freezing of the soil (Hopkins and Sigafoos 1951). Associated species, also in close niggerhead stands, are *Arctogrostis latifolia, Betula michauxii, Empetrum nigrum, Entreme edwardsii, Poa arctica,* and *Rubus chamaemorus.* Bog meadow occurs with *Carex* and heath plants. In upland meadows, *Dryas* is most important with *Arenaria capillaris* and *Festuca brachyphylla.* The floor of Anaktuvuk Pass, which connects the arctic slope with the central interior, is covered with the sedges *Carex bigelowii* and *C. aquatilis* and cottongrass. In drier areas are the heath plants *Cassiope tetragona, Ledum palustre decumbens, Rhododendron lapponicum, Empetrum nigrum,* and *Vaccinium vitis-idaea.* The lichen *Therefon* forms communities with moss (Rausch 1951).

Stone's caribou migrates north into the mountains, beginning in January, continuing through June, and returning from July to October. The species is most abundant in the White Mountains between the Yukon and Tanana rivers. Lichens, especially *Cladonia* and *Cetraria,* grasses, and sedges are important in their diet, and willow, birch, and *Vaccinium* are also utilized (Murie 1935). The wolf was originally the caribou's worst enemy. The Dall's sheep (Murie 1944), arctic hare, and snowshoe rabbit are present.

Both the tundra wolf and wolverine occur in the Brooks Range and elsewhere. The arctic fox is of irregular occurrence on the north slope, appearing in years following lemming abundance and very rarely on the south slope. The Alaska red fox preys on marmots. The Arctic grizzly bear *Ursus richardsoni* comes into the mountains to den in October, emerging in April. The food of this bear includes the vole *Microtus miurus,* ground squirrels, and lemmings, as well as *Equisetum,* sedges, cranberry, the hydric lichen *Therefon,* and roots of *Hedysarum* that they can dig out of loose soil.

The hoary marmot dens for the winter in September and comes out in June. *Saxafraga bronchialis, Pyrola grandiflora,* and lupine are its principal food, and it also eats the lichen *Therofron.* The location of these dens is indicated by the growth of an orange lichen in the nitrogenous soil around the entrances. Parry's ground squirrel, which eats a large amount of willow, commonly places its dens in *Cassiope*–lichen communities. The brown lemming feeds on grasses and sedges. The collared lemming feeds on *Dryas, Ledum,* and other vegetation. The red-backed mouse *Clethrionomys rutilus* is present up to 3000 or 3500 feet (900 or 1000 m). It selects areas covered with moss or *Cassiope.* The tundra vole is fond of sedge areas. The singing vole has been trapped everywhere except in *Cassiope*–lichen communities. There are also two shrews in the area (Rausch 1951).

HIGH ALTITUDE TUNDRA OF ALASKA AND ROCKY MOUNTAINS

Climate and Vegetation

The treeline varies from about 3500 feet (1000 m) in the extreme north to 5000 feet (1500 m) in the extreme south of Alaska and Yukon. On Mount McKinley in Alaska, the annual precipitation is about 15 inches (37.5 cm). The soil is frozen most of the year and solifluction terraces are common on north-facing slopes at an elevation of about 3900 feet (1180 m). A soft, carpet-like growth of moss, about an inch thick, often forms a ground layer, and other vegetation seldom extends over 10 inches (25 cm) high. The chief species are *Carex bigelowii, Dryas octopetala, D. alaskensis, Festuca altaica, Vaccinium vitis-idaea, V. uliginosum, Betula glandulosa siberica,* and *Empetrum nigrum. Hedysarum alpinum* and *Kobresia myosuriodes* are secondary dominants. *Oxytropis viscidula* and *Equisetum* are common.

There is a large area of alpine tundra in the Mackenzie Mountains of Yukon. Some alpine tundra of the eastern slope has contact with continental tundra. Alpine lichen–heath occurs on the more or less level, gravelly benches at 5500 feet (1667 m) which receive a heavy snow cover in winter. The heath is composed predominantly of *Stereocaulon paschale, Luzula arcuata, Salix pseudopolaris,* and *Cassiope tetragona saximontana* (Porsild 1951). Similar slopes facing east and south up to 6000 feet (1820 m) receive even more snow, but owing to the exposure, the snow melts earlier than on north slopes and vegetation completely covers the ground. The predominant plants are *Polytrichum,*

Cladonia alpestris, Salix reticulata, Betula glandulosa, Cassiope tetragona saximontana, and *Artemisia arctica; Salix arctica* and *S. pulchra* form small thickets three to five feet (1 to 1.5 m) high, occupying slight depressions in the terrain. Rocky ledge: and cliffs give a stony substratum, and only a few widely distributed species are present. On dry, windswept areas at 6500 feet (2000 m), lichen–heath occurs in saddles between peaks and consists of *Alectoria ochroleuca, Cetraria islandica, Luzula arcuata, L. confusa, Salix pseudo-polaris, Artemisia arctica,* and *Senecio yukonensis* (Porsild 1951).

East of Lapie River in Yukon, moist alpine tundra is characteristic of muskeg-covered mountain slopes Above timberline is a mossy alpine heath that contains the grasses *Arctogrostis latifolia* and *Hierochloe alpina,* the shrubs *Salix reticulata* and *Betula glandulosa,* and the forb *Polygonum bistorta plumosum.* Dry slopes and ledges have a number of species that obviously are growing outside their usual habitat: *Cassiope tetragona, Rhododendron lapponicum, Lupinus arcticus, Arctostaphylos rubra, Ledum palustre decumbens, Vaccinium uliginosum,* and *V. vitis-idaea minus* (Porsild 1951).

Major Influents

Caribou are among the important influents of the high arctic tundra, although they are commonly not present there in winter (Fig. 7-14). A favorite food of Stone's caribou on Mount McKinley is lichens, especially *Cladonia.* Grasses, sedges and willows are eaten in both winter and summer and *Hedy-*

Fig. 7-14. General view above timberline in the British Columbia Cassiar District east of Dease Lake, 50° N. Lat., 130° W. Long., with a caribou near the center of the photograph (from Bulletin 18, Bureau of Public Information, British Columbia, 1912).

sarum, a legume, in summer. Between 20,000 and 30,000 caribou are in Mount McKinley National Park. The calves are born out on the tundra rather than in the stream-skirting forest. Osborn's caribou in Yukon appears to have similar food, migration habits, and animal associates.

Grizzly bears are not abundant in the high tundra. Plant communities become modified, however, by the bear's digging, and the bears are instrumental in distributing seeds of the plants which they eat. Their food includes blueberries, soapberry, redberries, crowberry, and mountain cranberry. The bear also eats quantities of grass, especially *Calamagrostis canadensis*, horsetail, sorrel, spruce cones, and the sage *Artemisia ludoviciana*. Animal food includes marmot, caribou, Dall's sheep, ground squirrel, mice, ptarmigan, and a few insects. The bear exerts little effect on the populations of sheep and caribou. It dens from October to April, and, when active, its home range is about 10 miles (16 km) in diameter (Murie 1944).

There have been estimates of 5000 to 10,000 Dall's sheep in McKinley Park. When at a maximum, they come down from the rougher parts of the mountains and overbrowse willows, apparently causing them to die. Grasses and sedges are most frequently eaten, with horsetail and willow close behind. Sage, cranberry, crowberry, *Dryas*, *Hedysarum*, and *Oxytropis* are also important but lichens are not. Porsild (1951) finds *Equisetum arvense*, which has starchy tubers which make it an important sheep food, in moist spots up to 6000 feet (1800 m). The lambs are born in May (Murie 1944).

Mountain goats do not occur in Mount McKinley Park but are present in the highlands of southern Alaska and northwest British Columbia. They live in the roughest part of the high mountains. Little is known concerning their food (Fig. 8-7).

Fig. 7-15. A female willow ptarmigan from the mountains of British Columbia in breeding plumage (courtesy National Museum of Canada).

According to Murie (1944), there are between 40 and 60 wolves in the rough sheep country of Mount McKinley National Park. There were larger populations about 1880 and 1904 and small populations from 1916 to 1927. Mange and distemper contribute to their mortality. The aggregate of ground squirrels, mice, and marmots that they take for food equals that of sheep that they kill. Caribou is the largest item in their diet, but with 30,000 caribou and 10,000 sheep present, both with a high rate of reproduction, one would assume that the caribou and sheep would survive.

Minor Influents

A considerable number of small mammals are noted in connection with the food coactions of the major influents in Alaska. In general, these mammals also occur in Yukon. Of the birds, the rock ptarmigan is found at higher elevations than the willow ptarmigan (Fig. 7-15). The horned lark and Lapland longspur are also present (Rand 1946). No reptiles are to be expected and no amphibians were found.

There are very few records of invertebrates in high tundra. Winds and currents often carry insects up from lower elevations so that the mere presence of an insect in the high tundra does not indicate that it belongs there. The lepidopterons *Colias meadii, Boloria alberta,* and *Orenaia trivialis* are apparently high tundra species. Some of the common mosquitoes appear about a month later in the tundra than at lower altitudes (Frohne 1957).

Chapter **8**

The Northern Pacific Coast Rainy Western Hemlock Forest Biome and Mountain Communities

RAINY WESTERN HEMLOCK FOREST BIOME

The *hemlock–wapiti–deer–redcedar–Sitka spruce biome* lies adjacent to the Pacific Coast and extends from the middle of California to southern Alaska, about 2300 miles (3680 km) (Fig. 8-1). Its width varies from a few miles at the extreme north and extreme south to nearly 200 miles (320 km) near its center at the United States–Canadian boundary; for over 1500 miles (2400 km) it reaches about 100 miles (160 km) inland. The biome lies below 5000 feet (1500 m) in Oregon to below 2000 feet (600 m) in Alaska.

The mature dominant trees are very tall, 125 to 200 feet (38 to 60 m) high and 5 to 15 feet (1.5 to 4.5 m) in diameter; in the case of the redwood, 300 feet (90 m) or more high and 10 to 20 feet (3 to 6 m) in diameter. They form a canopy which makes a deep shade over 89 to 90 per cent of the ground. Secondary trees and shrubs find growth impossible except in canopy openings. In the mature forest, shrub and herb layers are poorly developed and consist of relatively few species. The layer of duff and organic soil is deep. The blue grouse and Sitka spruce occur scatteredly throughout. About A.D. 1550, when white man first appeared in the area, there was a large population of natives. They were salmon eaters and associated primarily with the streams. They did no extensive damage to the forest.

The biome may be subdivided into the hemlock–cedar forest on the western slope of the Cascades and throughout the coast ranges from northwestern California to 51° N. Lat. in British Columbia; the redwood forests, on various slopes and in pockets near the coast from near Monterey Bay, California, to southwestern Oregon, increasing in area northward; and the hemlock–spruce forest on the Pacific slope of the coastal mountains from 51° N. Lat. to Kodiak Island, Alaska.

211

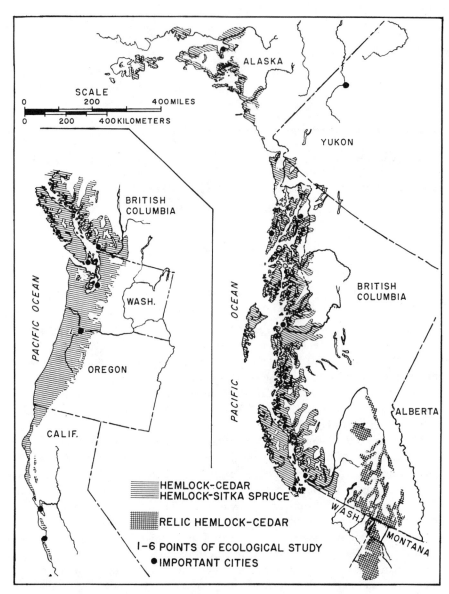

Fig. 8-1. Distribution of the hemlock–cedar, hemlock–Sitka spruce (after Shantz and Zon 1924; Halliday 1937), and the relict hemlock–cedar (called larch–pine by Larsen 1930; Alaska, Brooks 1906 and various other sources).

Climate

Climate is characterized by an annual rainfall of 50 to 100 inches (125 to 250 cm), coming chiefly in winter. Less than 20 per cent of the rain falls in the period April through September, and 0.0 to 2 per cent of it comes in July and August. Summer drought is a more critical factor than the rain that comes in the winter. North to about 51° N. Lat. near the north end of Vancouver Island, the driest month shifts from August to June, and this is accompanied by a change in the biota. On the Alaskan coast to Kodiak Island, June gets about 3 per cent of the rain with the heaviest precipitation coming in September; total rainfall is above 130 inches (325 cm). Rainfall is commonly lower on the coast than a short distance inland.

The frost-free period varies from 120 to 210 days. Hours of sunshine are low, and only about 50 per cent of possible hours of sunshine occur. Mean annual temperatures range from about 40° F. (4.4° C.) in Alaska to 56° F. (13° C.) in Oregon.

Community Constituents and Relations to Other Biomes

The western hemlock (Fig. 8-2) and western redcedar are the two outstanding plant dominants where the forest is best developed south of 51° N. Lat. The two species are most important in a narrow belt from 2500 to 5500 feet (760 to 1670 m). The grand fir occurs from sea level to 5000 feet (1500 m), the Pacific silver fir from sea level to 6000 feet (1800 m), and noble fir from 2000 to 5000 feet (600 to 1500 m). All have their greatest abundance and frequency in the Puget Sound region. The Douglas-fir ceases to be important farther north. The Sitka spruce is present throughout the length of the forest but is confined near the coast, notably in Washington. North of 51° N. Lat. due to a narrowing of the climatic area, hemlock and Sitka spruce are the predominant species in the vegetation (Clements 1920, Caverhill 1926,

Fig. 8-2. Old hemlocks in a row. They started growth on the trunk of a fallen tree in Olympic National Park along the Ho River trail (Department of the Interior, photo by George Grant, 1934).

Jones 1947). Black cottonwood and red alder occur throughout the biome.

This coastal forest biotic community is regarded as an entity separate from the transcontinental–montane forest. It has a distinct and unusual climate from sea level to the mountain divide, where there is also a sharp change in the forest character. For example, the western larch and western white pine are only on east-facing slopes in the Puget Sound area but occur generally with Douglas-fir in the Rocky Mountains east of Idaho (Fig. 6-2) (Larsen 1930). Moose, caribou, and mountain sheep do not occur regularly in the rainy forest (Osgood 1926). The lynx and fisher occur only as stragglers, and there is a distinct subspecies of marten limited to rainy coastal forest. Moose were reported to have come down the Skeena River in British Columbia in 1925, following fires and clearing, but apparently could not maintain themselves in the wet forest (R. W. Pilsbury, personal communication). The coyote and perhaps the red fox have invaded western Washington only as lumbering and clearing proceeded (Taylor and Shaw 1929). Early students of zoogeography recognized portions of the forest region as being distinct in their faunal maps (Kendeigh 1954). Agassiz's map in 1854 covers nearly all of it. Maps by J. A. Allen, dated 1892, recognize the northern and southern faciations. Drude's map in 1887 stops near the southern boundary of Oregon. Merriam's 1890 map is essentially correct. These maps are based on the presence of particular species and the absence of those found in adjacent regions. The dominant and influent animals of the area are related, however, to Rocky Mountain species.

The area of changes in the plant community, noted near 51° N. Lat., is near the northern distribution limit of the Roosevelt elk or wapiti. The mule deer is recognized as the Sitka subspecies from this area northward. The two subdivisions of the biome may be recognized as the hemlock–wapiti–cedar biotic association in the south and the hemlock–deer–Sitka spruce biotic association in the north. The wet belt plant association in southeastern British Columbia and Idaho is an ecotone between the several coniferous types of the coastal and Rocky Mountain regions.

The characteristic dense undergrowth which occurs throughout much of the biome consists largely of the salal, which grows to a height of 6 feet (1.8 m), vine maple up to 20 feet (6 m), salmon-berry 9 feet (2.7 m), and devil's club (Munger *et al.* 1926). Additional shrubs are Oregon grape, red bilberry, evergreen huckleberry, cream bush, *Salix scouleriana, Acer glabrum,* blackberry, stink currant, rusty leaf, twinflower, waxberry, and *Rhododendron macrophyllum* (Clements 1920, Dirks-Edmonds 1947). These shrubs often grow so profusely as to make travel on foot difficult.

HEMLOCK–REDCEDAR–WAPITI ASSOCIATION

Faciations

There are local and vertical variations in the percentage composition of trees making up the forest, which give a number of faciations. This has been shown for Vancouver Island (Caverhill 1926). From sea level to 2000 or 3000 feet (600 to 900 m), Douglas-fir constitutes 70 per cent of the trees; western red-cedar, 17; western hemlock, 6; Pacific silver fir, 2; western white pine, Sitka spruce, and others, 5. From 2000 to 4000 feet (600 to 1200 m) western red-cedar makes up 60 per cent; western hemlock, 22; Pacific silver fir, 11; Sitka spruce, 4; and all others, 3. Above about 4000 feet and locally at Quatsino Sound as low as 1700 feet (500 m), western hemlock is 60 per cent and Pacific silver fir, 39.

Mammal Influents and Dominants

The wapiti or Roosevelt elk was originally one of the most influent animals in the pristine forest. Lewis and Clark found them near the mouth of the Columbia River, and when pursued, the wapiti ran out of the woods in the same manner as the mule deer does. They are in part grazing animals but also browse on a great variety of bushes, leaves, twigs, branches of trees, and, to some extent, on tree lichens. They are especially fond of devil's club, rasp-berry, salmon-berry, willow, blueberry, vine maple, cherry, *Ceanothus, Holodiscus,* and wild rose. A great variety of herbaceous plants, including pea vines and clovers, are eaten as well as the rich mountain grasses. Deep snow within their range is of rare occurrence and of short duration, and they are able to withstand the winters (Bailey 1936).

The Columbian mule deer browses to a large extent and exacts a considerable toll from the vegetation, especially ferns, the young growth of shrubs, and herbs (Fig. 8-3). They are fond of the leaves, buds, and seeds of many species of *Ceanothus,* commonly called buckbrush, also of the evergreen barberry, willow, mountain mahogany, waxberry, blueberry, raspberry, salmon-berry, salal, and rose. There was no trace of grass in the stomachs of a considerable number examined by Bailey (1936). On Vancouver Island, Cowan (1945) concluded that the food potential in the climax cedar–hemlock forest was so low that it could support only one deer per square mile (2.6/km²). Most of the deer were in secondary communities where they had populations of 30 to 35 per acre (75 to 88/hectare). Trampling constitutes the chief effect of the mule deer upon the ground layer (Dirks-Edmunds 1947). When abundant, they make many trails through the forest and "beds" where they rest or bear their young.

The mountain lion affects the community primarily through its predation upon the mule deer. The bobcat also preys occasionally upon young deer. However, this animal is primarily a hunter of small game, including snow-shoe hares, birds, and mice. The wolves *Canis lupus fuscus,* south of 51° N. Lat., and *C. l. columbianus,* north of 51° N. Lat., are generally distributed through the pristine forest. They feed in large part on deer and wapiti but also utilize small animals when they are easier to obtain.

The black bear is only seasonally active within the forest, but they tear decaying logs and snags apart in search for food and consume huckleberries and other berries of the undergrowth (Dirks-Edmunds 1947). Many plant and animal foods are taken, including fish and the inner bark of pine, spruce, and hemlock belonging to forest climax and late subclimax. Bark injury and girdling, especially of young Douglas-fir, may be locally serious. The Klamath grizzly bear was originally fairly numerous on both sides of the mountains in Oregon and even more omnivorous than the black bear (Bailey 1936).

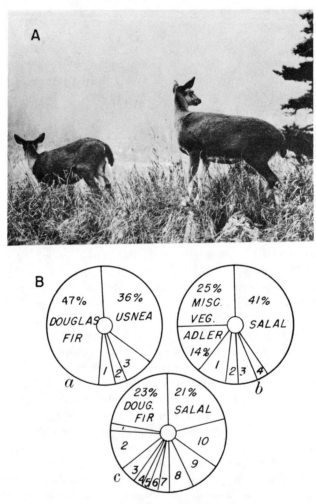

Fig. 8-3. A. Columbian mule deer standing in the grass *Bromus sitkensis* (E. Andrew). B. Food of the Columbian mule deer in southeastern Vancouver Island (from Cowan 1945): a, winter food, (1) miscellaneous vegetation, 5 per cent, (2) mushrooms, 4, (3) salal, 8; b, summer food, (1) black cap, 9, (2) bracken, 4, (3) willow, 6, (4) grass and sedge, 4; c, annual food, (1) cedar, 1, (2) miscellaneous vegetation, 11, (3) mushrooms, 5, (4) *Equisetum,* 2, (5) grass and sedge, 2, (6) thimbleberry, 2, (7) bracken, 3, (8) willow, 6, (9) alder, 10, (10) *Usnea,* 14 (courtesy Ecological Monographs).

Arthropod Dominants and Influents

Some of the insect tree destroyers are common also in the transcontinental–montane forest, but in a few cases they are subspecies of the transcontinental species. The hemlock looper *Ellopia fiscellaria lugubrasa* is a defoliator and tree killer of the first magnitude (G. J. Spencer, personal communication). This species destroyed considerable areas of hemlock and Douglas-fir forest near Grays Harbor, Washington, between 1928 and 1934. The black-headed budworm is almost as destructive to hemlock and attacks spruce and fir. Many Sitka spruce are killed by the Sitka spruce beetles *Dendroctonus obesus*. The Sitka spruce aphid *Neomyzaphis abientina* killed trees on Queen Charlotte Island and near Prince Rupert (G. J. Spencer, personal communication). Because of their effects, these four insects may be called dominants.

The hemlock sawfly *Neodiprion tsugae* is a serious influent, while the broad-headed borer *Trachykele blondeli* attacks the western redcedar, the larvae mining the trunk of living trees only. The bark beetle *Phloeosinus* bores into the bark. The redcedar is, however, relatively free from pests (Keen 1938). Other insect species are somtimes important (Fig. 8-4).

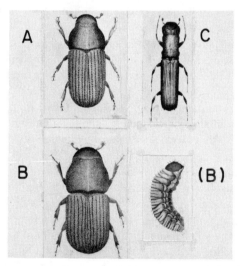

Fig. 8-4. A. *Dendroctonus pseudotsugae,* adult, an arthropod dominant in seral stages generally associated with Douglas-fir. B. *D. monticola* destroys lodgepole pine and western white pine but is most important in ponderosa pine. C. *Platypus wilsoni* bores deep into wood of dying conifers except pines and cedars (A-C after Swaine 1918).

Minor Influents

The red-tailed hawk, screech owl, and great horned owl feed extensively on mice and other small animals of the forest floor as well as on birds and an occasional insect from the higher strata. The great horned owl is especially noted as a hunter, its prey including rabbits, weasels, and skunks (Bent 1938).

The sharp-shinned and Cooper's hawks prey chiefly on birds, including wood-peckers and screech owls (Bent 1937).

The snowshoe rabbit is active the year round and is restricted to a vegetarian diet. When herbaceous plants are absent, they feed extensively upon the shrubs. They do not turn white in winter. The Pacific marten is common throughout the heavier forests of the Oregon region (Bailey 1936). It preys extensively on wood rats, squirrels, other mammals, and birds. The food of the western spotted skunk consists of insects, small rodents, small birds, amphibians, small reptiles, and occasional berries (Bailey 1936).

The diurnal pigmy owl in the Douglas-fir–hemlock community stays within about 30 feet (10 m) of the ground. This small owl has been known to kill chipmunks as well as mice, birds, and reptiles (Bent 1938). The pileated wood-pecker is active chiefly in the tree layer and feeds extensively on beetles and other insects (Bent 1939). Blue grouse scratches in the humus and soil among Oregon grape and salal bushes. It feeds also on young conifer needles, buds, and berries in the winter (Gabrielson and Jewett 1940). It is very abundant in British Columbia where it shows a preference for openings (Cowan, per-sonal communication). The band-tailed pigeon occurs as a seed and fruit-eating summer resident. The raven takes its toll of small birds, nestlings, and eggs, along with insects and carrion.

The deer mouse is the most common small mammal. Deer mice stay on the ground and lower strata, although they are good climbers. Their nests are built chiefly in and under old logs and stumps and around the bases of trees. They eat many kinds of seeds as well as numerous insects, insect larvae and eggs, berries, and some green foliage (Bailey 1936). In studies determining the effect of forest animals on seeds and seedlings on cut-over Douglas-fir land, Moore (1940) found that deer mice consumed more conifer seeds than any other group. He discovered also that they showed a decided preference for hemlock and Douglas-fir seeds. Hofmann (1920) states that the seeds buried by mice will germinate even after several years. After a fire, these seeds may be the means of regenerating the next forest of Douglas-fir. Seed crops of this species fluctuate from year to year and during years when the crop is light, the seeds may be entirely destroyed by the hymenopteron *Megastigmus spermo-trophus,* even before they are ripe.

The Douglas squirrel is abundant in most parts of the forest. Bailey (1936) states that they cut and store immature cones of Douglas-fir, Sitka spruce, hemlock, and pine. The bushy-tailed wood rat is occasionally observed. These animals are herbivorous, so far as is known, and they feed in the shrub and herb layers, making their homes mainly on the ground. Townsend's chipmunk lives in underground burrows and feeds extensively on roots, insects, and other animal life taken from that layer. Its diet also includes various berries and roots of shrubs.

The coast mole constructs subterranean burrows and heaps up mounds of soil. It takes insects, insect larvae, and other invertebrates for food. The little insectivore, Trowbridge's shrew, has been observed to feed readily upon

Douglas-fir seed. The Oregon creeping vole and the western red-backed mouse apparently do not occur in great abundance. So far as is known, both species are confined to the ground layer and are mainly herbivorous (Bailey 1936, Dirks-Edmunds 1947).

One of the interesting and characteristic birds of the community is the nomadic red crossbill which comes and goes rather irregularly; its characteristic noisy call is heard as it feeds on the conifer seeds in the tops of the trees, sometimes moving lower in summer. These birds are usually observed in flocks of 4 to 50 or more individuals. The pine siskin resembles the crossbill in feeding habits, but is observed less frequently (Table 8-1).

Table 8-1. *The Total Number of Individuals of the More Abundant Birds on One Hectare of the Douglas-Fir–Hemlock Associes*[a]

SPECIES	1935	1936	1937	1938
Gray jay	0	0	3	5
Steller's jay	1	0	4	6
Common raven	3	4	1	3
Chestnut-backed chickadee	4	12	2	18
Red-breasted nuthatch	2	5	4	6
Brown creeper	2	1	4	4
Winter wren	2	4	4	2
Varied thrush	4	3	8	2
Hermit warbler	5	..	7	4
Red crossbill	2	24	22	83
Total	25	53	59	133

[a] Observation dates, June 25, July 20, and August 21 each year (after Dirks-Edmunds 1947).

The western wood pewee and the hermit warbler are two insectivorous summer resident species of the tree layer. The hairy woodpecker, red-breasted nuthatch, and brown creeper are three permanent resident species that are almost exclusively insectivorous and appear to confine their activities to the trunks and larger branches of the trees. The omnivorous gray and Steller's jays frequent all layers in the community. The insectivorous chestnut-backed chickadee and the golden-crowned kinglet have been observed most frequently in young hemlocks. Two species that confine their activities chiefly to the lower layers of the forest are the varied thrush and the Swainson's thrush. The winter wren appears to be the only ground nesting bird (Dirks-Edmunds 1947).

Comparison of Faciations

QUINALT NATURAL AREA

The natural area in the Olympic National Forest near Quinalt Lake, Washington, at an altitude of 1500 feet (450 m), is at least 60 per cent cedar and 35 per cent western hemlock. The chief shrubs are salal and tall blue huckleberry. The very sparse herbs and low vegetation include western skunk cabbage, sword fern, and scattered representatives of several other genera. The ground was littered with fallen branches and trunks mainly of Douglas-fir when visited in 1945. A carpet of the mosses *Polytrichum piliferum, Funaria hygrometrica,* and *Ceratodon purpureus* (Jones 1947) covered everything on and near the ground.

Wapiti signs were abundant in 1945, especially tracks in the moss cover. The moss and decomposing trees were inhabited by the millipede *Harpaphe haydeniana* and young parajulids. Centipedes present were *Arctogeophilus melanonotus* and juvenile *Otocryptops;* only one small snail was found. Spiders were represented by *Hexura picea* and *Antrodiaetus hageni.* A large brown silverfish was among the most numerous insects. The camel cricket *Pristoceuthophilus* was less abundant than the ground beetle *Scaphinotus angusticollis velutinus.* Click beetles were conspicuous, including *Ctenicera protracta.* Larvae of click beetles, May beetles, and flies occurred in small numbers. Altogether the population of the ground stratum was much smaller than in deciduous forest at the same season.

The population of the herb and shrub levels was also low, only 4 to 10 per cent of the normal deciduous forest population. The Sitka bumblebee *Bombus sitkensis,* the yellow jacket *Vespula arenaria,* the larvavorid fly *Ursophyte migriceps,* and the boring beetle *Pidonia gnathoides* were conspicuous among a larger number of very small flies and spiders.

WIND RIVER NATURAL AREA

An area of climax in the Pinchot National Forest, Washington, contained 98 per cent western hemlock and 2 per cent cedar. On animal paths in the woods, the ant *Formica rufa melanotica* was the commonest insect, and tiger beetle larvae were present. The western toad *Bufo boreas* and the tailed frog *Ascaphus truei* were taken. Tadpoles of the latter clung to rocks in swift water. A deer mouse was trapped and the tenebrionid beetle *Iphthimus serratus* was much in evidence.

REDWOOD AREAS

Relatively small areas of redwood forests were visited in southwestern Oregon. Atmospheric humidity is high, and fogs are frequent and dense in this climax. The typical coastal forest is composed of Sitka spruce and grand fir on seaward slopes, redwood in the belt of heavy fog, and an occasional Douglas-fir. Tanoak, California-laurel, Pacific madrone, and alders are commonly mixed with the conifers; oaks in the areas of coastal forest tend to form distinct patches of oak woodland. The forest understory includes salal, vine maple, *Rhododendron,* and *Vaccinium.* The pileated woodpecker, olive-sided flycatcher, gray jay, chestnut-backed chickadee, brown creeper, hermit warbler, and red crossbill are listed by Miller (1951) as belonging to the coast forest.

The low broad-leaved shrub layer included seedlings of understory trees, bigleaf maple seedlings, Christmas fern, bracken, old man's beard, California blackberry, and hazel. The most characteristic shrub-nesting birds were the winter wren, varied thrush, hermit thrush, and golden-crowned kinglet. The shrubs were inhabited by a large population of spiders: *Linyphia pusilla, Anyphaena aperta, Theridion californicum,* and *Hyptiotes gertschi.* The only species taken on the low conifers were *Dipoena, Metellina curtisi,* and *Clubiona.* Larvae of Lepidoptera were numerous. The fly *Hylemya alacthoe* was conspicuous along with a few cantharid beetles *Malthodes.*

Herbs were scattered. The slim solomon and redwood sorrel were inhabited by the spider *Metellina curtisi,* the fly *Hylema alacthoe,* the beetle *Malthodes,* the sawfly larva Tenthrediniae, and the parasitic hymenopteron *Catastenus.*

The important ground nesting bird was the blue grouse. The ground and fallen tree trunks were inhabited by the ground beetle *Pterostichus vicinus,* beetle larvae, the camel cricket *Neduba,* the millipedes *Harpaphe haydeniana* and *Kepolydesmus,* and the centipede *Otocryptops sexspinosus.* The snails *Vespericola columbiana, Glyptostoma newberryanum,* and *Monadenia infumata* were present. Where some Douglas-fir was present and there was evidence of past fire, the invertebrates were somewhat different. *Diaea pictilis* was added to the spider list from the shrubs. The ground and fallen tree trunks were inhabited by *Pterostichus (Leptoferonia) humilis,* and the ants *Lasius niger* and *Phenolepis imparis. Zootermopsis,* a very large dampwood termite, was taken. A mollusk, not in the climax, was *Helminthoglypta tudiculata.* A single scorpion, *Uroctonus mordax,* was found.

Douglas-Fir–Western Hemlock Subclimax Community

On Saddle Mountain, a short distance west of McMinnville, Oregon, a representative hectare contained 75 Douglas-firs, height 200 to 250 feet (60 to 75 m), diameter 4 to 6 feet (1.2 to 1.8 m); 80 large hemlocks, height 130 to 175 feet (40 to 53 m), diameter 2 to 3 feet (0.6 to 0.9 m), their tops reaching the lower branches in the Douglas-fir crowns; and 2 noble firs, height 175 and 200 feet (53 and 60 m), diameter 4 feet (1.2 m) (Fig. 8-5) (Dirks-Edmunds 1947).

TREE TOP LAYER

On warm, sunny days numerous insects rise from the lower vegetation and fly among the crowns of the highest trees. Additional species from the high forest layers join them. These crowns are 80 to 135 feet (24 to 40 m) above the ground. The largest white pine butterfly *Neophasia menapia* is one of these species during August. Its eggs are attached to Douglas-fir needles near the tops of the trees; the larvae feed on the needles and lower themselves to the lower vegetation by silken threads in order to pupate. Also in the canopy are clubionid spiders, the mite *Bdella oblonga,* the collembolon *Entomobrya clitellaria,* predaceous larvae of serpent flies, psocids, the cercopid *Philaenus leucophthalmus,* the hemipteron *Catonia nemoralis,* the dipterons *Sciara* and *Mycetaulus,* the chalcids *Tetrastichus* and *Tridymus,* and the ichneumonid *Aplomerus.*

The dusky tree mouse *Phenacomys silvicola* is a canopy inhabitant, feeding on needles and twigs of the conifers in which it makes its nest. It seldom or never comes to the ground.

YOUNG HEMLOCK LAYER

There were 530 young hemlocks on a hectare, ranging in height from about two feet (0.6 m) to the lower crown of the large hemlocks. When clusters of trees reach 10 to 20 feet (3 to 6 m) in height, they produce a dense shade and eliminate lower layers.

More than ten species of spiders have been identified from this layer; of these *Linyphia diana, Hyptiotes gertschi, Meta curtisi,* and *Theridion sexpunctatum* occur in greatest numbers. Other families represented are the Agelenidae, Amaurobiidae, and Thomisidae. Oribateid beetle mites and predaceous representatives of the genus *Erythraeus* occur. In late summer the black mite *Platyliodes hoodi* is on the branches of the young hemlocks in great numbers.

Curculionid weevils, especially *Dyslobus lecontei, Nemocestes incomptus,* and *Sciopithes obscurus,* occur more frequently than any other beetles. Their larvae feed on roots; the adults principally upon buds, fruits, and foliage of

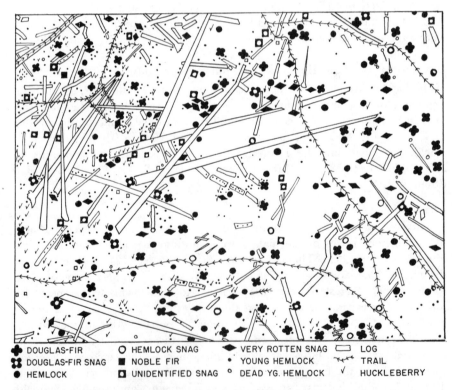

♣ DOUGLAS-FIR	○ HEMLOCK SNAG	◆ VERY ROTTEN SNAG	▭ LOG
♣♣ DOUGLAS-FIR SNAG	■ NOBLE FIR	• YOUNG HEMLOCK	⌇⌇ TRAIL
● HEMLOCK	▢ UNIDENTIFIED SNAG	° DEAD YG. HEMLOCK	∨ HUCKLEBERRY

Fig. 8-5. The condition of the forest floor on 0.75 hectare of Saddle Mountain, Oregon (1938). There were 62 hemlock and 53 Douglas-fir trees in the area, suggesting a large mortality in the development to the stage shown (after Dirks-Edmunds 1947).

shrubs and trees. Only one destructive bark beetle, *Pseudohylesinus,* was taken from the young hemlock layer. Gall midges, gnats, crane flies, the western syrphid *Syrphus opinator,* and adults of the Pacific brown lacewing *Hemerobius pacificus* are present; the lacewings feed on mites and aphids (Essig 1934). Smoke fly larvae feed on injurious caterpillars and stages of woodboring insects, while the bark louse *Peripsocus quadrifasciatus* feeds on the needles or bark of young hemlocks.

LOW SHRUB LAYER

The low shrubs consisted of widely scattered aggregations of the deciduous blue and red huckleberry alternating with the evergreen Oregon grape, salal, and sword fern. Arachnids constituted the largest single group of invertebrates, averaging 39 per cent of the total population during the summer, with a maximum of 54 per cent in August. The most abundant beetle, the cantharid *Podabrus piniphilus,* is reputed to feed extensively on aphids. The dipterons, including the mosquitoes *Aedes varipalpus* and *Theobaldia incidens,* are numerous during late summer and autumn. Tabinid deer flies and leptid snipe flies suck the blood of mammals. The big-eyed fly *Dorilas* is a parasitoid on leafhoppers (Essig 1934).

HERB LAYER

Vanilla-leaf, redwood sorrel, woodrush, and *Clintonia uniflora* are the common herbs during the aestival and serotinal seasons. The small pale green leafhopper *Empoasca* is a common insect among the herbs. Homopterons, chiefly aphids, are more abundant. Dipterons constitute approximately 20 per cent of the invertebrates, of which the brown snipe fly *Rhagio costalis* is characteristic.

GROUND LAYER

The forest floor was carpeted with mats of several species of mosses where dense growths of young hemlocks, shrubs, or herbs did not inhibit their growth. The two most common mosses were *Rhytidiadelphus loreus* and *Hylocomium splendens.* The mat averaged approximately six centimeters in thickness (Fig. 8-5). Beneath the mosses, there was a layer of duff or humus two or three centimeters in thickness. Hemlocks were growing commonly on fallen tree trunks (Fig. 8-2).

The large green slug *Agriolimax columbianus* and the large black-spotted green slug *A. c. maculatus* are conspicuous and active on cool, damp days during the summer and are a characteristic feature of the wet forest. The common snails are *Monadenia fidelis, Polygyra columbiana pilosa,* and the snail-eating *Haplotrema sportella* and *H. vancouverensis.* The snail-eating carabid *Scaphinotus angusticollis* was active every summer day. Millipedes, centipedes, mites, spiders, and a few phalangids constitute about 42 per cent of the invertebrate population. The carpenter ant *Camponotus herculeanus modoc,* that is found on dead trees, is the only common ant. Enchytraeidae or white earthworms average about 6 per cent of the invertebrate population.

During the dry season red or lumbricid earthworms are in the deeper soil and seldom occur above a depth of 50 centimeters; *Megascolides americanus* and *Plutellus* are known to occur (Macnab and Fender 1947). The Pacific giant salamander *Dicamptodon ensatus,* the salamander *Ensatina eschscholtzii,* and the western red-backed salamander *Plethodon vehiculum* occur under logs.

Aspection or Biological Changes with the Seasons (Fig. 8-6)

WINTER (HIEMAL) SEASON, NOVEMBER 1 THROUGH FEBRUARY 28

Weather data supplied by James A. Macnab for 1934 and 1937 indicate that minimum temperatures drop below the freezing point and snow often covers the ground. Maximum temperatures may approach 50° F. (10° C.). Humidity is high and precipitation is the heaviest for the year. Deciduous foliage is gone and herbs are dormant. Permanent resident birds are in flocks. Bears and chipmunks are in hibernation, but gnats, winter crane flies, a few spiders, and chipmunks become active during warm days.

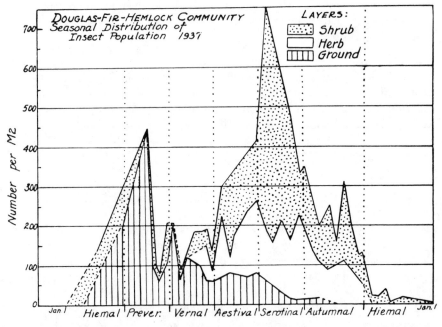

Fig. 8-6. Seasonal distribution of insects on Saddle Mountain, Oregon, in 1937 (data by J. A. Macnab, after Dirks-Edmunds 1947, courtesy Ecological Monographs).

EARLY SPRING (PREVERNAL) SEASON, MARCH 1 THROUGH APRIL 15

Minimum temperatures average above freezing, the mean approximately 40° F. (4.4° C.), and the maximum above 50° F. (10° C.). Precipitation decreases, but relative humidity attains its annual maximum; snow is frequent. Light intensity is low. Plants become greener, buds swell, and herbs sprout. The mosquito *Theobaldia incidens,* a few flies, moths, hymenopterons, and beetles are added to the insect population; jays, pileated woodpeckers, and juncos return to the area; chipmunks, bears, and amphibians become active.

SPRING (VERNAL) SEASON, APRIL 16 THROUGH MAY 31

The mean seasonal temperature rises above the annual mean of 50° F. (10° C.); temperature, precipitation, and humidity fluctuate widely as the general weather changes from cold to warm and from wet to dry. The herbs sorrel, yellow violet, and trillium and the shrubby huckleberry bloom and later also false Solomon's seal, vanilla-leaf, twayblade, Oregon grape, and devil's club. Herbs carpet the forest floor, shrubs leaf out, and coniferous trees produce new growth. Dipterons, coleopterons, and hymenopterons become very numerous, as do snails, slugs, and amphibians. Summer resident birds arrive and nest; mammals rear their young.

EARLY SUMMER (AESTIVAL) SEASON, JUNE 1 THROUGH JULY 15

Temperatures reach a maximum, precipitation is at its lowest point, humidity fluctuates widely, and light intensity reaches its maximum. Early weeks are marked by blossoms of the herb *Clintonia* and the shrub salal; the late weeks produce mature *Clintonia* and false Solomon's seal fruits. Dipterons, especially blood-sucking and predaceous types, are most numerous; amphibians are less active; chipmunks aestivate; and summer resident birds begin to depart. Table 8-1 shows that the summer bird population increased during the years of study.

LATE SUMMER (SEROTINAL) SEASON, JULY 16 THROUGH AUGUST 31

Temperatures reach the annual maximum and autumn rains begin. The last fruits of the herb and shrub layer mature and herbaceous foliage becomes yellow and ragged. White pine butterflies, hemlock looper moths, yellow jackets, wasps, and lacewings decline in numbers. Flocks of permanent resident and migrant birds comprise the avian population; mammals and amphibians become more active again.

AUTUMN (AUTUMNAL) SEASON, SEPTEMBER 1 THROUGH OCTOBER 31

Heavy fall rains alternate with "Indian summer" weather and frosts; maximum and minimum temperatures drop, the range of humidity decreases; and light intensity reaches its winter level. Mushrooms make their appearance; deciduous shrubs shed their leaves; and the herb layer becomes brown with decay. Winter crane flies, fungus gnats, and mycetophagous beetles appear; amphibians attain a second peak of abundance; the bird population is stable with only migrant flocks of juncos; hoarding activities of chipmunks, mountain beaver, and wood rats take place. Mating of deer and the prehibernating wandering of bears take place (Fig. 8-6) (Dirks-Edmunds 1947).

This aspection schedule represents only the bare outline of what takes place; a different way of dividing the annual cycle and much more detail of the biotic events in the forest of the Coast ranges have been worked out by Macnab (1958).

Coastal Promontories and Succession

GRASSY HEADLANDS

Headlands between fjord-like indentations tend to be dry, and their outer ends are covered with grass; the same is true of the small islands east of Vancouver Island. The chief grasses at one time were *Panicum pacificum* and *Festuca occidentalis,* but these are now largely displaced by *Bromus.* The communities of these most exposed areas are quite permanent and may perhaps be considered climax.

PIONEER STAGE ON ROCKS

In less exposed areas, the vegetation consists of two strata: a low herb stratum composed largely of grass which grows four or five inches (10 to 12.5 cm) high and has sheep sorrel scattered through it and a moss stratum composed of *Rhacomitrium canescens* that becomes thoroughly dry by the middle of June and has *Polytricum juniperinum* as its only important associate. The depth of the moss layer varies but becomes two or three inches (5 to 7.5 cm) deep in pockets of the rock. An inch or so of yellowish-brown soil accumulates in these pockets. Patches of lichen-covered rock, which moss has not yet invaded, are frequent (Table 8-2).

Table 8-2. Percentage Distribution of Individuals of Representative Abundant Species in Different Seral Stages

STATIONS	ROCK	GRASS	FOREST EDGE	DRY FOREST	MESIC FOREST
Machilis (thysanuron).............	87	13			
Short-horned grasshopper nymphs....	55	44	1		
Plagionathus obscurus (bug)..........		62	38		
Dicyphus agilis (bug)..............			55	11	34
Coenosia nivea (fly)................			17	22	61
Ischnorhynchus geminatus (bug)......				100	0
Mycomya littoralis (fly).............					100
Heleomyza fuscicornis (fly)..........					100

The abundant animal of the rocks and moss is the swift-moving, half-inch long thysanuron *Machilis* (Table 8-2). During a cloudy period from June 26 to July 5, 1928, red mites, carabid larvae, and collembolons, which are probably hidden in crevices during dry weather, were taken in the moss.

In the grass, small grasshopper nymphs are at times abundant; crab spiders *Latrodectus,* a few flies, the ant *Leptothorax texanus,* leafhoppers, and pentatomid nymphs in small numbers may be present. A small mite is sometimes abundant on the seeds of grasses.

FOREST-EDGE STAGE

The forest-edge may be viewed from the outside as a series of steps: ground, grass, shrubs, and trees; Douglas-fir and Rocky Mountain juniper are the tree dominants. The rose *Rosa nutkana,* the snowberry *Symphoricarpos racemosus,* and the gooseberry *Ribes divaricatum* make up the shrubs. The grass grows in patches alternating with the trees or shrubs but not under them. Brome grass and tufted hairgrass are the principal species. Forbs are yarrow, wild onion, chickweed, white hawkweed, tarweed, cat's-ear, tea vine, vetch, self-heal, and geranium.

The influent animals are Townsend's vole, deer mouse, screech owl, great horned owl, California quail, and the ants *Formica rufa obscuripes* and *Lasius niger americanus.* Animals present during the aestival season but possibly not at other times are chipping sparrow, white-crowned sparrow, short-horned grasshopper nymphs, the phylloxeron *Chermes* on fir needles, the beetle *Phalacrus ovalis,* the hemipteron *Plagionathus obscurus fraternus,* flies, a large variety of insects, and the sowbug.

DRY CONIFEROUS FOREST

Dry coniferous forest occurs on well-drained slopes. It is composed mostly of Douglas-firs of small diameters with 1 tree in 25 a lodgepole pine. The principal shrub is the ocean spray, which reaches a height of 15 feet (4.5 m), and the principal herb is the grass *Festuca subuliflora.* The forest floor may be slightly cluttered with fallen branches and trees, but the moss layer is deep. A large number of mordellid beetles may be found feeding on the blossoms of ocean spray. The grass was heavily infested in 1928 with the aphid *Macrosiphum granarium.* The spider *Linyphia,* incorporating as many as a dozen grass stalks in its web, was also conspicuous.

MESIC FOREST

The mesic forest is also dominated by Douglas-fir and lodgepole pine. Subordinate plants are the rose *Rosa gymnocarpa,* grand fir, western hemlock, ocean spray, and gooseberry. Herbs were bracken, a sword fern, salal, Oregon grape, Miner's lettuce, and trailing blackberry. The Columbia mule deer and raccoon were originally present. Mice and shrews still are. Perennial invertebrates are the springtail *Tomocerus flavescens americanus* and the slug *Agriolimax columbianus.*

Community Development

LAKE BOG SERE

Successions from open lake to mature bog are to be found on the Olympic Peninsula of Washington. The marginal flora of *Nymphaea polysepala, Typha latifolia, Menyanthes trifoliata,* and *Potentilla palustris,* and locally *Comarum palustre,* is followed by the sphagnum stage which floats on a soft and partially decayed and disintegrated organic matter. On the layer of sphagnum is developed the characteristic herbaceous flora of the bog, including *Carex, Eriophorum chamissonis, Sanguisorba micrecephala, Menyanthes crista-galli,* and *Trientalis arctica*. The chief woody plants in bogs are the swamp laurel *Kalmia polifolia*, Labrador-tea, sweet gale, Douglas spiraea, and the crab apple *Pyrus rivularis*. *Drosera rotundifolia* and *Vaccinium palustre* are characteristic of the more mature or older stage in bog development throughout the Pacific Coast.

Late stages of bogs are invaded first by western hemlock, then by western redcedar, lodgepole pine, and in coastal bogs by Sitka spruce, and finally but sparingly by Douglas-fir (Rigg 1925, Rigg and Richardson 1938). Trees grow very slowly in sphagnum, perhaps due to the toxicity of the substratum (Rigg 1919). Deciduous trees and shrubs are rare, except for red alder, cascara buckthorn, and western dogwood (Jones 1947).

PACIFIC COAST AND DUNES

The resemblance of dunes at Tahkenitch Lake, near Reedsport, Oregon, to dune communities about Lake Michigan is striking. The bare sand of blowouts is invaded by *Poa macrantha*, *Poa confinis*, and *Ammophila arenaria*. Grasses cover about 60 per cent of the area studied. The sand surface is traversed by the ants *Formica neogagates lasioides vetula* and *F. fusca subaenescens* and immature sand-colored spiders, totaling about 60 per square meter. The coast tiger beetle *Cicindela bellissima* is present in small numbers. The scarabaeid beetle *Polyphylla decimlineata* was dug from the sand. On the grasses are the green leafhopper *Laevicephalis siskiyou*, beetles, the tenebrionid *Coelus ciliatus*, the snout beetle *Trigonoscuta pilosa*, some melyids, the bee *Psithyrus variabilis*, and the same ants as on the ground.

A later shrub stage includes bearberry, hairy manzanita, bracken fern, bunches of the grass *Festuca rubra*, some *Polygonum paronychia*, and the strawberry *Fragaria chilensis*. About 25 per cent of the surface is still bare sand. The surface and top two inches (5 cm) of bare sand have few animal inhabitants, but in and under the vegetation the insect population is twice as large. Click beetle larvae, ants of the same species as in the earlier stage, three types of spiders, and the tenebrionid beetles *Eleodes* and *Coelus ciliatus* are present.

The young forest stage that follows the shrubs includes lodgepole pine, Douglas-fir, Sitka spruce, and hemlock. The taller undergrowth includes *Rhododendron macrophyllum*, salal, hairy manzanita, and the evergreen huckleberry *Vaccinium ovatum*, with much of the lower vegetation of the preceding stage in the more open spots. There are many lichens and mosses and about six centimeters of needles over the surface of the sand. Numerous ants, including *Formica fusca neorufibarbis*, *F. fusca subaenescens*, and *Camponotus sansabeanus vicinus* are in and on the fallen tree trunks. The termite *Zootermopsis angusticolles* and the yellow-margined millipede *Harpaphe* are also there. On both low and high undergrowth the insect and spider populations are small, including the click beetles *Ctenicera bombycinus* and *Athous pallidipennis*, a robber fly, some small flies, and the neuropteron *Agulla*.

HEMLOCK–SITKA DEER–SITKA SPRUCE ASSOCIATION

The center of the *hemlock–Sitka deer–Sitka spruce association* is on the Queen Charlotte Islands where Sitka spruce attains its maximum growth (Caverhill 1926). Northward, it extends on the islands and mainland to west of Kodiak Island, Alaska. It occurs in small coastal areas as far south as Oregon, where the subordinate plants and animals are similar to those of the Douglas-fir–hemlock community. Western hemlock commonly represents 25 to 40 per cent of the stands and Sitka spruce about 30 per cent. Associated species are western redcedar, Alaska-cedar, and grand fir. Grand fir is absent on the Queen Charlotte Islands (Caverhill 1926).

Temperature averages 5° F. (2.8° C.) colder in British Columbia than in the hemlock–redcedar association, and rainfall varies from 45 to 120 inches (112 to 300 cm). On the Queen Charlotte Islands, rainfall averages 52 inches (130 cm).

Major Permeant Influents

The Sitka mule deer (Fig. 8-7) is the most generally distributed of the influent mammals, although absent from some of the islands. The wapiti does not occur north of about 51° N. Lat. Swarth (1911) notes that deer are most abundant in some areas of southeastern Alaska. There are black bears and brown bears in the lowland forest, and they do much bark injury. The large Alaskan timber wolf occurs on Mount McKinley and along the coast, but its population has been reduced by trapping and poison.

Arthropod Dominants and Minor Influents

The green spruce aphid *Aphis abietina* kills spruce in the vicinity of Queen Charlotte Island (G. J. Spencer, personal communication). The hemlock looper and black-headed budworm act as dominants just as they do farther south.

The squirrel *Tamiasciurus hudsonicus* replaces *T. douglasii* of Oregon but retains its cone-gathering habits and characteristics. The red-tailed hawk, bald eagle, and great horned owl are important predators. The porcupine is common and does much damage by peeling trees. The smaller birds and mammals are mainly the same species as those in the cedar–hemlock association (Swarth 1911).

Community Development

Cooper (1923, 1939) has described the plant communities in certain areas of Glacier Bay, Alaska. About half of the species also occur on the Olympic Peninsula.

PIONEER STAGE

The first community on the areas most recently vacated by the glacier is characterized by a lichen, two xerophytic mosses, two species of *Equisetum*, the two grasses *Poa alpina* and *Trisetum spicatum*, saxifrage, the eyebright *Eu-*

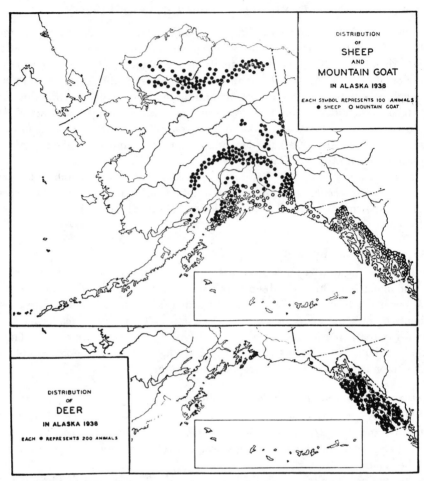

Fig. 8-7. The distribution and population of sheep and mountain goats in Alaska in 1936 (above); distribution and population of mule deer in Alaska in 1936 (Fish and Wildlife Service, Circular 3).

phrasia, Alpine pink, *Dryas,* three prostrate and six erect willows, bearberry, buffaloberry, and cottonwood. All occur in scattered areas and in a bewildering pattern (Figs. 8-8, 8-9).

WILLOW–ALDER STAGE

The thinleaf alder is dominant nearly everywhere in mature thickets. Its height averages 20 feet (6 m). Two willows of similar stature, *Salix alaxensis* and *S. sitchensis,* are also abundant, the latter species being the more important. The black cottonwood is commonly present. It is not numerous in individuals but is conspicuous because of its greater height.

The undergrowth of the alder thicket of Glacier Bay includes such mesophytic species as the ferns *Aspidium spinulosum, Polystichum braunii,* and *Asplenium*

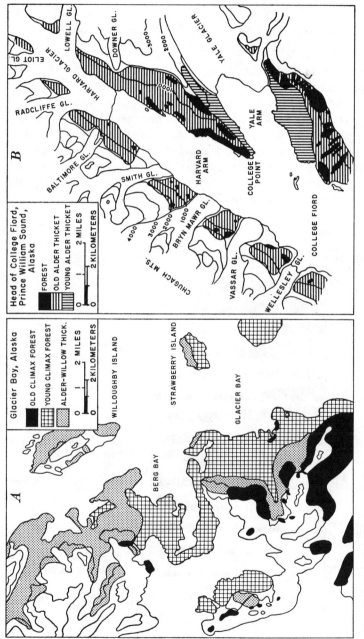

Fig. 8-8. The distribution of forest in the prince Williams Sound and Glacier Bay areas, Alaska. A. Inlets to the north from Port Well, a western segment of Prince William Sound near 148° W. Long. and 61° N. Lat. The black areas are forests of western hemlock, mountain hemlock, and Sitka spruce (after Cooper 1942). B. Distribution of forest about Glacier Bay showing relative age of climax (after Cooper 1923).

filixfoemina; the ground pine *Lycopodium selago;* the coral-root *Corallorrhiza mertensiana;* the colt's foot *Petasites frigida;* and the mosses *Rhytidiadelphus triquetrus* and *R. squarrosus.* The alder thickets become overtopped first by an occasional young spruce, then, as the spruces become more numerous and larger in size, merge insensibly into the conifer forest (Fig. 8-9).

LATE SUBCLIMAX AND CLIMAX FOREST

The coniferous forest which occupies the older portions of the ice-free region and adjacent islands is a nearly pure stand of Sitka spruce averaging 50 to 60 feet (15 to 18 m) in height. The symmetry of the trees, with leafy branches beginning near the ground, and the absence of standing dead trunks indicate that the forest is young. The oldest tree in a number that was examined was only 71 years of age. Occasional slender, narrow-topped cottonwoods and infrequent western hemlock and mountain hemlock accompany the spruces. Suppressed alders and the willows sprawl beneath the trees in more

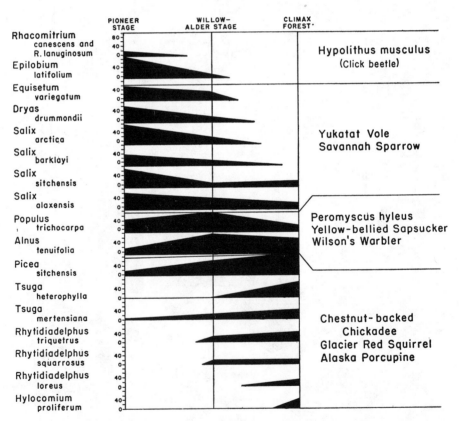

Fig. 8-9. Succession to the Sitka spruce–hemlock climax in southeastern Alaska. Percentage of frequency of important species of plants is shown for three stages (Cooper 1923). Associated animals are added as indicated by Grinnell (1909), Heller (1909, 1910), and Kincaid *et al,* (1910).

open places. Important shrubs include the salmon-berry, Pacific red elder, devil's club, and blue currant. The herbs are those of the Alaska lowland forest; the ferns *Aspidium spinulosum, Polystichum braunii,* and *Asplenium filixfoemina;* twayblade; baneberry; shinleaf; and one-flowered shinleaf. A rich moss carpet, covering the gravel and boulders many inches deep, consists of *Hylocomium proliferum,* three species of *Rhytidiadelphus, Mnium insigne, Dicranum rugosum, Drepanocladus aduncus,* and *Brachythecium albicans occidentale.* The mosses *Ulota crispa, Antitrichia curtipendula gigantes,* and *Plagiothecium piliferum* are partial to trunks and branches, especially of sickly alders and willows. This forest differs from the average coastal climax of southeastern Alaska only in the youthfulness of the spruces and the small proportion of hemlock (Fig. 8-9).

The climax undergrowth on Sitka Island, Alaska, includes Miner's lettuce, the miterwort *Tiarella trifoliata,* the yellow violet *Viola glabella,* and *Osmorhiza nuda.* The insect population of the undergrowth is especially rich in dipterons. The most abundant flies are *Bibio vestitus* and *Dilophus serraticollis.* The latter species occurs on salmon-berry blossoms when it is associated with nine species of syrphus flies, including *Cheilosia plutonia* and *Sphegina infuscata.* Fifteen species of empids or dance flies and eleven species of anthomyiid flies, including *Lonchaea albitarsis* and *L. deutchi,* are present. The common white *Pieris* is the only butterfly found, but 17 species of moths, mainly Noctuidae and Geometridae, are present. Few beetles, few ichneumonids, and no fossorial hymenopterons have been collected. The wasp *Vespa adulterina actica* has been taken at the blossom of the false azalea *Menziesia ferruginea.* Fallen trees contain ant colonies of *Formica fusca neorufibarbis, Lasius niger sitkaensis,* and *Myrmica breornodis sulcinodoides.* The bumble bees *Bombus californicus* and *B. flavifrons* are present, along with the honey bee *Psithyrus insularis* on blossoms of the salmon-berry (Kincaid *et al.* 1910).

Seral Communities South of Glacier Bay, Alaska

ROCKY SOIL

Vegetation on rocky soil usually consists of thickets of red alder, Sitka alder, Oregon crab apple, red elder, and salmon-berry. These thickets are found along streams, on landslides, and along most beaches. The commonest summer birds are rufous hummingbird, western flycatcher, robin, hermit thrush, Swainson's thrush, orange-crowned warbler, yellow warbler, fox sparrow, and song sparrow (Webster 1950).

POND SERE

Around open water ponds are open muskegs with much peat moss and numerous small shallow pools. The salamander *Ambystoma gracile* occurs in southeastern Alaska and probably uses such waters. Small shrubs occur, such as the crowberry *Empetrum nigrum,* mountain-laurel, and cranberry. Next come individuals, then clumps, of lodgepole pines and in smaller numbers, scrubby Alaska-cedars. There are various blueberries and huckleberries, particularly *Vaccinium parvifolium, V. caespitosum, V. membranaceum,* and *V. ovalifolium.* Western hemlock comes in later, along with a heavy under-growth of salmon-berry, black twinberry, devil's club, false azalea, and the various blueberries (Webster 1950). Occasionally there are small, very wet glades grown up to grasses and sedges instead of shrubs or trees. Here Lin-coln's sparrows occur sparsely. The common summer birds of the shrubs and small trees are yellow-bellied sapsucker, chestnut-backed chickadee, orange-crowned warbler, and Oregon junco (Swarth 1911, 1922).

Southeastern Alaska River Valleys and Lowlands

Webster (1950) reports that along the lower Stikine and other rivers on the mainland of southern Alaska, there is an associes of the cottonwood *Populus trichocarpa,* willow, red alder, and devil's club, that extends almost con-tinuously along the banks and covers practically all the islands in the winding river channels. The willows support the abundant flies *Bibio vestitus* and *Boletina groenlandica* and numerous horntails. The sawflies *Acantholyda marginiventris, Dolerus sericeus,* and seven species of another genus form a most conspicuous feature of the insect fauna. Only three species of hemipterons were seen, *Irbisia sericans, Scolopostethus thompsoni,* and *Corisa praesta.* The homopterons were more numerous; the seven species of leafhoppers captured are of wide distribution through North America. Ten species of arachnids and the two myriapods *Litiulus alaskanus* and *Geophilus alaskanus* were also taken (Kincaid *et al.* 1910).

More advanced in development are the dense stands of large cottonwoods, 40 to 60 feet (12 to 18 m) tall, with a heavy underbrush of willow, devil's club, and red alder, with a small Sitka spruce here and there. The common summer birds are ruffed grouse, hermit thrush, Swainson's thrush, robin, yel-low warbler, warbling vireo, and song sparrow. Two of these species, ruffed grouse and warbling vireo, occur in southeastern Alaska only in this forest. The western tanager and American redstart are limited to one or two of the river valleys. Hairy woodpecker, rufous hummingbird, golden-crowned kinglet, and hermit thrush nest in the climax vegetation. Most of the species also occur in the cedar–hemlock association. The small mammals of the cedar–hemlock community are represented here mainly by different subspecies (Webster 1950).

MOUNTAIN COMMUNITIES

The upper montane forest, immediately above the hemlock forest, is char-acterized by alpine hemlock, alpine fir, and lodgepole pine. The alpine meadow above the tree line is dominated by different species of *Carex, Poa* and *Agrostis* than occur in the Rocky Mountains. Mountain sheep are characteristic of high areas and goats of more rugged habitats without much regard to altitude (Dufresne 1942) (Fig. 8-9).

Alaska–British Columbia

The southeast Alaska mountain transect of Webster (1950) includes the alpine tundra (Arctic–Alpine Zone) above 2500 feet (760 m). The tundra area consists of islands on the higher peaks and on glacial strips that extend downward all the way to salt water in some cases. Away from glaciers, the tundra extends to the top of Mount Annahotz at 4700 feet (1425 m) on Baranof Island and to 15,300 feet (4636 m) on Mount Fairweather on the mainland. A permanent ice cap occupies all other altitudes above 5500 (1670 m) or 7000 feet (2120 m). The land above 2500 feet is never level; usually the slope is 45 degrees or more and the ridge tops are knife-edged, for this zone is above the rounded slopes which were overridden by Pleistocene glaciers (Webster 1950).

Most of the alpine vegetation is sedges and grasses. The low shrubs are willow, dwarf birch, the heathers *Cassiope mertensiana, Harrimanella stellariana,* and *Phyllodoce glanduliflora,* and alpine hemlock. Breeding birds are the rock ptarmigan, willow ptarmigan, white-tailed ptarmigan (only on the mainland), water pipit, gray-crowned rosy finch, and savannah sparrow. Mammals of wide distribution in southeastern Alaska and largely confined to this area are the collared pika, hoary marmot, and mountain goat (Webster 1950).

The upper montane forest (Hudsonian Zone) extends from 1500 to 2500 feet (455 to 760 m) on the western side of Baranof and Chichagof islands, but the proximity of glaciers and the direction of slope exposure make these limits variable. On Wrangell Island, the lower boundary comes about 1800 feet (545 m) and the upper at 2800 feet (850 m). On the mainland near Juneau and Wrangell, away from the glaciers, the limits are at about 1600 feet (485 m) and 2800 feet (850 m).

The climax community is a forest of mountain hemlock and Alaska-cedar, ranging in height from 30 to 40 feet (9 to 12 m) at its lower edge to only scrubby bushes at its upper edge. There is a little prostrate juniper, Sitka mountain-ash, and, on the mainland only, subalpine fir and Yukon birch. Sitka alder, alpine bilberry, and copper bush cover extensive areas of steep slopes with a subclimax of dense brush. A subclimax on boggy soil includes a few lodgepole pines but more often the alpine bilberry is succeeded by mountain hemlock. The common breeding birds are willow ptarmigan (upper edge only), ruby-crowned kinglet, Wilson's warbler, pine grosbeak, and Oregon junco. The common redpoll and golden-crowned sparrow are sparse along the mainland. The semipalmated plover breeds only at sea level near mainland glaciers (Webster 1950).

Mount Rainier

Mount Rainier, in Washington, has been extensively studied. It is volcanic in origin. At one time a crater three miles (4.8 km) in diameter was blown off; subsequent cinder deposits formed a rounded dome with an altitude of

14,363 feet (4352 m). There are 45 square miles (116 km²) of glacier on the mountain.

Under pristine conditions, permeant birds and mammals of the mountain were wapiti, Columbia mule deer, mountain lion, and black bear. Their smaller associates were raccoon, mountain sheep, and the mountain goat. These species ranged through all the forest and seral stages. The golden eagle and saw-whet owl also ranged widely (Taylor 1922).

Stupendous amphitheaters have been formed on the mountain, principally by the action of the frost. Some plants and animals appear especially adapted to rock slides, such as the mountain-ash and especially the pika.

Several species of insects, chiefly thysanurons, are found on the surface of the glaciers, worms of the genus *Mesenchytraeus* occur plentifully on the lower parts of the glaciers, and colonies of the bacterium *Protococcus nivalis* form patches of red snow. Pea fern, stonecrop, Sitka alder, and *Collomia debilis* grow on moraines in intimate relation to glaciers (Taylor 1922).

ALPINE MEADOW

The alpine meadow begins at 6500 feet (1970 m) altitude. The principal sedges are *Carex nigricans, C. pyrenaica, C. nardina,* and *C. illota,* with *C. festivella, C. atrata,* and others more seral in distribution. The chief grasses are *Agrostis rossae, Poa gracillima, P. arctica, P. epilis,* and *P. pringlei* (Clements 1920). Woody plants present are dwarf or snowy willow and crowberry. Forbs are numerous, particularly *Arenaria nardifolia, Draba aureola,* and the two Indian paint brushes *Castilleja oreopala* and *C. rupicola.*

During the winter, rocky slopes above timberline are deeply covered with snow except where the wind exposes the rocky ridges. Alternate freezing and thawing often coats all exposed surfaces with ice which renders food unobtainable and life for birds or mammals impossible. This icing is characteristic of high mountains in this region of heavy winter rainfall. During summer, four birds, the horned lark, gray-crowned rosy finch, water pipit, and white-tailed ptarmigan, and one mammal, the mountain goat, occur (Taylor 1922).

SUBALPINE OR UPPER MONTANE MEADOWS

Yakima Park on the northeastern side of the mountain is a meadow covered with the grass *Festuca viridula,* sedges, the rush *Juncus drummondii,* and about 20 forbs, including Indian paint brush, phlox, and lupine. Small subalpine fir and whitebark pines are scattered at wide intervals.

The boreal spider *Tarentula aculeata* and a good series of insects, including the carpenter ant *Camponotus herculeanus modoc,* were present in July, 1945. Among the birds listed by Taylor (1922) for the treeline area are the rufous hummingbird, horned lark, western meadowlark, Brewer's blackbird, savanna sparrow, and robin. Meadow mammals included the coyote, heather vole, creeping vole, long-tailed vole, water vole, Rainier pocket gopher, and Pacific jumping mouse.

UPPER MONTANE FOREST

The upper montane forest extends from 6400 feet (1940 m) down to 4500 feet (1364 m). On Mount Rainier, this forest is broken by meadows and early seral stages. Mountain hemlock, subalpine fir, whitebark pine, and Alaska-cedar are the principal trees. Important shrubs were delicious huckleberry, small red huckleberry, ovate salal, and juniper. Noteworthy herbs are white heather, red heather, Jeffrey shooting-star, smooth larkspur, avalanche lily, and dog-tailed lily.

The invertebrates have been but little investigated. The ground surface, stones, and fallen trunks of scattered trees are inhabited by the wingless orthopteron *Cyphoderris monstrosa,* the large ground beetle *Carabus taedatus,* the ants *Formica fusca neorufibarbis* and *Lasius niger sitkaensis,* and the crane fly *Tipula pseudotruncornum.*

The common birds are Clark's nutcracker, Cassin's finch, chipping sparrow, orange-crowned warbler, mountain chickadee, Townsend's solitaire, hermit thrush, and mountain bluebird, many of which are locally permeant.

The mammals include the heather mouse, creeping vole, water vole, Rainier pocket gopher, Cascade hoary marmot, yellow-pine chipmunk, Cascade golden-mantled ground squirrel, and pika (Taylor 1922).

Chapter **9**

The Summer Drought
or Broad Sclerophyll–Grizzly Bear Community

Sclerophyll vegetation extends from central Oregon through California into northern Baja California (Fig. 9-1), although it is not climax in central Oregon. The vegetation may be either forest, woodland, or chaparral. Leaves are commonly thick, coreaceous, highly cutinized, and shiny. Compound and lobed leaves are scarce. Sclerophyllous leaf characters have been acquired by plants belonging to more than a dozen families (Cooper 1922). Only 18 per cent of the dominant species are deciduous. The adaptations and adjustments of animals do not appear as striking as those of plants. The canyon live oak, interior live oak, and two subspecies of the California striped skunk originally covered much of the sclerophyll area, and grizzly bears were characteristic largely to the exclusion of the black bear (Grinnell 1933, 1938). The area has considerable rainfall in winter and a long dry period in summer. A similar climate prevails around the Mediterranean Sea, and it is of interest that this region has supplied many plants that are now cultivated within the boundaries of the broad sclerophyll area.

In 1600, the native Indian population was high, ranging from 7.7 to more than 19 persons per 10 square miles (3 to 7.5/km²) near San Francisco Bay. Agriculture was little developed because the seasonal distribution of rainfall was unsuited for corn. Food was obtained from salmon, acorns, and bulbs. The use of acorns led naturally to protection of the plant cover. Woody plants were used less by the Indians for firewood and shelter than in many localities because of the mild climate. However, what vegetation remains at the present time is either second-growth or badly disturbed by burning and grazing resulting from white man's activities. Native California grasses have been almost entirely replaced by exotic species which better tolerate heavy grazing. Fires are frequent, and much oak forest has been removed for fire-

238

wood and lumber. The grizzly bear, kit fox, wolf, wapiti, and some small mammals have been extirpated. On the other hand, deer have increased enormously in chaparral, woodland, and cut-over forest (Storer 1932).

CLIMATE

Total annual precipitation is significant but even more important is the length of the nearly rainless summer season. Annual rainfall in the northern end of the Sacramento Valley ranges from 16 to 38 inches (40 to 95 cm). In the chaparral area of Lake County, California, the annual rainfall is 21.6 inches (54 cm), but from June through September rainfall amounts to only 0.53 inches (1.3 cm); at Red Bluff in Tehama County, the corresponding figures are 22.9 and 1.8 inches (57 and 4.4 cm).

In the northward extension into Oregon of a poorer type of sclerophyll vegetation, the average rainfall from June through September is less than 2.1 inches (5.25 cm). North of the Umpqua Mountains in Oregon, there is 3.0 inches (7.5 cm), or more, of rain during the summer, and conifers replace the sclerophyll vegetation wherever there has been disturbance. This suggests an upper limit of 2.0 inches (5 cm) of rainfall from June through September for the occurrence of sclerophyll vegetation.

The southern coast ranges, southern Sierra Nevada foothills, and the Cuyamaca Mountains in California are regions largely covered by chaparral and have an annual rainfall of 21.7 inches (54 cm). June, July, August, and September usually have from 0.5 to 1.5 inches (1.2 to 3.8 cm) with most of this coming in September (Cooper 1922). Clements (1920) states that sclerophyll vegetation occurs in protected situations with as little as 10 inches (25 cm) annual rainfall and may occur on exposed slopes in northern California with 50 inches (125 cm).

TYPES OF SCLEROPHYLL VEGETATION

Sclerophyll vegetation varies both in life form and in taxonomic composition. Sclerophyll forest, woodland, and chaparral merge with one another without segregation into distinct regions. These three vegetation types, however, need to be recognized and described.

Broad Sclerophyll Forest

Aggregations of larger oaks with a grass ground cover occur in favorable spots within areas of chaparral and woodland but are most important in the northern part of the area (Cooper 1922).

Woodland

Woodlands are scattered trees with some other type of vegetation dominating the ground surface between and beneath the trees. This low vegetation may be grass, chaparral, or sagebrush (Jensen 1947). Woodland–grass surrounds the great inland valley of California, extends farther northeast along the coast range, and occurs in a few scattered outlying areas.

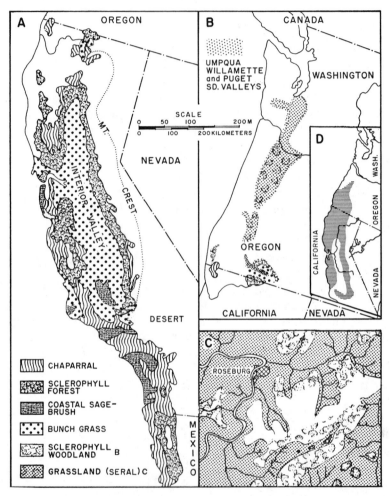

Fig. 9-1. A and B. Distribution of sclerophyll vegetation, sagebrush, and grassland in the valleys and coastal mountains (after Jenson 1947, Leopold *et al.* 1951, U.S. Forest Service maps of Oregon and Washington, and notes by several authors). C. Conditions around Roseburg, Oregon, in 1885 and 1935. In the survey of the 1880's the dotted area was called grazing land. Early photos of Mount Nebo near Roseburg show it entirely covered with grass, although now it is covered with oak–madrone scrub and Douglas-fir (J. A. Macnab, personal communication, 1951). The large high ridge in the 1880's was covered with woodland, but the blank area in 1935 was largely small Douglas-fir which is invading the woodland in the absence of excessive fire (T. T. Munger, personal communication, 1951). D. Distribution of the California yellow-legged frog *Rana b. boylei* (after Wright and Wright 1949, courtesy Comstock Publishing Company, Inc.).

Chaparral

Sclerophyll bush vegetation was formerly practically continuous near the coast from Baja California to the northern part of San Luis Obispo County of California; the strip was 50 to 60 miles (80 to 96 km) wide. It was interrupted by areas of coastal sagebrush which some writers regard as seral to chaparral. Still farther northward, as far as the Oregon border, chaparral occurred in scattered areas mixed chiefly with woodland-grass. It also occurred in discontinuous areas east of the interior valley of California.

PRINCIPAL CONSTITUENT SPECIES

Plants

Unity of the vegetation is indicated by woody species which occur as trees in woodland and forest but as shrubs in the chaparral. The canyon live oak and interior live oak are such species. A number of other oaks and three species of *Ceanothus* vary in form and stature in chaparral and forest but cover only a part of the area. The California coffeeberry or buckthorn behaves similarly but occurs also outside the area.

Animals

The major permeant dominant and influent animals were largely destroyed before their natural histories were recorded. It appears, however, that three species of grizzly bears, *Ursus californicus, U. tularensis,* and *U. colusus,* were important (Miller and Kellogg 1955). Some of the largest of these animals weigh 1100 pounds (4990 kg). The grizzly has broad "hands" and "fingers" terminated with long strong claws and is preeminently a digger. These same claws enable the bears to pluck fruits and nuts from trees and shrubs by the mouthful. The bears are essentially omnivorous, occasionally killing deer, wapiti, and pronghorn, and quite regularly securing and eating rodents and insects by digging. Roots and bulbs are utilized; one or two bears may dig over several acres of land, destroying all the ground vegetation (Grinnell 1938).

Herds of deer often range through sclerophyll forest, woodland, chaparral, and sagebrush. There are eight subspecies of the mule deer in and near the sclerophyll region (Cowan 1936); this is a species that ranges widely through the western mountains from southern Canada into Mexico. The population of deer in the less disturbed sclerophyll area varies from one to seven per square mile (0.4 to 2.7/km²) (McLean 1940).

In the Los Angeles National Forest the leaves of California scrub oak and Christmasberry make up as much as 95 per cent of the food of the deer during the summer and autumn. Chamise is utilized in spring (Cronemiller and Bartholomew 1950). A quantity of herbaceous material is eaten. Birchleaf mountain mahogany and hollyleaf cherry, both climax chaparral plants, are generally overbrowsed and not abundant, suggesting control by deer dominance.

The replacement of buckbrush by manzanita, due to the preference by deer for the buckbrush, is evidence of deer dominance. The deer is permeant, entering all successional stages and variations of the broad sclerophyll woodland and chaparral (Dasmann 1950, Leopold *et al.* 1951).

Deer were originally scarce in the Sierra Nevada Mountains (Leopold 1950a), being largely restricted to chaparral, woodland, and ponderosa pine which skirt the base of the mountains throughout about two-thirds the length of California. Following lumbering and fire in the mountain forests, however, woody shrubs and new tree reproduction afford deer unlimited food. A mule deer herd in Stanislaw National Forest now winters in large part on Jawbone Ridge, which extends southwest from the crest of the Sierras to Yosemite National Park (Jensen 1947, Leopold *et al.* 1951). The principal and favorite winter foods of deer in this area are buckbrush, 12 to 51 per cent, which is a regular chaparral constituent following fires; deer brush, which is important in ponderosa pine and chaparral; birchleaf mountain mahogany, a constituent of chaparral; mountain misery, 25 to 37 per cent, which is not usually listed as a chaparral plant; California black oak and other oaks, 13 to 14 per cent; and bearberry or manzanita, 11 to 13 per cent. The summer food is taken mainly from the higher altitudes. If the Jawbone Ridge deer originally stayed throughout the year at the lower altitude where it now spends the winter, half of the maximum population (7000) on twice the winter range area of 37 square miles (96/km²) would suggest 47 deer per square mile (18/km²) as the pristine number in the broad sclerophyll vegetation.

The mountain lion, according to Grinnell (1933), is primarily a constituent of the broad sclerophyll biotic communities. They are known to take deer. The California bobcat is most numerous in the foothill chaparral. They occasionally kill deer fawns. Wolves were originally present in sclerophyll vegetation but were not numerous. The coyote is still generally present and is an enemy of deer fawns. The golden eagle is also known to attack fawns. The striped skunk is generally distributed.

FOREST

Nowhere do the forests cover the landscape as a conifer forest commonly does. They occur, rather, in discontinuous patches alternating with woodland or chaparral. Species occur in a variety of combinations. Jensen (1947) recognizes tanoak–madrone, white oak–madrone, black oak–Oregon white oak, live oak, and blue oak–California white oak forests (Fig. 9-2).

Tanoak–Canyon Live Oak–Madrone Forest

The best examples of tanoak–canyon live oak–madrone forest are in Humboldt, Trinity, and Siskiyou counties in northwestern California. In addition to the three species named, California black oak, white or Garry oak, California-laurel, bigleaf maple, and a scattering of Douglas-fir occur (Cooper 1922). This forest merges into the coniferous biotic communities. It usually occupies south-facing slopes; north-facing slopes are frequently covered with a forest dominated by Douglas-fir, along with white fir and limber pine. An understory of broad-leaved sclerophyllous trees, in which *Castanopsis chrysophylla* is most abundant, is present; tanoak, madrone, and bigleaf maple also occur. *Corylus cornuta californica, Cornus nuttallii,* and *Ceanothus integerrimus* form a second lower stratum (Cooper 1922). There are serpentine soils in southeastern Oregon on which pines and cedars take the place of the understory sclerophyllous trees while the principal dominants remain the same but

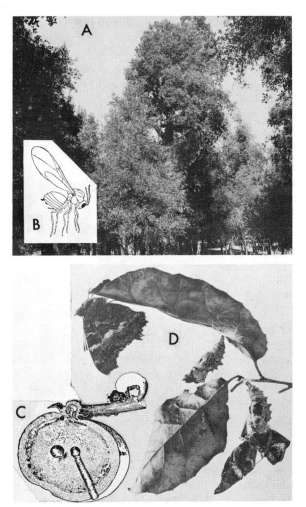

Fig. 9-2. A. Small stand of oak woods at about 3500 feet (1060 m) in the Descanso area. San Diego County, California, which is largely chaparral. B. Adult California gallfly *Andricus californicus,* which forms galls or oak apples on various oaks. C. Gall with larva cell and exit hole of the adult. D. California tortoise shell butterfly which feeds on various species of *Ceanothus* and sometimes on manzanita. (B, C, D from Essig's *Insects of Western North America,* courtesy of Macmillan Company).

dwarfed in size (Chap. 8) (Whittaker 1954). This forest is transitional to redwood forest. Cooper describes a forest of similar composition and relationships in the Sierras. Jensen's black oak–Oregon white oak forest appears to be a faciation of this forest southward.

Oregon White Oak–Madrone Forest

North of the Rogue River, Oregon, the tanoak and black oak become less important, leaving oak–madrone forest as an unstable faciation. Disturbances in the area have demonstrated that the climax is a coniferous forest (Munger, personal communication). Some of the hills in the Umpqua and Willamette river valleys have changed from oak–madrone forest interspersed with prairie (Lawrence 1926, Kuchler 1946) to Douglas-fir in the memory of the older residents (Fig. 9-1). There are small areas of oak-madrone forest near The Dalles in northern Oregon, east of the Cascade Mountains and below the ponderosa pine. In most areas north of Oregon, oak, madrone, and other broad-leaved trees are considered to be seral. In the Puget Sound area, oak–madrone alternates with lodgepole pine preceding other conifers in succession.

Lewis and Clark (1904-05) make frequent reference to oak and pine growing associated with prairie along the lower course of the Columbia River. Oak forest may typically occur between grassland and coniferous forest. It is popularly assumed that absence of the oak–madrone forest and presence of grass in the Willamette River Valley in Oregon is due to the prevalence of fires, resulting from the action first of aborigines and later of the early white settlers, who also removed the timber for other purposes.

A typical, probably second-growth, grove with 100 Garry oaks near McMinnville, Oregon, is on loose, dry soil overlaid with 21 centimeters of litter. The undergrowth consists of the hazel *Corylus cornuta*, 24 per hectare; *Rosa gymnocarpa*, 72 per hectare, a few black hawthorn, some blackberry *Rubus vitifolius*, and snowberry. The common herbs are cicely, rattlesnake weed, and Queen Anne's lace.

Lewis and Clark (1904-05) report deer as very abundant early in the nineteenth century; the wapiti, now extirpated, is mentioned less often. Signs of bear were seen. They state that mule deer prefer open ground and are seldom found in forest. When pursued, they flee to the hills. The wapiti behaves similarly. Their predators are the same as those that occur in the hemlock–cedar community.

The Garry oak is frequented by the western gray squirrel (Wight 1926). The bushy-tailed wood rat builds nests in shrub-covered stream margins and forest edges containing wild rose, spirea, and snowberry. Isolated thickets are inhabited by the brush rabbit, gray fox, and ruffed grouse.

A quantitative study of the grove described above gave 100 arthropods per square meter as compared with 1026 per square meter in Illinois deciduous forest at the same time. The Garry oak is subject to defoliation by the moth *Gracilaria pachardella;* ash on floodplains is defoliated by the noctuid *Homon-*

cocnemis fortis picina, and the red alder of floodplains is the preferred food plant of the western tent caterpillar *Malacosoma pluvialis.* Defoliations hinder reproduction and cause the death of some trees; this makes these insects important influents within the community.

California Live Oak–Madrone Forest

This forest extends on the outer coastal ranges in central California from the northern limit of live oak at 39° N. Lat. to the southern limit of madrone at 35° N. Lat. (Cooper 1922). It appears to be the southward extension of the tanoak–canyon live oak–madrone forest and occurs often on north-facing slopes. The deciduous California buckeye, California-laurel, and the deciduous bigleaf maple are frequently important, the latter especially in the more mesic localities.

California Live Oak–California White Oak Forest

The deciduous white oak and evergreen live oak are characteristic of the broad valleys and gentle foot slopes of the central coast ranges, notably in the San Francisco Bay region. The tall shrub layer includes coffeeberry, Christmasberry, blueberry elder, and the sumac *Rhus diversiloba.* The low-shrub layer includes *Rubus vitifolius, Symphoricarpos racemosus,* and *Solanum umbelliferum.* The ground is dominated by the trailing, evergreen, perennial yerba bucca. This forest merges into the woodland–grass community (Jensen 1947).

Only the mountain lion, black bear, and piñon mouse appear to show a preference for this forest (Longhurst 1940). The California oak moth *Phryganidia californica* defoliates large areas of live oaks (Fig. 9-3), but the white oak is less often damaged. The bark aphid *Symydobius agrifoliae* infests the twigs of live oak; the twig girdler *Agrilus angelicus* also appears to prefer live oak. The white oak is attacked by the oak bud aphid *Thelaxes californica,* while the live oak harbors the woolly oak aphid *Phyllaphis quercifoliae* which curls its leaves (Keen 1938, Essig 1934). In the litter and humus, arthropod populations are largest in March and smallest in autumn before the rains begin. Acarina, Corrodentia, Araneida, and Carabidae are perennial. Immature acarinons and collembolons begin to be abundant shortly after the autumn rains begin (Snell 1933).

Blue oak–white oak forest occurs as a variant of blue oak–white oak woodland with grass. There are doubtless several other local types of aggregated oaks which qualify as small areas of forest.

BLUE OAK WOODLAND WITH GRASS

Jensen (1947) listed woodland with chaparral beneath, woodland with sagebrush beneath, and woodland with grass. The last is widely distributed and important to industry. Various tree species are associated together in different localities, but only the predominantly blue oak communities have

Fig. 9-3. A. Oaks in a woodland grass area near Palo Alto, California, one defoliated, the other partially defoliated by the larvae of the California oak moth. B. An adult oak moth. C. Larvae killed by a wilt virus shown in D. E. Normal larvae. (A, B, C from Essig's *Insects of Western North America,* courtesy Macmillan Company; D by Dr. E. A. Stenhaus).

been investigated biotically. The associates of the blue oak in the Friant area of central California are the California white oak in moist areas, California buckeye and interior live oak on steep rough topography, and digger pine at the higher elevations (Klyver 1931, Hubbard 1941). The lower limit of digger pine agrees in this area with the lower limit of buckbrush, deer brush, and *Arctostaphylos mariposa*. The interior live oak is closely associated with rock outcrops (Talbot *et al.* 1942, soil map). The rocks have significance in controlling the occurrence and abundance of various animals. At its upper edge, the blue oak and digger pine form the pine–oak woodland recognized by Clements (1920).

The foothills in the central part of the Sierra Nevadas between Clavey River on the north and the San Joaquin River on the south have been studied extensively, as is evident in the biological transect of Grinnell and Storer (1924) (Figs. 6-9, 6-11), the forest transect of Hughes and Dunning (1949) (Fig. 6-9), studies in the San Joaquin Experimental Range by Talbot and Biswell (1942) and in the Friant Dam area by Hubbard (1941), and the vegetation transect of Klyver (1931) in the San Joaquin River region (Fig. 6-10). The exhaustive report on the San Joaquin Experimental Range, about 10 miles north of Friant, discusses the plants in regard to their utilization by cattle (Talbot and Biswell 1942). Coffeeberry was noted by these authors and by Klyver (1931). Their list of forbs of more than 30 species includes filaree, turkey mullein, tarweed, popcorn flower, red maids, and catch-fly as most numerous. Most of these are important in the food of the California ground squirrel. The grasses and grasslike plants include the common spike-rush, foxtail fescue, purple needlegrass, slender-leaf rush, and toad rush. Many grasses have been introduced from Europe, Australia, and elsewhere.

Major Permeant Influents and Dominants

The mountain lion, bear, and deer have evidently been eliminated from the San Joaquin Experimental Range, but all the other mammalian influents are still present. In the Jawbone area, deer are abundant in the woodlands during the winter. The golden eagle, mountain lion, black bear, and coyote are known to kill fawns. The winter population of the deer is often expressed as deer days per acre on the winter range. This ranges from 40 to 60 or more (100 to 150 deer days/hectare) in woodland–grass and woodland "brush" adjacent to Yosemite National Park (Leopold *et al.* 1951). This means that each acre had one deer on it 60 out of the 180 days of the winter period. In 1948, the mid-summer population of the Jawbone deer herd was estimated at nearly 8000. This was reduced to 6000 by autumn as the result of predation and hunting and to 3600 by spring 1949. A large summer increase resulted in a population of 4700 in the autumn of 1949. An annual increase of 32 per cent appears necessary to enable this herd to maintain itself, although elsewhere an increase of 25 per cent is usually sufficient.

The food of coyotes consists of insects, 8 per cent; birds, 2; reptiles, 4; and rodents and rabbits, 84. Ground squirrels are most frequently taken (Horn and

Fitch 1942). Control measures appear not to have reduced coyote populations noticeably.

Minor Influents

The California ground squirrel (Fig. 9-4) (Horn and Fitch 1942, Hubbard 1941) is conspicuous and diurnal. With the beginning of the autumn rains, green food is used extensively, including the entire plant of immature broad-leaf filaree, brome, and fescue grasses. Later, with the mature plants, only the tender leaves and fruits are selected. The impact on the herbaceous growth is greatest in the spring, with the result that the food supply of cottontails, formerly of pronghorns, and perhaps also the California wapiti is reduced. In the dry summer season, squirrels eat seeds, acorns, and tarweeds. The squirrels gather quantities of seed and cache them for future use. They also dig out quantities of subterranean fungi and roots of broad-leaf and red-stem filarees and foxtail and Bermuda grass.

The burrowing activity of the ground squirrels has a twofold effect. A burrow often represents a displacement of over 20 cubic feet (0.57 m^3) of earth. This soil is brought to the surface and spread over the surrounding vegetation, usually killing it. Digging operations continue from year to year. However, pulverized mineral soil, old nests, carcasses of squirrels, mice, snakes, lizards, toads, insects that have died in the burrow, feces and refuse plant material are all mixed together in the mounds so as to fertilize a more luxuriant vegetation. Large populations of ground squirrels bring a reduction of grasshopper populations because of the extensive digging.

The nocturnal kangaroo rat *Dipodomys heermanni* does not have well-developed storing habits as does the ground squirrel. However, during late spring, large surface piles of uncured grass may be present at the ends of its surface runways. During the winter, its food consists of filaree leaves, green grass (foxtail, slender oat, and bromes), and leaves of red maids and popcorn flower, pine nuts, and acorns.

The pocket gopher *Thomomys umbrinus* gathers its food at the ground surface and carries it underground for consumption. Broad-leaf filaree, popcorn flower, and miscellaneous green plant material are stored in loose piles. During the dry season, its food includes roots of tarweed.

Desert cottontails were scarce in 1934, increased until 1939, and then declined sharply. Abundant shelter, such as brush, rock piles, or thickets of tall vegetation interspersed with grassy openings, furnish favorable habitats. Breeding is limited to the late winter and early spring, and the young are well grown before the forage crop dries out during the summer. The seasonal peak of abundance occurs at the time when forage is maturing, which is the time when formerly it was also grazed most heavily by pronghorn. Moist area plants are used in the dry season. Clover, dock, rush, tarweed, and even turkey mullein are selected because of their succulence (Horn and Fitch 1942).

A less influent group of mammals includes the California raccoon near water-courses (Hubbard 1941). The striped skunk is found among the numerous rocks and is most abundant in the foothills. The California badger is on the experimental range but is not mentioned for the Friant area. Wood rats construct their nests at the base of oak trees in the Friant area. There is

Fig. 9-4. A. California ground squirrel feeding on filaree *Erodium*. B. The same, feeding on filaree root with coreaceous-leaved shrub behind. C. Young red-tailed hawks in nest in digger pine with the hind leg and tail of a squirrel at the left (all photos by H. S. Fitch).

a series of pocket mice, harvest mice, and deer mice, six or seven species in all, which function as part of the rodent food of predators.

The birds include a great number of migrants which regularly settle in the area temporarily in the course of their seasonal movements. Many of the birds that breed on the range, such as the kingbird, Bullock's oriole, and mourning dove, are not present during the winter. Raptorial species include the red-tailed hawk (Fig. 9-4C), Cooper's hawk, sparrow hawk, great horned owl, barn owl, long-eared owl, and screech owl, all of which nest on the area (Horn and Fitch 1942).

Twenty-six red-tailed hawks occupy 3500 acres (1400 hectares) each restricted to a definite territory. Hawk pellets contain insects, 8 per cent; birds, 3; reptiles, 20; and rodents and rabbits, 68. Nestling mortality is caused by the black fly, the larva of which occurs in running water. The great horned owls average 1 to each 80 acres (1/32 hectares). Their food is insects, 17 per cent; birds, 2; reptiles, 1; rodents and rabbits, 75. Found at the nest as food for young are wood rats, kangaroo rats, cottontails, ground squirrels, and pocket gophers (Horn and Fitch 1942).

The California quail requires brush, rocks, or small trees as escape cover from enemies. Small live oaks and some of the larger shrubs are highly important as roosting cover. Destruction or removal of such growths on range lands results in a reduced quail population. Types of cover used for nesting are grass or forbs, 32 per cent; fallen brush with grass, 19.8; and rocks with grass, 13.5.

The roadrunner is a common resident and is known to feed on young quails and reptiles (Storer *et al.* 1942). Primarily dependent also on the oak–grass woodlands are Swainson's hawk, golden eagle, sparrow hawk, mourning dove, eastern kingbird, western kingbird, Cassin's kingbird, yellow-billed magpie, mockingbird, loggerhead shrike, Brewer's blackbird, house finch, lesser goldfinch, and lark sparrow (Miller 1951).

Five lizards and six snakes are listed by Horn and Fitch as occurring on the experimental range. In the spring of 1939, the population of rattlesnakes was 1.0 per acre (2.5/hectare) and in 1940, 1.3 per acre (3.2/hectare). Rattlesnakes are sluggish and relatively inactive but wander widely. The time and extent of movements are exceedingly variable, but on the average the snakes move less than 10 feet (3 m) a day. Young rattlesnakes are born in the autumn and average 7.6 per litter; however, few snakes survive beyond the first year after birth. The 300 or more killed annually by humans account for only a small part of the total mortality. The rattlesnake is an important predator on rodents, especially ground squirrels, cottontails, birds, lizards, and the spadefoot toad. The effectiveness of rattlesnakes occupying ground squirrel burrows is increased through their striking any squirrel which comes within reach (Horn and Fitch 1942, Fitch 1948).

Other reptiles of the experimental range are striped racer, gopher snake, common kingsnake, the garter snakes *Thamnophis sirtalis* and *T. ordinoides,* night snake, side-blotched lizard, fence lizard, horned lizard, whiptail, and

Gilbert's skink. Most of these are taken as food by the various predators of the area. The few amphibians are the spadefoot toad, western toad, and the Pacific tree frog. The tree frogs breed in temporary pools after the first rains (Horn and Fitch 1942).

Feeding on the range grasses and forbs is the devastating grasshopper *Melanoplus devastator*. This species normally breeds in the grassy woodlands and often migrates into grassy valleys in destructive hordes. The valley grasshopper *Oedaleonotus enigma* occasionally also occurs in devastating numbers, while *Camnula pellucida* and *Sticthippus californicus* are present (Ray F. Smith, personal communication).

Biotic Studies in 1944

On the Hastings Reservation near Monterey, California, the native Junegrass, squirreltail, fescues, and pine bluegrass but no shrubs were present in a stand of blue oak. The only abundant insects were immature grasshoppers *Melanoplus*, and invertebrates in general were far less numerous than in an adjacent pure grass community. The partially insect-immune blue oak supported large numbers of the oak treehopper *Platycotis vittata*, the chrysomelid *Tanaops*, the ant *Crematogaster coactata*, the gall wasp *Synergus ochreus*, and dasytiid beetles.

On an adjacent hillside, blue oak, California live oak, California white oak, and madrone were present. The shrubs consisted of small oaks, Christmasberry, and some other species. The herbs were Indian milkweed and yarrow. Common insects found were the plant bug *Lygaeus kalmii*, the leafhoppers *Cochlorhinus pluto* and *Texananus oregonus*, and the bee *Ceratina nanula*. The shrubs were inhabited by the leafhopper *Paraphlepsius occidentalis* and large numbers of dasytiid beetles. The California oak moth *Phryganidia californica* was also present.

SCLEROPHYLL BUSHLAND OR CHAPARRAL

Chaparral, which Cooper (1922) considers to be climax, is the principal community over most of the southern coastal ranges. Northern slopes generally have some woodland and forest. The prevalence of the following species is indicated by the number of times they occur in 87 different localities in all parts of California: chamise, 75; Christmasberry, 26; scrub and leather oak, 23; interior live oak, 11; canyon live oak, 10; manzanita (all species), 50; buckbrush, 25; mountain mahogany, 19; and redberry, 10. Except for its local occurrence as an understory in ponderosa pine and in woodland, chamise has the same distribution as chaparral. A list of other common constituent species includes chinkapin, leather oak, bush-poppy, ribbon-wood, hollyleaf cherry, lemonade-berry, laurel-sumac, *Ceanothus crassifolius*, and eight other species of *Ceanothus* all of which are frequently called mountain lilac. Several of these species may be either shrubs or trees (Cooper 1922).

Faciations

Horton (1941) names four types of chaparral in the San Dimas National Forest: chamise chaparral, which occupies about two-thirds of the chaparral area and occurs in two faciations, one with *Ceanothus crassifolius* as the second dominant, and the other, on more xeric sites, with wild buckwheat; "chamise-sage" and black sage; oak chaparral, which consists of California scrub oak, Christmasberry, and hollyleaf cherry; and mixed chaparral, which is similar to oak chaparral but occupies dry sites and is dominated by *Ceanothus crassifolius, C. oliganthus,* Christmasberry, eastwood manzanita, and California scrub oak (Fig. 9-6).

Fig. 9-5. A. Chaparral; the light-foliaged plant is a species of manzanita. B. A deer feeding in chaparral on manzanita (photo by Joseph Dixon, courtesy California Division of Fish and Game).

Vertebrates

The chief mammals of the chaparral are the grizzly bear and mountain lion formerly, deer (Fig. 9-5), bobcat, coyote, gray fox, wood rat, and spotted and striped skunks. Brush rabbits are generally present. While rabbits feed to some extent on the climax chaparral shrubs, much of their food appears to be forbs. The Merriam chipmunk is confined to the chaparral in various localities, such as in the San Jacinto area. Small mammals peculiar to chaparral include the parasitic California mouse and five-toed kangaroo rats (Cogswell 1947, Grinnell 1915, Grinnell and Swarth 1913).

As was brought out by the studies in 1944, the commonest birds of the dry summer season are wrentit, common bushtit, and rufous-sided towhee. Scattered calls of the poor-will may also be heard at this season. In October, white- and golden-crowned sparrows, several races of fox sparrows, hermit thrushes, ruby-crowned kinglets, and thousands of Audubon's warblers are present. Many of these utilize the fruit of the *Rhamnus, Ceanothus,* and Christmasberry or toyon. The bushtit occurs in flocks of 15 to 40. Summer residents arrive in March, and nesting begins by the middle of the month. A host of northward migrants also passes through the chaparral in March.

Reptiles are numerous but occur without reference to vegetation. Amphibians appear to be wanting in the climax chaparral except for the tree frog which breeds in temporary pools formed by the first heavy winter rains.

Invertebrates

Collections were made during the dry season in the first two weeks of July, 1944, at 17 stations in the chaparral near Redding in northern California and Monterey and Glendora in southern California. No mollusks were obtained. Only the two centipedes *Taiyuna occidentalis* and *Idiothus shelfordi* and the one millipede *Xystocheir* were taken, although the winter wet season would have made many more available. A few spiders were ground inhabitants, and only three species were taken from the chaparral in all three areas: *Linyphia litigiosa, Misumenops californicus,* and *Oxyopes rufipes.* The most abundant insects were the hemipterons; specially noteworthy were the damsel bug *Nabis alternatus,* the leaf bug *Lygus hesperus,* and the shield bug *Homaemus variegatus.* The plant bug *Lygaeus kalmii* has a wide distribution elsewhere.

The shrubs and herbs in the northern area support a number of hemipterons and homopterons not found in the southern localities. These were the plant bugs *Rhyparochromus chiragra californicus, Malezonotus sodalicus,* and *Polymerus elegans,* the stink bugs *Tricholpepla aurora* and *Neottiglossa tumidifrons,* and the assassin bug *Zelus laevicollis.* In addition the treehoppers *Tortislilus pacifica,* the lantern flies *Didyssa chlathrara, Oecleus obtusus,* and *Scolops abnormis,* the cicada *Okanagana rubrovenosa,* and the two leafhoppers *Ballana paridens* and *Exitianus obscurinervis* were taken.

The hemipterons and homopterons at the Monterey and Glendora stations included the aphis killer *Anthocoris bakeri,* the jumping plant lice *Eurphyllura pruinosa* and *Euphalerus rugipennis,* the shield bug *Homaemus bijugis,* the leaf bugs *Phytocoris cunealis* and *Phymata pacifica,* the plant bug *Ligyrocoris sobrius,* the lantern fly *Dictyssa obliqua,* the stilt bug *Neides muticus,* and the squash bug *Ahyssus usingeri.* The following leafhoppers were taken: *Cochlorhinus pluto, Texananus oregonus, Paraphlepsius occidentalis,* and *Thamnotettex.*

Only a few beetles were secured. These included from the southern areas, snout beetles, the oak weevils *Thricolepis inornata* and *Curculio uniformis,* the buprestid *Anthaxia aeneogaster,* and the tenebrionid *Noserus plicatus.* The ground beetle *Amara insigni* was the only representative of this family. The buprestid *Acmaeodera gemina* was recorded only from the northern area. Three flies were recorded in the southern area: *Hyalomya aldrichii, Meromyza americana,* and *Trupanea femoralis.* As for orthopterons, only *Morsea californica* was taken from shrubs. Bees and wasps were not numerous; of the bees, *Diadosia ochracea* and of the wasps, *Leptochilus infussipennis* occurred near Redding. In the south, the parasitic *Euslandulum cyanea* was taken in chamise chaparral, and the ichneumonid *Anisobas bicolor* was taken in mixed chaparral at Hastings. The ants *Formica rufibarbis occidua* was abundant and found in both the northern and southern areas; *Formica r. rufibarbis* was taken only in the oak chaparral of the southern area. A species of *Camponotus* occurred quite regularly at all stations (Sparkman 1945).

In the San Dimas National Forest near Glendora, there were altogether 40 arthropods per square meter in the chamise–black sage chaparral, only 31 per square meter in the faciation without black sage, and only 27 per square meter in the mixed chaparral. The mixed chaparral at Redding showed only 11 arthropods per square meter. Insects are most abundant from March to May (Sweet 1930).

Biotic Seasons

Cogswell (1947) characterizes the period December through February as equivalent to spring, with its cool to warm and commonly wet weather, and March to May as summer with most of the flowers present and most of the birds nesting. The dry period from June through August is equivalent to autumn and September through November, when rains are increasing, to winter. This seasonal change is three to four months earlier than at similar latitudes to the east.

Status of the Association

The chaparral vegetation appears to constitute an ideal plant association, as far as growth form and taxonomic composition are concerned. However, it is not a biotic unit. The large mammals are not limited to the chaparral region even though the deer is a dominant. The distribution of the grizzly bears does not correspond with chaparral. The striped skunk, although confined to the area, can find a suitable niche in other vegetation than chaparral.

According to the distribution of birds (Miller 1951), the wrentit has essentially all its population limited to chaparral, while the green-tailed towhee and black-chinned sparrow have most of their populations in chaparral and a small part in sagebrush. The Allen's hummingbird, Anna's hummingbird, orange-crowned warbler, and Lawrence's goldfinch populations are largely in chaparral with a part extending into coast forest and woodlands, suggesting that these are northern species. On the other hand, the Costa's hummingbird, black-tailed gnatcatcher, California thrasher, rufous-crowned sparrow, and sage sparrow have a large part of their populations in chaparral and lesser parts in desert, suggesting that they are southern species.

Smith (1946) maps a dozen or more subspecies of lizards covering a part of the sclerophyll area with most of them apparently restricted to it. The meager insect records indicate some unity between northern and southern portions of the community, but also considerable disunity. The other two plant associations, woodland and forest, show even less biotic unity. Accordingly, the biota had best not be considered as a distinct biome in itself but as a series of biotic faciations of other biomes.

COMMUNITY DEVELOPMENT

Oak Forest

Along the west shore of San Francisco Bay, oak forest develops through a sere of salt marsh, willow, cottonwood, and boxelder; then the forest of coastal live oak and valley oak; but this forest finally succumbs to chaparral. The coastal area is flooded by creeks which unload considerable amounts of sediment eroded out of the adjacent coastal mountain ranges. This results in the gradually building up of the soil surface above the water table. Eventually the ground surface becomes too dry for the oaks, and chaparral comes in. This is largely a physiographic succession as the reactions of the organisms on the soil are of secondary importance (Cooper 1926). However, as Snell (1933) has pointed out, litter and humus accumulate, especially in the oak forest, and support a large number of animals. The pattern of this development resembles that of the Mississippi floodplain discussed in Chapter 4.

Chaparral Development on Washes

Three successional stages occur on the dry washes of Mill and Santa Ana creeks, at the southern base of the San Bernardino Mountains (Cooper 1922, Ingles 1929). The soil is mostly sand and silt with a great deal of humus formed chiefly from decaying leaves and logs. Excluding the cottonwood along the streams, the pioneer plants are the leafless composite scalebroom *Lepidospartum squamatum* and *Croton californicus*. This associes contains only earthworms, sowbugs, spiders, a single species of snail, a toad bug, and a few beetles. It is followed by coastal sagebrush with the third and final chaparral stage coming in with the establishment of chamise.

The sagebrush stage varies somewhat from the lower to the higher elevation; however, white sage, California sagebrush, wild buckwheat, scalebroom, and *Eriodictyon trichocalyx* are important throughout. *Yucca* and *Encelia* are present (Fig. 9-6). *Juniperus californica* is most numerous at the lower elevations. Black sage, white sage, and *Eriophyllum* are common.

Chamise is the most characteristic shrub in the chaparral. *Yucca whipplei* is present but is not as common as in the previous community. *Opuntia parryi* and *O. occidentalis* are present, although the latter is not common. *Eriodictyon trichocalyx* and *Mirabilis* are found in small groups throughout. Societies of

Fig. 9-6. A. Chaparral east of San Diego, California at 4000 feet (1200 m), showing the leaves and flower stalk of Spanish bayonet in the lower right corner. The low light-colored shrub is probably *Ceanothus verrucosus*. Considerable *Gutierrezia* is present. At the lower left corner is a photo of *Tegeticula maculata* which cross-fertilizes Spanish bayonet (courtesy U.S. National Museum, Washington, D.C.). B. The lower portion of a plant of California buckwheat with branch showing leaf arrangement; to the right, the leafless straight stems of white sage (altitude 500 feet, 150 m).

Lotus americanus and *Ceanothus crassifolius* occur scatteringly. There is a question whether this community, where drainage is good and flooding rare, may not later be invaded by *Adenostoma* and other species to form a more typical chaparral climax.

Of 157 species of animals found in the chaparral and sagebrush by Ingle (1929), 106 showed a preference for chaparral and 51 for sagebrush. Of the amphibians and reptiles there were 5 for chaparral and 2 for sagebrush; birds, 16 and 5; mammals, 3 and 2. There was a large group with a preference for the cottonwood–willow and a few for the scalebroom community.

The ground layer is inhabited by two lizards *Uta stansburiana hesperis* and *Eumeces skiltonianus,* which apparently are active throughout much of the year. With the exception of the Pacific tree frog, the snakes, toads, and frogs generally do not become active until sometime after the autumn rains begin. The California toad is first recorded in January, the canyon tree frog in February, and the gopher snake in April. Some birds, as the roadrunner, were present throughout the period of observation. Appearing for the first time in April are Costa's hummingbird, gray flycatcher, scrub jay, Bullock's oriole, white-crowned sparrow, rufous-crowned sparrow, and warbling vireo. The California quail nests in May. Mammals active the year around are the coyote, cottontail, and the pocket gopher *Thomomys umbrinus;* those that appear with the rains are a kangaroo rat and a microtine mouse.

According to Ingles (1929), the large centipede *Scolopendra polymorpha,* the scorpion *Anuroctonus phaeodactylus,* and the widely distributed annelid *Allolobophora iowana* occur under stones. The Jerusalem cricket appears after the autumn rains begin. Snails are unexpectedly numerous; *Helminthoglypta tudiculata* is present throughout the winter and spring. The beetle *Brennus (Scaphinotus) ventricosus* feeds on the adults of this snail. Other chaparral snails are *Vallonia cyclophorella, Sterkia hemphilli, Euconulus fulvus,* and *Zonitoides arboreus.* The last species is widely distributed over North America.

Of 19 species of carabid ground beetles, perhaps only 3 are found solely in the chaparral. Of the 15 species of tenebrionid beetles, only 3 are apparently confined to the cottonwood–willow community. Beetles appear in the autumn after the beginning of the rains.

Invertebrates in the herbs and shrubs during November are the grasshopper *Psoloessa texana pusilla,* the autumn bug *Homaemus bijugis,* the leaf beetle *Pachybrachis melanostictus,* the assassin bug *Sinea diadema,* and the stink bug *Chlorochroa ligata.* Appearing in April are the mud wasps *Ceratina subpunctigena* and *Pseudomasaris edwardsii;* the wasps *Podalonia luctuosa, Specodes arvensiformis, Halictoides mulleri,* and *Colletes californicus;* the moth *Euclidia cuspidea;* and the butterflies *Anthocharis lanceolata, Zerene eurydice, Colias harfordii, Coenonympha tullia california,* and *Lemonias gabbii.* Very few new species appear during the intervening months.

The period of greatest activity, represented by the presence of the greatest number of species, is March and April. This is probably determined in part by the greater soil moisture, as is the flowering of the plants. During the hot, dry summer months, many species of birds migrate north or into the mountains. Many invertebrate species aestivate. The snail *Helminthoglypta tudiculata* commonly becomes sealed in pockets in the ground by dried soil but becomes active on the day following the first rain. Centipedes are never seen during the dry season, but adults appear immediately after the first rain. Collembolons are never seen until after the rains start and then only in moist, well-aerated places, such as among the roots of club moss that grow under stones.

Fire Succession in Chaparral

The recognition of seral stages following fire is impracticable because of the diversified character of chaparral plants, the varying intensity of fires, and different effects produced by the season in which the fires occur. Chaparral species that sprout from the roots when the parts above ground are killed are chamise, eastwood manzanita, mission manzanita, interior live oak, leather oak, California scrub oak, canyon live oak, birchleaf mountain mahogany, chaparral whitethorn, and ribbon-wood. Nonsprouting species are common manzanita, hoary manzanita, Stanford manzanita, bigberry manzanita, Parry manzanita, whiteleaf manzanita, buckbrush, bigpod ceanothus, cupleaf ceanothus, hairy ceanothus, and wartystem ceanothus.

In ordinary or periodic fires few plants of sprouting species are killed. However, following burning, numerous annual grasses, broad-leaved herbs, and brush seedlings appear on most chaparral lands. On areas of sprouting chaparral, the herbaceous growth generally attains its maximum abundance during the first or second year and declines to a scattered stand by about the fifth year, owing to the effect of the very numerous bush sprouts that appear. On areas of nonsprouting bush, most of the invading herbaceous plants continue to increase in density well after the third year. Germination of some chaparral species is stimulated by heat, and the seeds of most chaparral species are more resistant to heat injury than those of grasses or broad-leaved herbs. Chaparral and its understory vegetation effectively protect soil from erosion. This protection favors a relatively high infiltration capacity of the soil for water (Sampson 1944a, 1944b).

Fire causes some increase in forage and browse for deer. Mule deer have a greater reproductive rate in open brush with herbaceous cover than in an eighteen-month-old fire burn or in dense mature brush (Taber 1953).

Coastal Sagebrush

Perhaps the naming of this community after specific plant dominants is not justified (Vestal 1938) because of the extensive variation in groupings of species that results from the presence of 47 genera and 102 species along with some 20 varieties. This bush community is limited to scattered areas from northern Baja California to San Francisco Bay. Clements (1920) states that it reaches through to southwestern Nevada, but this is out of the range of California sagebrush which he considers the most important species.

Communities dominated by one of the following five species are of general occurrence: California sagebrush, white sage, black sage, wild buckwheat, and purple sage. Chaparral chamise is a constituent in dry southern communities.

California sagebrush is generally distributed and occurs widely in pure stands along the coastal hills. Wild buckwheat and black sage are its most common associates. White sage is restricted to the southern part of the area and is subclimax in character. Wild buckwheat, including the variety *Eriogonum fasciculatus poliofolium*, has the widest range and forms communities in the *Larrea* desert. In San Dimas Canyon near Glendora, California, where all five of the commonest plant dominants as well as *Galium angustifolium* are present, the estimated arthropod population is 78 individuals per square meter, which is a large number for shrub vegetation. The average of six stands of chaparral here is 24 with the largest number, 43.

Sweet (1930) made a study of sagebrush in the fan of the San Antonio Canyon at Claremont, California. The fan is on an old established substratum composed of rocks, gravel, and sand. Three plant groupings were studied for their animal constituents, one included some *Lotus*, one *Rhus*, and one *Quercus agrifolia*. *Ribes indecorum*, *Rhus trilobata*, and *Rhamnus crocea ilicifolia* bloomed almost alone during the winter, but by May, the vegetation was gay with the varicolored bloom of annuals and perennials. By July, all fresh growth had vanished.

Of the hemipterons present, the noteworthy mirids along with the months of their occurrence were *Bolteria hirta*, March to June; *Europiella decolor*, March to June; *Irbisia sita*, March to May; *Lygus sallei*, October to December, June, July; *Nysius ericiae minutus*, October to July; *Orthotylus ferox*, March to May; and *Phytocoris canescens*, December, June. The homopterons were represented by the cercopid *Clastoptera lineatocollis*, April to December, and the scale insect *Amonostherium lichtensioides*, April 21 to June 13.

The beetles of the area were *Cantharis ingenuus*, April, May; *Diabrotica twelve-punctata*, May, June; *Monoxia debilis*, May to August; and *Pachybrachis hybridus* April, May. The lady beetles were *Hippodamia convergens*, April to July; *Hyperaspis lateralis*, April to October; and the weevil *Smicronyx obtectus*, May to October. Other species of beetles appeared less important. Lepidopterous larvae were only occasionally seen. There was apparently no preference for sagebrush on the part of any reptiles or amphibians.

Sage grouse, sage thrasher, vesper sparrow, and Brewer's sparrow are found chiefly in sagebrush. Populations of green-tailed towhee and black-chinned sparrow are divided between sagebrush and chaparral, while those of crissal thrasher and black-throated sparrow are divided between sagebrush and the desert (Miller 1951).

Chapter 10

Cold Desert and Semidesert Communities

The *shadscale–kangaroo rat–sagebrush biome* occupies the Great Basin portion of western Utah, Nevada north of 36° 30′ N. Lat., and Inyo County, California, west of Death Valley. It extends eastward to the canyons of the Colorado and Green rivers in eastern Utah and occupies much of the Green River Basin in southwestern Wyoming. A sagebrush faciation occurs in northeastern California, in central Oregon, and near the Snake River in Idaho (Fig. 10-1). The vegetation is largely shrubby. Cacti are scarce and the density of the shrubs varies. Species of the shadscale *Atriplex*, the sagebrush *Artemisia*, and the kangaroo rat *Dipodomys* occur throughout. The shadscale *Atriplex confertifolia* and sagebrush *Artemisia tridentata* are the principal plant dominants respectively of two associations within the biome (Clements 1920, Livingston and Shreve 1921, Billings 1945).

Areas dominated by sagebrush and shadscale are generally in contact with each other (Fig. 10-2). The shadscale community makes contact with the hot desert in southwestern Nevada. The sagebrush community has contacts with ponderosa pine forests and frequently grows beneath the trees in California and Oregon. The relations of sagebrush to pinyon–juniper woodland are similar. This woodland occurs above 5500 feet (1670 m) on numerous local ranges within the sagebrush region. In northern and central Utah, Colorado, and Wyoming, sagebrush mostly contacts bunch grass grassland. The Indian population was small in A.D. 1600 and depended largely on the pinyon–juniper woodland.

Climate and Soil

Soil differences are probably more important than differences in rainfall in separating the *sagebrush–rabbit* and the *shadscale–kangaroo rat associations*

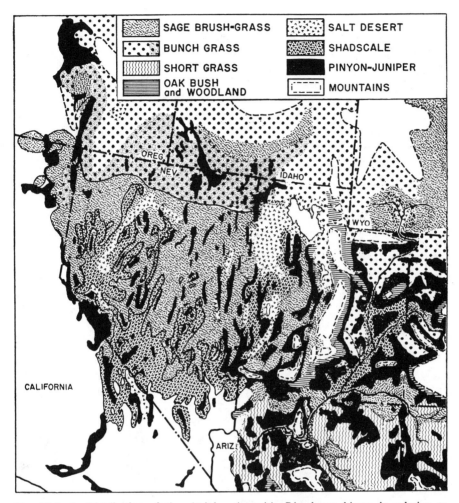

Fig. 10-1. A. Distribution of the *Atriplex–Artemisia–Dipodomys* biome in relation to pinyon–juniper, oak bush, and grassland biotic communities. The shadscale dominance mapped have areas of sagebrush–grass dominance adjacent at higher altitudes and below pinyon–juniper. Sagebrush also commonly occurs with conifers at low elevations. The mapping of Nevada is after Billings (1949), eastern Utah is after U.S. Geological Survey. Pinyon–juniper is reduced in areas as compared with the maps of Shantz and Zon (1924), and adjustments are made from the land classification map of Utah for 1931 and from Woodbury (1947). California is largely from Jensen (1947) and Arizona from Nichol (1937). The eastern edge of the Wyoming sagebrush–grass was definitely grass in 1870 (Great Plains Committee 1936). A light stippling covers areas that were probably bunch grass in 1600 (Weaver and Clements 1929, Clements and Shelford 1939).

within the biome (Hayward, personal communication). Due to the Great Basin having only internal drainage, much of the soil contains salt, and depressions containing greater accumulations have edaphic climaxes. *Artemisia tridentata* occupies alluvial fans and benchlands next to the mountains where the soil is relatively porous and drainage is good, so that most of the salt has been leached out. It shows a difference in growth form with differences in the soil (Fig. 10-7). *Atriplex confertifolia,* on the other hand, grows where the soil is fine sand or clay and contains some salt. There is often little difference in elevation between the two communities.

Mean annual precipitation in the shadscale area of Nevada and the Colorado River region is below 6 inches (15 cm). In western Utah and southwestern Wyoming, it is below 10 inches (25 cm). The maximum annual precipitation over a long period of years for western Utah has been 18.5 inches (46 cm), with a minimum of 3.5 inches (8.75 cm). The heaviest rainfall of the spring is usually in April or May; in some areas rainfall is also heavy in autumn.

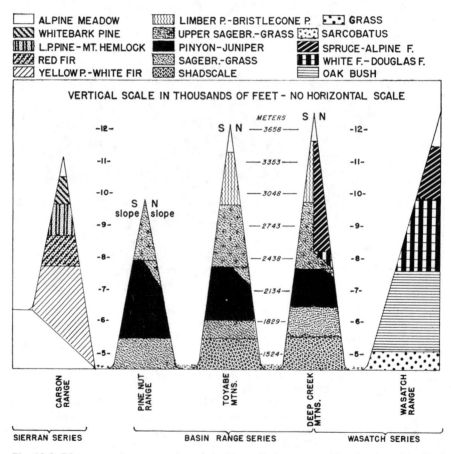

Fig. 10-2. Diagrammatic cross-section of the Great Basin communities showing altitudinal relations (after Billings 1951). The mountains are turned 90° so that their north slopes are on the right. The "alpine meadow" includes fell fields. Yellow pine includes *Pinus ponderosa* and *P. jeffreyi,* the latter more abundant. The *Sarcobatus* areas include some sage-grass.

The mean January temperature is between 29° and 39° F. (–1.7° and 3.9° C.) and the mean July temperature between 70° and 78° F. (21° and 26° C.). July temperatures in sagebrush may be 5° F. (2.7° C.) lower than in the shadscale areas (Billings 1949). The winters are not unlike those in central Iowa with regard to temperature, but snowfall is only about one-third as much. This justifies the title of cold desert given by Billings. Figure 10-3 compares the two most important climatic communities in Utah with the *Larrea* desert to the south (Billings 1949, Fautin 1946).

ASSOCIATIONS AND FACIATIONS WITHIN THE BIOME

Billings (1945) recognizes 14 plant communities, exclusive of cottonwood along rivers, and presents a figure showing the relations of 8 of them. In another paper (1949), he indicates that shadscale and six associated species, especially *Artemisia spinescens,* made one of the main community types or associations. *Artemisia tridentata* constitutes a second type or association. Other plant communities belonging to the hydrosere are the bulrush associes (tule), cat-tail associes (tule), alkali-grass associes, saltgrass associes, samphire (*Salicornia*) associes, and iodine bush associes.

Plant Constituents

The chief plants are deciduous, more or less spinescent, microphyll shrubs, chiefly *Atriplex,* and dense stands of tall sagebrush, chiefly *Artemisia.* The distributional limits of seven of the characteristic dominants center in western Nevada (Billings 1949). The height of the shrubs varies mostly between six inches and four feet (15 to 120 cm). Some of the principal plant dominants and influents are shadscale, winter-fat, bud sage, greasewood, Mexican tea, littleleaf horsebrush, sagebrush, and antelope brush.

In eastern Utah, Wyoming, and western Colorado, shadscale is largely displaced by saltbush and antelope brush by buckwheat bush. Black bush is very important in the upper Colorado River Basin in eastern Utah (Hayward, personal communication, and Edith Clements, personal communication of notes made by Shantz and Clements between Thistle and Monticello, Utah). The grasses, particularly rice grass, are about the same throughout. *Ephedra viridis,* littleleaf horsebrush, black sage, *Chrysothamnus,* and some others are limited to particular types of substrata.

The plants are widely scattered; in areas of shadscale usually only about 10 per cent of the ground is covered, but in some areas of sagebrush it reaches 85 per cent. In typical stands of shadscale, the ground between the shrubs may be nearly bare, or the perennial grass *Oryzopsis hymenoides* or the forb *Sphaeralcea* may be present. Minute native annuals may be present for a few weeks during a moist spring. The overwintering leaves of sagebush are lost when growth begins in the spring, and leaf fall is most rapid during periods of hot weather (Diettert 1933).

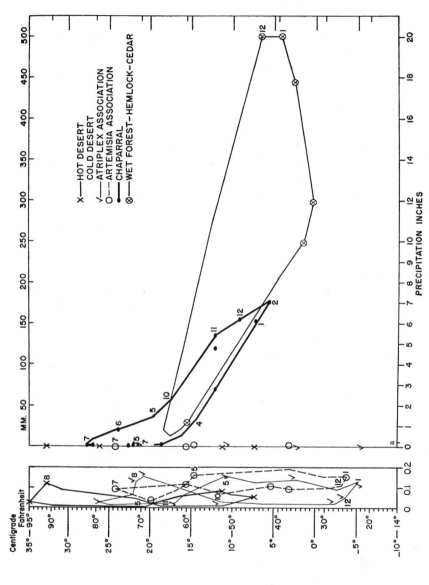

Fig. 10-3. Temperature and rainfall climograph of the three winter rain areas; the rainy hemlock–cedar figure is after Dirks-Edmunds 1947; the chaparral figure is for Ojai (near Ventura) and Redding, California, and the diagrams at the left for the cold and hot desert are after Fautin 1946. The point a on the large chart is at the 0.2 inch mark which is enlarged ten times at the left.

Animal Constituents

Outstanding and widely distributed animals are jack rabbits, kangaroo rats, kangaroo mice, pocket mice, grasshopper mice, and antelope ground squirrels. In western Utah, populations range from 14 to 41 per hectare. All except the jack rabbit are burrowers. The burrowing habit also extends to the carnivores, badger, kit fox, and coyote. Several lizards do some burrowing and moving of soil. The pronghorn antelope was originally distributed throughout the area. Deer are local and present only in winter.

PERMEANT INFLUENTS AND DOMINANTS

The number of influent species and the size of their populations are small. Major influents are Great Basin coyote, badger, pronghorn, locally the mule deer, kit fox, bobcat, great horned owl, prairie falcon, Swainson's hawk, golden eagle, red-tailed hawk, and bald eagle.

The total population of pronghorns in Utah in 1922-24 was estimated to be 670 and in Nevada, 4273 (Nelson 1925). In 1942, California had 3752 (McLean 1944), but this may be double the population present in 1924. Assuming 1876 for California sagebrush, the total number of pronghorns in the cold desert in 1922-24 was 6800. There may have been more than 600,000 under pristine conditions, but this is uncertain since drinking water is a limiting factor.

Pronghorns commonly occurred in bands of 20 to 1000 individuals. They constitute the food supply of such predators as the wolf, coyote, and bobcat, especially during the winter when snow impedes their movements. At the present time, pronghorns tend to scatter during the summer and remain in the low hills, but during the winter they congregate more commonly in the valleys (Fautin 1946).

The abundance of pronghorns on the short grasslands east of the Rocky Mountains indicates their preference of grass for food. In areas where grass is available, it occurs in 40 to 60 per cent of the stomachs examined but was mixed with other plants. In the Great Basin, however, pronghorns browse largely on shrubs; sagebrush constitutes from 35 to 40 per cent by volume of their food in autumn and spring and 95 per cent in winter; bitterbrush or antelope brush, 10 to 13 per cent and 0.1 per cent, respectively. Other shrubs are used in spring to the extent of 9 per cent. Forbs make up 35 to 52 per cent in autumn and spring but only about 2 per cent in winter. Grasses constitute only 1 to 2 per cent throughout the three seasons studied (Ferrel and Leach 1952).

Coyotes range throughout but are few in number in the open communities of shadscale, winter-fat, and little rabbitbrush, where the short plants afford little cover. The number of scats and calls in one study indicated that they were some 11 times more abundant in greasewood and sagebrush, where their chief food, rabbits, was also most abundant. Coyotes mate from late January to early March, and the pups are usually born in April and early May. The size of the litter varies from 4 to 17 with 7 about the average. One of three dens discovered in a greasewood community was an enlarged badger den (Fautin 1946).

A study of 8263 stomachs from 17 western states by C. C. Sperry (Fautin 1946) shows that rabbits and other rodents comprise more than 50 per cent of the food taken. Reptiles occur in 3 per cent of the stomachs, *Sceloporus* being taken most frequently and rattlesnakes next. Blowsnakes, horned toads, uta lizards, garter snakes, and racers are also eaten. The number of insects consumed is usually small, but some stomach contents are nearly all insects, especially grasshoppers. The remains of "cactus pears" or *Opuntia* fruits are found occasionally.

Four dens of the desert fox were found in open spots, one in a winter-fat area and the others in shadscale communities. The fox is largely nocturnal, staying in its den during the day. The remains of 19 kangaroo rats, 2 kangaroo mice, 1 pocket mouse, 1 deer mouse, and rabbit bones were near the dens. The kangaroo rat is so important in the diet of the desert fox that in some areas its population is found to be very closely correlated with that of the kangaroo rat.

Burrows of the badger are traceable by the excavations they produce in search of ground squirrels and kangaroo rats. Fresh diggings were found in every community except winter-fat. In the sagebrush community, there was estimated to be about two badgers per square mile (0.8/km²) and in shadscale and greasewood communities, one badger per two square miles (0.2/km²). One badger often digs out three or four ground squirrel burrows per day. They leave an area when the food supply is depleted. Since they hibernate, they influence the community only during eight or nine months of the year. Their excavations are utilized as dens by coyotes and are places of refuge for burrowing owls, cottontails, lizards, snakes, insects, spiders, and scorpions (Fautin 1941, 1946, Bailey 1931).

Bobcat tracks were found in the shadscale about three miles (4.8 km) below the base of a mountain; it is only an occasional visitor in these lower communities (Fautin 1946).

MINOR PERMEANT INFLUENTS

The burrowing owl occupies badger burrows in all communities except sagebrush and are active both day and night. The owls are most abundant in shadscale areas where the vegetation is sparse and open. In most cases their burrows are from a fourth of a mile to a mile or more apart (0.4 to 1.6 km). The three to seven young per brood are able to fly during mid-July. Kangaroo rats and insects taken at night are important in their food.

The ferruginous hawk, according to C. L. Hayward (personal communication), is the most characteristic bird of the cold desert. It nests on ledges, in juniper close to shadscale communities, and in the adjacent valleys. The habits of the golden eagle are similar. Swainson's hawk is a common summer resident throughout the western part of Utah. Two nests of this species, located in April, were in the juniper *Juniperus osteosperma*. One nest was about 10 feet (3 m) and a second, 13 feet (4 m) above the ground. The second nest was shared by a "pack rat" whose abode was in the lower part. The hawk feeds on insects, chiefly grasshoppers, as well as chipmunks and ground squirrels. It is most abundant in the sagebrush community. Red-tailed hawks and marsh hawks are also present.

Prairie falcons were recorded in White Valley, Utah, on 35 per cent of the trips taken in 1939 and on 28 per cent in 1940. This species is widely distributed and ranges throughout all the biotic communities. Although it nests in the low cliffs of the adjacent mountains, it does most of its foraging in the valleys. The falcon preys upon a variety of animals, small birds probably being taken more often than by any of the other hawks. The remains of lizards found below its nest indicate that they are taken. The small permeant birds will be noted in the faciations where they are important.

Coactions and Reactions

Ord's kangaroo rat is most abundant in the winter-fat, greasewood, and sagebrush communities in White Valley and in small-leaf horsebrush on sand at Dugway, Utah (Vest 1955 and personal communication). It generally occurs on sandy soils and the Great Basin kangaroo rat, on gravelly ones. Both occur throughout the major community west of the Wasatch Mountains, while Ord's kangaroo rat also ranges into Colorado and Wyoming. There may be 13 mounds per hectare (Fig. 10-4). The breeding season begins early in the spring, and two to four young are born in June.

In one study 71 per cent of the rats that were carrying food materials had leaves of the perennial shadscale in their cheek pouches, which also included winter-fat, 3 per cent; ephedra seeds, 3; young Russian thistle seedlings, 7; globe mallow leaves, 1; and grass seeds, 15. They are active at night, gathering and storing food materials. Food is stored in side pockets connected with the burrow and in small pits, 1 to 2 inches (2.5 to 5 cm) in diameter and about 1.5 inches (3.75 cm) deep, excavated on the surface of the mound. Within the greasewood community, such pit caches were found to consist of leaves of *Atriplex nuttallii;* the volume of one amounted to about two quarts (2 liters). Another cache in the shadscale community, consisting of seeds of *Ephedra nevadensis,* had a volume of 1.5 quarts (1.5 liters).

The storage of seeds may aid in the distribution of plants. Seeds of the annual *Bromus tectorum* stored in the surface caches are known to sprout and grow (Billings, personal communication). Kangaroo rats also influence the growth of plants by altering the chemical and physical composition of the soil as a result of their burrowing activities. The burrows facilitate the penetration of water and oxygen to greater depths. The bringing of the subsoil to the surface allows it to mix with the topsoil and to increase in moisture equivalent and water-holding capacity (Greene and Murphy 1932, Grinnell 1923, Taylor 1935). Due to the production of excreta and the decomposition of stored plant materials in their burrows, the soil near the dens contains greater quantities of soluble salts, especially calcium, magnesium, bicarbonate, and nitrate ions,

Fig. 10-4. A. Kankaroo rat mound in shadscale vegetation in White Valley; note several burrow entrances. B. Partially excavated burrow of a banner-tailed kankaroo rat in shrub grassland of western New Mexico (photo by E. Rigg).

than away from the immediate location of dens (Greene and Reynard 1932). In certain winter-fat areas, plants growing in the vicinity of kangaroo rat mounds are almost twice as tall as those in adjacent areas (Fautin 1946). Where there is a continuous hardpan close to the surface, winter-fat is unable to establish itself, and large areas are covered by the subdominant little rabbitbrush. However, when there are pocket gopher and kangaroo rat burrows interspersed through the little rabbitbrush, small circular "islands" of winter-fat occur. A transect through such an island shows that the hardpan has been broken by the rat burrow, allowing the roots of the winter-fat to penetrate to greater depths and to obtain sufficient moisture for support (Fautin 1946).

Kangaroo rat burrows are used by reptiles, especially lizards, for escape from predators and as places in which to hibernate. When abandoned by the rats, the burrows are utilized by many spiders, tarantulas, scorpions, and insects, especially tenebrionid beetles. Kangaroo rats also constitute a large part of the food supply of desert foxes, burrowing owls, badgers, and coyotes. Because of their nocturnal habits they are seldom taken by hawks. Rattlesnakes and gopher snakes often prey on them by entering their burrows. Because of these various biotic interrelationships and because the kangaroo rat is so widespread and abundant, it is obviously a dominant species.

The antelope ground squirrel frequents rocky areas and at times has a population of 10 per hectare. Its food consists chiefly of fresh plant material. The Townsend's ground squirrel is distributed throughout Utah and Nevada and is a local dominant. There are several other species of small resident burrowing rodents that supplement the soil moving of the kangaroo rats: the little pocket mouse (see Chap. 15), the dark kangaroo mouse, and the western harvest mouse (Fautin 1946). There are also a few wood rats and many deer mice. The pocket gopher *Thomomys umbrinus* occurs in the various shadscale communities and is largely nocturnal in its foraging above ground. The pocket gophers are very important in soil formation; Grinnell (1923) estimates that they cast up 7.2 tons of earth per square mile (2500 kg/km²) in Yosemite Park, California, each year. The earth-moving activities of the lizards have not been evaluated.

SHADSCALE–KANGAROO RAT ASSOCIATION

Billings states that the shadscale community is a climatic climax because it occurs on mildly saline and even nonsaline soils under conditions of low rainfall. However, in western Utah, with more rain than in either Nevada or eastern Utah, the salts in the soil produce optimum conditions for this association. Faciations are almost endless because of the varying availability of moisture in the soil and differences in the impact of solar radiation (Shreve 1924).

The shadscale vegetation in western Utah lacks *Sarcobatus baileyi* and the desert thorn *Lycium cooperi*. Pure stands of shadscale occur on the heavy silty soils of the drier valleys (Billings 1951), and the species varies from first to fifth in abundance in all the many faciations and facies. The Great Basin kangaroo rat makes up 22 to 80 per cent of the burrowing rodents in all the faciations. Ord's kangaroo rat was present in most of them in smaller numbers during Fautin's period of study (1939-40). C. L. Hayward (personal communication) has found, on the other hand, that over several years Ord's rat is usually the more abundant and widely distributed of the two species.

Shadscale–Rice Grass–Ord's Kangaroo Rat Faciation

This is the most widely distributed grouping of species. An area was studied by Shelford and party in June, 1945, and a count of plants made. The area had been protected from livestock grazing for ten years, but had not completely recovered its original condition (Table 10-1).

Animal dominants recognized on the area were the two kangaroo rats, the little pocket mouse, the jack rabbit, and the pronghorn. The common birds were the horned lark, vesper sparrow, western kingbird, and loggerhead shrike. Horned toads, western whiptails, northern side-blotched lizards, and leopard lizards were seen but not censused. The Great Basin rattlesnake was abundant.

Table 10-1. The Abundance and Frequency of Plants on Six 20-square-meter Plots in the Shadscale–Rice Grass–Ord's Kangaroo Rat Faciation at the Desert Range Experiment Station, Milford, Utah

Vegetation	Number	Frequenc
Shadscale	192	6
Indian rice grass	73	6
Winter-fat	67	4
Black sage	49	6
Rabbitbrush	23	2
Galleta grass	1068	5
Globe mallow	21	4
Wild tobacco	56	4
Galium	37	3
Cactus (*Opuntia*)	1	1
Miscellaneous forbs	2	3
Miscellaneous grasses	3	2

Invertebrates included the spider *Metepeira foxi* which appeared to be characteristic. The most strikingly abundant insects were leafhoppers; Fautin, who identified *Eutettix insanus* and *Aceratagallia cinerea,* also found this true. The grasshoppers *Trimerotropis pallidipennis* and *Melanoplus occidentalis* were present in moderate abundance. A few years later, the latter species occurred in "outbreak" proportions. *Cordillacris occipitalis* and *Areochoreutes carlinianus strepitus* were present in 1945. Fautin had earlier found *Xanthippus lateritius* and *Aeoloplides tenuipennis.*

Present are the nocturnal burrowing cricket and the tenebrionid beetle *Eleodes* which is generally distributed in arid areas. Tenebrionids take the place of the carabids of forested areas. As to ants, *Pogonomyrmex occidentalis utahensis* is widely distributed, and *Myrmecocystus navajo* and *Dorymyrmex pyramicus* are often present.

The character of the vegetation at a contact with hot desert is evident at Daylight Pass near Beatty, Nevada, at an elevation of 4300 feet (1300 m). *Sphaeralcea ambigua, Larrea divaricata, Eriogonum inflatum, Aster tortifolius,* and *Mendora spinescens* are present along with *Atriplex confertifolia.* The Great Basin kangaroo rat also occurs here.

Horsebrush–shadscale–Ord's kangaroo rat faciation. Of its four important plant dominants, littleleaf horsebrush is most outstanding in regard to density and ground coverage, followed by shadscale, little rabbitbrush, and galleta grass. The faciation is not common in Nevada. The principal animal dominants are the two kangaroo rats with the Ord's rat more abundant than in the shadscale faciation above. The black-throated sparrow is a relatively abundant nesting bird. Nearly all the reptiles in the shadscale faciation are occasionally present. On the shrubs, Fautin (1946) found that the mirids, notably *Lygaeus elisus, Psallus,* and two species of *Nysius,* were the most abundant hemipterons. There were the spider *Misumenops celer,* the homopteron *Ceratagallia artemisia,* and the beetle *Eupagoderes varius.* On the ground were the tenebrionid beetles *Eleodes hispilabris* and *E. obscura.*

Greasewood–Harvest Mouse Faciation

This widespread community (Fig. 10-5) is largely dominated by grease-wood, salt-blite, shadscale, saltgrass, black-tailed jack rabbit, and harvest mouse. The jack rabbit reaches its maximum abundance here. The mouse is restricted to this community in White Valley, Utah, and to the sagebrush community. Ord's kangaroo rat is more abundant than the chisel-toothed

Fig. 10-5. A. Indian rice grass in shadscale vegetation. B. Greasewood flat in White Valley, Utah (photos by R. W. Fautin, courtesy Ecological Monographs).

kangaroo rat, and the deer mouse is very common. The loggerhead shrike is a permanent resident, and was found nesting in no other community. Northern side-blotched lizards are restricted to sand dunes, and rattlesnakes are rarely encountered. The sagebrush swift was found in no other community in White Valley but was relatively common here (Fautin 1946).

Spiders are very abundant. *Sassacus papenhoei* appears to be restricted to this community. Hemipterons, notably the stink bugs *Thyanta rugulosa* and *Chlorochroa sayi,* are conspicuous. Homopterons make up half of the insect population, largely due to the treehopper *Enchenopa permutata.* The chrysomelid beetle *Pachybrachis* is present.

Winter-Fat–Rice Grass–Kangaroo Mouse Faciation

This faciation (Fig. 10-6) is largely dominated by winter-fat, Indian rice grass, shadscale, globe mallow, the kangaroo mouse *Microdipodops pallidus,*

Fig. 10-6. A. *Atriplex confertifolia;* note compact leaves. B. Horned lark nest at the base of white sage or winter-fat (photos by R. W. Fautin). C. The segregation of lizard populations in three of the faciations.

the little pocket mouse, and the northern grasshopper mouse. The minor influent birds are much the same as in the shadscale community with the horned lark the only nesting species. Reptiles and insects are practically the same as in the shadscale with probably fewer numbers of most species.

The little rabbitbrush is a subclimax species that is widespread in the desert. It is not found on soil with a high mineral content or on heavy soils where the water table is high. Consequently, it is most common in areas where shadscale and winter-fat are dominant (Fautin 1946).

Black Sage–Wood Rat Associes

This subclimax community occurs on black lava rock knolls which attain an elevation of 5000 feet (1500 m). The principal dominants are shadscale, black sage, and galleta grass. The desert wood rat and Nuttall's cottontail reach their maximum populations here.

Subordinate species of the horsebrush *Tetradymia glabrata* and *T. spinosa*, Indian rice grass, broomweed, gray molly, the rock inhabiting *Laphamia stansburii*, the long-tailed pocket mouse, and the canyon mouse are restricted to this community. No ground squirrels were seen and jack rabbits were scarce.

The birds are practically the same as in the horsebrush–shadscale community. Visitants are the gray flycatcher, violet-green swallow, and cliff swallow. The reptiles are also the same as in the horsebrush–shadscale community; the western whiptail and the leopard lizard occur in smaller numbers than elsewhere, but the collared lizard reaches its greatest abundance (Fautin 1946).

Edapho-climatic Communities

Greasewood forms edapho-climatic climaxes on heavy salty alluvial soils of the finest types in the bolsons and playas where ground water is not far below the surface. Its chief associate is lenscale, but in many cases greasewood is present in pure stand. Scattered clumps of seepweed and poorly developed patches of desert saltgrass occur in south-central Oregon, southwestern Wyoming, New Mexico, Texas, and California (Shantz and Zon 1924, Billings 1945, Fautin 1946). Billings states that myrmicine ants bury *Atriplex* seeds under the friable surface soil crust to a depth of five to ten centimeters. The greasewood community is inhabited by the Sonoran deer mouse, the Bonneville Basin kangaroo rat, and the white-tailed antelope squirrel.

There are a great many variations in the grouping of plant and animal species due to variations in the salt and water conditions of the soil. At a point 66 miles (106 km) southwest of Ely, Nevada, near the center of the biome, the following plants were noted in July, 1944: *Sarcobatus vermiculatus, Atriplex parryi, Allenrolfea occidentalis, Chrysothamnus graveolens,* and *Artemisia.* There was an unusual array of grasses: *Hordeum jubatum, Sporobolis airoides, Agropyron spicatum,* and *Agrostis.* Insects were numerous, including the beetles *Collops flavicinctus* and *Systena,* the cicada *Okanagana utahensis,* the leafhoppers *Parabolocratus viridis* and *Exitianus obscurinervis,* and the ants

Iridomyrmex pruniosus and *Pogonomyrmex californicus.* One corizid bug
Harmostes angustatus was taken.

Relations of Animals in Different Faciations and Facies

Differences in habitat preferences and potential influence of animals in
different types of vegetation are shown in Table 10-2. The density of 5.0 jack
rabbits per 10 hectares is an average of records varying from 3 to 22. The
rabbits come into the greasewood for shade during the day, giving high mid-
day counts; in the mornings and evenings they scatter out in the more open
areas to feed. Pellet counts indicate that rabbits are considerably more abundant in winter-fat and shadscale communities that are protected from grazing
and contain grass than they are in the same type of communities which are
subjected to grazing. Pellet counts are very high in grass. As a consequence of
its feeding, the jack rabbit is a very important influent of the communities in
which it occurs and should be ranked as a dominant in some of them (Fautin
1941).

*Table 10-2. Abundance of Animals per 10 Hectares and Per Cent Occurrence in
Sample Plots in Different Types of Vegetation*

VEGETATION	BLACK-TAILED JACK RABBIT		NUTTALL'S COTTONTAIL	
	FREQUENCY PER CENT	AVERAGE NUMBER INDIVIDUALS	FREQUENCY PER CENT	AVERAGE NUMBER INDIVIDUALS
Shadscale...........	17	0.7	0	...
Horsebrush..........	40	1.0	85	2.8
Greasewood..........	100	5.0	0	...
Sagebrush...........	100	4.3	37	0.5

Nuttall's cottontail was found regularly only in the horsebrush and black
sage communities. It relies less on its speed for protection than the jack rabbit
and seeks safety by darting into a crevice or hole. It occurs among large rocks
or occupies old badger excavations.

Bird populations are relatively low in all communities. The average population from April to September varies from 10.6 birds per 10 hectares in
horsebrush and 12 in shadscale to 25.7 in greasewood. Populations are high in
the shadscale and horsebrush communities during spring and early summer,
but as the temperature increases and the horsebrush begins to shed its leaves
the birds move out. The permanent resident population is much the same in
all communities. The permanent birds include the sparrow hawk, horned lark,
common raven, and loggerhead shrike (Fautin 1941).

Horned larks are present in all communities, being most numerous in shadscale during the breeding season but occurring in greatest abundance in the
greasewood community during late summer and autumn. The high frequency

of occurrence of this species in sagebrush and greasewood is due to the presence of large open areas where it is always found except during the heat of the day when it seeks the shade of the tall shrubs. It nests in the shadscale community. Nests are placed on the ground in a shallow excavation beneath low-growing shrubs. Nesting pairs are well spaced, the minimum distance between nests being 52 meters.

Snakes are more important than lizards. The crepuscular Great Basin gopher snake is encountered frequently, especially during May, and is present in all communities but is most abundant in the shadscale community (Hayward, personal communication). It preys to a great extent on ground squirrels, locally called "gophers." Its stomach contents include ground squirrels, antelope ground squirrels, and kangaroo rats. The Great Basin rattlesnake occurs in all communities except sagebrush. This species is also largely crepuscular in its activities. The striped whipsnake is widespread through all the communities, but is much less abundant than the other species. Its food includes checkered whiptails, uta lizards, grasshoppers, and tenebrionid beetles. Figure 10-6C shows the segregation of lizards into different faciations.

Hydroseral Stages

Salicornia rubra occurs close to the water in playas. Where the ground is a little higher, iodine bush and sometimes salt grass occur. This is usually near greasewood.

In Smoky Valley near Millett, Nevada, there is a series of springs, varying in size from small, water-filled holes only a few inches across to large open pools many feet in diameter. The ponds resulting from these springs are fairly permanent and contain an abundance of vegetation as well as animal life. They are inhabited by more kinds of ducks, herons, and coots than any other aquatic habitats within the area. Tadpoles of the spadefoot toad *Scaphiopus hammondi* and the toad *Bufo boreas* occur. Silver buffaloberry is a predominant plant in a belt one-half to two miles (0.8 to 3.2 km) wide around the alkali flat southeast of Millett. These thorny plants occur singly or in dense thickets ten feet (3 m) or more in height. The thickets make excellent retreats for birds and mammals. The abundant fruit is food for the sharp-tailed grouse, chipmunk, deer, and some 12 species of small birds (Linsdale 1938).

A slightly deepened playa pond at the Desert Range Experiment Station contained water in the last week of June, 1945. There was none of the usual aquatic plants present, but a very few greasewood, *Atriplex nuttallii* shrubs, and a scattering of herbs occurred. Tadpoles of the spadefoot toad were numerous and of large size. Midge fly larvae of *Tanytarsus nigricans*, *Tendipes decorus*, *Procladius*, and *Spaniotoma* were present in large numbers. The mayfly nymph *Callibaetis*, two species of corixid water-boatmen, and dytiscid beetles were also present in small numbers. The wet mud margin was occupied by three species of the ground beetle *Bembidiom*, including *B. henshawi* and *B. fuscicurum*, and the shore bug *Salda interstitialis*. Mud-using wasps were conspicuous visitors.

Farther away from the pond the vegetation was even more scattered. Gray molly was most abundant, followed by *Sarcobatus vermiculatus, Atriplex nuttallii,* and *A. canescens* in equal abundance. Here one square meter of vegetation held 55 minute insects of which 19 were leaf bug nymphs.

The Humboldt River in Nevada is a muddy stream with a bed of alkali silt. From Golconda to Humboldt Sink, it meanders through a treeless valley. It is bordered by dense thickets of mostly the coyote willow *Salix exigua,* called pin willow by the cowpunchers because it seldom gets larger than an inch in diameter. At this size it dies and remains erect among the younger sprouts to make a thicket scarcely penetrable except along animal trails. Eight or more species of dragonflies occur, including *Gomphus olivaceous nevadenses* and *G. intricatus,* some in considerable numbers (Kennedy 1917). Salt grass sod areas, as well as a wide area of greasewood, occur near the river (Shantz and Zon 1924). The cottonwood *Populus fremontii* dominates the only arborescent community to enter the Carson Desert region below 4500 feet (1360 m). It forms gallery forests along the lower Truckee, Carson, and Walker rivers (Billings 1945).

Sand Areas

Sand dunes occur in the northwestern part of Utah and in many other areas. Dean Vest (1955 and personal communication) reports a dense growth of plants in the sand areas due to the high water and low salt content. There appear to be two types of sand dunes, one in the foothills with *Juniperus osteosperma* as the most abundant plant and the other one in the valleys with horsebrush as the most important plant. Both of these areas have a larger population of rodents and a greater variety of species than in surrounding communities at the same level. The most abundant rodent in both types of sand dune vegetation is a pallid-colored Ord's kangaroo rat; the second in abundance in the juniper area is the white-tailed antelope ground squirrel. The kangaroo mouse is absent from the foothill sand. While all successional stages in the development of the sand community are available, the invasion by plants and animals has not been described. There is a *Dalea* community on stabilized dunes in western Nevada which contains *Atriplex canescens* and *Tetradymia comosa* (Billings 1945).

Rock Areas

Rock from lava flows occurs in various areas, notably in parts of the Snake River plain. The dwarf golden bush *Haplopappus nanus* and *Drymocallis pseudo-rupestris* appear early on laval rock in the Craters of the Moon National Monument but decrease as the rock gets older. On the other hand, sagebrush, antelope brush, rabbitbrush, and *Gayophytum ramosissimum* increase in density with the age of the flow. The same is true of the grasses *Agropyron spicatum inerme* and *Poa secunda,* which are characteristic dominants of the Palouse prairie. Another variety of *A. spicatum* and the galleta grass also increase with the age of the flow. Weaver and Clements (1929) hold the view that with fire and grazing removed, this area would become bunch-grass prairie.

SAGEBRUSH–RABBIT ASSOCIATION

Sagebrush is usually the outstanding plant dominant. Grasses regularly present are bluebunch wheatgrass, Sandberg bluegrass, porcupine grass, squirrel-tail, foxtail barley, and alkali sacaton. Because of the grasses, the several faciations are all properly referred to as sage–grass communities. There is climax sagebrush in xeric areas, especially in the bunch-grass grassland area and in Montana east of the mountains, but it is safe to assume that in 1600 it may have amounted to as little as 50 per cent of the area mapped by Shantz and Zon (1924). The sagegrass is very characteristic of the Great Basin area. On the residual soils above 4000 feet (1200 m) in the western Great Basin, the sagebrush association is the climatic climax (Billings 1945). This relation-ship was very evident in the area studied by Fautin (1946) and has been re-ported by Stewart, Cottam, and Hutchins (1940) and by Shantz and Piemeisel (1924) for other parts of the Great Basin. This community is characterized by the sagebrush *Artemisia tridentata* and the joint fir *Ephedra viridis*. On rocky knolls, *A. tridentata* is replaced by *A. nova* (Billings 1945, Fautin 1946).

The low, stocky form and the arrangement in clumps with intervening bare ground make sagebrush favored living quarters for many kinds of animals. It provides shelter from the wind, pursuing predators, and the sun. The black-tailed jack rabbit and the cottontail *Sylvilagus idahoensis* are prominent species, but *S. nuttalli* is apparently absent. Sagebrush also provides perching places and nesting sites for birds. Shreds of sagebrush bark and small stems make good nest material (Linsdale 1938). A few animals, notably the pronghorn, use the sagebrush for food. In the western half of the biome, sagebrush is in 89 to 100 per cent of all pronghorn stomachs and balsam root in 10 to 45 per cent. Green grass occurs in 40 to 60 per cent of the stomachs and dry grass in 15 to 93 per cent. Pronghorns are probably more numerous in the sagebrush community than in shadscale. Shadscale is used commonly as a food during the winter but not during the summer. Antelope brush was found in 7 to 60 per cent of the stomachs (Ferrel and Leach 1950, 1952). Deer are important at the western edge of the region, especially in winter. Other resident dominant mammals are the Ord's and Great Basin kangaroo rats.

Sagebrush–Jack Rabbit–Wheatgrass Faciation

This is the faciation of sagebrush that covers much of Utah and adjacent states to the east and southeast. Fautin (1941, 1946) found bluebunch wheat-grass, needle-and-thread grass, and balsam root important species besides the sagebrush. Resident mammalian influents in this community are the sage-brush chipmunk and the Great Basin pocket mouse. The western harvest mouse is restricted to this community and to greasewood. Both species of kangaroo rats were present but less abundant than in the shadscale.

The minor influent birds are nearly the same as in the greasewood community. However, the summer residents include mourning dove, common nighthawk, Say's phoebe, and poor-will.

The reptiles are mainly the common species of other communities but are in lesser numbers. The sagebrush swift is restricted here and to the greasewood community (Fautin 1946, Woodbury and Woodbury 1945).

The number of spiders per collection was greater in this community than in any other that Fautin studied, especially during the latter part of July and the first part of August; a great many were immature. The most abundant species are usually *Philodromus* and *Dendryphantes,* which are found also in the greasewood community. The species *Xysticus cunctator* and *Icius* appear to be restricted to sagebrush.

Six of the 15 species of hemipterons collected were restricted to sagebrush. Species of mirids occur in greatest abundance, of which *Chlamydatus uniformis* and *Tuponia* are most numerous. The damsel bug *Nabis alternatus* occurs with greater frequency in this community than elsewhere. Homopterons are more abundant than the hemipterons. The cicada *Neoplatypedia constrictá* occurs on the main branches. The spittle bug *Clastoptera brunnea* is restricted to the community. Cicadellids are usually far more numerous than in other communities, the most abundant species during June being *Empoasca nigra typhlocyboides.* The fulgorid *Hysteropterum cornutum* is usually abundant. The blister beetle *Epicauta maculata* appears restricted here. Orthopterons and coccinellids are very abundant; six species of the latter appear to be restricted to sagebrush. On the other hand, lepidopterons, dipterons, and hymenopterons are usually scarce. Only five species of ants and four species of parasitic hymenopterons were taken. The ants most frequently found are *Camponotus* (*Myrmentoma*), *Monomorium minimum, Iridomyrmex,* and *Formica obtusopilosa* (Fautin 1946).

Near Austin, Nevada, in 1945, at about 6500 feet (1970 m), the author found that the characteristic shrubs of western Utah were sagebrush, rabbitbrush, black sage, rice grass, and wheatgrass. Three bugs, *Thyanta rugulosa, Harmostes reflexulus* and *Pseudopsallus tanneri,* were common.

Sagebrush–Pygmy Rabbit–Bitterbrush Faciation

This community occurs between 4000 and 7000 feet (1200 and 2100 m) and is characterized by sagebrush, horsebrush, joint fir, bitterbrush or antelope brush, hop-sage, gooseberry, and the two rabbitbrushes *Chrysothamnus viscidiflorus puberlus* and *C. graveolens.* In the western Great Basin, the characteristic grasses are *Sitanion hystrix* and *Poa secunda,* and the forbs are *Delphinium andersonii, Zygadenus venenosus,* and numerous others (Billings 1945). Klyver's (1931) transect extended barely across the upper Owens River near the present Lone Valley Reserve. In the Owens Valley area there is sagebrush with and without pinyon–juniper from 5000 to 7000 feet (1500 to 2100 m). Most of these species are not important in Utah, Idaho, and eastward. Large areas of this community are also found in northeastern California (Jensen 1947, Grinnell *et al.* 1930) (Fig. 10-7).

The bobcat, sagebrush chipmunk, Klamath chipmunk, Columbia kangaroo rat, a couple of harvest mice, and piñon mouse are characteristic animals. The pygmy rabbit is abundant north of Lake Tahoe. The desert cottontail takes its place south of Lake Tahoe to the Owens River area (Fig. 6-10). The pronghorn and also the bison in the northern part of the area occurred under pristine conditions. Robins and evening grosbeaks use the western

Fig. 10-7. A. Sagebrush–bitterbrush vegetation five miles northeast of Reno, Nevada, at an altitude of 4800 feet (1450 m) (courtesy A. E. Weislander, U.S. Forest Service). B. Sagebrush in eastern Utah blending into pinyon–cedar. The different-colored and smaller plants on the lower slopes are the same species as those in the foreground but are growing under different soil and moisture condition.

juniper. Bitterbrush or antelope brush was defoliated on 50,000 acres (20,000 hectares) in California in 1943 by the Great Basin tent caterpillar *Malacosoma fragilis,* which may make it an arthropod dominant. The species occurs throughout the eastern front sagebrush.

High Altitude Sagebrush–Cottontail–Cream Bush Faciation

Included in this grouping of species between 6000 and 9500 feet (1800 and 2880 m) is a small area of sagebrush with scattered ponderosa pine noted on a volcanic tableland adjacent to the Owens River in California (Klyver 1931). This is probably still a pristine community, but its occurrence in any large areas has not attracted notice. Adjacent to much of the high sagebrush reaching from Mono Lake north through California into eastern Oregon are large areas of ponderosa pine, mainly near 6000 feet (1800 m). Pinyon–juniper or juniper alone ranges from 5500 to 8000 feet (1670 to 2400 m) but is commonly scattered in dense stands of sagebrush at its lower altitudinal limit. The conifers are locally absent from slopes, but sagebrush is continuous up to 9500 feet (2880 m), although there is a gradual change of most other associated plant and animal species. A plant grouping, which includes cream bush and often mountain mahogany in Nevada and snow bush in California, can be recognized (Billings 1951). Cream bush is also reported in the Inyo Range east of Owens Valley, California, and mountain mahogany is recorded mixed in sagebrush up to 9500 feet (2880 m) near the eastern end of Klyver's (1931) transect. Little is known about the animals in this faciation, except that Nuttall's cottontail occurs throughout.

Permeant Influents Along the Western Margin

Mule deer occupy the slopes of the Sierras south of Lake Mono to Owens Lake; their winter range lies in the Great Basin sagebrush community and probably to a minor extent in the hot desert. They leave the highland west of Owens River with the first snowfall. Some of them move down the river and across it at a definite point each year. They then continue across the Inyo and other desert ranges into the cold desert to the east (Cowan 1936).

Likewise, north of Mono Lake and into Oregon mule deer come out of the mountains into the sagebrush at the beginning of winter. In northern California they form a winter herd of 20,000 to 28,000 animals on 335,000 acres (134,000 hectares) and are often concentrated on about 100,000 acres (40,000 hectares). Their forage has become badly overbrowsed. Figure 10-8 shows by months the kinds of food used. The mountain mahogany and snow bush, characteristic of the higher altitudes of the dry mountains, are utilized during the migration. In much of the winter range, juniper is absent, and the principal shrubs are sagebrush and bitterbrush. Throughout the area there are many grasses such as big squirreltail, Thurber needlegrass, *Poa, Agropyron,* and a number of introduced species.

The seasonal migrations of the pronghorn in northeastern California are similar to those of the deer, but the two species are not conspicuously associated. Their winter concentrations are farther south and in different valleys than those of the deer. The autumn, winter, and spring food is that to be found in the Great Basin habitat. Their summer food in the mountains has not been reported.

Previous to 1825, small herds of bison moved south through Surprise and Alturas valleys in northeastern California, which is an area frequented by pronghorns and deer. It is probable that the bison fed on grass in the well-watered Surprise Valley, but they were also known to browse.

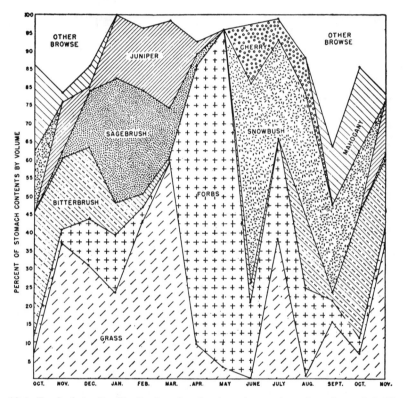

Fig. 10-8. Food of the Devil's Garden deer herd at the northern boundary of California (from the Interstate Deer Herd Commission 1951).

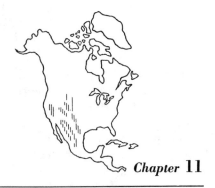

Chapter 11

Ecotone Woodland and Bushland Communities

The communities discussed in this chapter occur in the foothills of both the eastern and western cordilleras from Montana into Hidalgo, Mexico, on the east, from Oregon into Zacatecas on the west, and on nearly all the low mountains in between. They are absent from the ranges near the Pacific Ocean in Oregon and through most of California. There are two types of vegetation involved: woodland, which is an orchard-like but irregular arrangement of low, short-trunked trees with a scattered growth of shrubs and herbs beneath, chiefly pines, junipers, and oaks in various combinations, and a usually dense growth of shrubs thinning out into grassland or desert at low elevations.

The area extends 2000 miles (3200 km) from north to south, and covers about 800,000 square miles (over 2,000,000 km²). No single species of plant or animal appears to be present throughout. The pinyon *Pinus edulis,* the alligator juniper *Juniperus deppeana,* and the oak *Quercus arizonica* probably have the widest distribution, but rarely do all three species occur in the same area. There appear to be no animals which are distributed throughout any considerable part of either the woodland or bushland. The annual rainfall ranges from 12 to 25 inches (30 to 62 cm) ; the winters vary greatly in amount of snowfall and in minimum temperature.

In 1600, the native population was small; the Indians were primarily maize growers but utilized pinyon nuts for food and trees for firewood and housebuilding. They apparently did not destroy the vegetation over areas of any considerable size.

The following regions may be recognized (Fig. 10-1) :

(1)   *The western juniper–buckbrush region.* Buckbrush, sagebrush, or bitterbrush occupies the interspaces between the western junipers. This vegetation occurs in northeastern California, northwestern Nevada, and southwestern

Oregon; its eastern limit is not known. This community has already been discussed in Chapter 10.

(2) *The pinyon–juniper region.* Except for the northwest portion, pinyon–juniper vegetation is scattered over the Great Basin, extends in a minor way into southern California, occurs on the Rocky Mountain plateaus between northern Wyoming and central Arizona and New Mexico, and has an important extension southward through Texas into Mexico.

(3) *The oak–juniper region.* Oak–juniper is common in eastern Arizona, New Mexico, and west Texas, and extends southward on the slopes facing the Mexican Plateau between 500 and 6000 feet (150 and 1800 m) elevation. The vegetation contains *Agave* in Mexico, but *Agave* is not important in Arizona. Oak–juniper alternates with pinyon–juniper in the Davis Mountains, and constituent species of both communities occur together in the Chisos Mountains of Texas.

(4) *The oak woodland region.* Oak woodland occurs at the lower altitudes in Arizona and on the Pacific slope of Sonora Province in Mexico.

(5) *The oak bushland.* This vegetation, sometimes erroneously called chaparral, occurs in the Rocky Mountains and merges into oak woodland in Arizona. It is generally characterized by the presence of wavyleaf oak and hairy cercocarpus. Bushland occurs below pinyon–juniper over the eastern two-thirds of the pinyon–juniper range.

Most woodland dominants vary from trees as much as 50 feet (15 m) high to shrubs as low as 10 feet (3 m) and, in some cases, to bushes of lesser height. Various authors have considered the woodlands a distinct plant formation and have combined the Rocky Mountain oak bushland with the California chaparral into another distinct formation. However, in the Schimper (1903) sense, a "formation" is a distinct large area of specialized vegetation, and this cannot be applied to these broken attenuated stands of vegetation as they occur in North America. Since the animal life is also not distinctive, woodland and bushland cannot qualify as biomes (Figs. 10-1, 11-5) (Hayward 1948), although the California chaparral can (Cronemiller 1942).

PINYON–JUNIPER WOODLAND

The pinyon–juniper woodland is the most clearly defined of the several communities and is best known scientifically. The two dominant genera, *Pinus* and *Juniperus,* are regularly associated, are similar in character and requirements, and give a uniform physiognomy. This woodland occurs most frequently as a narrow belt, from less than a mile to a few miles wide, from the front range of the Rocky Mountains to the eastern slopes of the Sierra Nevada foothills. Its northernmost limit with both pinyon and juniper present is in the Big Horn Mountains of Wyoming. It is absent north of Pikes Peak, Colorado, in the main ranges of the Rocky Mountains. Its most southerly known areas are along the Pan-American Highway between 220 and 232 kilometers north of Mexico City. Here, the juniper predominates. In the hills 320 kilometers from the city, the pinyons are most conspicuous. The greatest width of the area occupied by the community is about 900 miles (1400 m) along 35° N. Lat. Woodland composed of *Juniperus occidentalis* and *Pinus edulis* is essentially limited to the three Pacific Coast states and Baja California. Over the remainder of the region, the woodland consists of *J. osteosperma, J. scopulorum, P. edulis* and *P. cembroides,* although the latter species is restricted to the Rocky Mountains. The only large continuous areas of woodland are on the plateaus in the northern half of Arizona, although smaller areas of significant size occur in eastern and southern Utah and in parts of southwestern Colorado.

Climate

The climate of the vast attenuated area occupied by the pinyon–juniper woodland can be suggested only by scattered data. In the rough topography, the community occurs at elevations of 4500 to 9000 feet (1360 to 2720 m). The annual rainfall is 10 inches (25 cm) in Oregon, 14 inches (35 cm) at Santa Fe, New Mexico, and 16 inches (40 cm) at Mesa Verde, Nevada, and in eastern Utah. Mean annual temperatures range from 48 to 51° F. (8.9° to 10.6° C.). The winters are cold, with 0° F. (–17.8° C.) occurring from time to time in some areas.

Kaibab–Zion Area

PLANT RELATIONS

On the Kaibab Plateau in northern Arizona, pinyon–juniper woodland forms a characteristic zone (Fig. 11-1) between ponderosa pine forest and short-grass plains on the east, cold desert sagebrush (Fig. 11-2) on the north, and rabbitbrush and shadscale on the south. The plateau contains some oak brush on the west side. Areas of cliff rose and other broad-leaved shrubs occur on the northeast (Fig. 11-2). The pinyon–juniper is found mostly between 5500 and 6800 feet (1670 and 2060 m) in a zone from 4 to over 12 miles (6 to 19 km) in width. On exposed southwestern slopes, the upper limits extend frequently to 7250 feet (2200 m), while on cool north and northeastern slopes they come at only 6500 feet (2000 m) and the lower limits go down to 5000 feet (1500 m) (Rasmussen 1941).

The low branched trees, reaching a height of only 20 to 30 feet (6 to 9 m), do not ordinarily form solid stands but are scattered, with the intervening spaces covered with grass, sagebrush, or other shrubs. The relations of pine and juniper are shown in Figure 11-1. Toward the upper edge of the woodland zone, the low trees gradually give way to ponderosa pine. In some local areas, there is a narrow belt of oak bush between the pinyon–juniper and ponderosa pine.

In the canyon of Zion National Park, Utah, 80 miles (128 km) northwest of the Kaibab area, the principal dominants are the Utah juniper and the singleleaf pinyon, with *Juniperus scopulorum* and *Pinus edulis* in the more moist situations (Woodbury 1933).

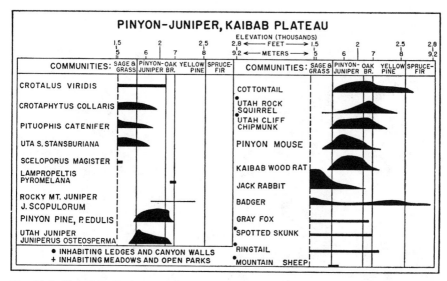

Fig. 11-1. The altitudinal distribution and relative abundance of woodland influents and dominants on the Kaibab Plateau. The confinement of amphibians and reptiles to the low elevations is shown in the upper two-thirds of the first column; woodland trees are shown in the lower third. The distribution of mammals in the woodland area is shown in the second column (after Rasmussen 1941). Compare with Figure 6-14.

The cliff rose occurs throughout the pinyon–juniper on the Kaibab Plateau (Table 11-1) and is important in Zion National Park, but less extensive in other areas. Although ordinarily a shrub 5 to 10 feet (1.5 to 3 m) high, it also forms small trees from 15 to 20 feet (4.5 to 6 m) in height. It is frequently damaged by the feeding of deer during the winter (Fig. 11-2). Toward the northern edge of the Kaibab woodland area the most important shrub is sage-

Table 11-1. *The Relative Abundance of Secondary Plants at Different Altitudes of the Pinyon–Juniper Woodland[a]*

VEGETATION	LOWER	MIDDLE	UPPER
SHRUBS	*Number per Acre (0.4 Hectare)*		
Sagebrush	1340	320	140
Prickly pear	80	20	0
Joint fir	80	0	0
Desert barberry	20	0	0
Cliff rose	60	140	220
Serviceberry	0	40	80
HERBS	*Per Cent of Ground Covered*		
Blue grama	30	10	0
Broomweed	3	30	10

[a] Rasmussen 1941.

Fig. 11-2. A. Impact of deer on the Kaibab woodland: the pinyon on the left and the juniper on the right both show deer browse lines; the juniper has a broken down limb; damaged sagebrush is in the foreground and broken stems lie scattered on the ground. B. Undisturbed cliff rose, pinyon, and sagebrush in the northeast corner of the Grand Canyon National Game Preserve, which the deer could not be induced to use in winter Grasses and herbs cover about half of the ground. C. Condition of an area similar to B but between seven and ten years after the maximum population of deer. In the center is a juniper with browse line; at right center is a dead cliff rose; to the right of the tree is a regenerating cliff rose; and at the left is another which was tall enough to show a browse line. The sagebrush in the foreground has regenerated; the ground is littered with broken stems.

brush. In addition to the shrubs listed in Table 11-1 other less important species are bitterbrush or antelope brush, mountain mahogany, small sagebrush, Apache plume, black bush, and *Yucca baccata* (Rasmussen 1941). In Zion National Park, there is some undergrowth of Mormon tea, cliff rose, Arizona mahogany, and silverberry (Woodbury 1933).

The important Kaibab herbs are *Solidago petradoria, Pentstemon* (several species), *Calochortus nuttallii, Sphaeralcea, Cryptantha, Hymenopappus lugens, Opuntia basilaris, Mammillaria arizonica,* and *Erodium cicutarium. Pentstemon* and *Sphaeralcea* were also noted in Zion and Mesa Verde national parks.

Finger-like projections of cold desert shrubs extend from lower elevations up into the area of pinyon–juniper community. Sagebrush is dominant in the better soils with the shallow-rooted woodland trees dominant on the ridges and in more rocky situations. As altitude increases, sagebrush gradually decreases until it practically disappears at the upper limits of the woodland community.

MAJOR PERMEANT MAMMAL DOMINANTS AND INFLUENTS

The mule deer is a dominant. This is true even though the montane and subalpine forest and the oak bush are essential for breeding and passing the summer. The deer spends the winter almost exclusively in the pinyon–juniper, and this has proven to be a critical period in the annual cycle of the species. On the Kaibab, the species migrates only to an area west of the center of the plateau. This is a tradition for the species which various management efforts have failed to change. Its favorite winter food plants in the northeast and east remain untouched. The mountain lion is its chief predator, but the deer population remains large in spite of losing one or two individuals per week per mountain lion (Rasmussen 1941).

The permeant mountain lion and coyote are also most abundant in the woodland during the winter. Both should be considered dominants. The mountain lion has maintained a large population on the Kaibab Plateau in spite of the destruction by man of relatively large numbers (600 in 12 years, 1906 to 1918). There are perhaps two litters of three kittens each per year. In the roughest country, where the species is most numerous, there may be one animal to every 200 to 225 deer (Rasmussen 1941).

The coyote is present throughout the year on the Kaibab Plateau. The summer population is the lowest, estimated at 5.5 to 7 per 10 square miles (2.1 to 2.7/10 km²). This is to be compared with 10 to 11 per 10 square miles (3.8 to 4.2/10 km²) at higher elevations. In the autumn, there is a considerable downward movement out of the mountains so that the coyote population during the winter is two or three times its numbers during the summer. These high populations occur both in areas of deer abundance and in areas where deer are scarce or absent.

In the late autumn of 1931, several coyotes could be seen by close observation during one day's ride. Tracks and droppings were in evidence in great number. Within the droppings, one could identify deer hair, rabbit fur, deer mice remains, and grass. Some droppings from late summer were composed primarily of vegetable matter: juniper berries, serviceberries, and a considerable amount of grass. Other droppings held the remains of deer mice and chipmunks. Thus within this community, the coyote's main food consisted of the small rodents that are found there.

The deer population increased from 4000 in 1906 to a maximum 100,000 in 1924. The cause of the great increase has been ascribed to the withdrawal of predator checks. Both the mountain lion and coyote were removed in large numbers. However, this period came at a time when solar radiation and weather were favorable to high fecundity, and the population increase could have resulted in part from higher rates of reproduction (Shelford 1952a, 1952b).

Another resident predator which ranks as an important influent on the Kaibab Plateau is the bobcat. The lair of the bobcat is, as a rule, on the cliffs and ledges, and it is found wherever such topography occurs. However, their numbers do not equal those of the coyote. The food of the bobcat consists of small rodents, some birds, and an occasional fawn (Rasmussen 1941).

MINOR MAMMALIAN INFLUENTS

The desert wood rat is very characteristic of the pinyon–juniper woodland where it is active the year around. It is present both in the Kaibab Plateau and in Zion National Park. Its nest is placed about the stump of a juniper or pinyon or under the protection of the spiny leaves of a yucca. The houses which contain the nests are made up of whatever the "pack rat" can collect. One such house contained approximately ten bushels of materials, of which 85 per cent was sticks and twigs of pinyon, juniper, the barberry *Berberis fremontii*, and some miscellaneous other species; 5 per cent, empty pinyon cones; 4 per cent, the bones of cattle, horses, deer, and jack rabbits; 2 per cent, rocks; 1 per cent, opuntia cactus; 1 per cent, mushrooms; and 2 per cent, deer hair and hide. The nest also contained about one-half pound (225 gm) of pinyon nuts and a small collection of juniper and algerita berries. The wood rat has a wide choice of foods: pinyon nuts which it stores, juniper berries, fruits, and seeds of a variety of plants, herbs in summer, and a considerable quantity of *Yucca baccata* in winter. In level portions of the pinyon–cedar woodland, there was about one house to each one and one-fourth acre (2/hectare).

The nocturnal piñon mouse is perhaps the most abundant mammal within the pinyon–juniper, outnumbering the deer mouse. It is most common in regions where cliffs and broken country provide suitable breeding sites but is not limited to such areas, and mouse droppings may be seen about hollow stumps of pinyon and juniper throughout the woodland. Juniper berries, grass, and herb seeds are included in its diet. Mice in general appeared to be more abundant in this community than in others. The mice are active all year, although the breeding season is in the summer. A very rough estimate of 12 deer mice per acre (30/hectare) was made on the basis of trapping during October, 1931 (Rasmussen 1941).

The cliff chipmunk is active throughout the year. It is more abundant where cliffs and ledges are present. It feeds on and stores pinyon nuts, juniper berries, acorns, fruit of the cactus, and seeds of composites.

The black-tailed jack rabbit is abundant in the area surrounding the Kaibab Plateau and extends upward with the sagebrush through the pinyon–juniper to the ponderosa pine. Its abundance, combined with its voracious feeding habits, gives it considerable influence in the community. Vorhies and Taylor (1933) have shown that 13 rabbits ordinarily eat as much as one sheep (Rasmussen 1941). The size of the rabbit population is related to the amount of sagebrush present, apparently due to the shelter provided by the bushes. The rabbit's diet is practically 100 per cent grass. Nuttall's cottontail occurs throughout the woodland but is not as abundant as the jack rabbit.

The rock squirrel has a local distribution, depending on the presence of rocks and canyon walls, but nevertheless is probably the most characteristic mammal of the pinyon–juniper woodland. There may be several per acre where conditions are favorable.

The porcupine occurs in small numbers in the woodland. There seems to be an increase in abundance during winter, resulting from a limited downward migration from higher elevations. The pinyon is its main food in this association, but some cliff rose may also be taken.

Small numbers of gray fox, ringtail, spotted skunk, and badger occur, but these carnivores are more abundant in adjacent communities.

BIRDS

The red-tailed hawk is a year-long resident over the entire Kaibab Plateau. In summer, there is about one pair per square mile (0.4/km²), but this number is doubled in winter. The golden eagle is also present, nesting in the rougher terrain (Rasmussen 1941).

Small birds are more numerous in climax woodland than in any other communities on the Kaibab Plateau, although even here their numbers are not large due to the scarcity of water and the absence of subclimax stages. There were eight singing males on five acres (2 hectares) during May, 1931. This area was three to five miles (5 to 8 km) from the nearest permanent water (Rasmussen 1941).

The plain titmouse is largely restricted to the pinyon–juniper, being the least migratory of the native species. It forages about the smaller limbs and twigs of trees and larger shrubs. Observations indicate that the abundant leafhoppers are eaten in great numbers during summer, along with a large assortment of other small arthropods. There are about 40 birds per square mile (15/km²).

The scrub jay is a permanent resident in this community and the oak bush, both on the Kaibab Plateau and at Zion National Park. Prominent among items eaten are pinyon nuts, ground beetles, grasshoppers, and ants. The jay occurs in numbers up to 30 birds per square mile (12/km²).

The mountain chickadee is usually present. The red-shafted flicker appears in all wooded portions of the plateau in populations of 18 to 20 per square

mile (7 to 8/km²). Extending into higher elevations are the mountain bluebird, robin, and the sporadic Steller's jay. Juncos are common and conspicuous
during the winter. In the Book Cliff area of eastern Utah, the piñon jay, plain
titmouse, and bushtit are characteristic species (Hardy 1945).

REPTILES

Small numbers of reptiles occur in the pinyon–juniper woodland, increasing
in numbers at lower altitudes. The horned lizard, however, becomes more
abundant in the open ponderosa pine forests. The insect-eating sagebrush swift
is the most common species, averaging six to eight per acre (15 to 20/hectare).
The collared lizard and northern side-blotched lizard have been recorded, and
the Great Basin rattlesnake is found in limited numbers, particularly along the
rocky canyons extending upward from Kanab Creek and Grand Canyon and
in Zion Canyon. The population was large in 1935, following several dry years.
Woodbury (1933) also found the kingsnake and red racer at Zion National Park.

INVERTEBRATES

Ants predominate in and on the ground. *Formica fusca subaenescens,
Leptothorax texanus,* and a very dark red variety of *Myrmica scabrinodis* occur
on both the ground and the vegetation. *Monomorium minimum* and *Crematogaster lineolata* occur about fallen logs, tree trunks, and under rocks. Nests
of *Camponotus sansabeanus vicinus luteangulus* were located under rocks,
but no activity of this species was noted during the middle of the day. The
termite *Reticulitermes tibialis* makes galleries in pinyon and, to a lesser extent,
in juniper wood wherever dead limbs rest on the ground. The tenebrionid
Eleodes and carabid *Cymindis blanda* are common. The poisonous spider
Latrodectus mactans was found under rocks on exposed ridges; it is fairly
common. An asilid *Ospriocerus abdominalis* and a cicada *Platypedia putnami*
are also characteristic species of this community (Rasmussen 1941).

At Zion National Park numerous cicada nymphs feed on roots as much as
120 centimeters below the ground surface. Tenebrionid larvae feed on fungi
and appear in numbers when they become adults. Some of the predaceous
insects are also subterranean, for example, the free-ranging larvae of the robber
fly and bee fly. The scorpions *Hadrurus hirsutus* and *Vejovis punctipalpus* and
the centipedes *Lamyctes pinampus* and *Otocryptops sexspinosus* seek shelter
under rocks in the daytime and range out in search of insect prey at night.
The tarantula *Eurypelma steindachneri* usually uses holes in the ground
(Woodbury 1933).

Eighty-three species of invertebrates were collected quantitatively in the
vegetation. The total population averaged 20,000 per acre (50,000/hectare)
with peaks in late May and again in late summer following the rain season.
This population was composed, by percentages, of spiders, 25; Chermidae, 18;
ants, 12; other hymenopterons (mostly ichneumonids), 8; dipterons, 10; cicadellids, 9; hemipterons, 6; coleopterons, 5; orthopterons, 3; and others, 4. The
most abundant species were the immature crab spiders *Misumenops asperatus,*

the jumping plant louse *Paratrioza cockerelli,* and the ant *Formica fusca subaenescens.* All three were taken from tree-shrub and herb layers. The spider and plant louse were most abundant on the trees and shrubs. The ant was present in the ground collections as well.

The herbs supported the beetle *Monoxia,* which was the most abundant species and occurred on *Gutierrezia.* The banded-winged locusts *Trimerotropis inconspicua, T. cincta,* and *T. cyaneipennis* usually rested on the bare ground in the spaces between trees. They foraged, however, on both grasses and shrubs. In July, 1931, there was considerable local damage by orthopterons to cliff rose, sagebrush, and grass in the western portion of the Kaibab community, in the same area where the deer had most severely damaged their range. *Trimerotropis cincta,* collected from cliff rose, was probably important (Rasmussen 1941).

In the shrub and tree layers of the Kaibab forest, the most abundant species were the spiders *Dendryphantes* and *Oxyopes,* two species of leafhoppers, the jumping plant louse *Psylla brevistigmata acuta,* the beetle *Anthonomus,* and a cecidomyiid fly. On the shrubs antelope brush and wavyleaf oak, as well as on the pinyon and juniper, were the spiders *Theridion, Metaphidippus,* and *Oxyopes rufipes,* the black mirid bug *Psallus nigerimus,* and leafhoppers. The leafhopper *Exitianus obscurinervis* was present on both the herbs and the woody plants. The chrysomelid beetle *Coscinoptera axillaris* was present on the shrubs.

Great Basin

Near the Desert Range Experiment Station in southwestern Utah, the juniper comes down to 5400 feet (1636 m), and occurs scatteringly in the cold desert (Fig. 10-7B). Shadscale and rabbitbrush are the principal dominants, but black sage is important and there is an unusual amount of *Poa secunda* and Indian rice grass. In 1945 the grasshoppers *Cyphoderris monstrosa* and *Aulocara elliotti* and the plant bug *Melanotricus coagulatus* were taken. They were not found in the cold desert where the juniper was absent.

Pinyon–juniper at its lower edge mixes with sagebrush through much of Utah and Nevada. In California and western Nevada, *Pinus monophylla* is present and is associated with *Juniperus californica* and/or the western juniper. *J. californica* occurs also in lower California, where it is associated with *P. edulis, P. monophylla,* and *P. quadrifolia.* The pines all drop out north of the Reno, Nevada, area, leaving juniper woodland (*J. occidentalis*) northward into Oregon (Chap. 10).

The wintering deer is a permeant dominant in the juniper and juniper–sagebrush–pronghorn areas. The Interstate Deer Herd Commission (1951) shows junipers and antelope brush are browsed very much as they are on the Kaibab Plateau. There are similar coactions with deer and the other dominants in and along the eastern California mountains where the pinyon–juniper woodland occurs.

Birds preferring pinyon–juniper woodland in California are piñon jay, Scott's oriole, and ladder-backed woodpecker. Also occurring here are broad-tailed hummingbird, Cassin's kingbird, gray flycatcher, plain titmouse, common bushtit, Bewick's wren, cactus wren, and blue-gray gnatcatcher (Miller 1951).

Rocky Mountains

MESA VERDE NATIONAL PARK, COLORADO

Studies were made at 7200 and 6000 feet (2180 and 1820 m). The trees at both stations were *Pinus edulis* and *Juniperus osteosperma*. At the higher elevation there were above 5 pinyons to 1 juniper, at the lower elevation about 103 pinyons to 1 juniper. Antelope brush, scrub oak, *Yucca, Lupinus, Pentstemon,* and globe mallow were present at both stations in late June, 1944. The soil was inhabited by dipterous and other insect larvae, as it was in Zion National Park. One red velvet ant was taken. Tenebrionids were numerous, and the millipede *Oriulus venustus* was common. Sparkman (1945) found *Liometopum apiculatum lactuosum* to be the most characteristic ant of the pinyon–juniper. It builds carton nests under stones. *Camponotus laevigatus* and *C. sansabeanus vicinus nitidiventris* are likely to be found regularly. Other species of ants taken are *Lasius niger, Solenopsis molesta, Formica neogagates lasiodes vetula,* and *F. fusca argentea.* The termite *Reticulitermes tibialis* is present in fallen tree trunks.

The herbs yield 30 arthropods per square meter. Hemipterons are represented by *Teleonemia,* a brown mirid *Deraecoris bullatus, Coquilletia insignis,* and numerous tingids. The homopterons are leafhoppers, including *Ballana recurvata, Dikraneura carneola,* and *Exitianus obscurinervis.* These plant sapsuckers are unusually free of spider enemies. One parasitic hymenopteron is present with moth larvae. A number of moths frequent the flowers of the yucca, and a few gall wasps *Rileya tegularis* are present. The tree layer is well inhabited by sap-sucking insects.

On five acres (2 hectares) there were found 15 chipping sparrows, 4 black-throated gray warblers, 10 mountain chickadees, 3 mountain bluebirds, 3 violet-green swallows, 2 crows, 5 towhees, and 3 vireos. The first four species are generally distributed in woodland.

ALAMOGORDO AREA, NEW MEXICO

Pinyon pine, alligator juniper, one-seed juniper, and associated species occur on exposed slopes between 6300 and 7200 feet (1900 and 2180 m), and on sheltered slopes between 5800 and 6800 feet (1760 and 2060 m). Below this belt is the desert and above, the montane coniferous forest. The rock squirrel, piñon mouse, and Mexican wood rat reach their greatest abundance here. The mouse and wood rat occur also in the montane forest, and the rock squirrel ranges from the desert to the summit of the mountains. The jack rabbit and the cottontail of the desert range up into the pinyon–juniper, and montane forest inhabitants, notably the western chipmunk and tawny deer mouse, range down into the pinyon–juniper belt (Dice 1930). Few mammal species seem to be characteristic of the pinyon–juniper woodland on these mountains, and no form is restricted to it.

DAVIS MOUNTAIN AREA, TEXAS

Rainfall here in western Texas is about 16 inches (40 cm). The pinyon–juniper community is found on north-facing slopes and mesas above an altitude of about 6000 feet (1800 m). Pinyon is the most abundant dominant; *Juniperus monosperma* and *Quercus grisea* are important on mesas and rocks but may be scarce or lacking on the slopes. They constitute a definite oak–juniper community mainly on south-facing slopes below 6000 feet (1800 m). Mammals present in both woodlands are black bear, spotted skunk, mountain lion, bush mouse, black-tailed jack rabbit, and mule deer. The harvest mouse is recorded from the pinyon–juniper woodland but not from the oak–juniper (Blair 1940).

BIG BEND AREA, TEXAS

Pinyon–juniper woodland occupies the Chisos Mountains above 4800 feet (1450 m) except for small areas of coniferous forest (Muller 1939 and personal communication). Mexican pinyon, alligator juniper, drooping juniper, one-seed juniper, emory oak, and gray oak are important constituents. Additional trees are madrone, evergreen sumac, mountain mahogany and Gregg ash. Several grasses, especially *Muhlenbergia,* give character to the community. This is evidently a transition between pinyon–juniper and oak–juniper. It is the southern limit of continuous pinyon–juniper (Taylor *et al.* 1944). The pinyon–juniper area on the outer eastern slope of the Mexican Plateau near 20° N. Lat. is small and apparently isolated.

The gray fox, brush mouse, white-tailed deer, whip-poor-will, Colima warbler, and rufous-sided towhee are the principal mammals and birds. The alligator lizard and the black-tailed and rock rattlesnakes occur.

OAK–JUNIPER AND OAK WOODLANDS

The oak–juniper woodland is centered in southern Arizona, New Mexico, and northern Mexico. It extends north locally in the mountains of the eastern half of Arizona and the western half of New Mexico to the thirty-fifth parallel. There is no pinyon–juniper south of 33° 30′ N. Lat. in Arizona (Nichol 1937). In New Mexico and Texas only an arbitary boundary can be drawn for oak–juniper, due to its relations to the topography. Oak–juniper occurs commonly in the mountain ranges of trans-Pecos, Texas, and is found scattered in the canyons and escarpments of the Staked Plains and the Edwards Plateau. Here the pinyon–juniper woodland is commonly at higher altitudes and on north-facing slopes than it is in the Davis Mountains. More mammals, including the ringtail, occur in the oak–juniper (*Quercus grisea–Juniperus monosperma*) than in the pinyon–juniper (Blair 1940).

On the Edwards Plateau escarpment near Kerrville, Texas, the dominant trees are Ashe juniper, live oak, Spanish oak, shin oak, or blackjack oak, depending on soil characteristics, particularly depth of soil and moisture relations. The grasses are little bluestem, some *Muhlenbergia lindheimeri,* silver beardgrass, and side-oats grama. Taylor and Buechner (1943) describe the deterioration of the community through overgrazing, such as could result from very large bison populations. There were few jack rabbits, and the game animals were turkey, white-tailed deer, bobwhite, mourning dove, and fox squirrel. The deer had populations of three to seven per square mile (1.2 to 2.7/km²). Buechner (1946) adds the scaled quail as an extirpated species. He listed 42 permanent resident birds, including the golden-fronted woodpecker. Four other woodpeckers, common in the north, pass the winter here. Smith (1946) maps at least 16 species or subspecies of lizards as occurring in this region; snakes are not much fewer in numbers. The invertebrates in a pastured area during July are numerous, including the orb-weaver spider *Thiodina sylvana, Oxyopes,* and immature *Habrocestum.* A grasshopper found on *Euphorbia* is *Hadrotettix trifasciatus.* The grasshoppers *Spharagemon bolli, Chortophaga viridifasciata,* which feeds on grasses, and *Psoloessa texana* are probably regular inhabitants. A hemipteron present is the black shield or stink bug *Mormidea lugens.* The chief coreid is *Mecidea longula.* Immature leafhoppers are abundant. The commonest ant is *Crematogaster laeviuscula,* occurring in the trees and on the shrubs and herbs. The sapromyzid fly *Pseudocalliope variceps* is present. Two species of tree snails, *Bulimulus alternatus* and *Helicina orbiculata,* occur in the trees and shrubs.

On the Carmen Mountain, just south of the Rio Grande, a virgin vegetation characterized by several kinds of black oak, live oak, and junipers, some three or four feet (1+ m) in diameter is reported. The forest floor is well covered with forbs. Gloyd and Smith (1937, 1942) found ten species of reptiles.

Farther south in Mexico, below 25° N. Lat., the eastern slope montane forest is an orchard-like woodland consisting principally of oaks, none of which occurs in the United States: *Quercus clivicola, Q. canbyi, Q. porphyrogenita,* and *Q. virginiana fusiformis.* Where trees are widely spaced, *Agave americana* thrives. The sparseness of the forest cover is also responsible for an abundance of grasses: *Stipa mucronata, Aristida purpurea, A. arizonica, Bouteloua radicosa, B. hirsuta,* and *B. filiformis.* At higher altitudes the woodland blends into oak–pine forest (Chap. 18).

Emory oak, silverleaf oak, and Arizona oak, 10 to 20 feet (3 to 6 m) tall, occur in the Santa Catalina Mountains of Arizona and are associated with occasional junipers and the Spanish bayonet *Yucca.* The undershrubs are fragrant sumac and manzanita. The bear grass *Nolina,* agave, cacti, other grasses, and low herbs occur on the gravelly clay soil. Dice and Blossom's (1937) record of the number of mammals taken or seen of each species gives some idea of their relative abundance: rock squirrel, 3; cliff chipmunk, 3; Bailey's pocket mouse, 1; rock pocket mouse, 1; white-throated wood rat, 2; brush mouse, 5; antelope jack rabbit, 1; Arizona cottontail, numerous; mule deer, 1.

A study was made in July, 1944, of the invertebrates in an area northwest of Nogales, Arizona, containing *Quercus emoryi,* a few acacias, wild grape, and the subordinate plants *Hesperidanthus linearifolius, Mimosa biuncifera,* and the grasses *Aristida* and *Festuca.* The invertebrate constituents included the four species of ants *Pogonomyrmex californicus, Camponotus acutirostris primipilaris, Novomessor albisetosus,* and *Xiphomyrmex spinosus;* the ground beetle *Discoderus robustus;* the leaf beetle *Pachybrachis;* the dermestid *Cryptohopalum blateatum;* the cercopid *Clastoptera arizonana;* the leafhopper *Aligia turbinata;* the arachnid *Thrombidium magnificum; Arctosa chamberlini;* and *Xysticus cunctator.*

In western Mexico, from about 5000 to 6000 feet (1500 to 1800 m) in the north and somewhat higher in the south, agaves, junipers, and oaks with grasses, *Yucca, Nolina, Dasylirion,* and *Opuntia* underneath give a parklike landscape. From 6000 to 7000 feet (2100 m), there is a transition belt, and above 7000 feet, an oak–pine community. The change toward higher altitudes is from juniper through oak to pine of the ponderosa type.

Oaks are dominant in the Mexican foothill communities where *Quercus emoryi, Quercus grisea,* and other species are found. Their appearance varies from chaparral through savannah to forest. The juniper *Juniperus deppeana* occupies sunny slopes and dry ridges, singly or in open groves, and commonly interspersed with oaks and agaves. The agaves grow singly throughout.

On the lower slopes in Sonora, Mexico, the transition from the grassland to oak–agave–juniper contains relatively little yucca, bear grass, and sotol, but some desert plants occur. The lower limit of oaks in Sonora is considerably below that on the Chihuahuan side of the mountains, being common at elevations below 3000 feet (910 m) (Brand 1936).

Dalquest (1953) reported for San Luis Potosí the spectacled squirrel, bush mouse, white-throated wood rat, and ringtail; the latter two species also occurring in the desert. The widely distributed white-tailed deer is found in the area as was also the mule deer in primitive time.

BUSHLAND

The Rocky Mountain bushland consists almost exclusively of deciduous shrubs, ranging in height from 2 to 20 feet (0.6 to 6 m), but most commonly from 5 to 10 feet (1.5 to 3 m). Wavyleaf oak and common chokecherry often become small trees (Fig. 11-3). Squaw bush may form a gigantic growth 20 feet (6 m) high and 25 to 30 feet (7.6 to 9 m) in diameter, while mountain mahogany is usually a slender, erect shrub. Bushland is scattered because of the diverse topography which it occupies. The number of plant dominants is large with several species appearing regularly in each community (Clements 1920).

Precipitation comes at all seasons. In central Utah mean precipitation over an eleven-year period was 17.6 inches (44 cm), mean temperature about 43° F. (6.1° C.), with the maximum 97° F. (36.1° C.), and the minimum –30° F. (–34.4° C.) (Hayward 1948).

The *Cercocarpus–Quercus* community reaches its best development in Colorado, southern Wyoming, northern new Mexico, and eastern Utah (Fig. 10-1), at elevations between 5000 and 8000 feet (1500 and 2400 m). It usually lies below the pinyon–juniper belt. There is a climax bushland along the eastern slope of the Sierra Nevadas composed of desert mahogany, the snow bush *Ceanothus cordulatus,* and greenleaf manzanita, with huckleberry oak and shrub-sized canyon live oak occurring locally (Klyver 1931). Farther north *Ceanothus velutinus,* also called snow bush, often makes dense stands. The shrubs may occur mixed with sagebrush or alone and may be found among pinyon and juniper trees or junipers alone. This vegetation is more closely related to chaparral than to the Rocky Mountain bush.

Fig. 11-3. View toward the south at 8500 feet (2600 m) in the Sandia Mountains near Albuquerque, New Mexico, showing a bush community of wavyleaf oak and Mexican locust. At the right are ponderosa pines and at the left Douglas-firs. This is probably a seral community resulting from fire (author's photo 1910, legend from Watson 1912).

The relative frequency of occurrence of the plant dominants on the eastern front of the Rocky Mountains is shown in Table 11-2. Chokecherry, which occurs along streams, has the highest basic water requirements of these five dominants, followed by squaw bush, oak, serviceberry, and cercocarpus. C. L. Hayward (personal communication) states that *Cercocarpus ledifolius* is nearly equivalent to the oak in water requirements. *Fallugia paradoxa, Cowania mexicana stansburiana,* and *Cercocarpus montanus* are all slightly more xeric. *Arctostaphylos pungens* and *Ceanothus cuneatus* are seral in the southwest. *Acer grandidentatum* is also important.

Communities of sumac and oak also occur as narrow belts for hundreds of miles along the bluffs of the upper Missouri River and its tributaries. They are important to animals with forest-edge habits such as the plains grizzly bear,

Table 11-2. Relative Frequency of Plant Dominants as Indicated by the Number of Localities in the Rocky Mountains Where Found[a]

VEGETATION	CENTRAL	SOUTHERN	NORTHWESTERN
Wavyleaf oak............................	126	26	0
Hairy cercocarpus......................	107	16	0
Saskatoon serviceberry...................	74	13	15
Squaw bush.............................	58	13	15
Chokecherry............................	58	4	15

[a] Clements 1920.

wapiti, plains white-tailed deer, and others, but their relations to the Rocky Mountain bushland are obscure.

North Central Utah

Biotic studies of the bushland near Provo, Utah, indicate that it is an ecotone (Fig. 11-4). Twenty-six of its plant species also occur in the montane forest and 28 in communities on the Great Plains. Confined to the bushland are *Quercus gambelii, Vicia americana,* and *Cercocarpus ledifolius* (Tidestrom 1925, Hayward 1945, 1948).

PERMEANT MAMMALIAN INFLUENTS

The majority of the vertebrate inhabitants of the bushland are either transitory, seasonal, or no more confined to this community than they are to any one of several others (Fig. 11-5). Of all the original mammalian inhabitants of the bushland, the mule deer is probably the most important and dominant. Its behavior and migratory relations are similar to what they are on the Kaibab Plateau. The deer becomes concentrated in the bushland in winter, utilizing the same plant species as in the Kaibab. The autumn migration of deer into bushland from the montane forest is related to snowfall, occurring mainly during November. In the spring, the return movement to higher elevations follows roughly the retreat of snow, but animals are found in large numbers on favorite winter feeding grounds on the west face of Mount Timpanogos as late as the last of May. The extent to which the Gambel oak is used as browse is somewhat variable, depending upon the availability of more favored foods, but in some areas browsing of the oak is severe (Hayward 1948).

The coyote, mountain lion, wolf, black and grizzly bears, and the bobcat were originally abundant. The wolf and grizzly bear are now gone, and the other species are greatly reduced in numbers.

SMALL PREDATORS AND RODENTS

The long-tailed weasel is common throughout the bushland community. Tracks are frequently seen in the snow, and the animals themselves are often encountered in field work. Both the spotted skunk and the Great Basin skunk occur. Nearly all of the rodents are active through the year, but the smaller

Fig. 11-4. Mountain bush on the west front of Mount Timpanogos. The dominant plants are *Quercus gambelii* and *Acer grandidentatum*. The parklike character is shown by the patches of bush beginning in the center of the figure below the solid bush cover and extending to the foreground where cottonwoods occur along the Provo River. The small area in the near foreground at the right suggests the original bunch grass grassland of the lower altitudes (photo by D. E. Beck, courtesy of C. L. Hayward).

species, at least, probably have maximum populations in late summer. Mammals which hibernate include the jumping mouse, marmot, Uinta ground squirrel, and golden-mantled ground squirrel. The latest date for activity in the jumping mouse is September 9; in spring, specimens have been taken as early as May 3. The hibernation period of the marmot varies considerably with the elevation. The rock squirrel and the least and cliff chipmunks do not enter into true hibernation but become quiescent for short periods in severe weather. Tracks and indications of feeding activities of all three species are evident in the snow throughout the winter. All these mammals are montane forest species that extend down into the bushland, and several occur also in the pinyon–juniper woodland on the Kaibab Plateau (Hayward 1948).

IMPORTANT BIRDS

The 107 species of birds may be divided into permanent residents, including seven hawks, owls, and eagles, 26; winter visitants, 20; summer residents, 30; and migrants or local transients, 31.

The impact of overwintering birds is important. Thousands of Oregon juncos, for instance, winter in the bushland of the Wasatch Mountains where they spend at least nine months. During the rest of the year, the junco migrates and nests from eastern Oregon and Washington, northern Idaho, and north-western Montana northward into Canada. Birds overwintering in the bush-land are largely limited to westward and southwestward slopes, where large areas of ground are almost continuously exposed. During the few days after a rare heavy snowfall the shrubs become more directly important in providing galls, fruits, seeds, and buds as food. Various psyllid galls filled with nymphs subsist on dead leaves, many of which remain on the shrubs throughout the winter. These are utilized especially by the scrub jay which breaks open the hard galls with the beak and removes the food. Steller's jays and flickers probably also use this food. The fruit of the hackberry, which remains on the shrubs throughout the winter, is important food, especially at the lower edge of the bushland, for robins, Steller's and scrub jays, Townsend's solitaires, flickers, Bohemian and cedar waxwings, and rufous-sided towhees. Mountain and black-capped chickadees are the chief twig-feeding species found in the bushland in winter, but they appear to depend more upon dormant insects and eggs that are attached to the surface or under loose bark. At the upper limit of the bushland, large flocks of robins and smaller numbers of Townsend's

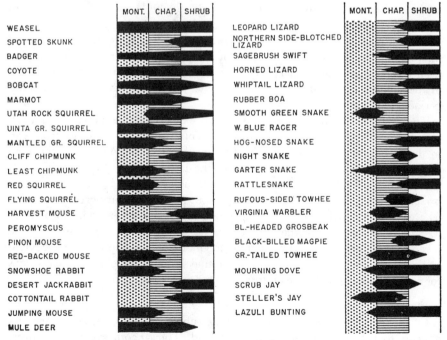

Fig. 11-5. The bushland described in Figure 11-4 is the central community CHAP., with montane forest MONT. above, and sagebrush and grassland SHRUB below.

solitaires and flickers feed upon the dry berries that remain attached to the twigs of the serviceberry. Buds of the shrubs are eaten by the ruffed and blue grouse, pine grosbeak, and evening grosbeak.

By far the great majority of wintering birds obtain their food directly from the surface of the ground. Rufous-sided towhees, pheasants, and quail feed largely under the protection of the shrubs, especially oak. Juncos, pine siskins, house finches, white-crowned sparrows, rosy finches, and goldfinches feed in open spaces and utilize the shrubs mainly as perches or for protection from predatory mammals. The food of these ground feeders is largely seeds of the herb layer. Some dormant invertebrates are available, but the great majority of the invertebrates spend the winter under rocks or deep in the ground where they are not available. Careful watching of the feeding process of these fringillids indicates that there is plenty of food on the surface of the ground easily accessible without much effort on the part of a bird. On the few occasions when there is a complete coverage of snow there are usually sufficient herbs, particularly sunflowers and grasses, projecting above the snow to provide food until the ground is bare again. Acorns, shed by the oaks in early autumn, are used mainly by mammals but are taken also by jays.

Aggregation among wintering birds constitutes a conspicuous feature of the winter aspect. Although there are occasional individuals that show little tendency to associate with their own kind or other species, flocks commonly range in size up to a thousand or more birds. In the Wasatch bushland, there is a general tendency for the flocks to occur in the mouths and at the bottoms of canyons and to avoid open flats or ridges. The flocks are almost invariably mixtures of several species, characteristically juncos of several subspecies, pine siskins, robins, jays, and chickadees. The highest concentration of birds occurs in the night roosts, and flocks tend to move out during the daytime feeding period into surrounding areas.

The presence of summer residents, which extends from about the last week of May to mid-July, coincides with the period of greatest activity of invertebrates in the community and the growth of young birds corresponds with the time of highest population of invertebrates in the latter part of June and the first three weeks of July. The Virginia warbler nests mainly in bushland but is present in fairly large numbers for only about 18 weeks. During the remaining 34 weeks, it is either in its wintering range in Mexico or en route to or from it. Studies of the food of the bushland birds clearly show that in summer the invertebrates form the major item of food of the avian population. Several of the more common birds utilize the ground layer under oak or other shrubs for nesting sites but make use of the shrubs for other purposes.

REPTILES

Of the eleven species of reptiles only the rubber boa which lives under rocks appears to belong primarily to the bushland. The smooth green snake overflows from the lower montane forest into bushland, and the other species, more characteristic of grassland or cold desert, here find their upper limits.

The sagebrush swift feeds on arthropods living near the ground; it hibernates from before October 1 to April 1 (Hayward 1948).

PLANT–INVERTEBRATE RELATIONS

At the upper elevations of the bushland, a number of montane shrubs enter into the community and mingle with the oaks to form a dense continuous thicket (Fig. 11-4). These include serviceberry, bigtooth maple, chokecherry, snow bush, and the rose *Rosa woodsii*. At the lower border, the climax bushland is in scattered patches with extensive areas of herbs and low shrubs occupying the intervening spaces, giving to the whole a parkland appearance. Various species of the sagebrush–grass community occur in this transition and in the bushland proper. The predominant tall shrub is the Gambel oak. At the lower edge of the bushland the fragrant sumac, smooth sumac, and the hackberry *Celtis reticulata* mingle with the oak or more often form independent patches or clumps. The first appearance of leaves at the lower border is usually about the middle of May but not until the first of June at the upper limits. By the last of October the leaves have again fallen, leaving the dull gray branches exposed for the remainder of the year.

Another conspicuous feature of the Wasatch bushland is the presence of patches of mountain mahogany (*Cercocarpus ledifolius* and *C. montanus*), antelope brush, and cliff rose, all of which tend to be evergreen and constitute important winter browse for deer. These shrubs, however, occupy rocky ridges and represent a stage in the xerosere rather than climax vegetation. *Ceanothus* is also evergreen. Juniper appears as scattered individuals and nowhere forms a woodland as occurs in other portions of the state (Hayward 1948).

The invertebrates of the tall shrub layer are characteristic of the layer, not commonly occurring in either the herb and low shrubs or on the ground. Many parasitic and gall-forming species are confined to one or another of the shrubs. Most of the quantitative sampling of this study is confined to the deciduous oaks. The greatest numbers appear in June in contrast to the herb–low shrub layer in which populations are highest in August. There is a good deal of variation in the populations of various species. In 1944, the bud-eating weevil *Thricolepsis inornata* appeared in very large numbers about May 26 and continued to about June 20, when it disappeared. At this same period a species of lace bug was also very abundant, and large numbers of parasitic hymenopterons, mainly of the subfamily Tetrastichinae, were common. The following year the weevil appeared only in small numbers, but the lace bugs and parasitic hymenopterons were again abundant at about the same time.

The parasitic hymenopterons are the most consistently characteristic invertebrates of the tall shrubs, particularly oaks, although other species often appear in greater numbers for short periods. Two cynipid wasps, *Acraspis villosa calvescens* and *A. hirta packorum*, are known to form leaf galls on *Quercus utahensis* (Kinsey 1929). Oak twig galls, presumably formed by cynipids of the genera *Andricus* or *Neuroterus*, are conspicuous objects. From oak galls, Hayward (1948) has bred a species of *Synergus*, which is an inquiline with the gall maker, and the following chalcidoideans which are either parasitic or hyperparasitic on the gall insects: *Torymus rubenidis, T. warreni, Eupelmus allynii, Ormyrus,* and *Eurytoma.* Hackberry supports a number of conspicuous and important leaf galls. These are formed by psyllids or jumping plant lice of the genus *Pachypsylla. P. venusta* has been reared from large galls on the leaf petioles, the galls causing the leaves to remain on the twigs all winter. Nymphs occur in great masses in these galls and are eaten extensively by the scrub jay and probably other birds. *P. celtidis-vesicula* has been bred from leaf galls, and *P. celtidis-gemma* is known to form leaf galls. Adults of these psyllids appear in May or early June. A number of hymenopterons belonging to the Chalcidoidea have been reared from galls, apparently parasitic on larval psyllids. These wasps include *Amblymerus, Eurytoma,* and *Psyllaephagus pachypsyllae* (Hayward 1948).

The herb–low shrub layer is well defined. At the lower elevations it occupies the greater part of the total area, occurring between the scattered clumps of tall shrubs. With an increase in altitude the tall shrubs come to cover more and more of the surface area, leaving only small patches available, so that the herb layer is limited to those few species that are able to grow in the shade of the tall shrubs.

Under pristine conditions, the wheatgrass *Agropyron spicatum* was the dominant herb species at lower elevations, with *Poa secunda* also playing an important role. Both of these species are important dominants of the Palouse Prairie. Other native species in this layer include *Poa longiligula, P. curta, Oryzopsis hymenoides, Stipa lettermani, S. columbiana,* and *Hesperochloa kingii.* They belong primarily either to the grassland below or the forest above (Hayward 1948).

Populations of invertebrates in the herb–low shrub layer matrix show a progressive increase in numbers from May until August and a general decline in September. At high altitudes, the highest populations are reached in July. Ants are by far the most abundant group in the herb–low shrub layer, where they constitute about 41 per cent of the total invertebrate population during the summer. The minute black ant *Monomorium minimum* appears very frequently in samples, the odorous ant *Tapinoma sessile* is common, and the acrobat ant *Crematogaster lineolata* often occurs in considerable numbers. At the higher elevations, ants make up only a little more than 5 per cent of the total invertebrate population; only *Tapinoma sessile* was found.

Leafhoppers form a conspicuous part of the invertebrate population in the herb–low shrub layer throughout. At the lower elevations they amount to 14 or 15 per cent of the total invertebrate population. There are 5 species represented at 6800 feet (2060 m) and 17 species at 5200 feet (1575 m) (Hayward 1948).

Locusts and grasshoppers form a conspicuous part of the constituents of the herb–low shrub layer, especially on the forbs, and are probably favored by cattle grazing of grasses. The following species occur at two elevations: 6800 feet (2060 m), *Scudderia furcata, Melanoplus mexicanus, M. biliturittus,*

M. bivittatus, Arphia simplex, Steiroxys pallidipalpus; 5200 feet (1575 m), *Arphia simplex, Xanthippus corrallipes, Melanoplus mexicanus, Mermiria bivittata, Hesperotettix viridis, Trimerotropis cyaneipennis, Schistocerca shoshone, Circotettix undulatus,* and *Leprus interior* (Hayward 1945).

MICROSCOPIC ORGANISMS OF THE SOIL

Bacteria, fungi and minute arthropods of the soil were investigated, along with the larger invertebrates, in bushland on Mount Timpanogos, Utah, from April 1 to June 1, 1945. There were 1,000,000 bacteria and actinomycetes per gram of soil in the open areas, and 500,000 under the oak, and the organisms were different in these two situations. Fungal molds on the other hand were five times more abundant in the oak soil (50,000/gram) than they were in the open soil. In the oak soil *Penicillium* was predominant, with *Mucor* also present; in open areas *Rhizopus, Aspergillus,* and *Fusarium* were the most common genera represented (unpublished studies of Desma H. Galaway referred to by Hayward 1948).

INVERTEBRATES

With respect to invertebrates, samples of one-tenth square meter to a depth of five centimeters showed that populations were consistently greater and more variable in soil under oak than in the open. Populations in the oak soil rose slowly in May and June, rose rapidly in July, and began falling off in August.

In open soil, Acarina form about 47 per cent of the average summer populations of all invertebrates, Formicidae about 43 per cent, and Araneida about 3 per cent. In the soil under oaks, Acarina is again the most abundant, forming about 77 per cent of the summer population of invertebrates. Formicidae comprise about 9 per cent and Collembola about 7 per cent. The small thief ant *Solenopsis molesta validiuscula* is very characteristic in this habitat. The shrubs and the herbs contribute much in the way of plant parts that become incorporated into the litter and soil and are the food base for the existence of many of these small organisms. The mites, collembolons, many spiders, insect larvae, millipedes, and centipedes require continuous darkness or semidarkness throughout their entire lives. On the other hand, some of the common inhabitants, especially ants, may invade the higher layers, even in daylight, for a portion of their food and carry materials for the construction of their nests into the soil.

Most of the large invertebrates of the ground layer live under the protection of rocks. By far the most common of these animals are ants. At the higher elevations, 85 per cent of rocks turned over have ant colonies under them. The following species have been identified: *Formica neogagates lasioides, F. rufa obscuripes, F. fusca gelida, Lasius latipes, L. niger sitkaensis, L. niger neoniger, Tapinoma sessile, Camponotus sansabeanus vicinus, Aphaenogaster uinta, Pheidole, Monomorium minimum, Leptothorax nitens, Crematogaster lineolata,* and *Solenopsis molesta validuiscula* (Hayward 1948).

Ten species of spiders live under the protection of rocks. The millipede

Taiulus tiganus is common under rocks throughout the bushland along with the scorpion *Vejovis boreus.* The most common centipedes are *Pokabius utahensis, P. socius, Scolopendra polymorpha, Yobius haywardi* and *Bothropolys permumdus.* Isopods are not at all common; only colonies of *Porcellionides pruinosus* are found in the lower part of the bushland. Mollusks are likewise quite uncommon; only a few specimens of *Oreohelix strigosa depressa* have been found. The most common beetles under rocks are *Discoderus amoenus, Iphthimus sublaevis, Canthon corvinus, Coniontis uteana,* and *Galeruca externa.* Paper wasps quite frequently utilize the underside of rocks for the attachments of their nests. *Polistes fuscatus utahensis* and *Mischocyttarus flavitarsis* are the most common (Hayward 1948).

Eastern Slopes of the Rocky Mountains

The bushland on the eastern front of the Rocky Mountains near Pikes Peak and Rocky Mountain National Park, Colorado, is composed principally of mountain mahogany, although *Prunus* occurs in moist places. Grasses, some of them common in grassland, occur as an understory with the shrubs *Rosa, Rhus, Berberis,* and others. In Engelmann Canyon, the shrubs are *Quercus undulata, Rhus trilobata, Cercocarpus montanus, Holodiscus discolor, Prunus virginiana, Symphoricarpos, Yucca glauca,* and *Elymus ambiguus.* The following animal groups are common in the bushland near Pikes Peak: ants, *Leptothorax acervorum canadensis, Lasius niger neoniger, Formica fusca subaenescens, Tapinoma sessile,* and *Formica neogagates lasioides vetula;* the flea beetle *Phylotreta pussila,* which is numerous on forbs at all altitudes; and the hemipterons *Ischnorhynchus geminatus, Lygaeus elisus, Phytocoris palmeri, Harmostes reflexulus* which are common on forbs, and the red-shouldered bug *Thyanta cursitor* which feeds on forbs and grasses. Few flies are present; only *Dioxyna picciola* was taken. The hymenopterons include the parasite of gall wasps *Eudecatoma varians.* The leafhopper *Neokolla hieroglyphica* is very common and *Cuerna lateralis* less so. Only one mature spider *Misumenops asperatus* was taken. The small chipmunk *Eutamias quadrivittatus* occurs in Engelmann Canyon bushland. White-tailed jack rabbits are seen. There is evidence of ground squirrels.

Near Loveland, Colorado, on July 29, 1936, there were 40 grasshoppers per square meter including *Melanoplus regalis, M. confusus, M. foedus, M. flavidus, M. bivittatus, Pediodectes stevensonii, Brachystola magna, Hadrotettix trifasciatus, Hesperotettix viridis, H. gillettei, Dactyltum bicolor pictum, Mermiria maculipennis macclungi,* and *Amphitornus coloradus.* Mantis nymphs were present. The mountain mahogany was stripped of its leaves, and the leaves and stems of forbs were badly damaged.

Nuevo León, Mexico

The Nuevo León bushland lies between the more mesic mountain communities and the hot desert at about 2000 feet (600 m) elevation on the west side of the mountain range which skirts the east side of the plateau. Evergreen broad-leaved sclerophyll species *Ceanothus, Quercus, Cercocarpus, Garrya, Arctostaphylos, Yucca, Arbutus,* and *Rhus* are prominent almost everywhere, but a considerable proportion of the shrubs is deciduous. The outstanding difference between this vegetation and that of the desert plateau below is the small number of thorny shrubs present.

The woody plants of the Nuevo León bushland include seven species of *Quercus,* two species of *Ceanothus,* two species of *Cercocarpus* and two species of *Garrya* that are not found in the Rocky Mountain bushland. Only two species, *Rhus trilobata* and *Arctostaphylos pungens,* appear to be common to the Colorado and Nuevo León communities. Most important species are evergreen, but they vary in importance. In one area *Quercus flocculenta* may be widely dominant. In another, *Pinus cembroides* associated with *Rhus virens* may entirely replace *Quercus.* In yet another, *Ceanothus lanuginosas, Arctostaphylos pungens,* and *Quercus cordifolia* are dominant. This vegetation type includes a greater number of important woody species than any other type in the Mexican region (Muller 1939).

There are no animal studies for the area. Mule deer are abundant where not heavily hunted, and black bear descend from adjacent montane forests (Muller, personal communication).

Chapter 12

The Marginal Contacts of the Temperate Grassland

The main body of temperate grassland lies in the interior of the continent. A detached segment occurs in central California. Grassland approaches the sea only along the coast of Texas and northeastern Mexico and in California. It has contacts with deciduous forest in the east; aspen parkland in the north; lodgepole pine, ponderosa pine, and Douglas-fir forests in the eastern foothills of the Rocky Mountains; woodland and bushland from Wyoming to Texas; shrubby areas in eastern Texas and Tamaulipas; and cold and hot deserts in Utah and Arizona.

In dry regions there is commonly a gradual transition at these contacts, with mixing of grass and woody plants. In moist regions, however, an abrupt change occurs from grassland to forest, and the assemblage of plants and animals is largely different from that of either grassland or forest. These forest-edge communities are in a state of constant fluctuation back and forth with changes in climate. The dynamics of the forest edge are perhaps better explored than those of any other type of community.

GRASSLAND–DECIDUOUS FOREST CONTACT

The contact between grassland and deciduous forest (Fig. 12-1) lies on a plain that varies only a little more than 500 feet (152 m) in elevation. It extends from 30° to 48° N. Lat., a distance of more than 1200 miles (1920 km). The distance from the easternmost limit of large areas of prairie in Indiana to the western limit of climax red oak–basswood forest is nearly 600 miles (960 km) (Aikman 1926), and from important small areas of prairie in Indiana and Ohio to the western limit of bur oak forest along streams is nearly 1000 miles (1600 km). Climax forest and climax prairie are interspersed

306

Fig. 12-1. A map showing the central portion of the deciduous forest grassland and forest edge. A. Continuation of Missouri River in North Dakota. B. Other grassland areas in Indiana and Ohio.

through a region of not less than 360,000 square miles (932,400 km²). The actual area covered by forest-edge vegetation within this general region is estimated at 10,000,000 acres (4,000,000 hectares).

Climate

Rainfall during May and June ranges from 5.8 inches (14.5 cm) at Mankato, Minnesota, to 9.2 inches (23.0 cm) at College Station, Texas, and from 7.47

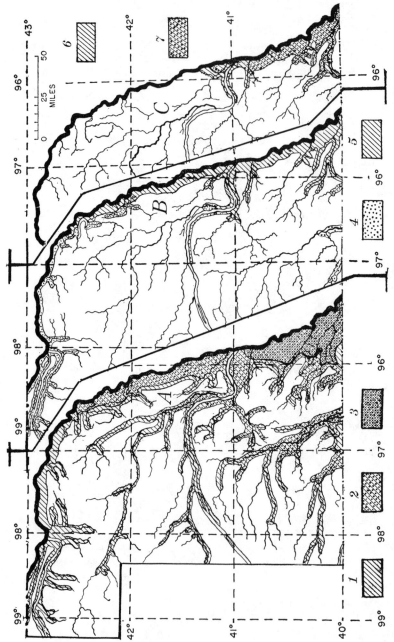

Fig. 12-2. The distribution of forests along the Missouri River and its larger tributaries in eastern and northeastern Nebraska (after Aikman 1926). A. Xeric seral communities partially covering the entire forested area, greatest width 50 miles (80 km): 1, Bur oak forest; 2, bitternut hickory and bur oak; 3, bitternut hickory and bur oak with chinkapin oak. Not all the land is covered with these types; some of types B and C are included due to the complex distributions shown in Figure 12-3. B. Climax communities which enter only a part of the area shown under A: 4, basswood; 5, basswood and northern red oak. C. The more mesic seral communities do not cover the entire area outlined: 6, shagbark hickory; 7, shagbark hickory and black oak.

inches (18.7 cm) at Plymouth, Indiana, to 8.06 inches (20.2 cm) at Columbus, Nebraska. Annual precipitation varies from 24.06 inches to 38.50 inches (60.2 to 71.2 cm), north to south, and from 35.31 inches to 27.59 inches (88.3 to 69.0 cm) east to west. Precipitation decreases from east to west most rapidly immediately west of the Ozark uplift in Oklahoma. Mean annual excess of precipitation over evaporation of 0.0 to 10 per cent is fairly characteristic of the region under consideration (Visher 1946), except in the extreme north and in the prairie peninsula which extends across Illinois (Transeau 1905) where evaporation is less.

Forest-Edge Vegetation

Forests along streams and occasional groves in grassland are similar in biological constituents. Groves are called mottes (Texas) and bluffs (Canada), while shrub patches are called ruffs (Illinois) or shinneries (cheniers) in Oklahoma; where groves are numerous, the country is called parkland or savanna. Gleason (1912) reports on a grove in eastern Illinois in which the most important tree in size and number is the bur oak. Black walnut, hackberry, and slippery elm are also present. The arrangement of species on the forest edge in Nebraska is shown in Figure 12-2. Weaver, Hanson, and Aikman (1925) state that climax red oak–basswood stretches westward and north-westward along small streams in Nebraska in ribbon-like bands that decrease in width to not more than 30 meters (Figs. 12-2, 12-3). Basswood goes farther west than red oak. Ironwood is generally present. Throughout the best de-

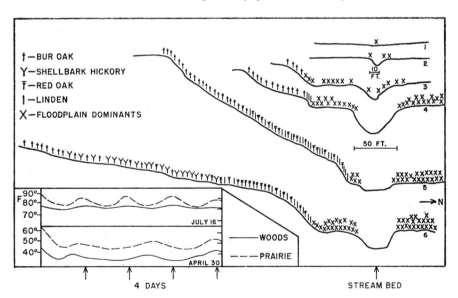

Fig. 12-3. Profiles of Weeping Water Valley, where transects 1-6 were made; numbers 3-6 show only the south slopes bordering the valley (Weaver *et al.* 1925). Insert contains selected four-day records of soil temperature at a depth of two inches (5 cm) in the prairie and woodland stations at Lincoln (after Pool *et al.* 1918).

velopment of the red oak–basswood climax, the sparse undergrowth includes honeysuckle, black raspberry, prickly-ash, Virginia creeper, and the wild rose *Rosa setigera.* The larger shrubs are represented by pawpaw and redbud (Aikman 1926). The other oaks and the hickories, secondary species in the oak–basswood climax, extend farther west than does the basswood; bur oak continues to at least the 100th meridian along the Missouri River and makes a forest of one principal dominant on the low uplands. On small tributaries, bur

Fig. 12-4. South-facing slope of the Missouri River near Yankton, South Dakota, covered with patches of rough-leaf dogwood, coralberry, snowberry, and smooth sumac. Bur oak occurs in the ravines (photo by J. E. Weaver).

oak forest is in contact with the floodplain forest. The outermost tree species in a wide stream-skirting forest makes up the entire forest along the smaller streams (Weaver *et al.* 1925).

At Lawrence, Kansas, at the time of settlement (1860), the forest was a half mile to a mile (0.8 to 1.6 km) wide on each side of the Kansas River. The University of Kansas Natural History Reserve now contains a forested area that is four miles (6.4 km) from the river, and the forest is of very different composition than in 1860 (Fitch and McGregor 1956). This indicates that there has been an active invasion of grassland by the forest edge.

The forest edge, either of the stream-skirting forest or of the scattered groves, consists of low trees, shrubs, and forbs, 10 to 60 feet (3 to 20 m) wide (Fig. 12-4). The most stable and perhaps climax edges are those of bur oak forests. Hazel and sumac are common shrubs in the edge over much of the region.

In northern Illinois (Vestal 1914a), low trees make a zone of variable width, outside of which are zones of shrubs and forbs. Edges with full exposure to the sun show very distinct zonation and are wider than north edges (Fig. 12-5). The low tree zone is composed of wild crab, either *Malus ioensis* or *M. coronaria,* plum, or hawthorn and is usually from 25 to 45 feet (82 to 148 m) in height. The shrub zone may be dominated by the dogwood *Cornus obliqua* with hazel and elder locally abundant. The dogwood is usually about 3 feet (1 m) high; the hazel, 6 feet (2 m); and the shrub zone, 3 to 10 feet (1 to 3 m) wide. The grape *Vitis vulpina* covers some of the outer shrubs and trees, and smooth sumac may occur on south exposures. The outermost zone is usually an almost pure growth of the sunflowers *Helianthus decapetalus* and *H. divaricatus.* The goldenrod *Solidago missouriensis,* tall tickseed, and species of *Verbesina* are present locally. The height of the sunflowers is 2 to 3.5 feet (0.6 to 1 m), and the zone is 2 to 5 feet (0.6 to 1.5 m) wide. The prairie

Fig. 12-5. Forest invasion of prairie and forest edge. A. Prairie looking west along the sinuate south edge of the forest skirting the Des Plaines River near Riverside, Illinois, June, 1907. Goldenrod and sunflower are evident in front of the shrubs, largely *Crataegus,* in the distance; rosin-weed and compass-plant are present in the foreground. B. Approximately the same location in 1947. The trees at the right have died or been removed. The prairie has become invaded by shrubs, and only a small open spot remains. Compass-plant is missing and grasses have been largely displaced by forbs. The erect stems are those of three species of goldenrod. The highest branches at the left are those of an old cherry.

just outside the sunflower zone is almost identical with that farther from the forest. Low spots in the prairie often contain temporary spring ponds. Patches of shrubs in the prairie away from the edge indicate invasion (Short 1845).

In eastern Nebraska the shrub edge is usually in contact with bur oak forest. Hazel occurs next to the oaks, then wolfberry and coralberry, and finally smooth sumac the outermost. Gooseberry and, less commonly, raspberry occur along with grape and other vines. The shade under the shrubs at a station near Lincoln, Nebraska (Pool *et al.* 1918), was so dense that most of the annual herbs as well as the herbaceous forest perennials were extremely small. The number of herbs on 16 square meters were the following: *Amphicarpa bracteata comosa*, 3; *Acalypha virginica*, 16; *Andropogon scoparius*, 19; *Monarda fistulosa*, 26; *Aster*, 71; *Fragaria virginiana*, 27; and *Helianthus hirsutus*, 108.

The absence in Nebraska of the zone of small trees that was found in Illinois is evident. The forb zone is also much less definite in Nebraska and in Kansas than in Illinois. On south-facing edges, the grasses may grow inward under the trees for a short distance, and small trees and shrubs may be scattered among the grasses. In heavily shaded north-facing edges, forbs occur in patches between the shrubs and the grasses. In a small piece of natural forest edge in eastern Kansas, located at the top of a ridge above a north-facing slope, stunted *Andropogon* grass grows among and under the shrubs, especially where there is no shade from trees. Dogwood alternates with smooth sumac as the principal shrub. Both of these along with several other plants, including the goldenrod *Solidago ulmifolia*, are scattered out into the grassland. In an eastern Nebraska area scattered trees cover a ridge that has forest on its north-facing slope. Sumac is scattered among the grass with other edge shrubs occurring in patches or singly. The sunflower *Helianthus decapetalus* is present as in Illinois. Bluegrass penetrates into the woods and predominates on the ridge beneath scattered bur oaks. Ravines of small intermittent streams that are too small to support trees are sometimes filled with a mixture of forbs and edge shrubs. In northern Oklahoma, low-growing trees in small ravines are surrounded by *Symphoricarpos* (Carpenter 1939). Near the 100th meridian in Kansas (Griswold 1942, Carpenter 1939), forest-edge shrubs are in contact with floodplain forest of cottonwood, willow, elm, and hackberry. A coralberry–sumac edge occurs around the cottonwood–elm floodplain forest in Kansas, the width of the edge varying from 9 to 300 feet (3 to 90 m) depending upon the amount and type of recent soil disturbance.

From central Kansas southward, the forest climax changes from red oak–basswood to oak–hickory, with post oak and, to a lesser extent, blackjack oak as characteristic species (Aikman 1935). In the contact with oak–hickory, the coralberry is the common representative of the genus *Symphoricarpos*. Dwarf oaks, including live oak, become important, and the edge extends much farther east than is common at more northern latitudes. Fire burns off and kills the tops of persimmon trees which then, through sprouting, form dense thickets.

Tharp (1926) states that in east Texas, north of 30° N. Lat., the forest margin is characterized by scattered clumps of post and blackjack oaks without shrubs. In Waller County, Texas, live oak and post oak both invade the prairies. South of 30° N. Lat. the contact is between the Gulf Coast prairie and magnolia forest and has been little studied.

In Oklahoma shin oak is often present, along with post and blackjack oaks (Bruner 1931). Outside of the trees, coralberry occurs, followed by the sumac *Rhus copallina* and *Rhus glabra*. Outermost and growing among grasses is dwarf chinkapin oak. Any one or two of these species may be absent in a particular locality. In ravines, the coralberry and sumac are accompanied by wild plum. Next to floodplain forest, persimmon, buttonbush, dogwoods, and the two sumacs form the edge.

Forest-Edge Animals

The width of the stream-skirting forest is important to large forest animals which require extensive home ranges for grazing, browsing, or hunting. Under primeval conditions, black bear, turkey, white-tailed deer, gray fox, and gray squirrel declined in numbers to the north and west along the Missouri River and its tributaries until finally the area of forest became too small to support them at all. Luttig (1920), on his journey in 1812, noted the last black bear at the northwestern corner of Missouri and turkey near the mouth of the

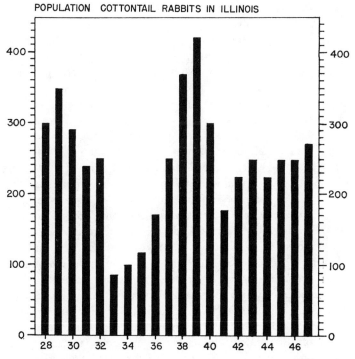

Fig. 12-6. Variations in Illinois cottontail rabbit population indicated by arbitrary numbers (modified from Yeatter and Thompson 1952).

Platte River. According to notations made by Lewis and Clark in 1804, forest deer decrease in abundance near the mouth of the Platte River and by degrees farther upstream; turkeys and squirrels are reported as far north as Sioux City.

The cottontail is a characteristic coactor in forest edges and forest openings west of the Appalachian Mountains. Its population in Wisconsin fluctuates from 2.5 to 5.2 times the lowest minimum (Grange 1949). Figure 12-6 shows fluctuations of cottontails in central Illinois. Pellet counts may be used as an index of abundance (Brown 1947). The usual summer range for cottontails is about 1.3 acres (0.5 hectares) and the winter range is about twice this (Schwartz 1941). In western Kansas, the population becomes concentrated in the lowlands during the winter. The home range is intersected by paths which are kept open by nibbling. The impact of the species on the vegetation varies sharply, but there is little or no quantitative information available. There are three or four litters per year, depending on the latitude and weather conditions; nests of young are hidden away in small pockets in the soil in the tall grass (Seton 1929).

The closest space competitor of the cottontail is the bobwhite. It is one of the most universally present of the forest-edge birds. Under original conditions in Illinois, the bobwhite is reported to have nested chiefly in the forb zone outside the shrubs of the forest edge. The bobwhite covey varies from 6 to 30 or more individuals. The roosting huddle is a compact circle with heads facing outward, a distinct advantage for quick flight. The rounded bodies of the birds slip through the shrubs easily, where predators are deterred; grass and forbs offer obscure nesting and feeding places.

Bison crashed through river-skirting forests and forest edges, often on regular beaten trails, to reach drinking water. Large migrating groups caused considerable damage to the vegetation at every river crossing. The animals doubtless used the stream-skirting forest for shelter in severe weather.

Wapiti divided their time between forest, bushland, and grassland. Lewis and Clark found the wapiti so emaciated in winter that they were unfit for food. The white-tailed deer and mule deer frequented the forest edge, the latter largely limited to areas west of the 98th meridian. Seton (1929) states that the mule deer mother usually hides her fawn at different points in the thicket from day to day. This practice continues until the young are seven or eight weeks old. In the western part of the plains, deer come down from the mountains to pass the winter. The white-tailed deer ranged nearly across the plains along the large rivers. They made paths in the tall thickets and woodland through which they traveled and bedded down at night in the groves—sometimes in groups of considerable number. Both species use home ranges of 100 acres (40 hectares) but sometimes wander distances of four to six miles (6 to 10 kilometers). In winter, both species of deer and the wapiti feed on shoots and buds of trees and shrubs. During the summer they feed on herbs and twigs. In the summer of 1804, Lewis and Clark report seeing them in great abundance on the prairies near the forest edge. At that time their population may have been as high as 50 per square mile (20/km²) of river-skirting forest.

The raccoon is a common inhabitant of the stream-side forest in the east and frequently wanders out onto the grassland. It feeds commonly on plums and other fruits and thereby helps to distribute their seeds. Raccoons were seen by Lewis and Clark and other early explorers along the Missouri River in central Missouri as far as the mouth of the Grand River. Hayelden reported them as abundant near Council Bluffs in 1862 (Bailey 1926). There is considerable evidence that they have extended their range westward with settlement, through tree planting and releases by hunters. They are abundant in eastern Nebraska at the present time. In the southern portion of the forest-edge region the opossum occupies a nearly equivalent niche.

The red fox is a regular resident of the forest edge and utilizes both the forest and adjacent grassland in search of food (Scott 1947). Many of the forest-edge plants and animals are used for food, from deer down to ants and from plums to ragweed. The occurrence of red foxes was noted by Lewis and Clark.

The coyote prefers open, sunny places for making dens and rearing young but uses the forest edge for part of its activities. In former days it doubtless destroyed young deer, rabbits, bobwhite, ground squirrels, mice, and song birds, which divided their time between thicket and grassland.

Striped and spotted skunks dig and scratch in the soil to secure insects and other invertebrates, and burrow to considerable depths for shelter. The spotted skunk eats a greater variety of food than the striped skunk does. In addition to insects, its diet includes fruits, birds' eggs, nestlings (including bobwhite), salamanders, lizards, and crayfish.

The meadow jumping mouse is an important forest-edge animal that is found in lowland thicket communities (Blair 1938). It constructs globular nests and eats relatively large quantities of grass seeds and small fruits.

The Franklin's ground squirrel occurs only in the northern two-thirds of the forest margin area, with its southern limit in Kansas. Its habitat preference appears to be for grass and forbs with invading shrubs. This species is omnivorous, utilizing a variety of food from strawberries to mice and from cacti to ants. Its den is a tunnel or network of tunnels dug a few inches into the ground. The burrowing of the squirrel opens up the soil with only minor disturbance of vegetation and soil fauna. These breaks in the prairie sod admit seed into the soil. The animals are sometimes locally gregarious. Scattered individuals have home ranges of about 1.6 acres (0.6 hectares). Mice, shrews, and moles are not especially important in the forest edge.

The sharp-tailed grouse inhabits open areas in forest and grassy areas that contain bushes and is distributed widely over North America. The greater and lesser prairie chickens are parkland species, although the lesser prairie chicken will breed where shinnery and thickets occur. The sage grouse flourished in the stream-skirting forest and bush-covered edges in eastern Montana and Wyoming.

Small birds are widely distributed throughout the forest margin area. Species nesting principally in shrubs and small trees are brown thrasher, catbird, American goldfinch, yellow-billed cuckoo, field sparrow, indigo bunting, and cardinal. In the northern part of the area, the lark sparrow, chipping sparrow, and American redstart are added to the community. They are replaced in the southern part by the mockingbird, Bell's vireo, and loggerhead shrike. Tree sparrows and slate-colored juncos are the common species during the winter. Birds with a nesting preference for the marginal trees and for securing their food from grassland or edge are red-headed woodpecker, common crow, robin, ruby-throated hummingbird, and Cooper's hawk. Young song birds are fed animal food, chiefly insects and spiders. Adult song birds also feed on animal matter and some species on seeds and plant materials as well. Lepidopterous larvae are important as food for young birds; adults are used but little. Larvae are relatively large, easy to secure, and easily digested. Those available near the forest edge are the snowberry clear-wing moth, Virginia creeper sphinx, catocala, *Heliothis armigera,* and eight-spotted forester moth larvae, all of which feed on grape and Virginia creeper. Unfortunately nothing is known of the abundance of these species.

Spiders generally available near the forest edge include the nursery spider *Dapanus mira,* the crab spider *Xysticus gulosus,* and the tube spider *Clubionia tibialis.* The large orb weaver *Araneus raji* is usually two meters above ground in shrubs or low trees. Muma and Muma (1949) list five eastern Nebraska spiders which occupy both grass and shrubs: *Oxyopes salticus, Tetragnatha laboriosa, Agriope trifasciata, Misumenops celer,* and *Dictyna sublata.*

A great many nymphs of orthopterous insects are fed to nestling birds. Commonly available species are the broad-winged katydid, the narrow-winged katydid, a candle head bug, and the differential grasshopper. This last species frequents shrubs and, when abundant, may defoliate them.

Spittle bugs and leafhoppers are capable of weakening plants so that they fail to reproduce or to extend their territory vegetatively, or they die. The spittle bug *Lepyronia quadrangularis* feeds on the principal shrubs of the forest edge, *Corylus, Cornus, Rhus,* and *Symphoricarpos,* and on a few native forbs (Doering 1942). The froghopper *Clastoptera proteus* also feeds on *Cornus.* R. H. Beamer and P. J. Christian of the University of Kansas, list the following leafhoppers feeding on forest-edge plants (personal communication): on hawthorn, *Idiocerus provancheri, I. crataegi,* and *Typhlocyba duplicata;* on sumac, *I. dolorus* and *Empoasca;* on hazel, *Typhlocyba;* and on dogwood, *T. putmani* and *Eurythroneura corni.* The large and small milkweed bugs, *Lygaeus kalmii* and *Oncopeltus fasciatus,* are usually present, along with two stink bugs, *Euschistus tristigmus* and *Cosmopepla bimaculata,* on blackberry and with *Corizus bohemanii* on *Cornus.*

In Nebraska, the leafhopper *Erythroneura* may be abundant in August and *Dikraneura abnormis* from June to December. The flea beetle *Chalcoides fulvicornis* is abundant in spring. Some chrysomelids have free-living, leaf-eating larvae which birds may utilize. It is surprising that the young of some birds are fed weevils possessing extremely hard outer coverings. Among the Hymenoptera, ants are numerous, and few species are characteristic of the forest edge. The most abundant ant occurring in the litter is *Leptothorax curvispinosus ambiguus* (Fichter 1954).

The zone of crab apple, hawthorn, plum, and cherry is important as a source of food wherever it occurs, due to fruits and flowers of the trees. The fruit fly *Drosophila* feeds on decaying plums and other fruits; four species are regularly recorded in the forest margin area. Larvae of other flies are parasitic in the spittle of the common spittle bugs.

Both the primitive Indian and the early white man sought the shade and shelter of the forest edge. When in a forested area, they removed the trees from considerable areas to create open spaces. When in parkland, they usually selected the edges of the wooded areas for building their homes. The forest provided logs for cabins and wood for fuel. The early white settlers located their towns on the edges of forests. According to Kroeber (1939), the Indian populations in both the oak (parkland) forest edge and in the aspen (parkland) forest edge were small and similar to population densities on the plains, that is, 2 to 5 persons per 100 square kilometers.

Dynamics of the Forest Edge

Evaporation in hazel and sumac thickets of the forest edge is nearly the same as in the forest and is much lower than on the prairie (Weaver and Thiel 1917). This is correlated with the higher relative humidity in the tall shrubs compared with the prairie. Peculiarly, mean summer temperatures in the grassland are commonly lower than among the dense shrubs. During the middle of a representative day, however, the prairie may have an air temperature of over 36.7° C. (98° F.) ; the shrubs, 30.7° C. (87.2° F.), and the bur oak forest, 25.3° C. (77.5° F.) ; while the corresponding soil temperatures would be 26.6° C. (79.8° F.), 23.6° C. (74.5° F.), and 21.4° C. (70.6° F.). Except at the beginning of the season, the water content and acidity of the soil during the growing season is lower in the prairie than in the shrub and forest (Aikman 1928). Some forest ground animals occur under the shrubs, but forest herbs are largely absent.

Dry and rainy periods of longer or shorter duration have alternated over thousands of years. During long, wet periods, forests expand from groves and stream-skirting strips to take possession of prairie areas. Some forests in Iowa and Illinois are now growing on black prairie soil that is 20 to 36 inches (50 to 90 cm) deep. During long dry periods the process has been reversed. Grasses invade wooded areas and kill the shrubs and trees probably by monopolizing the water supply through a superior system of deep roots. Most of this competition between forest and prairie communities goes on in the shrubby edge that separates them.

Animal reactions on the habitat and coactions with other organisms are important in this struggle between communities. Germination of seeds may be improved by passing first through the alimentary tracts of animals. Thirty-two per cent of sumac seeds taken from rabbit pellets germinated while only 19 per cent of the hand-picked seeds did so (Brown 1947). Rabbits also scatter seeds from the sand dropseed on open ground where they may better germinate and provide grass for food and shelter. The rabbit thus tends to maintain and extend its natural habitat conditions. The fox, raccoon, and opossum feed on fruits and scatter their seeds.

The fruits and seeds of forest-edge trees and shrubs usually make up about one-third of the kinds of food taken by bobwhite, and seeds of grasses and legumes constitute about one-fourth. Good cover is required during feeding and probably influences the food taken (Stoddard 1939). The germination of sumac seeds is improved by passage through the digestive tracts of bobwhites (Krefting and Roe 1949). The mobility of the bobwhite probably causes it to spread seed farther than does the cottontail rabbit. Song birds cause wide dispersal of seeds. Sumac seeds have been found in nearly all seed-eating birds. Passage through the robin improved the germination of the seeds of the black cherry, blackberry, elder, and white mulberry; passage through the catbird improved the germination of gooseberry, blackberry, and elder.

Seeds in bird and mammal excrement germinate where they are dropped (Brown 1947). Bird plantings are common at least in the eastern part of the range of the shrub edge. Professor F. C. Gates (personal communication) has observed seedlings of dogwood and sumac in cow tracks in Kansas. Bur oak acorns blown or carried into the grass were readily pushed into the soft soil by ungulates under pristine conditions, and this aided their germination.

Plants are also injured by animals. Cottontails damage rose, chokecherry, and sumac by eating the bark, sometimes girdling the plants (Brown 1947). The larger animals, such as deer and wapiti, may browse shrubs severely and destroy seedlings.

The soil surface may be disturbed by the dust-bathing of bobwhite. This usually takes place under a thicket and may cover a considerable area. The deer and wapiti disturb the soil and vegetation where they sleep.

Hazel and sumac invade tall grass in the southerly post oak area by means of rhizomes or root sprouts (Clements *et al.* 1929). Forbs may crowd out shrubs due to their low water requirements. In the presence of moisture adequate for shrub growth, they shade out and destroy the grass. Hazel may shade out the less tolerant sumac, only to be destroyed in turn by the bur oak which invades shrubby areas readily. A seedling of this tree develops a 22-centimeter tap root before it opens its leaves. The development of a tap root is a critical stage in the establishment of a tree among grasses even when it is planted by squirrels or ungulates (Weaver and Kramer 1932).

The drought of 1933-37 reversed the succession of trees into grassland (Albertson and Weaver 1945). The dying of trees left the way open for grasses to invade forested areas. The climax oak–basswood forest suffered most from the drought; 80 per cent of the red oak and basswood in eastern Nebraska within 25 kilometers of the Missouri River died. Shrubs suffered near Hayes, Kansas, 240 kilometers west of the limit of the climax oak–basswood. Snowberry and ill-scented and smooth sumac, when growing in close thickets, suffered severe mortality. All but occasional plants growing in dense thickets died on south-facing slopes; more survived on north-facing slopes. The foliage of trees was often further reduced and in many instances destroyed by grasshoppers, webworms, and leaf-eating larvae of other insects, for which the dry weather appeared favorable. Trees along the Missouri River and elsewhere were sometimes defoliated two or three times during a single season. Smooth sumac was not only defoliated, but bark was also removed from its stems.

ASPEN FOREST EDGE OR PARKLAND

The ranges of cottontail, bobwhite, and white-footed mouse end in the north between 48° and 51° N. Lat. They are replaced by the snowshoe rabbit, ruffed grouse, and red-backed mouse. Oak tends to be replaced by aspen; and balsam, poplar, and paper birch appear scatteringly. Hawthorn becomes scarce and occurs only in widely separated areas; crab apple is absent. Chokecherry mostly replaces wild plum, snowberry replaces coralberry, and smooth sumac becomes scarce.

The grassland is separated from the coniferous forest in the north by quaking aspen forest (Fig. 12-7). Outlying islands of aspen occur farther out in the grassland, and the aspen forest may likewise be interspersed with islands of grassland. This intermingling of vegetative types has led to the term "aspen parkland." Conifers invade the aspen forest locally and sometimes occur as islands among the aspen. This community stretches from northwestern Minnesota to near Edmonton, Alberta, a distance of 900 miles (1440 km), and then turns south, extending into Montana. Northeast from Regina, Saskatchewan, and southeast from Edmonton, the parkland is nearly 200 miles (320 km) wide.

Plants

Near Birtle, Manitoba (Bird 1930), the dominant plants are quaking aspen, 5 to 12 feet (1.5 to 3.6 m) in height and 2 to 3 inches (5 to 7.5 cm) in diameter, and snowberry, hazelnut, chokecherry, and the rose *Rosa blanda*. In some locations, *Viola canadensis* was found in considerable abundance. A quadrat of 25 square meters gave the following number of plants:

Healthy aspens	31
Aspens girdled by rabbits in 1928 and killed	14
Aspens girdled by rabbits before 1928 and killed	6
Healthy hazelnut	0
Hazelnut killed by rabbits	18
Healthy rose	0
Rose killed by rabbits	17
Healthy chokecherry	0
Chokecherry killed by rabbits in 1928	6
Chokecherry killed by rabbits before 1928	2

Vertebrate Animals

The snowshoe rabbit is an outstanding influent. It is subject to great fluctuations in its abundance (Seton 1929, MacLulich 1937). At times of minimum numbers, there may be as few as ten individuals to the square mile (4/km²), but maximum populations a few years later may reach into the hundreds. The habitat preference of the species is for the young aspens about the edges of groves and of second-growth in burned-over areas, especially where there is considerable shrubby undergrowth. During the summer while there is an abundance of green food and protection, they multiply rapidly. From four to five

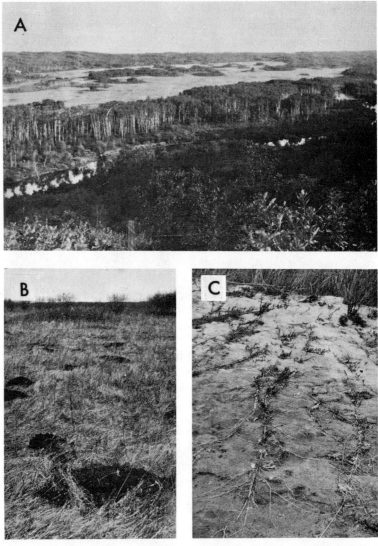

Fig. 12-7. A. Typical view of the poplar savanna found along the Birdtail River near Birtle, Manitoba. The xeric westerly slope is still largely prairie but is being rapidly invaded by poplar groves. The dark patches are principally shrubs such as snowberry; the larger ones have been invaded by aspens. B. Pocket gopher mounds; this disturbance of the soil favors the seeding of snowberry. C. Roots of *Salix interior,* exposed by a spring flood, sending up shoots.

broods of two to five young each are produced. In the winter when food is scarce, they subsist on the twigs and bark of practically all the woody plants, a great many of which become stunted or killed (Fig. 12-8). In one representative quadrat, all the hazelnut, rose, and chokecherry were killed back to the roots, and 64.5 per cent of the poplars were girdled and killed. Some shrubs are preferred to others; the oleaster *Elaeagnus commutata* and oak seedlings are cut back year after year until they die. It appears that oak might be more abundant and that the aspen would invade more quickly into the prairie, were

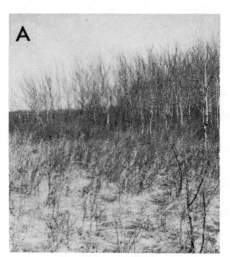 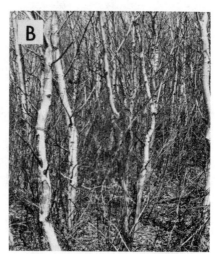

Fig. 12-8. A. Edge of a poplar grove with snowberry below and in front. Dark stems are young poplars killed by rabbits during the preceding year. B. Young white-barked trees in foreground are dead as a result of girdling by rabbits; a thick growth of snowberry is below the trees.

it not for the rabbits (Bird 1930). In winter a great many rabbits are killed by predators, especially by the great horned owl, goshawk, weasel, and coyote. Both white man and Indian killed many rabbits for food and fur.

Bird (1930) quotes Alexander Henry that where bison are abundant the grass is shorter. Some tree-covered areas are deprived of their undergrowth and trees are polished to the height of the bison because of their being used as rubbing posts by the animals.

Birds are abundant in the shrubs and aspen trees. Characteristic species are the American goldfinch, which nests in the young aspens and feeds on the sunflower and thistle seeds in the neighboring prairie; the yellow warbler, which is 97 per cent insectivorous; and the red-eyed vireo, which feeds 85 per cent on insects and spiders. Where there are patches of chokecherry and saskatoon serviceberry, the rufous-sided towhee, catbird, robin, and brown thrasher, all of minor importance, occur.

The coyote, weasel, and skunk of the prairie community also hunt in the aspen parkland. The red-backed mouse, yellow-pine chipmunk, and Franklin's ground squirrel are characteristic minor subinfluents. The pocket gopher and badger penetrate into the edge of the aspen groves. Their diggings supply seed beds for the grassland-invading shrubs.

Invertebrates

Insects are few in the aspen parkland compared with the prairie and mature poplar forest. No species were found to be exclusive except those confined to specific food plants, such as the long-horned beetle *Ropalopus sanguinicollis* on cherry and *Saperda bipunctata* on the stems of *Amelanchier*. Another species of long-horned beetle was also found boring considerable numbers in the stems of young aspens. In the autumn, the ground surface layer of this community is considerably increased by an influx of hibernating invertebrates from the adjoining prairie and down from the trees and shrubs of the aspen itself.

Hydroseral Community Development

Occurring mainly in moist and wet habitats around the numerous ponds, the two willows *Salix petiolaris* and *S. interior* differ but slightly as to their plant and animal associates. *S. petiolaris* is absent only where there is a high percentage of alkaline salts. It constitutes a facies of the willow community of wide distribution, which leads to aspen forest if conditions are hydric, or to prairie if xeric. A slough surrounded by *S. petiolaris* often has water covering the bases of some willows. The willow grows in a close bushlike form from just above the average high water level of the spring to about six inches (15 cm) below the average low water of the autumn. The rich black humus soil is generally saturated with water, if not actually flooded. The sedge *Carex vesicaria* is the principal subordinate plant. *Fragaria virginiana, Aster, Scirpus, Pyrola, Spiraea alba,* and *Potentilla anserina* also occur. A dense growth of moss may cover the surface locally to a depth of almost an inch (Bird 1930).

The sandbar willow *S. interior* facies is also of wide distribution throughout the parkland and represents the first successional tree stage between running water and the forest. It is found on the mud banks and sand bars of all the rivers. When the rivers are sluggish, it may be preceded by the grass *Glyceria grandis*. It spreads rapidly by means of long roots parallel to the surface of the ground, from which frequent shoots are sent up (Fig. 12-7C).

A few boxelders may be present, and scattered herbs consist of the following species in order of abundance: *Ranunculus cymbalaria, Carex, Mentha arvensis, Potentilla anserina, Scirpus,* and *Agrostis*. Moss is scarce and litter does not cover the surface because of seasonal flooding. In neither willow communities are there well-developed seasonal socies.

Animal constituents of the willow community include several species of Chrysomelidae; other insects of lesser importance are the collembolon *Sminthurus* and numerous leafhoppers of the genus *Deltocephalus*. Spiders also are quite numerous. Frogs breed in the sloughs in great numbers and spread from them into the neighboring prairies and woods. The willows and sedges also give shelter and breeding places for large numbers of coots, ducks, and other aquatic birds. Terrestrial birds are not confined to any particular stratum. The redwinged blackbird and common grackle feed on both animal and vegetable matter and often wander some distance from the sloughs in search of food. The yellowthroat is almost entirely insectivorous. The yellow warbler ranks close to the yellowthroat in number and importance, and the song sparrow is common.

Muskrats are seen in both willow facies, and a pair built a nest in the *Salix petiolaris* study area. Voles are also found in both facies but particularly in *S. petiolaris* after the surrounding prairie had been burned. They make numerous runways and nests in the moss at the base of the willows but leave them as soon as an abundant supply of grass returns in the prairies.

Populations of invertebrates are found to be much higher in *S. petiolaris* than in *S. interior* communities. The ground population of the former is especially large on account of the accumulation of moss and surface debris which affords food and shelter. In the *S. petiolaris* community the total invertebrate population in March, before the breakup of hibernation, was over 9.5 million per acre (24 million/hectare) but steadily decreased to an autumn minimum of 0.5 million (1.25 million). The invertebrate population of the *S. interior* community was 3.5 million per acre (8.7 million/hectare) in early spring, fell to a minimum of 1.5 million (3.7 million) at the end of May; rose to 6.75 million (16.9 million) on June 22 because of the presence of many minute dipterons, and then dropped to the autumn minimum of 1.0 million (2.5 million).

The sawfly *Pontania* makes large round galls on the leaf of the willow, and the cecidomyiid *Phytophaga walshi* makes rosette galls on the terminal twigs, particularly of *S. interior*. There may be as many as 15 to 20 on a single stem, but they appear to have but slight effect on the health of the tree.

Insects of seasonal importance are the collembolon *Sminthurus,* the leafhoppers *Deltocephalus mollipes* and *D. noveboracensis,* and the snails *Vertigo ovata, Succinea,* and *Lymnaea*. Beetles belonging to the families Staphylinidae and Carabidae and the frogs *Rana pipiens* and *R. sylvatica cantabrigensis* are abundant on the ground. The frogs feed on insects and snails. The soil population is made up entirely of white worms of the family Enchytraeidae and of the larvae of insects. During the brief period when the willows are in flower, they are visited during the warm part of the day by many small cecidomyiids, chironomids, other dipterons, and andrenid bees, insects which assist in their pollination. During the winter, snowshoe rabbits wander into the willow communities from neighboring aspen groves and eat large numbers of willow shoots.

The biotic interaction and food chain relationships of the willow community are intimately interwoven with those of surrounding communities, particularly that of the *Salix discolor* facies. The predominant of the latter facies is the beetle *Galerucella decora,* which in years of abundance spreads from this food plant to *S. petiolaris, S. interior,* and quaking aspen. Snowshoe rabbits come in from the aspen groves to feed on the willows and the garter snake *Thamnophis* to feed on frogs.

OTHER GRASSLAND CONTACTS

The aspen parkland extends southward from the Edmonton area along the base of the mountains where the adjacent forest is composed of lodgepole pine and Douglas-fir (Fig. 6-2) (Lewis *et al.* 1928). Apparently there is little of a special character between this forest and grassland (Daubenmire, personal communication).

Ponderosa Pine Forest

In the contact of grassland with ponderosa pine, shrubs are few, and forbs and grasses cover the forest floor (Larsen 1930). In many places such as in Montana, southern Canada, Idaho, and near Flagstaff, Arizona, trees often grow singly or in small groups in the grassland.

The Rocky Mountain bushland probably does not extend north of central Wyoming. Usually as the grassland is approached, the shrubs become scattered and smaller. At Manitou, Colorado, the contact between grass and bushland is a graded mixture of grass, mountain mahogany, and other shrubs (Vestal 1917). Near Loveland, Colorado, there was a similar gradient in which grasshoppers were observed to destroy all shrub and forb foliage in 1936. This shrub vegetation drops out in Colorado near the Arkansas River where the pinyon–juniper woodland begins. Where both are present, the bushland is commonly both above and below the woodland.

Woodland and Bushland

In open spaces near the margin of the pinyon–juniper woodland, species occur that are characteristic of the adjoining grassland. In northeast New Mexico (Emerson 1932), the open spaces show 25 to 33 per cent bare ground, although bare ground occupies only about 0.4 per cent of the adjacent typical short grassland.

From southern Colorado south through Texas, cane cactus grows a long way out onto the short grass plains. Between it and the oak–juniper woodland occur Apache plume and barberry (Fig. 12-9) (Cottle 1932).

Oak bushland lies in contact with bunch grass in Nevada and Utah (Fautin 1946, Linsdale 1938). On the western front of the Rocky Mountains in Utah there is an oak bushland that is frequently called chaparral (Clements 1920, Hayward 1948). At its lower edge it was originally made up of bunch grasses with the shrubs occurring in groups or mottes (Fig. 11-4) (Hayward 1948).

Texas–Tamaulipas Area

Before grazing was introduced, grassland extended south of the Rio Grande as a triangular area reaching into Mexico almost to the latitude of Tampico. This triangle was bounded on the west by low hills and mountain ridges. Contact in Tamaulipas is made largely by two or more acacias (e.g., *Acacia amentacea* and *Mimosa emoryana*) which then merge into more mixed shrubs.

Near the Texas shore, groups of shrubs with ericaceous leaves is suggestive of the magnolia–holly forest. This simulant is further emphasized by the extensive distribution of live oak, particularly the variety *Quercus virginiana fusiformis*. In some areas live oak occurs as mottes or single trees scattered on the prairie.

Fig. 12-9. The grassland edge at the base of an oak woodland near Alpine, Texas. The foreground is short-grass grassland which is prevalent generally: in 1930 the grasses were (1) *Bouteloua curtipendula, Andropogon scoparius, A. saccharoides,* and *Sporobolus wrightii*. In 1954 these species were restricted to scattered individuals among the oaks, and the grassland in the foreground was *Tridens pilosus* and *Aristida longiseta*. Shrubs in the foreground are (2) the cane cactus *Opuntia imbricata*, (3) *Berberis* (*Mahonia*) *trifoliata,* and (4) *Fallugia paradoxa;* the trees are (5) gray oak, (6) Emory oak, (7) pinyon, and (8) juniper (photo and data from B. H. Warnock, personal communication).

Cold Desert

The contact of bunch grass with sagebrush in the Great Basin and on the Columbia Plateau of Oregon and Washington is not a striking line that has attracted attention. There is a gradual increase of sagebrush and decrease of grass. Species of *Rosa* and *Symphoricarpos* form numerous thickets on protected upland slopes, especially near the edge of timbered foothills. Here, a high percentage of the nonforested area is covered by scrub rather than by grass. The thickets are dense and vary from a half to one and one-half meters in height. On the most mesic sites, these shrubs are accompanied by taller species such as *Prunus virginiana, Crataegus douglasii,* and *Amelanchier florida*. Characteristic forbs in the community are *Agastache urticifolia, Veratrum speciosum,* and *Epilobium angustifolium* (Daubenmire 1942). In shrubby areas on rocky slopes in the Blue Mountain area of southeastern Washington occur the rattlesnake, bullsnake, two or three predatory birds, deer mouse, and cottontail as residents (Dice 1916).

Hot Desert

The contact between mesquite grassland and the hot desert in Arizona, New Mexico, and Mexico is also diffuse. Shantz points out that the margins of this grassland in Arizona are often characterized by dense growths of cactus, es-

Fig. 12-10. A. Mesquite or desert grassland east of Tucson, Arizona. This was grass in 1890 but had been invaded by cacti in 1936, the time that this photograph was taken. B. The white-throated wood rat in its nest with storage of mesquite bean, after the top of the house, a pile of twigs and pieces of plants, was removed (from Vorhies and Taylor 1940).

pecially when grassland is being invaded by the cactus. The margins and over-grazed areas of desert grassland often have a heavier growth of cacti than that which characterizes the desert. East of Tucson, Arizona, there is an area which was mesquite grassland about 1890. In 1936, mesquite was abundant here along with some *Larrea,* and cacti were very numerous (Fig. 12-10). *Opuntia fulgida* grows fruit each year.

Wood rats are abundant in these areas; they store fruits in their dens and eat only the pulp. Ground squirrels enter the dens and eat the seeds which the rats have removed from the pulp. Desert hackberry and mesquite also supply food for the rodents.

In one of these marginal areas, 18 persons set 9 wood rat traps on two hectares just before dark on July 10, 1936. Each two persons then strip-cruised another area with a gasoline lantern for about one-half hour. In this period they saw or caught 12 pocket mice, 12 wood rats, 4 common toads, 2 lizards (*Cnemidophorus tigris gracilis*), 1 kingsnake, and 2 racers. At the end of the cruising period each of the nine traps had caught a wood rat. The following day, the 18 persons again cruised one to two hectares in the area and reported 61 wood rat dens, 76 kangaroo rat mounds, 5 holes presumably made by pocket mice, and 1 jack rabbit form; lizards, ants, grasshoppers, and robber flies were present in numbers. On one hectare there were two gnat-catchers, two mourning doves, one mockingbird, and one kingbird.

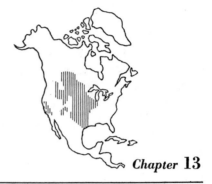

Chapter **13**

The Northern Temperate Grassland

The grassland biome occurs both in the tropics and in temperate climates. One type of grassland occurs in small areas in the middle western and southern parts of Central America, usually covering the tops of hills or low mountains. From maps of Antillian vegetation, grassland apparently occurs under similar conditions in the West Indies. Another type occurs between streams in the lowlands of Central America, including Panama. Most of this country is savanna. Grassland also occurs above timberline, where it is called alpine meadow, and in mountain meadows where fires have destroyed the forest. Several such areas occur in Yellowstone National Park.

On the other hand, in temperate North America the *needlegrass–pronghorn–grama grass biome* covers large areas (Fig. 13-1). It is uniformly characterized by the presence of perennial grasses and originally by a population of grazing and burrowing animals, both the grasses and the animals being dominants. Grassland is the largest biome in North America. It stretches from Edmonton, Alberta, almost to Mexico City, a distance of 2400 miles (3840 km), and from the Pacific Coast to western Indiana. Needlegrass (*Stipa*) and grama grass (*Bouteloua*) occur throughout; both bison and pronghorns were formerly present in about 75 per cent of the area, and one or the other was found over the entire grassland. The northern part of the grassland discussed in this chapter is moist and cool. The southern portion of the grassland, which will be considered in the next chapter, contains scattered shrubs and is drier.

The northern temperate grassland lies mainly north of a line at 34° N. Lat., extending from the California coast to north-central Texas, and then southeast to the Gulf Coast at about 27° N. Lat. The grassland extends north into Alberta and from western Indiana into California. The altitudinal range is from

328

sea level in southern Texas to 6000 and 7000 feet (1800 to 2100 m) in the Rocky Mountains and California.

On the open plains the vegetation stretches as far as the eye can see, green in the rainy season and brown in the dry season. A common morning scene contains a herd of bison or pronghorns in the distance, jack rabbits returning to their forms, a wolf or a coyote trotting to its den, several small birds flying overhead and singing, and locally a prairie dog or a ground squirrel sitting upright at its burrow.

In *:.*D. 1500 the horse had not yet been reintroduced into North America, and the native human population of the Great Plains was small. When the horse became common, a hundred years or so later, the population of Indians increased, especially in the smaller areas west of the Rocky Mountains, but nowhere equaled the population that occurred in the woodland or forest (Kroeber 1939).

Regions

Although Shantz (Shantz and Zon 1924) maps and names 20 faciations in the United States, and there are at least two more in Canada and three in Mexico, these may be combined into only three regions of importance biotically. The *combined tall-grass and mixed grassland* extends from the deciduous forest to about 104° W. Long., near the western boundaries of Kansas and the Dakotas. This boundary is much farther west than that shown by Aikman (1935) after the worst years of the great drought. It includes the tall grass of Shantz (1924) which comprises the subclimax prairie and true prairie of Clements (1920), the tall-grass and coastal prairies of Clements (Clements and Shelford 1939), and the mixed prairie as defined by Clements (1920) and mapped by Shantz (1924) (Fig. 13-1). Its limits are well represented by the original distribution of the greater prairie chicken. The *short-grass grassland region* lies between the mixed prairie and the base of the Rocky Mountains and includes certain areas among the mountain ranges. Large populations of pronghorns once occurred here. The *bunch-grass region* lies between the northern Rocky Mountains in the United States and the Pacific coastal mountains, and between the Sierras and the coastal mountains farther south.

Climate

The range of climatic conditions within this community is greater than in any other North American biome. On the east from Texas to Indiana, its boundary lies close to the isohyet of 40 inches (100 cm), while farther north precipitation drops to 25 inches (62.5 cm) in correlation with a decreased rate of evaporation, and in the west to 12 inches (30 cm). A dry season prevails from early or late summer to autumn or early winter. In the tall-grass area in the east, precipitation equals or exceeds the evaporation (Visher 1946), but nearly everywhere else the rainfall is less than the evaporation. The strong wind that prevails over the plains is a factor that induces high rates of evaporation. The considerable variability of the water supply from year to year causes

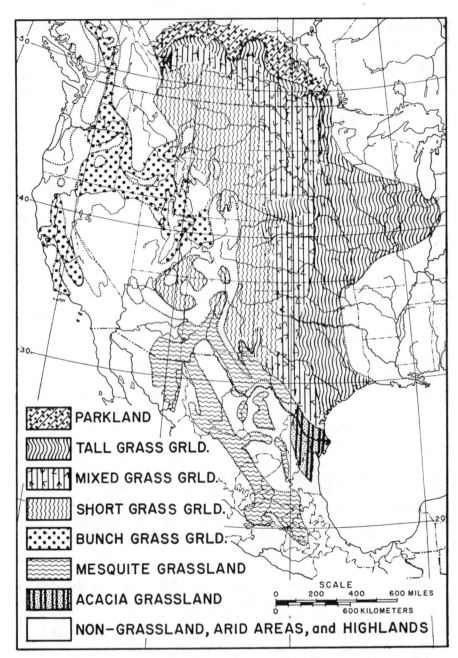

Fig. 13-1. The main types of grassland in North America, often recognized as associations. Boundaries between tall grass, mixed grass, and short grass have shifted from time to time due to changes in rainfall, grazing by bison, and other factors (University of Chicago base map).

trees to be absent and makes subterranean habits advantageous for small animals. The extremes of temperature are even more striking than are those of precipitation, varying from a frost-free season of only three or four months in Canada to one of practically an entire year in southern Texas.

Life Forms and Life Habits of Important Constituents

The grassland is a community dominated by herbs. The herbs are divisible into grasses and forbs, the grasses being the dominants. The community and habitat relations of such sedges as *Carex filiformis* and *C. stenophylla* are essentially the same as short grasses.

Dominants and influents which give unity to the biome by their ubiquitous presence are the grasses, Junegrass, blue grama, side-oats grama, hairy grama, needle-and-thread, green needlegrass, sheep fescue, little bluestem, buffalo grass, and the animals, bison, pronghorn, badger, jack rabbit, and the grasshopper *Melanoplus mexicanus*.

Grassland vegetation is layered both below and above ground (Fig. 13-2).

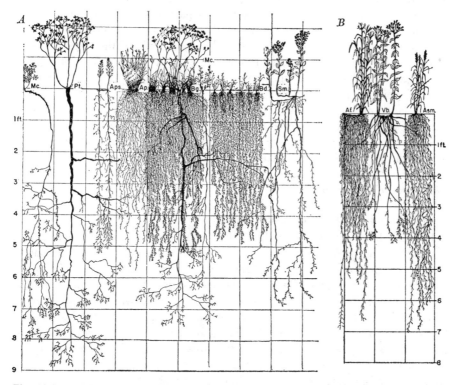

Fig. 13-2. A. Diagrams of short-grass plant tops and roots: Mc., *Sphaeralcea coccinea;* Pt., *Psoralea tenuiflora;* Aps., *Ambrosia psilostachya;* Ap., *Aristida purpurea;* Bg., *Bouteloua gracilis;* Bd., *Buchloë dactyloides;* and Sm., *Solidago mollis.* Note the very definite root layer formed by the grasses and the deeper one by coarse forbs. B. Af., *Andropogon gerardi;* Vb., *Vernonia baldwini;* Asm., *Agropyron smithii* (after Albertson 1937).

In most grasslands the grasses are of two or more heights. In the tall-grass prairie, the grasses are 50 to 150 centimeters high; in the mixed prairie both relatively tall and short grasses occur, respectively, 30 to 60, and 20 to 50 centimeters high; in the short-grass plains, 5 to 40 centimeters. The bunch-grass prairie may include both tall and short grasses, but the bunch grasses are commonly 60 to 100 centimeters high.

Life habits of animals may be classified as subterranean, cursorial, and arboreal. Life habit ratios may be computed for different taxonomic groups by computing the percentage of species breeding in each of the three strata. For example 53 per cent of the grassland birds breed in nests built on the ground, contrasting with 20 per cent of forest birds. On the other hand, fallen or standing trees are used by 60 per cent of the forest birds. A great preponderance of subterranean breeders characterizes grassland mammals, exclusive of bats (Clements and Shelford 1939).

Craig (1908) points out that few forest birds sing on the wing, but grassland birds such as the horned lark, bobolink, Smith's longspur, chestnut-collared longspur, lark sparrow, lark bunting, and Sprague's pipit often do.

Eyesight is keen in some grassland species, particularly the pronghorn, and continuous watch for enemies from vantage points takes the place of the secretive habits (Bailey 1931, Seton 1929) characteristic of the larger animals in the forest. Grassland animals are commonly swift runners. The pronghorn will travel 40 miles (64 km) per hour over considerable periods and 54 miles (86 km) per hour for short distances.

The large prairie dog towns, large herds of pronghorns (400 animals or more reported by early explorers), and the enormous numbers of bison, often in herds of 100,000 to 2,000,000 animals (Seton 1929), indicate the gregariousness of grassland animals. Likewise prairie chickens are gregarious on their drumming areas in contrast to the solitary habit of their forest relative, the ruffed grouse (Craig 1908).

Major Permeant Dominants and Influents

In 1600 there were probably 45,000,000 bison occurring in all parts of the grassland of North America, except California. The size of herds in the bunch grass and in the short grass of New Mexico were smaller than those of the plains and the eastern tall grass. Bison had penetrated forested areas and were apparently responsible for maintaining grass-covered areas where the substratum was unfavorable for trees, as near Crab Orchard, Tennessee. They also invaded the forest in Virginia and Pennsylvania in small numbers. The animals made trails over the grassland and through the stream-skirting forest in their quest for water. The bison was a dominant in the mixed prairie and the western edge of the tall-grass areas and was an influent elsewhere (Fig. 1-1).

The pronghorn had a population similar to that of the bison (about 15,000,-000 on the plains, 15,000,000 in California, and 15,000,000 in the arid southwest and Mexico). It occurred in all types of grassland, although its activities were limited to the western part of the tall-grass community. Pike found them abundant in eastern Kansas in 1806 (Cockrum 1952). They tended to occupy drier portions of the grassland than did the bison.

The white-tailed jack-rabbit of the northern grasslands and the black-tailed jack rabbit of the southern grasslands have their ranges overlap roughly between 36° and 42° N. Lat. They made definite trails across the plains when populations were large. Jack rabbits prefer short grasses with borders or mid grasses for shelter.

The cottontail, wapiti, and deer are secondary influents in the grassland wherever there are considerable stream-skirting forests. Predators operate throughout the grassland. Some of the predatory birds are restricted to the vicinity of the stream-skirting forests. The wolf, coyote, and kit fox are able to live in pure grassland (Criddle 1925).

The badger is universally present, digging for rodents. Burrowing rodents near the center of the grassland area include ground squirrels, prairie dogs, pocket gophers, two species of pocket mice, and a kangaroo rat, all of which have large populations. Moles and shrews disturb the surface of the ground, and large numbers of grasshopper mice live in holes made by other species. There is also a number of predators which burrow into the ground, such as the black-footed ferret and kit fox, which are characteristic of grassland, and wolf, coyote, spotted skunks and hog-nosed skunk, whose geographic distribution is more extensive.

The greater prairie chicken, the burrowing owl, and perhaps the sage grouse are species whose entire lives are spent in open grassland. The bobwhite, sharp-tailed grouse, and several predatory species requiring more cover exercise considerable influence on the grassland. Small minor influent and permeant birds are the horned lark, lark sparrow, lark bunting, and vesper sparrow. Their impact on the community has not been adequately measured, but they feed their young large numbers of insects.

Of reptiles, the bullsnake or gopher snake is found throughout and feeds on rodents. Grasshoppers and ants are the outstanding groups among grassland insects. There are more than 250 species of grasshoppers originally restricted largely to the grassland. The grasshopper *Mermiria neomexicana* occurs in nearly all the grassland communities.

The effect of plants and animals on the habitat is marked by the very deep root systems of the prairie grasses (Fig. 13-2) and by the immense numbers of burrowing mammals and insects. The abundance of thirteen-lined ground squirrels with their extensive network of burrows means that an enormous amount of soil material must be moved by this species alone and in a relatively short time. The prairie dog commonly digs to a depth of 12 to 15 feet (3.6 to 4.5 m) and brings enormous amounts of soil material to the surface. Since the prairie dog is concentrated in towns, its effect on the habitat is not as general as with the ground squirrels.

Another reaction of great significance is the tramping of the soil by the large numbers of bison, pronghorns, and jack rabbits. The effect on the soil and grass cover of the passage of a large herd of bison must have been great but has not been adequately described.

An outstanding coaction is the grazing or clipping of grasses by the large ungulates, prairie dogs, ground squirrels, grasshoppers, and other herbivores. Great quantities of animal droppings fertilize the soil. Early settlers used dry bison chips as their principal source of fuel. These chips were what was left after the droppings had afforded sustenance and shelter for insects.

TALL AND MIXED GRASSLAND

Tall and mixed grasslands reach from east-central Texas to the aspen park-land in Canada and in broad bands between the deciduous forest in the east and the short-grass plains at about the 100th meridian (Fig. 13-1). In North Dakota the boundary swings to the west across the northeastern corner of Montana. At the western edge of the region, stream-skirting forests have dwindled to mere cottonwood floodplains. Tall grasses predominate next to the forest margin but become more and more mixed with short grasses in drier habitats westward.

Along its eastern boundary from Oklahoma to Illinois, annual precipitation approaches 40 inches (16 cm) but drops to about 20 inches (8 cm) in southern Manitoba. Along the western limit of the region, precipitation varies from 20 inches (8 cm) in Oklahoma to 25 inches (10 cm) in Nebraska, 20 inches (8 cm) in North Dakota, and 15 inches (6 cm) in the extreme northwest. Evaporation and temperature vary in a similar manner. Drought periods are less frequent and less severe near the forest than in the more westerly parts.

Plant Dominants

The dominant plants are porcupine grass, prairie dropseed, little bluestem, side-oats grama, Junegrass, western wheatgrass, plains muhly, panic grass, and the sedge *Carex pensylvanica*. In mixed prairie, additional species include green needlegrass, needle-and-thread grass, sand dropseed, slender wheatgrass, galleta, and purple three-awn. There are numerous species of forbs throughout. Match weed or broomweed, scurf-pea, sunflowers, goldenrods, and the ragweed *Ambrosia psilostachya* occur from Texas into Canada.

Near Hays, Kansas, within the mixed prairie region, little bluestem occurs in a faciation similar to that in Illinois; big bluestem grows with western wheat-grass; and short grasses occur on the top of knolls. In Saskatchewan the mixed prairie contains species of *Stipa* and *Bouteloua* (Coupland 1950). In and near the Wichita National Wildlife Refuge in Oklahoma, there is much little blue-stem adjacent to short grass.

Animal Dominants and Influents

There is no evidence that animals recognize the difference between tall-grass and mixed-grass prairies. Many animal ranges terminate near the western edge of the combined areas. On the other hand, the pronghorn attains its maximum abundance farther west on the short-grass plains and extends eastward only to the western quarter of Missouri. The white-tailed jack rabbit barely crosses the Mississippi River, while the black-tailed jack rabbit stops in Missouri. The bison appears to have been most numerous in the mixed-grass area. Bison wallows in central Oklahoma and their paths to rivers there and in Illinois can still be seen. Bison were plentiful, however, also in the tall-grass area. In 1679 at the beginning of winter, LaSalle saw a bison stuck in a marsh near South Bend, Indiana. In 1680 he found the prairie near Morris, Illinois, occupied by numerous bison. This animal disappeared east of the Mississippi River about 1800; according to an Indian tradition, most of the Illinois herd was killed by a blizzard about 1775 (M. B. Shelford 1913). Bison had declined to a few isolated bands in the Dakotas by 1886. In the mixed prairie, extensive grazing by bison and prairie dogs tended to change *Andropogon* tall grass into short grass. The boundary between the two grassland types probably shifted back and forth over 100 miles (160 km) during periods of years when the size of animal populations or the amount of rainfall varied.

Osborn and Allen (1949) describe the extinction of a prairie dog town as tall grasses encroach the area. It appears that grazing by bison and other animals is required to keep out the tall grasses and permit the occurrence of this species.

The gray or buffalo wolf was originally described from the tall-grass prairie country of Iowa. The coyote is fond of retreating into the forest margins; it still occurs in the area. Its principal food consists of ground squirrels, mice, birds, grasshoppers, and a limited amount of fruit. The badger formerly ranged eastward into the Indiana–Ohio–Michigan prairie peninsula and still occurs sparingly in Illinois and Indiana. Snead and Hendrickson (1942) found its food to consist of ground squirrels, mice, cottontails, and insects in ratios of 7:5:3:3.

The black-tailed jack rabbit eats 12 kinds of grasses, 27 forbs, and 5 shrubs (Brown 1946, 1947). Those most heavily utilized were broomweed, soapweed, cactus, and chalk lily. The stem just below the branches of the flower stalk of broomweed and the long needle-like leaves of the soapweed are heavily grazed during the autumn and winter months. The upper portions of the tap roots and the seedling leaves of the chalk lily are heavily utilized after the first damaging frost in the autumn. The leaves and the leaf sheaths surrounding the inflorescence of sand dropseed and the green leaves of western wheatgrass and sedge are the parts generally eaten, the sedge within two to three inches (5 to 7.5 cm) of the ground. Blue grama and buffalo grass were utilized lightly. In the Fort Hays region of western Kansas, where this study was made, jack rabbits averaged 185 per square mile (71/km^2) in 1946.

In a study of seeds in pellets, Brown found that jack rabbits deposited 7.7 pounds of seed per acre (8.6 kg/hectare) during the month of October. The seeds from the pellets compared with hand-picked seeds germinated in the following percentages: buffalo grass, 100 and 57.5; sand dropseed, 4.2 and 3; Hooker's dropseed, 25 and 14. White-tailed jack rabbits are scarce in the Fort Hays area, but were originally generally distributed over the northern plains east to the Mississippi River in Iowa. The species decreased with the plowing of the natural grass areas; the black-tailed jack rabbit, however, apparently increased.

Six species of small mammals had large populations on the Fort Hays study area (Wooster 1935, 1936, 1938, 1939). The plains harvest mouse, between 1931 and 1936, averaged 716 per square mile (275/km²). They eat the seeds of switch grass, buffalo grass, and sunflower and grasshoppers. The deer mouse averaged 1728 per square mile (665/km²). They utilized the seeds of 7 grasses and 17 forbs, including sunflower. Insects, however, especially grasshoppers, were preferred. The prairie vole averaged 2462 per square mile (947/km²). They stored their nests with seeds of two short grasses, two mid grasses, and some perennial ragweed. The thirteen-lined ground squirrel was present in numbers of 1952 per square mile (751/km²) during a part of the year. They utilized buffalo grass and wild onion bulbs along with grasshoppers, crickets, beetles, and spiders for food and short and mid grasses for nests. The remainder of the population was made up of the short-tailed shrew which had 337 per square mile (130/km²) and the least shrew with a population of 614 per square mile (236/km²). Prairie dogs, pocket gophers, and pocket mice had been extirpated.

Important predators present were the coyote with two per square mile (0.8/km²), Swainson's hawk, rough-legged hawk, and ferruginous hawk. There was a large excess of food available for the predators. With the wolf, badger, and foxes eliminated, the population sizes indicated for the prey species probably do not represent their pristine numbers.

The greater prairie chicken was abundant in 1832 and during the next 15 years. Flock formation occurred in the autumn, and several thousand often entered the oak forests along the streams (Nice 1931). Population size is affected by weather conditions (Shelford and Yeatter 1955). Young prairie chickens, 8 to 10 weeks old, are fed the short-horned grasshopper *Melanoplus differentialis,* long-horned grasshoppers including *Neoconocephalus robustus,* the leaf-feeding beetles *Calligrapha similis* and *Cryptocephalus venustus,* the imbricated snout beetle *Epicaerus imbricatus,* the June beetle *Phyllophaga,* spiders, robber flies, miscellaneous beetles, bees, and wasps. Grass seed, goldenrod heads and leaves, and rose hips are also utilized (Yeatter 1943). Other strictly prairie birds are the meadowlark, dickcissel, and horned lark. All of these nest on the ground, consume large amounts of seed, and feed their young with quantities of prairie insects.

The reptiles of the tall-grass and mixed prairies are limited to snakes, except in the south. The garter snake *Thamnophis radix* feeds principally on earthworms, frogs, and toads. The chief rattlesnake east of the Mississippi River is the massasauga, which lives mostly on wet ground; the western rattlesnake and blue racer occur throughout, and the bullsnake is characteristic and generally distributed. The bullsnake commonly feeds on striped ground squirrels and meadow voles. Lizards, toads, and frogs appear not to breed in the climax but occur in some seral stages.

Insects are abundant in prairies, both in numbers and in biomass (Fig. 13-3)

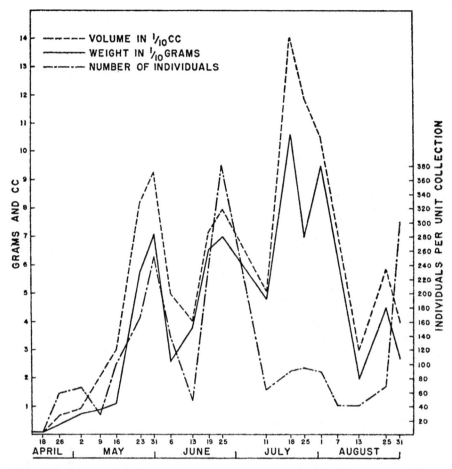

Fig. 13-3. The relation of volume, weight, and numbers of individuals in samples of arthropod populations in south-central Oklahoma in the summer of 1937 (redrawn from Shackleford 1939).

(Adams 1915, V. E. Shelford 1913, Hendrickson 1930, Shackleford 1929, Weese 1939, Fichter 1954). By late August they sometimes reach 10 million individuals per acre (25 million/hectare).

The Orthoptera are important and numerous (Criddle 1933, Whelan 1938). The grasshopper *Encoptolophus sordidus costalis* is characteristic of the tallgrass prairie; *Melanoplus bivittatus* is important in Nebraska (Fichter 1954). *Melanoplus keeleri luridus* and *Conocephalus saltans* are generally associated with the grass *Andropogon. Melanoplus dawsoni* occurs in tall grass from Alberta to Missouri. Some orthopterons feed on forbs rather than the grasses.

Insects feeding on grasses also depend for shelter on grasses of particular life form. In east-central Texas, the slender-bodied *Mermiria m. maculipennis, Syrbula admirabilis,* and related species are found among the *Andropogon,*

Sporobolus, and other tall grasses; while *Ageneotettix deorum* and *Trachy-rhachys kiowa fuscifrons* and two other small flattened acridians frequent low, matlike grasses, especially buffalo grass (Isely 1938). Although other plants are eaten, *Thelesperma trifidum* is the chief host-plant of the wingless *Paraidemona punctata.*

Gutierrezia dracunculoides is the host-plant of *Hesperotettix v. viridis,* and *Helianthus annuus* the primary host of *Hesperotettix speciosus. Melanoplus texanus* eats a number of plants including *Gaillardia pulchella. M. plebejus* prefers grasses. *Engelmannia pinnatifida* is one of the possible host-plants of *M. flabellatus,* a very wide-ranging species. *Rudbeckia hirta* is the food plant for *M. impiger. Hypochlora alba* is definitely associated with the western mug-wort. Caged *M. discolor* shows a definite preference for *Salvia farinacea.* Three species, *M. differentialis, M. mexicanus,* and the flightless *Brachystola magna* are pests and general feeders. Katydids frequently utilize forbs.

Spharagemon collare cristatum and *Melanoplus angustipennis* reach their greatest numbers only in light sandy soils where they lay their eggs; alluvial soil species are *M. differentialis* and *Dissosteira carolina,* eroded soil species, *Trimerotropis pistrinaria,* and calcareous soil species, *Xanthippus corallipes pantherinus* (Isely 1935, 1937, 1938).

Among the Hemiptera, the lygaeid *Ligyrocaris diffusus,* the plant bug *Lygus oblineatus,* the predatory bug *Orius insidiosis,* and the damsel bug *Nabis ferus* are in all faciations and subclimax communities. The common stink bug *Euschistus variolarius* is also in all prairie faciations but feeds on the forbs *Astragalus* and *Ratibida.* The predatory ambush bug is widely distributed on composites, most often found lurking on goldenrod (Hendrickson 1930). The sucking hemipterons and homopterons deplete the grasses, as is illustrated by the ravages of chinch bugs, original inhabitants of the tall-grass prairie. *Agallia constriata* is one of the abundant and widely distributed tall-grass leafhoppers.

Among the Coleoptera of the grassland is the twelve-spotted cucumber beetle and various others that are forb eaters. Decaying material and bison excrement support the scarab *Aphodius distinctus,* while the lady beetle *Hippodamia convergens* is predatory. The common and abundant species of insects often do not feed on the grasses, notably leaf beetles of the genus *Pachybrachis* (Whelan 1936). Lepidoptera are usually not abundant and are not grass feeders. The only lepidopteron regularly found in the Iowa prairies is *Cercyonis alope olympus* (Hendrickson 1930).

In the mixed prairie of the Canadian provinces, the pale western cutworm *Agrotis orthogonia* feeds on wild grasses. The loose soil around bison wallows and in the diggings of other animals formerly provided favorable conditions for their egg laying (C. W. Farstad, personal communication). When rain brings the larvae to the surface, they are attacked by birds, parasites, and other organisms.

Among Diptera, several species of long-legged flies of the genus *Colicops* are common or abundant. Syrphus flies, whose larvae feed on aphids, and robber flies, such as *Asilus* and *Promachus,* abound; the latter prey on flying insects, picking them up with great dexterity and often taking forms larger than themselves. Hendrickson found robber flies most abundant in an *Andropogon* community.

Hymenoptera are represented chiefly by the ants which effect important coactions in their gathering of both plant and animal food and important reactions in their moving of soil. Species differ sharply with changes in grass composition (Hendrickson 1930). Halictid bees are usually present in the grasslands. The grass stem sawfly *Cephus cinctus,* which is found on wheat, is a native of this community. It feeds on the stems of wheatgrass, on two other species of *Agropyron,* and on *Hordeum jubatum.* It is distributed in the mixed prairie south of 36° N. Lat. (Seamans and Farstad 1938).

Biotic Changes Northward

In southwestern Manitoba, the little bluestem and big bluestem grasses may be climax on knolls (Bird 1930). A few grasses drop out near the international boundary; *Agropyron dasystachum* and the two half-shrubs, *Eurotia lanata* and *Artemisia frigida,* remain. These species are generally distributed over the northern high short-grass plains. The prairie dog is replaced by the Richardson's ground squirrel north and east of the Missouri River. This ground squirrel has habits similar to those of the prairie dog except that it does not occur in groups or towns. Arnason (1941) names 35 species of insects, including four grasshoppers, near Saskatoon which are characteristic of grass and which do not attack cultivated crops.

Overgrazing

Grazing brings changes in both the plant and animal composition of grassland. In the Niobrara Game Refuge near Valentine, Nebraska, on July 24, 1945, one ungrazed meadow was dominated by grasses, although a considerable series of forbs were present. The common species were two gramas, fescue, western wheatgrass, *Stipa comata,* pepperwort, lead plant, and *Hymenopappus filifolius,* as well as other species. A second area overgrazed by bison contained buffalo grass, pepperwort, sedge, lead plant, *Hymenopappus filifolius,* and *Opuntia polyacantha.* There was also a little wormwood and rabbitbrush present. Ninety-six sweeps with the insect net in each area showed a greater number of animals on grass in the ungrazed area and on the forbs in the overgrazed area. A froghopper, a cercopid, and three species of shield bugs found in the ungrazed area were absent from the overgrazed area. Nymphs of a long-horned grasshopper were more numerous in the ungrazed area, but nymphs of the short-horned grasshopper were more abundant in the overgrazed area.

Weese (1939) estimated the effect of overgrazing on insect populations by quantitative daily collections from moderately overgrazed and ungrazed grassland in the Wichita Mountains Refuge in Oklahoma, between June 6 and July 3. The protected area was characterized by a heavy growth of tall grasses, principally little bluestem, while the overgrazed area supported a rather sparse vegetation of buffalo grass and *Coreopsis.* The flies were equally numerous in the two areas. The other groups ranged from 2.3 to 9 times as abundant in the overgrazed area as in the ungrazed grassland.

Effects of Excessive Rainfall and Drought

The drought began at Hays, Kansas, in 1933 after a period of six years with average annual precipitation five inches (12.5 cm) above normal; this was followed by seven years with precipitation nearly five inches below normal (Albertson 1941). The native vegetation was greatly modified first in one, then in the other, direction. During the wet period, the more mesic grasses, such as the wire grasses *Aristida purpurea* and *A. longiseta*, side-oats grama, little bluestem, switch grass, and Indian grass advanced a considerable distance within the more xeric areas occupied by the two short grasses, buffalo grass, and blue grama. The wire grasses, side-oats grama, and little bluestem invaded the short-grass area. Mid grasses formed an upper story over large areas of the short grasses. Even the tall grasses were found growing through the short-grass community in buffalo wallows and other depressions. The perennial ragweed *Ambrosia psilostachya* of the hillsides and ravines invaded the short-grass areas as an upper story of the vegetation.

During the ensuing drought period, the basal ground cover of the short grasses became reduced in 1940 from between 80 and 95 per cent to 20 per cent. In the dry upland soils, little bluestem withered early in June, but the deeper-rooted big bluestem persisted a longer time (Weaver and Albertson 1939, 1940). *Stipa spartea* and *Bouteloua gracilis* were also more resistant, rolling their leaves and assuming a condition of drought-dormancy. The grasses, however, often failed to bloom. The forb population also steadily decreased. In 1937 the still resistant forbs were red false mallow, few-flowered psoralea, the spiny haplopappus, common thistle, blazing star, and narrow-leaved four-o'clock. However, even they were almost completely gone by the fall of 1939, as were the native grasses. This decrease in competition allowed the growth of pepper grass, stick-tights, and various species of *Chenopodium* and *Amaranthus*. The root systems of some of these species can be compared in Figure 13-2 and in Albertson (1937).

The cactus *Opuntia macrorhiza* is a constituent of the grassland on the Great Plains. It increases in drought periods and almost disappears in wet periods. Many individuals are eliminated by the cactus moth which thereby lessens its own chances for existence (Bugbee and Reigel 1945).

Among the forbs, resistance to drought is closely correlated with the relative development of the root system. Species with roots penetrating 8 to 20 feet (2.4 to 6 m) into the moist subsoil are little affected by the progress of drought. Where the root systems are shallow, the decrease in water content of the tissues is pronounced.

Wooster (1935) found that the black-tailed jack rabbit population increased about threefold following the summer droughts of 1933 and 1934, although there was no mass movement of the jack rabbits. The prairie vole decreased sharply in numbers. The deer mouse persisted in the usual numbers. The bird dickcissel became scarce or absent.

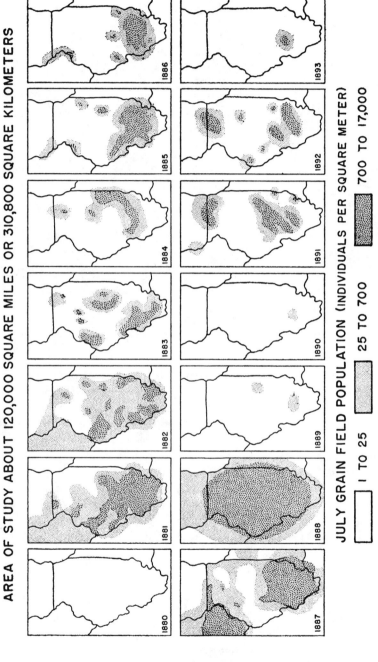

DISTRIBUTION OF DENSE POPULATIONS OF THE CHINCH BUG
(BLISSUS LEUCOPTERUS SAY)
AREA OF STUDY ABOUT 120,000 SQUARE MILES OR 310,800 SQUARE KILOMETERS

JULY GRAIN FIELD POPULATION (INDIVIDUALS PER SQUARE METER)

1 TO 25   25 TO 700   700 TO 17,000

Fig. 13-4. Variation in abundance of chinch bugs in tall-grass area. These are measured on field crops, but populations vary in size also in natural grass (after Shelford and Flint 1943).

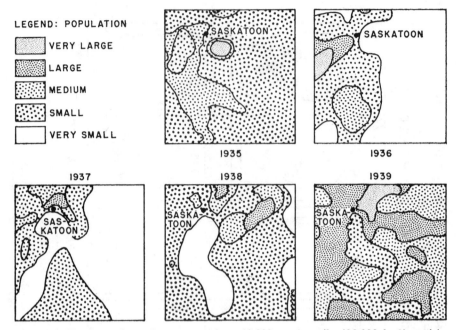

LEGEND: POPULATION

▨ VERY LARGE

▨ LARGE

▨ MEDIUM

▨ SMALL

☐ VERY SMALL

1935

1936

1937

1938

1939

Fig. 13-5. Numbers of grasshopper eggs in a 10,000 square mile (26,000 km²) prairie, largely cultivated area in Saskatchewan, 1935 through 1939. The successful hatching of the eggs depends upon favorable rainfall (from maps by K. M. King, Dominion and Provincial Departments of Agriculture, Canada).

There appear to be few or no reports on the effect of drought on the general population of insects except for agricultural pests. Dry weather favors large populations of chinch bugs in Illinois, but heavy rainfall in May and June kills the young (Fig. 13-4). Grasshoppers are favored by dry weather. Native grasshoppers in Saskatchewan which show correlations between population size and rainfall (Fig. 13-5) are the red-legged grasshopper, the two-striped grasshopper, and the clear-winged or roadside grasshopper. In addition to these about a dozen other kinds usually occur in numbers, including Packard's grasshopper, the narrow-winged grasshopper, and Bruner's grasshopper. In drought years, grasshoppers become numerous and make flights from the high plains into the Mississippi Valley. Such flights were recorded in 1876 and in the 1930's.

Aspection

The sedge *Carex pensylvanica* is first to bloom in the prevernal season in South Dakota (Harvey 1908). No prairie grasses flower in May, although some forbs do. In the aestival season from mid-June to mid-July *Koeleria cristata* blooms; in the serotinal season, so do buffalo grass and three species of *Bouteloua*. The bluestems come into flower in the autumn. Different species of insects reach peak of numbers at different times during the year; for instance at Norman, Oklahoma (Carpenter 1939), during the prevernal aspect, the bug *Homaemus aeneifrons;* vernal, the bug *Strongylocoris stigicus;* aestival, the bug *Solubea pugnax;* aestival-serotinal, the stink bug *Mecidea longula;* serotinal, the leafhopper *Idiocerus crataegi;* and autumnal, the bug *Ortholomus scolopax.* Many other examples could be given.

Coastal Prairie

The tall- and mixed-grass faciation blends into the tall-grass faciation of the Gulf of Mexico coast south of 31° N. Lat. This faciation lies largely in Texas but extends about 75 miles (120 km) inland and eastward into Louisiana, where it appears not to be a climatic climax under present conditions, and only a short distance south of the Rio Grande. Its distribution agrees well with that of the Attwater subspecies of the prairie chicken. The annual rainfall ranges from 29 inches (72 cm) at Matamoros to 50 inches (125 cm) at the Sabine River to 55 inches (138 cm) in Louisiana. This is the highest rainfall for any major part of the grassland.

Silver beardgrass is a characteristic dominant north to Kansas. The little bluestem appears to be uniformly distributed. Buffalo grass, big bluestem, *Paspalum plicatulum,* Texas needlegrass, smut-grass, and hairy tridens are also dominants (Clements and Shelford 1939, Tharp 1926). Purple three-awn grass is apparently subclimax (Tharp 1926). Some of the grasses not common in tall grassland generally, such as *Aristida purpurea, Bouteloua rigidiseta, B. trifida, Tridens pilosus* and *Hilaria berlangeri,* are among the commonest grasses in the Tamaulipas area; several of them go into the tropics. Silver beardgrass, *Hilaria berlangeri,* and *Heteropogon contortus* are important at altitudes of 3700 to 4700 feet (1120 to 1425 m), in the desert grassland of Arizona, New Mexico, and western Texas (Whitfield and Beutner 1938). The faciation merges with the desert grassland to the west and the Tamaulipas grassland to the southwest.

Certain subspecies of the pygmy mouse, harvest mouse, and black-tailed jack rabbit appear to be characteristic. Mammals that were dominant or influent over the biome generally were present in former days. The bison occurred in considerable numbers, and pronghorns were numerous. The badger and subspecies of spotted skunk, pocket gopher, red wolf, and coyote take the place of northern races.

Among birds, pristine populations of Attwater prairie chicken were high. A subspecies of horned lark and the boat-tailed grackle are characteristic (Bailey 1905). Most of the common species occur well south into Tamaulipas, Mexico.

The common tall grassland orthopterons, such as *Melanoplus differentialis, Chortophaga viridifasciata, Ophulella pelidna, O. speciosa, Orchelimum concinnum,* and *Scudderia texensis* are usually common. *Amblycorypha huasteca* reaches north in the tall-grass prairie as far as Kansas. The spider *Aysha gracilis* occurs at Victoria, Mexico. Most of the spiders in this area are also common in Illinois.

The coast is bordered by a continuous subclimax marsh which varies from five to ten or more miles (8 to 16 km) wide. The dominants of this coastal marsh are *Spartina spartinae, S. patens,* and *Sporobolus virginicus.* In the marshes the wood rat and swamp rabbit are the chief mammals. Other seral stages of this prairie are the gulf bayous, mudflats, and sand areas. The mole is a sand area species.

Outside of the marsh area there are sandy beaches with zones similar to those shown in Figure 3-11 except that the beach vegetation merges into the *Spartina* vegetation with no woody species. The lowest zone is the wet beach with flesh flies and tiger beetles; the second zone has beach crabs and the burrowing wasps *Bembex.* The third zone, observed west of Sabine Pass, Texas, in July, 1942, had *Salicornia, Gaillardia pulchella, Aster,* and a little *Spartina,* and supported a large group of insects $(50/m^2)$, including the tree cricket *Oecanthus quadripunctatus,* the narrow bug *Ischnodemus badius,* the spittle insect *Clastoptera xanthocephala* of wide distribution, and about ten other species.

SHORT-GRASS GRASSLAND

The principal grasses of the short-grass grassland or the *blue grama–pronghorn association* are *Bouteloua gracilis* (blue grama), *B. hirsuta, Buchloë dactyloides,* and *Hilaria jamesii.* The shrub *Artemisia frigida* occurs in the northern part of both the short-grass and mixed-grass prairie (Fig. 13-6). In the southern parts, *H. jamesii* is important. Eight short-grass faciations have been recognized, with *B. gracilis* prominent in five of them. Short grass reaches from the southern boundary of New Mexico to points more than 100 miles (160 km) within Alberta and Saskatchewan.

Aspection is shown by the flowering of forbs. Characteristic species representative of each aspect in the central part of the association are the common pasque-flower in the prevernal; *Senecio aureus, Sphaeralcea coccinea,* in the aestival; and the composites *Aster ericoides, Artemisia glauca, Liatris punctata,* and *Solidago missouriensis* in the serotinal aspect.

Major Permeants, Dominants, and Influents

The bison was formerly a dominant animal. Herds of a million or more were reported by the early explorers. These aggregations were of greatest magnitude during migration as the animals moved north a distance of 200 or 300 miles (320 to 480 km) in early summer and south again in early winter. Bison and other animals maintained the mixed prairie and the little bluestem bunch grass in a short-grass condition by their grazing and tramping (Larson 1940). Wallowing caused considerable disturbance of loose soil. The herds were harassed by the buffalo wolf, but the bison were able to defend themselves with skill and vigor—they tossed wolves with their horns and trampled them with their hoofs (Seton 1929).

The pronghorn antelope occurred in great numbers in this association. It preferred rolling topography and sought temporary shelter in ravines and cottonwood-covered valleys during storms. During the winter, it shifted to areas with thin snow; otherwise the bands remained in the same general locality. The species was an important influent throughout its range and locally a dominant (Pendergraft 1946, Rand 1947).

Lesser Influents

The white-tailed jack rabbit occupies about the northern two-thirds of the grassland, and the black-tailed jack rabbit, the area south of Nebraska. Their

Fig. 13-6. A. Short grass and the sagebrush *Artemisia frigida* at the upper end of Bull Lake, Wyoming, at an altitude of about 6000 feet (1800 m). The mound is that of the agricultural ant (photo by Elliot Blackwelder, ca. 1920). B. Mountain short grass in northeastern Arizona with burrows of the Zuni prairie dog; *Bouteloua eriopoda* is conspicuous, *Hilaria jamesii* is commonly one of the dominants (photo by H. C. Hanson).

trails are often conspicuous in the grama grass during the middle of the summer. The rabbits were originally preyed on by coyotes, wolves, and foxes. These predators also fed on rodents, a few insects, and fruits. The badger is a rapid digger and excavates its food supply of ground squirrels and other rodents with great facility. Other important mammals in the short-grass community include the desert cottontail and the spotted ground squirrel. The kit fox occurred throughout; it has been nearly exterminated by poison placed on bison carcasses to kill wolves (Bunker 1940). The plains area is likewise covered by one or another of several species of skunk and by a weasel. The Richardson's ground squirrel sometimes eats grasshopper eggs. Grasshopper mice, pocket gophers, and pocket mice are also present. The distribution of several of these species is bounded by the western edge of the mixed prairie.

The birds which are most conspicuous over the plains are the horned lark, McCowan's longspur, which feed on grasshoppers, chestnut collared longspur, lark sparrow, lark bunting, Sprague's pipit, Brewer's sparrow, grasshopper sparrow, and the western meadowlark. The lesser prairie chicken was one time locally abundant. In general, hawks and owls are not numerous, but the burrowing owl is usually present.

Among the reptiles the plains garter snake, the western rattlesnake, and the bullsnake make up the principal constituents. These species take a heavy toll of ground squirrels, pocket mice, and harvest mice. Farther south, horned lizards, collared lizards, fence lizards, and utas occur.

Arthropod Influents and Dominants

One of the outstanding and abundant insect groups is the grasshoppers. Seven common or abundant species characteristically feed on grasses or herbage. Notable in this group is the high-plains grasshopper *Dissosteira longipennis*. Bruner (1891) states that they feed on grama grass and buffalo grass, in preference to other grasses, and have at times destroyed these species over large areas. Most of the burrowing mammals of the high plains destroy their eggs. *Melanoplus mexicanus* occurs at all times and is sometimes migratory (Fig. 13-2). Orthopterons which occur in both the climax and subclimax mixed prairie and in short grass are *Acrolophitus hirtipes, Amphitornus coloradus, Cordillacris o. occipitalis, Ageneotettix deorum, Psoloessa delicatula, Aulocara elliotti, Arphia pseudonietana, Encoptolophus sordidus costalis, Spharagemon equale, Metator pardalinus,* and *Philibostroma quadrimaculatum*.

Steep banks of small ravines that are without plant cover support tiger beetles, such as *Cicindela denverensis*. Both the larvae and the adults of these species burrow into steep banks in preference to flat areas. These ravines also afford convenient places for wolves and coyotes to dig their dens.

New Mexico and Arizona Grassland Areas

There are a number of areas of short grass largely surrounded by mountains. One of these, an area of some two hundred square miles (520 km²), lies east of the Kaibab Plateau and is known as Houserock Valley (Rasmussen

1941). The vegetation is dominated by blue grama grass, which forms a continuous sod over much of the area. Another prominent grass is galleta, and sand dropseed occurs on sand areas. Small areas of *Munroa squarrosa* and Rothrock grama are present. Within the memory of a very old man of the Kaibab Indians, trips were made there to hunt pronghorns. Pits were dug in which the hunters concealed themselves until the animals approached near enough to be shot with bows and arrows. The area contains the typical smaller vertebrates of the short-grass plains. The kangaroo rats *Dipodomys ordii* and *D. microps* also occur.

COMMUNITY DEVELOPMENT IN THE TALL-GRASS, MIXED-GRASS, AND SHORT-GRASS ASSOCIATIONS

Marshes

The most important seral stages occur in the ponds and marshes. Temporary ponds are usually dominated by *Spartina,* though in many cases rank forbs are present. Prairie ponds differ from forest ponds in the lack of shrubs and trees on their margins (Shackleford 1929). Prairie marshes, especially those that became dry in the autumn, are extensive, especially in Illinois and Indiana, and are the favorite haunts of numerous vertebrates, such as the massasauga and the numerous birds. Notable among the invertebrates of such areas is the eastern lubbery locust. The sword-bearer grasshopper places its eggs between the stem and lower leaves of big bluestem, which is a late seral species.

According to Ball (1932), leafhoppers in the central Iowa grassland fall into a rough series relative to the amount of available moisture. *Flexamia stylata, F. abbreviatus,* and *Latalus configuratus* are associated with cord grass on low wet ground; *F. sandersi, F. reflexa,* and *F. strictus* are associated with big bluestem; and *F. visenda, F. albida,* and *Laevicephalus collinus* occur in the late successional and moisture series just below grama grass.

Several species show preference for the xeroseral plants of bare soil. The chinch bug belongs primarily to the tall-grass prairie but has a preference for scattered grasses on bare open ground. Its chief places of hibernation are in *Andropogon* bunch grass and in the margins of groves and stream-skirting forests. Its abundance varies greatly from year to year (Fig. 13-4).

Sand areas are large and scattered over the grassland. Shantz maps 15 areas lying chiefly between southern South Dakota and southwest Texas. The largest area is in Nebraska. There are other areas of small size such as that in the Illinois River area studied by Vestal (1913). This vegetation is usually referred to as sand grass–sage (*Calamovilfa longifolia–Artemisia filifolia*) vegetation. There may be as much as 35,000 square miles (91,000 km²) of sand grass on the plains.

Nebraska Sand Grassland

In Valentine Migratory Waterfowl Refuge about 35 miles (56 km) south of Valentine, the grasses at the time of study were rather widely spaced and included *Calamovilfa longifolia, Panicum virgatum, Stipa comata, Muhlenbergia pungens,* and *Eragrostis.* The cover was about half forbs, including *Helianthus petiolaris, Lygodesmia juncea, Petalostemum,* and occasional small cherry shrubs. *Artemisia filifolia,* usually a characteristic shrub, was absent. The surrounding climax was mixed prairie.

The present animal inhabitants of the sand tall-grass areas (Beed 1936) are striped skunk, spotted skunk, long-tailed weasel, yellow pocket gopher, deer mouse, prairie vole, and meadow jumping mouse. Mammals of small and medium size usually found in grassland but absent here include the prairie dog, thirteen-lined ground squirrel, kangaroo rat, and three species of pocket mice.

In July, 1945, the social digger wasp *Microbembex* was present in numbers in blowouts where vegetation was absent. The ant *Lasius niger neoniger* was plentiful. A burrowing camel cricket had a hole 37 centimeters deep; the spider wasp *Pompilus scelestus* was present. The large tiger beetle *Cicindela formosa generosa* was present around the blowouts. In the sparse grass and forbs the only vertebrate found was the lizard *Holbrookia maculata.* Seven species of tall-grass grasshoppers were present. Of the homopterons, the froghoppers *Lepyronia gibbosa* and *Philaronia bilineata* were numerous; of the beetles, the milkweed long-horned beetle *Tetraopes vestitus* was present. The shield bug *Homaemus bijugis,* the stink bug *Prionosma podopoides,* the assassin bug *Apiomerus spissipes,* and the buffalo treehopper *Tortislilus inermis* represented the hemipterons.

Ponds and small lakes are common in the area. Near their margins, grasses such as *Calamagrostis inexpansa* and *Spartina pectinata,* the bur-reed *Sparganium multipedunculatum,* and the sedge *Carex* grow in dense stands. There are considerable differences in the invertebrate constituents of the vegetation compared with the sandy upland.

Illinois Sand Grassland

Important bunch-grass species are *Koeleria cristata,* with large, compact, flat-topped tufts; *Stipa spartea,* tall, loose, few-leaved; *Panicum pseudopubescens,* with short, broad leaves, forming very flat bunches; and species of *Bouteloua.* The sedge *Cyperus schweinitzii* forms sparse, open bunches; little bluestem and big bluestem form very large bunches. Typical broad-leaved perennials are *Aster linariifolius, Lithospermum gmelini, Aster sericeus,* and *Tephrosia virginiana.* In addition *Rhus aromatica illinoensis* forms dense masses which often build up small dome-shaped dunes. The mat plants are *Opuntia humifusa* and a species of *Antennaria.* The interstitial plants are usually annuals, such as *Oenothera rhombipetala, Ambrosia psilostachya, Linaria canadensis,* and *Cassia fasciculata* (Vestal 1913).

The herb layer supplies food both to nonselective plant feeders, such as the grasshoppers *Mermiria bivittata* and *Melanoplus angustipennis* and the leafhopper *Neokolla hieroglyphica,* and to feeders that select particular plants or parts of plants (Vestal 1913). Truly subterranean plant eaters are not abundant but some beetle larvae compete with the pocket gopher. A few strong flying predaceous animals such as dragonflies secure food from the air, while predaceous animals on plants merely hide or rest and await the prey. The predaceous animals of the ground surface and subsurface are apparently much more abundant than the plant eaters. Surface predaceous animals include the spider *Lycosa,* the tiger beetles *Cicindela formosa generosa* and *C. scutellaris lecontei,* the hognose snake *Heterodon nasicus,* the lizard *Cnemidophorus sexlineatus,* and the western meadowlark (Vestal 1913). The parasitic bunch grass animals are represented by the mite *Eutrombidium rostratus,* which attacks a grasshopper, and the larvae of *Systoechus vulgaris,* which attack grasshopper eggs.

Texas–New Mexico Sand Grassland

Portales Valley in Texas and New Mexico contains a dune area in a broad series of shallow basins. The area was excavated by a stream that has since been captured by the Pecos River. Porous materials deposited by the former stream partly fill the basin and constitute the present shallow ground-water horizon, from 3 to 35 feet (1 to 10 m) beneath the soil surface. Some level areas within the dune belt have a mature soil and a caliche substratum, others are mere sandy flats without a caliche layer and above a shallow water table. The level areas probably originally supported a climax mixed prairie biota of the high plains. *Hilaria, Buchloë,* and *Bouteloua,* with *Mimosa, Acacia,* and *Prosopis* now cover them. The sandy flats are seral communities of an edaphic climax vegetation (*Andropogon hallii, Andropogon gerardi*) (Hefley and Sidwell 1945).

The vertebrate influents and dominants originally included the bison, pronghorn, and possibly deer. Lesser influents, the conspicuous box turtle *Terepene ornata,* kangaroo rat, bullsnake, garter snakes, rattlesnake, lark bunting, scaled quail, and the western meadowlark, were present. Grasshoppers of several species also exert considerable influence on the community.

In the absence of an underlying layer of caliche, upward passage of moisture from the shallow ground water affords moisture to initiate a succession beginning with *Yucca* and followed by *Andropogon.* After this, sumac takes possession. Sumac is deep rooted and is spread by kangaroo rats which gather and store the fruit. There are numerous burrowing insects present, both in the vegetation and in blowout areas, which include mutillids, digger wasps, and tiger beetles (Hefley and Sidwell 1945).

The establishment of sumac, bluestem, and yucca brings a semistable condition of the substratum, and subsequently sand sage appears along with scattered bunches of little bluestem. Moist dunes probably reach this condition two or three decades after their initiation. Decline in sand sage and increase in the control of the community by tall grasses bring in the subclimax. Sagebrush and the tall grasses furnish protection for animals, and their numbers increase. Coyotes and cottontails are more numerous here than elsewhere in the area.

Dry dunes differ from moist dunes by being located above a relatively impervious stratum of caliche with a mature prairie soil interposed between their bases and the caliche. The climax is apparently a faciation of the mixed-grass prairie. The dominant grasses appear to be the western little bluestem and the alkali sacaton *Sporobolus airoides,* with *Sporobolus cryptandrus* present in many places almost to the exclusion of other mid grasses. Subdominants are *Hilaria jamesii, Buchloë dactyloides,* and *Bouteloua.* Forbs present are *Stillingia sylvatica* and *Yucca,* with sand sage locally (Hefley and Sidwell 1945).

PALOUSE PRAIRIE

The Palouse Prairie is an area of bunch grasses and lies in the basins of the Columbia and Snake rivers, largely surrounded by the Cascade and various ranges of the Rocky Mountains. The grass cover is often luxuriant. The summers are hot and dry. Most of the precipitation falls during the months of November through February; July and August are often nearly rainless. The Palouse Prairie extends a little way into the Great Basin of California and Nevada and eastward and northward into Wyoming, Montana, and British Columbia. It is separated from the California Prairie by mountains.

Plant Dominants

In addition to *Agropyron (A. spicatum, A. smithii, A. dasystachum),* the following are important: *Poa nevadensis, Stipa occidentalis, S. elmeri, S. thurberiana, S. viridula, Poa secunda, Festuca idahoensis, F. occidentalis, Oryzopsis hymenoides, Hordeum brachyantherum,* and others. *Festuca idahoensis* is mixed with *Agropyron* at somewhat higher elevations. The forbs *Lupinus, Astragalus, Balsamorrhiza, Solidago, Agoseris,* and *Aster* are the major contributors to the aestival and serotinal aspects. Weaver (1917) figures the root system for several of these species that occur in southwestern Washington. The area today is largely covered by the sagebrush *Artemisia tridentata.* Northcentral Utah originally had some good bunch grass without sagebrush (Fig. 13-7, 13-8). An examination of a number of old cemeteries containing relatively undisturbed vegetation indicated that parts of the original prairie may have contained sagebrush plants scattered 10 to 25 feet (3 to 8 m) apart. Evidence from such relict areas, old scientific records (J. C. Merriam 1899), stockmen, and grazing exclosures (Fig. 13-9) suggests that the prevalence of sagebrush has been produced by overgrazing since A.D. 1600. Sagebrush is often accompanied by saltbush, rabbitbrush, or antelope brush. Good bunch grass occurred in the British Columbia fruit-growing valleys in 1930 (Tisdale 1947).

Permeant Dominants and Influents

In the National Bison Range of Montana, shrubs are of little importance; nearly pure grass covers hills and knolls which have only an occasional ponderosa pine. The bison afford a special habitat of importance in their excrement which is utilized by the beetles *Cymindis planipennis, Percosia obesa,* and *Eleodes pimelioides,* the snail *Succinea,* the spider *Phidippus altanus,* the ant

Fig. 13-7. A. Bunch-grass grassland in the cemetery at Mona, Utah. The principal grass is *Agropyron spicatum; Gutierrezia* and *Lygodesmia* occur. Old residents say no sagebrush has been removed. B. The solid sagebrush area in the background outside the Mona cemetery where grazing by cattle occurs.

Lasius niger alienus americanus, centipedes, and fly larvae (Mohr 1943).

In the Sheldon Antelope Refuge of northwest Nevada, the principal grasses are *Stipa thurberiana, Poa secunda* (abundant), *P. nevadensis, Festuca idahoensis, Elymus condensatus, Koeleria cristata,* and others. The shrubs include the black sage, the chief food of the pronghorn antelope, and sagebrush, which is also utilized. In addition there are the shrub *Tetradymia* and the rabbitbrush *Chrysothamnus viscidiflorus.* The herbs are *Oenothera tanacetifolia, Stenotus acaulis, Pentstemon breviflorus, Astragalus haydenianus, Eriogonum orendense, Delphinium andersonii megacarpum,* and *Crepis occidentalis.* The food of the pronghorn in early spring is new grass; in late spring, herbaceous vegetation; in the summer, sage; in August and September, the antelope brush *Purshia tridentata;* and in October and November, grass. Deer and wapiti come down from the mountains and enter the area locally in the winter. The bison has now been extirpated. The black-tailed jack rabbit is confined to the higher areas. The little cottontail and pygmy rabbit are generally distributed. The pocket gopher *Thomomys talpoides* and golden-mantled ground squirrel are usually evident; six closely related subspecies of the latter are present. The chipmunk *Eutamias minimus* runs in the sage area, the rock squirrel is seen on the tops of lava rock outcrops, the white-tailed antelope squirrel and deer mouse dwell in the meadows, and pygmy mice and wood rats are present.

Fig. 13-8. Sagebrush on the Snake River Plain near Bliss, Idaho, with an old badger hole occupied by a coyote.

The bobcat sometimes chases pronghorns. The coyote is present, and the badger is occasionally seen. The wolf was probably present in small numbers in 1600.

The sharp-tailed grouse was formerly an important bird. The sage grouse covered the same area and also extended more to the east (Scott 1942). The resident small birds in the sagebrush are Brewer's sparrow, sage thrasher, loggerhead shrike, green-tailed towhee, and horned lark. The usual carnivorous birds are present. The short-eared owl nests on the ground, its numbers fluctuating with the rodent population. The burrowing owl and marsh hawk nest in the grassland.

Invertebrates

The common garden spider *Argiope trifasciata* is found in the Montana bison pastures, as also are the Mexican grasshopper, pellucid grasshopper, two-striped locust, Carolina locust, big-headed grasshopper, the grass-hoppers *Spharagemon collare, Hesperoleon coquilletti, Pseudopomala bra-*

Fig. 13-9. Grazed and ungrazed grassland near Steamboat Springs, Colorado, in the transition area between bunch grass and mixed grass. Outside the fenced area, sagebrush is predominant and continues up the slope where it makes contact with oak bush. The extensive grass in the fenced area includes bluebunch fescue and blue grama, and only scattered wormwood or sagebrush. The insert is a close view of the grass (photo by H. L. Andrews 1936).

chyptera, and *Amphitornus coloradus,* and the cricket *Acheta assimilis.* Hemipterons are numerous, including the plant bugs *Lopidea nigridea* and *Pseudopsallus tanneri,* the shield bug *Aelia americana,* and the two bugs *Homaemus aeneifrons* and *Eurygaster alternata.* The buffalo treehopper *Stictocephalus inermis,* which damages cultivated grasses, and the cercopid *Philaronia bilineata* are also found. Hymenopterons include the wasp *Ectemnius dilectus,* the bee *Perilampus hyalinus,* and the ants *Myrmica brevinodis discontinua, Tapinoma sessile, Formica sanguinea puberula,* and *F. cinerea.*

Seral Stages

The southeastern quarter of the state of Washington is bunch grass or bunch grass–sage. The eastern and east-central portion of this area is arid, and soils poor and sandy. There are large outcrops of rock dissected by numerous deep valleys. It is in these places that shrubs and forbs, which quickly invade when the grasses are weakened, persist (Dice 1916, Munger *et al.* 1926, Merriam and Stejneger 1891, Grinnell 1933). A few miles west of Moss Lake in a sandy area of sparse grass, *Bromus tectorum* is present along with *Oryzopsis hymenoides* and *Stipa*. The shrubs and herbs are *Chrysanthamnus graveolens* and *Psoralea lanceolata*. The principal grasshopper is *Melanoplus foedus;* associated with it are the wingless notable shield bearer, the three-lined grass katydid, and the long-legged cricket.

Farther east near Ritzville in an area of level rock and compact soil, the grasses are similar but more *Festuca idahoensis* and *Agropyron* are included. The same grasshoppers and, in addition, *Ageneotettix deorum, Amphitornus coloradus,* and *Melanoplus m. mexicanus* occur.

The rough, rocky talus land in the Big Bend country of Washington is called scabland, especially around such areas as Grand Coulee. The sagebrush is in part the scabland sage, the favorite food plant of the coulee cricket. The cricket eggs hatch in spring, and the nymphs find shelter under the fallen leaves of the sage. When about one-tenth grown, they start migrating by walking in vast hordes and devouring all vegetation in their paths. The Mormon cricket has similar habits in the area near Great Salt Lake. These crickets commonly attract birds. Near Great Salt Lake, gulls have been effective in reducing the cricket population to a low point. Gillette (1905) mentions a migrating horde as 900 feet (270 m) wide and a half mile (800 m) long. On the eastern Wasatch highland and mesas in Wyoming, eastern Utah, and western Colorado, the grassland communities are primarily bunch grass. Ungrazed grassland near Steamboat Springs, Colorado, included *Bouteloua gracilis* and *Festuca idahoensis* in 1936. Most of Wyoming west of the divide is sagebrush with its associated species, their presence probably the result of overgrazing.

CALIFORNIA PRAIRIE

Mountain ranges and desert isolate the California Prairie from other grasslands. Until recently, however, it has been connected in northern California with an extension of the Palouse Prairie. South of Mount Shasta and from the coast to the foothills of the Sierra Nevada this prairie under pristine condition probably covered more than one-fourth of the land, including the valleys of the Sacramento, San Joaquin, and areas south of the Cross Ranges, parts of the lateral valleys, and the adjacent Pacific slopes and foothills. It also reached into southern California. In the coastal area originally, the south-facing slopes were often bunch grass, and the north-facing slopes chaparral.

The climate is characterized by winter rainfall, the maximum amount falling in December, January, and February. Except in the neighborhood of the coast, the summers are hot and the winters mild with little or no snow. The annual rainfall varies from a minimum of 6 inches (15 cm) in the upper San Joaquin Valley and in Antelope Valley at the west end of the Mojave Desert to 25 to 30 inches (62 to 75 cm) along the coast.

Plant Dominants

Most of the dominants are strikingly different from those of the plains to the east but have the bunch-grass life form. The absence of grama grasses is noteworthy. The principal dominants include *Stipa pulchra, S. lepida, S. coronata, Melica imperfecta, Elymus triticoides,* and *E. glaucus.* Also important are *Poa scabrella, Festuca idahoensis* and *Hordeum brachyantherum.* As a consociation or in mixture with *Melica, Poa,* or *Koeleria, Stipa pulchra* occupies a larger area than all other dominants combined. *Elymus glaucus* is characteristic of the oak savanna and *Stipa lepida* and *S. coronata* of upper slopes leading to chaparral (Clements and Shelford 1939).

The perennial forbs are either vernal plants with bulbs or corms (*Brodiaea, Calochortus,* and *Allium*) or belong to genera typical of mixed- and tall-grass prairie. California poppy is limited to the southwest. Great masses of annuals, representing more than 50 genera and several hundred species, are present. Their densities fluctuate widely with the amount and distribution of rainfall and with temperatures.

Animal Dominants and Influents

Although bison is not known to have occurred in California, the pronghorn was originally very abundant in the San Joaquin Valley, as in other bunch-grass areas, occurring in herds of two or three thousand. Stephens (1906) states that pronghorns fed largely on grasses, only occasionally on twigs or leaves of shrubs. The tule elk and wolf have been gone for a century. The coyote preyed largely on jack rabbits, although they also ate grasshoppers, other insects, and fruit (Stephens 1906).

The black-tailed jack rabbit originally occurred in great numbers, which were further increased by the destruction of carnivores. In 1893 a drive at Fresno resulted in the destruction of 20,000 rabbits (Palmer 1897). Nearly all drives were in counties originally containing large areas of climax grassland.

The badger was not abundant in California in 1906 (Stephens 1906); it has a distribution in California somewhat similar to that of the grassland.

The ground squirrel *Citellus beecheyi* constitutes one of the most characteristic species of this grassland but extends into grass-covered woodland in the foot-hills and into mountain parks (Grinnell and Dixon, 1918), as do the various dominant grasses. The seeds of grasses and forbs are used by the ground squirrel for food and the grasses for lining nests.

The pocket gopher *Thomomys umbrinus* is important in these areas. The San Joaquin pocket mouse *Perognathus inornatus* apparently belongs to the grassland; kangaroo rats are not important. The birds of the valley and coastal grasslands include the horned lark and western meadowlark; predatory species enter from the adjacent woodlands. The western rattlesnake and the gopher or bullsnake occur.

Remnants of grassland near Davis, California, were found occupied in 1944 by the pellucid and the devastating grasshoppers. Both of these species occur at times in large numbers and migrate from place to place. The homopteron *Xerophloea viridis* feeds on native grasses. Little is known about the remainder of the invertebrate population.

Chapter 14

The Southern Temperate Grassland

The desert or mesquite grassland or the *grama grass–pronghorn–curly mesquite grass association* of the temperate grassland biome lies in the highlands of the southern half of Arizona and New Mexico, in Texas, in Mexico south to 20° N. Lat., and in the lowlands of Texas and Tamaulipas between 29° and 22° N. Lat. (Fig. 14-1). The landscape is usually short grass with a scattering of shrubs, although there may be extensive invasion of the grassland by mesquite and other shrubs under heavy grazing.

Bison formerly occurred in the moister parts of the grassland, and prong-horns were generally present. The grassland was not disturbed by Indians before A.D. 1500 while they still lacked the use of the horse.

The large *plateau region* includes the valleys and slopes of the Mexican Plateau, the transverse ranges south of the Colorado River, and the southern Rockies at altitudes from 900 to 5000 feet (275 to 1520 m). It occurs in Arizona, New Mexico, Texas, and Mexico, and originally covered an area of approximately 300,000 square miles (780,000 km²). The range of the banner-tailed kangaroo rat coincides well with this region. At low altitudes (approximately 400 to 1200 m), tobosa grass and black grama are characteristic. The foothill grassland (1020 to 1520 m in the United States, 1500 to 2300 m in Mexico) is without trees or shrubs, and curly mesquite grass and blue grama are important constituents. On the western mountain slopes in central Sonora (approximately 275 to 1000 m) the species composition has not been reported, but there is tall grass with scattered oaks (Fig. 14-4A).

The *Gulf of Mexico lowland region,* 164 to 1640 feet (50 to 500 m) contains *Acacia, Bouteloua trifida,* and curly mesquite. It occupies eastern Nuevo León, eastern Tamaulipas, and southern Texas.

In addition to these areas, there is a nearly continuous belt of varying width extending south from New Mexico at elevations ranging from 1000 to 2300 meters (Shreve 1924). This grassland continues south through Durango (1800 to 2000 m), Zacatecas (near 2400 m), into Aguascalientes, Jalisco, and Michoacan. According to Arias (1942) and other authors, its southern end broadens out to cover the entire intermountain plateau and thus connects with grassland of the eastern ranges of the plateau. Some of this grassland may be seen northeast of Mexico City between 2400 and 2500 meters. Areas of grassland on the west-facing mountain slopes in Sonora and on the eastern slopes of the coastal mountains of southern California and Baja California will not be discussed because of lack of studies made in them.

Between the Sierra Madre Occidental and the Sierra Madre Oriental are many bolsons or areas of internal drainage. These vary from a few hundred to

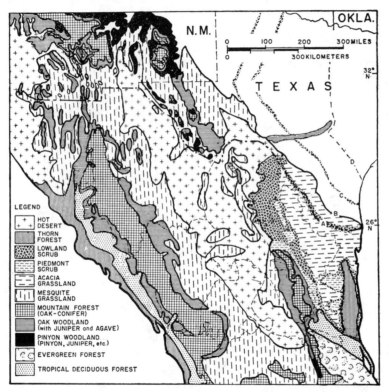

Fig. 14-1. Probable distribution of mesquite grassland, *Acacia* grassland, and associated communities about A.D. 1600. Settlement by the Spanish apparently greatly accelerated or initiated the invasion of *Acacia* grassland by the lowland scrub (Bray 1906). In Texas it is believed that this scrub was present in the Rio Grande floodplain and adjacent areas (A) while mesquite extended to the line B. This vegetation is an impenetrable thicket of thorny plants and cacti, frequently under mesquite trees. By 1923, the thorn shrubs had reached line C and the mesquite had reached line D. Acacias without mesquite invade grassland in various places, especially in Mexico. The separation into pinyon–juniper and oak woodland and bush is arbitrary outside Arizona.

several thousand hectares in size. The northern part of the Mexican Plateau is occupied by the extensive Bolson de Mapimi. The centers of bolsons that receive drainage from mountains are occupied by alkaline flats and temporary lakes. Bolsons in the center of the plateau receive limited drainage and are without a central playa or lake bed. In these the floor of the bolson is nearly level and the soil is deep and fine in texture. Collectively, these bolsons cover a large expanse of northern Chihuahua, the largest being the Llano de los Gigantes and Llano de los Cristianos.

Rainfall at Tucson, Arizona, averages 28 centimeters and at Silver City, New Mexico, 44 centimeters (Whitfield and Beutner 1938). In Chihuahua, it is also 44 centimeters (LeSueur 1945). Most of it falls in July, August, and September. In Arizona and Sonora, west of the mountains, rainfall is also heavy in December. Precipitation in the Mexican grassland lies between approximately 40 and 50 centimeters on the west side of the central basin, and between 50 and 60 centimeters on the east side of the mountains of northern Coahuila (Shreve 1942). The annual evaporation at Roosevelt, Arizona, is 205 centimeters and at Wilcox, Arizona, 218 centimeters, or five to seven times the rainfall.

In Mexico and the southwestern United States, limestone soils do not support heavy stands of grass, and the desert invariably extends to much higher elevations on limestone than on other types of rock and their derived soils (Shreve 1942). In the Santa Rita Reserve area, the soil is sandy and the slopes are not steep. In the Santa Catalina areas, loose rocks and gravel are a conspicuous element of the soil, but there is an abundance of sand and clay.

DOMINANTS AND INFLUENTS OF GENERAL DISTRIBUTION

The following short grasses predominate throughout the southern grassland area: blue grama, side-oats grama, and *Bouteloua chrondrosioides*. Poverty three-awn grass is a close fourth in abundance and frequency. Silver beardgrass occurs locally. There are in addition about 30 other species belonging to familiar genera (Shreve 1942). The shrubs include various species of *Prosopis, Acacia, Quercus, Yucca, Nolina,* and *Dasylirion.*

Only a few species of animals occur through all the various faciations and associes. A generally distributed permeant is the pronghorn (Nelson 1921). The bison have never exercised any important influence in the southern grassland, especially in the plateau area. Mule deer frequently enters grassland areas at higher altitudes and adjacent to woodland. The black-tailed jack rabbit occurs nearly throughout.

The wolf has dens mostly in the open ponderosa pine, pinyon–juniper, and oak woodland, which lies just above the grassland. However, they killed pronghorns under pristine conditions and doubtless utilized rabbits and other small mammals of the grassland. The coyote is the chief control over the number of jack rabbits. It also eats some fruits and cactus and dens at higher elevations. The bobcat is present and helps in reducing populations of rabbits and wood rats.

The badger is generally distributed in all the grassland faciations. It travels about considerably, digging out rodents which are its food. It is perhaps the least stationary of the earth-moving animals.

The banner-tailed kangaroo rat (Fig. 10-4) is one of the most characteristic animals of the mesquite grassland. Its ecological equivalent, the Nelson's kangaroo rat, occurs on the eastern side of the Mexican Plateau. These rats are nocturnal, silent for the most part, active, and somewhat social; they progress by leaping and signal by a drumming or thumping on the ground with their hind feet. The breeding season extends from January to August. The number of young per litter varies from one to three and averages two. They do not hibernate but store food for use during adverse seasons. Storage in each den varies in quantity from 5 to 5750 grams. Materials stored are mostly seeds and forage plants such as various species of *Bouteloua* and *Aristida,* with Rothrock grama the most important. The dens vary from 6 inches to 4 feet (15 to 120 cm) in vertical height and from 5 to 15 feet (1.5 to 4.5 m) in diameter. They are a tortuous network of burrows, with many storage and some nest chambers, the whole arranged so as to be two to four stories high (Vorhies and Taylor 1922).

The white-throated wood rat ranges from southern Arizona to Mexico City. It is nocturnal and very active; it signals by tapping with the hind feet and keeps in close vicinity of its den. It breeds from January to August. The litters average two young, but there may be three. The den of the white-throated wood rat is a conspicuous mound of available loose debris, beneath which a partially subterranean portion contains the nest of softer material (Fig. 14-4). Dens are usually located beneath a shrub, tree, or cactus, which thus furnish both shelter and food. Quantitatively their food consists of cactus, 44 per cent; mesquite, 30; grasses, 5; carpetweed, 2; miscellaneous forbs, 18; and animal food, 1. A considerable variety of invertebrates, chiefly insects, inhabits the dens along with the wood rats (Vorhies 1945, Vorhies and Taylor 1940).

FACIATIONS

Two plant faciations have been described by Whitfield and Beutner (1938), and Dice and Blossom (1937) made collections of animals in the two faciations, but otherwise they have been little studied. In addition there is the agave grassland.

Low Altitude Faciation

Black grama and tobosa grass predominate, with other climax plant dominants in order of abundance being Rothrock grama, bush muhly, poverty three-awn, purple three-awn, and sand dropseed (Figs. 14-2, 14-3). Creosote bush and other desert shrubs were formerly scattered throughout most of the

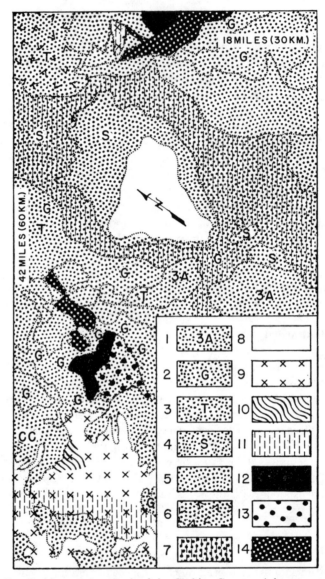

Fig. 14-2. Details of mesquite grassland in Cochise County, Arizona, extending in a northeasterly direction with the most westerly corner on the San Pedro River northwest of Benson. The stippled area includes much grass: 1, Arizona three-awn grass; 2, black, blue, hairy, and Rothrock grama grasses; 3, tobosa grass; 4, sacaton grass; 5, pure grama grass grassland; 6, tobosa grass with small scattered shrubs; 7, grama grasses with considerable mesquite; 8, Willcox Playa; 9, catclaw desert; 10, creosote bush desert; 11, mesquite near river; 12, oak woodland; 13, oak woodland with grass; 14, oak woodland with shrubs (after Darrow 1944).

area. The low altitude grassland occupies about 85 per cent of the area in the southwestern United States.

The Santa Rita Range Reserve (800 to 1400 m) is in the lowland faciation. Here the scattered shrubs are mesquite, palmilla, cholla, bisnoga, and cat claw. On the slopes of the mountains in Sonora, the grassland goes down to 900 meters as compared with 1200 meters in Chihuahua. Yuccas, bear grass, and sotol are few. *Parkinsonia,* mesquite, and acacia are well represented.

Fig. 14-3. A. Mesquite grassland with *Yucca* near Bowie, Arizona (photo by H. L Andrews). B. Mesquite grassland in western Texas (photo by Edith Clements). The shrub (A) is *Rhus microphylla;* the plant (B) is *Yucca,* probably *elata* (identification by J. L. Gardner).

Jack rabbits are important animal influents or dominants. The antelope jack rabbit is a common influent in the Sonoran grassland and in the more arid northern parts of the Chihuahuan grassland. Its food consists of 45 per cent grass, but there is no evidence that they eat *Heteropogon, Andropogon,* or *Hilaria.* Mesquite makes up 36 per cent and cactus about 8 per cent; the remaining 11 per cent includes some of the shrubs above noted. These rabbits have their nests in depressions just as cottontails do. (Vorhies and Taylor 1933).

The black-tailed jack rabbit is primarily a desert species but it enters the grassland in numbers where there is overgrazing. One animal per four hectares would be a maximum population. It consumes only about one-half as much grass as does the antelope jack rabbit, but mesquite composes 56 per cent of its diet. The black-tailed jack rabbit appears to render areas more suitable for itself by reducing the grass and favoring the forbs which furnish it with shelter (Vorhies and Taylor 1933). Succulents, especially cacti, are consumed in greater amounts as drought conditions increase (Vorhies 1945). The enemies of the jack rabbit include the golden eagle, great horned owl, and the red-tailed and ferruginous hawks. The desert cottontail is common in many grassland localities (Vorhies and Taylor 1933), and the presence of the grasshopper mouse, Merriam's kangaroo rat, banner-tailed kangaroo rat, and round-tailed spermophile are noteworthy. Many of the species listed in Table 14-1 are represented by other subspecies or ecologically equivalent species in the Davis Mountains of Texas.

The scaled quail is reported to occur in the grass and mesquite plains in northeastern Sonora (Van Rossem 1945). Gambel's quail is dependent on shrubs and trees and is not known to utilize grass, except the sprouts of *Panicum*

Table 14-1. *The Relation of Mammals to Altitude Within the Mesquite Grassland*

LOCALITY	CATALINA MOUNTAIN[a]		SANTA RITA RESERVE[b]
Elevation...............................	1200-1400 m	1150 m	900 m
Cactus mouse............................	1	0	0
Deer mouse.............................	1	0	0
Ord's kangaroo rat......................	1	0	0
Rock squirrel...........................	1	0	0
Mule deer..............................	3	0	0
Western harvest mouse...................	4	0	0
White-throated wood rat.................	17	4	4
Black-tailed jack rabbit..................	8	0	+
Desert cottontail.......................	++	0	3
Antelope jack rabbit.....................	3	0	11
Desert pocket mouse.....................	1	0	1
Southern grasshopper mouse..............	2	1	5
Harris' antelope squirrel.................	1	2	1
Bailey's pocket mouse....................	9	29	8
Arizona pocket mouse....................	0	1	0
Rock pocket mouse......................	0	5	0
Merriam's kangaroo rat..................	0	1	18
Bobcat.................................	0	0	1
Round-tailed spermophile................	0	0	1
Banner-tailed kangaroo rat...............	0	0	3

[a] The Catalina Mountain area contains curly mesquite grass and blue grama, with oaks at the upper edge and palo verde.
[b] The Santa Rita area contains tobosa grass and Rothrock grama with mesquite.

hallii which are taken in abundance. Small birds generally distributed over the grassland are horned larks; black-throated, lark, grasshopper, and other sparrows; loggerhead shrike; nighthawks; and Scott's orioles.

The earless lizards *Holbrookia maculata approximans* and *H. texana* occur in the southeastern plains area of Arizona (Gloyd 1937). *Cnemidophorus perplexus* is present in the same locality and is mentioned in connection with grass. The large kingsnake *Lampropeltis getulus splendida* and the Mexican garter snake *Thamnophis eques* occur in the same region. A rattlesnake and a bullsnake are credited with destroying rabbits (Vorhies and Taylor 1933).

Island-like areas covered by grass are called llanos. Llanos within bolsons are predominantly *Hilaria mutica* with some *Bouteloua gracilis* and *B. eriopoda* (Shreve 1942). There are a few shrubs: *Florestina tripteris, Viguiera phenax,* and *Xanthocephalum gymnospermoides.* In the northern San Luis Potosí, the llanos are occupied by the Mexican prairie dog, spotted ground squirrel, and burrowing owl. Some bolsons have very large numbers of kangaroo rats, probably *Dipodomys ordii.*

Several grasses (*Hilaria mutica, Bouteloua gracilis, B. curtipendula, B. hirsuta, Setaria macrostachya,* and *Trichachne californica*) characterized a border area which was visited between the low altitude and foothill faciations, 186 kilometers north of Chihuahua City. The principal forb was *Gutierrezia.* Shrubs were very scattered. There were also some scattered yuccas.

Grasshoppers taken from the grasses in this area were *Ageneotettix deorum,* probably a strict grass feeder, and immature *Eremiacris,* probably the species *E. acris* which is associated with *Hilaria mutica. Ageneotettix* and adult short-winged katydids lay eggs on grasses and feed on both grasses and forbs. The hemipterons found were the squash bugs *Mecidea longula, Liorhyssus hyalinus,* and *Arhyssus lateralis.* The plant bugs were the false chinch bug and *Nysius californicus* which feed on grasses. The damsel bugs were represented by *Nabis alternatus;* the stilt bug *Jalysus wickhami* was present. Homopterons present were the fulgorids *Oliarus pima* and *Hysteropterum sepulchralis* and the cicadellids *Neokolla curcubita* and *Exitianus obscurinervis.* In the grasses also were the leaf beetles *Promecosoma virida* and *Pachybrachis nigrofasciatus;* the snout beetles *Centrinaspis, Pantomorus albosignatus,* and *Mitostylus setosus;* and the melyrid *Collops flavicinctus.* The hymenopteron *Euparagia maculiceps* was present, and the ants were *Pogonomyrmex barbatus, P. californicus, Dorymyrmex pyramicus,* and *Myrmecocystus melliger.* Flies were not numerous, except for *Trupanea pseudovicina.* The spiders of the grasses were *Misumenops coloradensis* (also in shrubs), *Tibellus chamberlini,* and immature specimens of several genera. The millipede *Orthoporus chihuanus* was taken from the shrubs. The Texas horned lizard and the whiptail were present.

There was a dense stand of shrubs, one-half to two meters high, nearby, which included *Acacia constricta, Flourensia cernua, Condalia lycioides, Eysenhardtia, Rhus microphylla,* and *Larrea divaricata.* The grasshopper *Bootettix argentatus* was taken from the creosote bush. This grasshopper has a striking resemblance in color to its obligate host plant (Ball *et al.* 1942). Mirids were represented on the shrubs by *Phytocoris vanduzei* and *P. ramosus*. The stilt bug *Jalysus wickhami* was present.

In western Texas, only one area of the typical lowland faciation is known, occurring at 1200 meters on an old lake bed. Here tobosa grass and tar bush were found. The transition from the western shrub or desert grass to the eastern lowland type is broken by desert in the Pecos Valley. To the east from that point is the curly mesquite grass, buffalo grass, and opuntia cactus (Cottle 1931). This vegetation has probably advanced northward as the result of grazing.

Foothill Faciation

This community is characterized by curly mesquite and blue grama and extends from approximately 1040 to 1520 meters. There are indications that it was formerly more widespread. In addition to curly mesquite grass, blue grama, and hairy grama, are the New Mexican feathergrass, side-oats grama, and the following dominants from lower altitudes: bush muhly, black grama, tobosa grass, and several species of *Aristida*. The forbs are *Castilleja integra, Psoralea obtusiloba,* and *Scutellaria wrightii*, which on dry slopes under pristine conditions extended up to 1820 meters. Match weed and *Eriogonum wrightii* are sometimes present. Scattered shrubs from the semidesert are mesquite, cat claw, *Acacia constricta*, ocotillo, and *Yucca baccata* (Whitfield and Beutner 1938). Throughout both this and the low altitude faciations where there is greater available water, occurs a mixture of side-oats grama, silver beardgrass, tanglehead, bullgrass, Junegrass, and cotton top (Whitfield and Beutner 1938).

Most of the discussion of the two faciations above applies especially to the grassland north of the United States boundary. In northern Chihuahua, Mexico, LeSueur (1945) maps areas of *Bouteloua gracilis* at higher elevations and does not emphasize *Hilaria*. In Shreve's (1942) description of the grasslands on the Mexico Plateau, he does not stress curly mesquite grass but mentions the Mexican *Hilaria cenchroides*. While there are obvious differences with altitude, lines of separation hardly exist. Throughout the extensive grass areas of Chihuahua and Durango, and farther south on the eastern slope of the Sierra Madre Occidental, species of *Bouteloua* predominate. The commonest of these is *B. gracilis,* which is estimated to form at least 80 per cent of the grassland cover. In Zacatecas and Jalisco, hairy grama, purple grama, and *Hilaria cenchroides* or *Sporobolus trichodes* alternate or associate as the most common species. Throughout the grassland area large coarse grasses occur sporadically or in isolated colonies, including *Andropogon saccharoides, A. barbinodis, Stipa eminens, S. clandestina, Sporobolus airoides, Elyonurus tripsacoides,* and *Trichloris crinita*. Relatively moist areas are heavily carpeted by buffalo grass. Dry localities with shallow soil at low elevations are thickly covered with fluffgrass or well-defined colonies of burro grass. As altitude increases toward the west in the desert area of Chihuahua, the grasses become mixed with creosote bush, mesquite, and tar bush. Creosote bush drops out at higher elevations and yuccas, sotol, agaves, and bear grass come in. The chief grasses are the gramas along with curly mesquite grass, tobosa grass, and dropseed. This best grass area is from 1200 to 1500 meters in elevation, but grasses continue in the open woodland 300 meters farther up (Fig. 14-4). This is the area in which the large herds of pronghorns and their animal associates, noted above, occur. The white-sided jack rabbit ranges over the southern half of the grassland area while Gaillard's jack rabbit occupies the northern half and extends into New Mexico.

Fig. 14-4. Tall grass under oaks ten miles west of Huachinera, Sonora, on the upper Bavispe River, June, 1953 (photo by Allen Philips).

In the northern part of this grassland area, about 72 kilometers north of Chihuahua and at an elevation of 1500 meters, the roadside in July, 1944, was less heavily overgrazed than elsewhere. The grasses were *Bouteloua gracilis* and *Aristida*. The principal forb was the yellow-flowered *Baileya multiradiata*. There were scattered yuccas. Kangaroo rat mounds were present. Here, there were juvenile broad-winged grasshoppers *Trimerotropis* and *Platylactista azteca*, characteristic of dry, bare areas (Ball *et al.* 1942). The predatory ground mantis *Yersiniops solitarium* was present. Hemipterons were represented by *Nysius californicus* and *Arhyssus lateratis* and the homopterons by the brown leafhopper *Exitianus obscurinervis* and *Aceratagallia*. The only adult spider, an enemy of the leafhoppers and bugs, was the abundant *Misumenops coloradensis*. Two species of snout beetles, *Centrinaspis picumnus* and *Pantomorus albosignatus*, were present. The small red tiger beetle *Cicindela lemniscata* was common on the bare ground of paths. There were a number of the fly *Olarius aridus* and of chalcids. The ants were *Pogonomyrmex imberbiculus*, *P. desertorum*, *Dorymyrmex pyramicus niger* and *Pogonomyrmex barbatus*.

Again east of Tucson near Benson, Arizona, at an altitude of 1450 meters in July, 1944, *Yucca elata*, *Zinnia acrosa*, and very few *Fouquieria* were present and the prostrate herb *Allionia incarnata* was characteristic. The grasses consisted of two species of *Bouteloua, Aristida longiseta,* and *Panicum hallii.* Mammals or signs of mammals were pocket mice, kangaroo rats, wood rats, foxes, and jack rabbits. Birds observed were the mockingbird, horned lark, various sparrows, and the nighthawk. The horned lizards *Phyrnosoma coronatum* and *P. solare* and checkered whiptails were seen or specimens taken.

Grasshoppers are abundant, including *Psoloessa texana pusilla,* which feeds on germinating grass seeds. *Amphitornus coloradus ornatus* is a characteristic grass-eating species of the foothill faciation. *Eremiacris virgata* is commonly associated with certain species of *Hilaria.* The toad lubber *Phrynotettix tshivavensis,* which is found on *Euphorbia,* and *Hadrotettix trifasciatus* are forb-eaters. Additional grasshoppers taken were *Xanthippus c. corallipes, Mestobregma plattei rubripenne,* and *Trachyrhachys mexicana,* which are associated with grama grass. They are all either forb or grass-feeders (Ball *et al.* 1942). Tinkham (personal communication) states that *Trimerotropis melanoptera* and *T. laticincta* are also present. The ground beetle *Discoderus robustus,* tenebrionids, the millipede *Orthoporus productens,* and the centipede *Scolopendra polymorpha* occur on the ground under yucca debris, and the termite *Amitermes* was found in a dead yucca. Hemipterons are represented by the white coreid bug *Mecidea longula* and the two lygaeids *Geotomus parvulus* and *Phlegyas annulicornis.* The common homopterons are the leafhopper *Prairiana subta* and the fulgorid *Fitchiella melichari.* The spiders from the grasses and forbs are *Latrodectus mactans, Habronattus brunneus,* and *Tibellus chamberlini.* The common ants are *Pogonomyrmex sanctihyacinthi* and *Novomessor cockerelli.*

Agave–Bear Grass–Hairy Grama Faciation

Agave lecheguilla, Nolina erumpens, and *Bouteloua hirsuta* make up a faciation that occurs in the Rocky Mountain highland of Texas and in the Mexican states to the south. There is considerable variation in its composition. On Mitchell Flat at 1420 meters in the Davis Mountains of Texas, blue grama, hairy grama, and black grama are the principal dominants. Curly mesquite grass is not mentioned. Mesa muhly and *Stipa* are unimportant. Another station, 105 meters higher, includes some pinyon and juniper. The bunch grass *Nolina texana,* blue and hairy grama, cat claw, and *Yucca elata* occur. Other areas contain allthorn, Mexican tea, and opuntia cactus (Blair 1943).

Farther south, the upper limit of this grassland comes at about 1520 meters, and its lower limit varies from about 1220 meters on the north side of the Chinati and Chisos mountains to 920 meters in the southern part of the Big Bend area to as low as 700 meters in some places east of the Dead Horse Mountains. It forms a zone around the north, northeast, and northwest sides of the mountains. By far the most common dominant is the chinograss *Bouteloua breviseta.* Other important dominant grasses are *Bouteloua hirsuta, Aristida wrightii,* and *Aristida adscensionis.* Much of the area contains sotol, a most conspicuous and important subdominant. Important secondary species in certain places are the western wheatgrass, false grama, bear grass, *Agave lecheguilla,* blood of the dragon, candelilla, and hechtia.

The animal influents and dominants of the grass–sotol community are the Texas antelope squirrels, harvest mice, brown towhees, meadowlarks, little striped whiptails, western patch-nosed snakes, bullsnakes, and glossy snakes (Taylor *et al.* 1944). The deer mouse is commonly associated with grass and occurs in the foothill country of Coahuila, in west Texas, and in eastern New Mexico, particularly in the foothill areas near Alamogordo.

In Coahuila, there is a grass belt between the desert and the woodland at 1200 and 1500 meters. The dominant grasses are similar to those in Texas, noted above, as are also the shrubs with the addition of *Viguiera stenoloba.* The yellow-nosed cotton rat is the most characteristic species, with Nelson's kangaroo rat and the white-throated wood rat also occurring. The black-tailed jack rabbit is present. In the grassland of southeastern Coahuila and Nuevo León, the cover is rarely more than 50 per cent grass, the remainder being other herbaceous perennials. There are frequent mottes of *Quercus cordifolia* from 1.5 to 2 meters high, which also contains other shrubs and semisucculents in greater frequency than in typical grassland. The grasses are blue grama, large-flowered tridens, *Hilaria cenchroides* (confined to Mexico), wolftail, and side-oats grama. Associated forbs are *Zinnia anomola, Dichondra argentea, Dyschoriste decumbens,* and *Houstonia rubra* (Shreve 1942). The mammals are the black-tailed jack rabbit, Mexican prairie dog, spotted ground squirrel, Ord's kangaroo rat, and deer mouse (Baker 1956).

Community Development

South of Samlayuca, Chihuahua, there is a very extensive area of dunes, the sands of which are 94 per cent silica. The height from the apex of a dune to the bottom of a blowout is usually about 15 meters. Successional stages, although very irregular, appear to be: (a) low annuals, *Portulaca pilosa, Nama demissum, Euphorbia labomarginata, Dalea glaberrima,* and others; (b) erect ruderals, *Acanthochiton wrightii, Abronia fragrans, Cordylanthus wrightii,* and others; (c) grass, *Sporobolus cryptandrus, Panicum propinquum, P. havardii;* and (d) grass and sedge, *Bouteloua barbata, Aristida purpurea, Eleocharis, Cyperus* (LeSueur 1945).

Hydroseral succession in a valley west of the city of Chihuahua follows the general pattern: (a) submerged plants, *Chara* and, occasionally, *Myriophyllum;* (b) floating plants, *Spirodela polyrhiza;* (c) water's edge plants, *Pectis aquatica, Lilaea subulata, Sagittaria graminea, Juncus interior, J. balticus;* and (d) climax plants, *Bouteloua gracilis* (LeSueur 1945).

The valley grassland north of Chihuahua, where overgrazed, undergoes the following series of changes: (a) decrease of gramas and increase of other grasses; (b) increase of ruderals (chiefly composites), *Aster, Erigeron, Cosmos, Aplopappus, Gutierrezia,* and others; (c) addition of five species of *Dalea,* and *Nolina;* (d) increase of brush, *Ephedra trifurca, Eysenhardtia spinosa,* and *Opuntia;* (e) shrubs, *Acacia greggii, Mimosa dysocarpa,* and *A. constricta;*

and (f) the final stage, which on dry flats and ridges is *Larrea divaricata* and *Acacia constricta,* in canyons and arroyos *Mimosa,* and on deep soil with water available *Prosopis juliflora.* With overgrazing by cattle or by native rodents such as kangaroo rats, desert vegetation tends to replace grassland (LeSueur 1945).

Acacia Grassland

This is called acacia grassland because acacias partially displace the mesquite (Fig. 14-1). It occupies the lowlands of southern Texas, northeastern Coahuila, Nuevo León, and Tamaulipas. This grassland has been modified by invasion of the shrubs which make up the Tamaulipan thorn scrub of Muller (1947). Shreve (1917) recognized the vegetation as unique and called it Texas semi-desert. On his map it is bounded on the northeast by the central branch of the Nueces River, on the northwest by the escarpment of the Edwards Plateau, and on the east by the Gulf. This grassland area corresponds to the distribution of the scaled quail. It may, therefore, be called the curly mesquite–scaled quail–acacia faciation of the grassland biome.

The Rio Grande Plain lies below the Edwards Plateau and stretches from the central branch and main trunk of the Nueces River to the San Fernando River and beyond in Mexico. North of the Rio Grande, the Edwards Plateau is the only important high area. The base of its escarpment is at about 300 meters. At the time of settlement this grassland occurred 50 to 75 miles (80 to 120 km) north and east of the Rio Grande. In Mexico, the plain is broken by the Sierra de San Carlos and, further south, becomes narrow between the Sierra de Tamaulipas and the Sierra de las Rusias, which lies near the coast. The grassland extends up to and slightly above 500 meters, as for example north of Monterrey, and to less than 100 meters at China, Nuevo León.

Annual rainfall ranges from 43 to 75 centimeters, with two rainy seasons. May and June are usually the wettest months with 11 to 22 centimeters of rain. The second rainy season during September has from 8 to 17 centimeters. Rainfall decreases north and northwest from central Tamaulipas. Near Tampico, there is merely a general increase of rainfall during the summer months, with September having 33.2 centimeters. Laredo has the highest annual temperature, 24.6° C.; Uvalde, near the base of the Edwards Plateau, has the lowest, 21.2° C.

The Spanish settlement of Monterrey was founded in 1560, and Saltillo, which lies in a pass through which pronghorns could have traveled, was founded in 1586. After the Spanish had had about 300 years experience in raising cattle on the range, the land near Corpus Christi was still very fertile. Groups and belts of timber are found near the coast; but after this is left, a vast undulating prairie extends from the Nueces to within three or four miles of the Rio Grande.

The grass with the highest frequency in the open areas is *Bouteloua trifida.* Purple three-awn grass is second, followed by curly mesquite and reverchon three-awn. Less frequent are *Aristida longiseta, Sporobolis buckleyi,*

Tridens texanus, Setaria macrostachya, and *Hilaria mutica.* In northeastern Coahuila, there also occur *Andropogon scoparius, A. saccharoides, Chloris virgata,* and *Buchloë dactyloides* (Muller 1947). This vegetation often assumes the characteristics of well-developed grassland, but less frequently in Coahuila than in Nuevo León. The scattered shrubs are mesquite, *Cordia boissieri, Acadia amentacea, A. farnesiana, Mimosa emoryana, Leucophyllum frutescens, Yucca tréculeana, Opuntia lindheimeri,* and *Karwinskia humboltiana.* Near Linares *Salvia ballotaeflora, Selloa glutinosa, Lantana involucrata* are also found.

Pronghorns are numerous north of the Rio Grande; they were abundant in Coahuila in the 1890's. They feed on cacti in arid regions. Bison formerly occurred in the northwestern portion of this grassland (Young and Goldman 1944). Roe (1951) states that Franciscan monks in 1602 encountered numerous herds of bison near Monterrey; every year they moved north in April and May and returned in September or October. A wolf was also present in the early days.

The following permeant mammals occur on the Tamaulipan plain and are likely to be influent in the community (Dice 1937): coyote, black-tailed jack rabbit, desert cottontail, and collared peccary. The peccary roots in the earth for earthworms, insects, reptiles, bulbs, and roots much like the domestic hog. Nuts and succulent plants are also utilized. The general effect of the peccary has not been evaluated, but it is supposed to be favored by the presence of shrubs.

Near China, Gral Bravo, Monterrey, and Sabinas Hidalgo in Nuevo León, the principal mammals recorded in 1942 were the black-tailed jack rabbit, cottontail, southern plains wood rat, which is an inhabitant of half-open country (Bailey 1905), and builds its nest under the more thorny shrubs and cacti, and numerous burrowing mammals. The Mexican ground squirrel, hispid pocket mouse, fulvous harvest mouse, and white-footed mouse were noted by Dice. The pygmy mouse is a grass inhabitant, and the Mexican spiny pocket mouse is an important grass seed eater.

The scaled quail occupies the entire area of the grassland, the chachalaca occurs in thickets, and the bobwhite in forest edges. The white-winged dove breeds in the area. The turkey vulture, black vulture, and various hawks and owls are present. The birds noted on six hectares in early August, 1942, combining lists of three two-hectare areas, were black-throated sparrow (40 individuals), lark sparrow (64), black-chinned sparrow (2), mockingbird (23), white-winged dove (12), hooded oriole (4), doves (29), fish crow (8), turkey vulture (3), painted bunting (10), curve-billed thrasher (3), verdin (1), and pyrrhuloxia (2).

The reptiles recorded for the area near Sabinas Hidalgo, Nuevo León, are the eastern spotted whiptail, blue spiny lizard, Texas rose-bellied lizard, mesquite lizard, and crevice spiny lizard. *Holbrookia texana* has been taken at both China and Sabinas Hidalgo. It occurs in both the Arizona–Mexico and the Texas–Tamaulipas areas. The bullsnake and other plains snakes are known to occur. The pink coachwhip has been taken at Monterrey. The Mexican milk snake barely crosses the Rio Grande into Texas.

In July, 1942, the spiders *Misumenops dubius, Argiope trifasciata,* and *Phidippus texanus* appeared restricted to grasslands and forest edges. On the other hand, *Oxyopes salticus,* the black widow spider, and *Mimetus hesperis* are associated with a combination of shrubs and grass in warm areas, sometimes in semidesert. *Peucetia viridans* is widely distributed and usually associated with shrubs and trees. The grasshoppers usually present are *Mesochloa abortiva, Ophulella speciosa, Trachyrhachys kiowa fuscifrons,* and *Opeia obscura,* which feed on short grass, and *Syrbula fuscovittata,* which prefers tall grasses (Isely 1938). The swift-flying *Trimerotropis pallidipennis* is nearly always present. The long-horned *Arethaea* and *Dichopetala brevihastata,* with preferences for forbs and shrubs, are regular residents. The latter occurs throughout the shrub grassland. With the exception of the last, none of these species was captured in the shrub areas near Victoria and Linares, but the forb-eating *Schistocerca obscura* was taken near Linares and is a well-known species in Kansas. Some of the ants taken here, especially *Pogonomyrmex barbatus,* are widely distributed through the mesquite grassland. *Dorymyrmex pyramicus* is similar in its distribution. *Mecidea longula,* a stink bug, is widely distributed in both the eastern and western portion of the shrub grassland. The lygaeids *Phlegyas annulicornis* and *Geocoris uliginosus* occur both in the Tamaulipas and Arizona grasslands. The leafhoppers *Carneocephala flaviceps* and *Xero-phloea viridis* are of wide distribution, occurring in both Tamaulipas and Texas. The same is true of the cercopid *Prairiana subta.* Flies and the beetles were poorly represented. The shrub-inhabiting tree snail *Bulimulus alternatus* is quite generally present.

Much of this grassland has become greatly modified by the invasion of shrubs. Bartlett (MS) noted an area of grassland north of the San Fernando about seven kilometers wide in 1930 that was covered with *Bouteloua trifida* and widely spaced shrubs. This grassland area was absent in 1943. Goldman (1944) returned after four years' absence to a collecting spot in Tamaulipas to secure additional specimens of grass-nesting birds and had difficulty in finding the spot or the birds because the increase of shrubs had almost crowded out the grass. North and east of Monterrey, especially near China, there was a marked decrease in grass between 1938 and 1944. The Rio Grande plains probably contained woody vegetation for several hundred years previous to 1750 in spots unfavorable for grass; but otherwise the plain was grassland (Bray 1905). In the memory of many men, hundreds of square miles now brush land were formerly open prairie, for example, between San Antonio and Del Rio.

Cattle eat the mesquite beans but do not digest the seeds which are thus scattered into areas where the grass is overgrazed. Certain changes take place in the composition of the grasses when mesquite enters; for example, at Austin, the Texas needlegrass forms close sod under mesquite trees to the exclusion of other grasses. Other woody species follow the mesquite, such as *Condalia obtusifolia* in central Texas; *Lippia ligustrina, Opuntia leptocaulis,* and others in southern Texas, follow until the former pasture land becomes covered with a dense, impenetrable growth (Bray 1905). This process appears to have been the basis for Shantz's mapping of two parallel bands of vegetation running through

Fig. 14-5. A. Typical *Acacia* grassland at Sabinas Hidalgo, Nuevo León, Mexico. B. Tamaulipan scrub with *Yucca* which encroaches on grassland along the Pan-American Highway south of Linares 1941 (author's photos). C. *Holbrookia texana* (photo by Anna Allen Wright). D. Tree snails in January. Tamaulipan scrub: E. In the summer of 1942. F. Approximately the same spot in winter, large shrubs 10 to 15 feet (3 to 4.5 m) high with scattered grass beneath and in small patches.

west-central Texas from the southwest corner of Oklahoma to the Gulf Coast, parallel with the Pecos and Rio Grande rivers. The western one was thornbush, and the eastern, mesquite. It appears that invasion of grassland has progressed northward from the Rio Grande since settlement by white man. Lines A, B, C, D, on Figure 14-1 indicate the recent progression (Price and Gunter 1942). The dense vegetation in the lower Rio Grande Valley is frequently marked as desert, although the rainfall is fairly high.

At a point 893 kilometers north of Mexico City (north of Linares), the shrubs *Cordia boissieri, Acacia amentacea, Forestiera angustifolia,* and *Prosopis juliflora* are found, and in places *Lantana involucrata, Salvia ballotaeflora,* and *Leucophyllum frutescens* occur. These shrubs are four to six meters apart. At 740 kilometers the shrubs are about 3.5 meters apart, and one or two additional species are present. In both places, the grasses *Tridens texanus, Sporobolus buckleyi, Aristida purpurea,* and *Bouteloua trifida* are scattered in the openings between the shrubs along with the forb *Euphorbia prostrata.* A third study station at 651 kilometers was very similar to that at 740 kilometers and the same grasses occurred. The leaves fall from most of the shrubs in winter, but the tree snails remain in place on the stems and twigs (Fig. 14-5). At 651 kilometers, *Rhus* and *Selloa glutinosa* were present although not seen at the other two stations. The shrubs become larger and denser both with increasing distance southward and with increasing altitude.

The pygmy mouse and the spiny pocket mouse are recorded from very near the 740 kilometer station, and the southern plains wood rat is known from the southern tip of the Tamaulipas grassland, although it does not occur west of the mountains. The cottontail and the squirrel *Sciurus negligens* is recorded from Linares. The birds noted on six hectares, August, 1942, at 740 and 893 kilometers north of Linares were mourning dove (1 individual), indigo bunting (4), olive sparrow (17), white-winged dove (4), turkey vulture (1), black vulture (2), long-billed thrasher (6), tufted titmouse (3), white-eyed vireo (8), cardinal (19), and boat-tailed grackle (5).

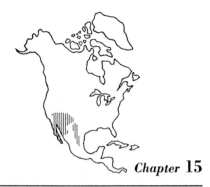

Chapter 15

The Hot Desert

A desert is a tract of bush-covered land with much bare ground exposed. The vegetation is adapted to conditions of little and irregularly occurring rainfall. Animals are adjusted physiologically or in habits of life to tolerate the arid environment. Deserts that are generally warm throughout the year and are very hot in summer are called *hot deserts;* deserts that become very cold in winter are called *cold deserts,* as were discussed in Chapter 10. The hot desert is not, however, frost-free. Minimum temperatures not infrequently drop to freezing in Arizona, Sonora, and Baja California or occasionally to 0° F. (–18° C.) on the Mexican Plateau. Annual rainfall ranges only between 3½ and 13 inches (9 to 32 cm).

The North American hot desert is divisible geographically into two parts (Fig. 15-1). A western portion lies astride the Colorado River and Gulfo de California, chiefly in lower California, Baja California, Arizona, and Sonora. An eastern part is astride the Rio Grande and Rio Conchos, chiefly in New Mexico, Texas, Chihuahua, Coahuila, Durango, Zacatecas, and San Luis Potosí. The two deserts have been connected to varying extents in the past through some of the valleys of New Mexico (Tinkham, personal communication).

Creosote bush is nearly always present in all parts of the desert. Two important associates of the creosote bush are the white bur sage in Arizona, California, and Sonora and the tar bush in Texas, Chihuahua, and southeastern Arizona. Other species occurring in both deserts are burroweed, mariola, and burrobush. Merriam's kangaroo rat also occurs in all parts of the desert, so that the desert community may be designated the *creosote bush–kangaroo rat biome.* The desert pocket mouse and the grasshopper *Trimerotropis pallidipennis* are also of general occurrence.

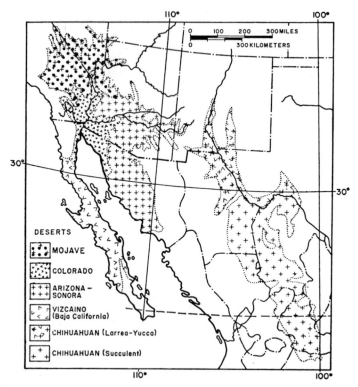

Fig. 15-1. Subdivisions of the North American hot desert (modified after Shreve 1942, Brand 1936, Muller 1947, Nichol 1937, Jensen 1947, and Benson and Darrow 1944, courtesy University of Arizona Press).

The dominant plants are from 10 to 30 feet (3 to 9 m) apart. The crowns rarely touch each other. The low water content of the soil necessitates a large area and wide spacing of the plants for adequate water absorption by the roots. Half-shrubs or undershrubs in many places grow in the open spaces among the larger plants. It appears that they must fit into the general distribution of roots in the soil. Probably they crowd the larger plants a little farther apart.

The diversity of habitat in the areas occupied by desert communities probably exceeds that of any other region (Figs. 15-2, 15-3). There are differences between north- and south-facing slopes, between sun and shade (Shreve 1924, 1931), and between upland and lowland. Soils are mainly of coarse texture, loose rock fragments, gravel, or sand (Fig. 15-4) (Benson and Darrow 1944). There is great variation in the amount of soluble salts in the surface materials. The sharp differences in the habitat result in great diversity and complexity of plant distribution. Because one type of vegetation changes so slowly into another, succession is very difficult to observe and may not occur except on floodplains. The vegetation on most areas may be considered essentially climax (Shreve 1942).

Some plants bloom whenever the requisite amount of moisture becomes available. In the western desert, when there is sufficient rain in February or March, the desert becomes brilliant with the blooming of winter annuals, among which the yellow California poppy is important. The same blooming occurs in the eastern desert when the heaviest rainfall comes in late spring or midsummer. This includes four-o'clocks, milkweeds, asters, senecios, milkworts, mallows, lupines, and astragalus. These plants may occasionally bloom also in the autumn when they get adequate moisture (Tharp 1939). A large group

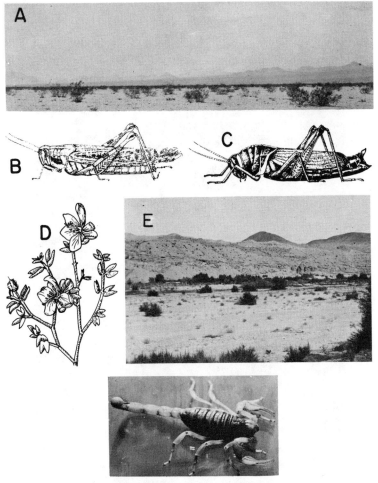

Fig. 15-2. A. Mojave Desert near Kelso, California, in 1910, with only scattered *Larrea* present. B. The gray bird locust *Schistocerca vaga* which sometimes feeds on *Larrea*. C. The desert grasshopper *Taeniopoda eques* which frequents shrubs along water courses, notably mesquite (courtesy of Macmillan Company). D. A branch of *Larrea divaricata* in bloom (after Jepson 1925 and others, redrawn by Alice Boatwright). E. The sparse vegetation in and along the Mojave River, 1910. F. A hairy scorpion.

of arthropods also remain drought-locked in the hard soil until the soil is sufficiently softened by rain to enable them to dig their way out.

Large ungulates are almost absent on the desert. Carnivores are small and usually nocturnal and permeant. The principal dominants are nocturnal burrowers, particularly kangaroo rats and pocket mice. Their moving of soil, bringing of earth materials to the surface, and burying of seeds and other or-

Fig. 15-3. Two general views of the hilly Arizona desert southwest of Picket Post Mountain near Superior, Arizona. A. Ocotillo or coachwhip, barrel cactus, burroweed, and teddy bear cholla. B. Saguaro or giant cactus, foothill palo verde, Engelmann prickly pear, and brittle bush *Encelia farinosa,* the small shrub in the foreground (photos by H. K. Gloýd).

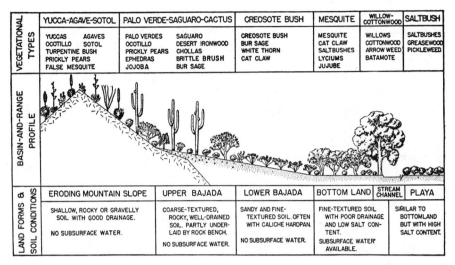

	YUCCA-AGAVE-SOTOL	PALO VERDE-SAGUARO-CACTUS	CREOSOTE BUSH	MESQUITE	WILLOW-COTTONWOOD	SALTBUSH
VEGETATIONAL TYPES	YUCCAS AGAVES OCOTILLO SOTOL TURPENTINE BUSH PRICKLY PEARS FALSE MESQUITE	PALO VERDES SAGUARO OCOTILLO DESERT IRONWOOD PRICKLY PEARS CHOLLAS EPHEDRAS BRITTLE BRUSH JOJOBA BUR SAGE	CREOSOTE BUSH BUR SAGE WHITE THORN CAT CLAW	MESQUITE CAT CLAW SALTBUSHES LYCIUMS JUJUBE	WILLOWS COTTONWOOD ARROW WEED BATAMOTE	SALTBUSHES GREASEWOOD PICKLEWEED

BASIN-AND-RANGE PROFILE

	ERODING MOUNTAIN SLOPE	UPPER BAJADA	LOWER BAJADA	BOTTOM LAND	STREAM CHANNEL	PLAYA
LAND FORMS & SOIL CONDITIONS	SHALLOW, ROCKY OR GRAVELLY SOIL WITH GOOD DRAINAGE. NO SUBSURFACE WATER.	COARSE-TEXTURED, ROCKY, WELL-DRAINED SOIL. PARTLY UNDER-LAID BY ROCK BENCH. NO SUBSURFACE WATER.	SANDY AND FINE-TEXTURED SOIL. OFTEN WITH CALICHE HARDPAN. NO SUBSURFACE WATER.	FINE-TEXTURED SOIL WITH POOR DRAINAGE AND LOW SALT CONTENT. SUBSURFACE WATER AVAILABLE.	SIMILAR TO BOTTOMLAND BUT WITH HIGH SALT CONTENT.	

Fig. 15-4. Benson and Darrow's idealized profile (1944) through a range of hills in the relatively moist Arizona desert showing vegetation types and soil conditions (courtesy of University of Arizona Press).

ganic materials is a process which impresses all students of desert animals.

In A.D. 1500 the largest human population was in the Sonora–Gila–Yuma region (Kroeber 1939). There was land cultivation on the river floodplains which would have disturbed the early seral stages, but the upland desert was little modified. In this area populations reached 19 individuals per 10 square miles (7/10 km^2).

The hot desert may be subdivided into regions of ecological importance based upon the presence or absence of succulent plants and associated animals, differences in taxonomic composition of the biota, and differences in the abundance and uniformity of occurrence of the principal species (Brand 1936, Nichol 1937, Benson and Darrow 1944, Jensen 1947, Shreve 1925 to 1944).

In the western section, the *Mojave Desert* consists of a series of undrained basins separated by low mountains in California, Nevada, and northwestern Arizona. Death Valley is one of these basins. The annual rainfall is only about 3.5 inches (8.75 cm) and comes erratically. Between 1909 and 1921, there were three rainless periods at the Bagdad station of 24, 25, and 32 months duration. The vegetation consists of shrubs, yucca, and the Joshua tree. The *Colorado Desert* lies below 1000 feet (300 m) in the basins of the Gila and Colorado rivers. The Salton Sea Basin is included within it. There are many sand dunes. It has the same low annual rainfall as the Mojave Desert, and frosts seldom occur. There are fewer tall cacti here than in the Arizona–Sonora Desert to the east (Benson and Darrow 1944). The *Vizcaíno* or *Baja California Desert* has more variable and uncertain rainfall than even the Mojave Desert. A number of plants not found in the Colorado Desert occur here (Shreve 1926). The *Arizona–Sonora Desert* with its characteristic tall cactus or saguaro lies between 1000 and 4000 feet (300 to 1200 m). The term "Sonoran Desert" as formerly applied to the combined Colorado and "Arizona" deserts should be abandoned. The deserts in Arizona and the eastern half of Sonora have similar climatic and biological characteristics and should be grouped together. Rainfall is somewhat higher than in the other desert regions above described.

In the eastern portion, the *Chihuahuan Yucca Desert* lies on both sides of the Rio Grande south to the Rio Conchos in Chihuahua between elevations of 2000 and 5000 feet (600 to 1500 m). Studies are too inadequate for its accurate mapping. The *Chihuahuan Succulent Desert* makes up the remainder of the eastern desert, reaching south into San Luis Potosí of the eastern Sierra Madre of the Mexican Plateau. It lies between 2000 and 6000 feet (600 to 1800 m) elevation.

COLORADO DESERT

The Colorado Desert (Fig. 15-1) is characterized by frost-sensitive plants such as smoke tree, desert ironwood, and certain daleas in the climax and by the dense growths of willow, cottonwoods, and mesquite on the early floodplain. These species are not, however, important dominants. Certain reptiles are characteristic of this region (Gloyd 1937) and some mammals are restricted to floodplain areas (Burt 1938). In the desert near the Colorado River, between Needles and Yuma, creosote bush occurs in varying abundance from the second terrace of floodplains to the tops of the highest hills. Only the most rocky slopes and the periodically eroded wash-bottoms lack this plant altogether. On some upland mesa areas, creosote bush grows to the entire exclusion of all other ligneous vegetation. In places it invades the saltbush community (Grinnell 1914).

Parts of the desert mesa are swept clean of fine sand by the prevailing winds, and the resulting surface consists of packed gravel or windworn pebbles, called "desert pavement." Other parts of the desert have a sandy soil where a common small shrub is the white bur sage. On stony ground often no woody plant is present except the creosote bush.

Merriam's Kangaroo Rat–Creosote Bush Faciation

This is Grinnell's "creosote association (mesa)." The two principal dominants are the creosote bush, in general very abundant in the area, and Merriam's kangaroo rat, which is at its maximum abundance here. Other important mammals are the yellow-backed pocket gopher and the little pocket mouse which occurs particularly on sandy soil. The black-tailed jack rabbit may be seen in numbers returning to their daytime forms soon after sunrise. At night the desert coyote and kit fox forage practically everywhere. Species of lesser abundance are Harris' antelope squirrel, white-tailed antelope ground squirrel, long-tailed pocket mouse, rock pocket mouse, and spiny pocket mouse, all in stony habitats, and round-tailed ground squirrel, cactus mouse, and desert pocket mouse on sandy soil. The bats are represented by the cave bat, big brown bat, and leaf-nosed bat, but their relative abundance in the different faciations of the biome is not well known.

Grinnell found only two breeding birds, the lesser nighthawk and the black-throated sparrow, in the creosote bush faciation proper; neither of them is abundant. A few additional transient or wintering species also occur. Common lizards are the southern whiptail and desert spiny lizard.

The invertebrates have not been adequately investigated. The author estimates arthropod population at one spider, three ants, and one beetle or hemipteron per square meter. Six ant hills per hectare was a common number. *Bootettix punctatus* and its associate, the katydid *Insara covilleae*, which Ball *et al.* (1942) state are obligate on creosote bush, are present as adults in late summer.

Saltbush–Desert Kangaroo Rat Faciation

Desert saltbush occurs alone or mixed with creosote bush, and there is some mesquite on fine loam soil. Ocotillo and brittle bush generally associated with giant cactus are present in small areas only. Locally there may be some desert trumpet and palo verde (Grinnell 1914, Nichol 1937).

As compared with the creosote bush faciation, ground squirrels and four of the five species of pocket mice are absent. The cactus mouse and desert kangaroo rat are abundant; Merriam's kangaroo rat, black-tailed jack rabbit, and desert cottontail are present in small numbers. The pocket gopher and the bats are replaced by other species. The badger is almost limited to this area in the desert. Resident birds include small numbers of the Gambel's quail and the roadrunner.

A collection of invertebrates by the writer in midsummer, 1944, yielded ten specimens per square meter. These included the leaf beetle *Pachybrachis mellitus* and the ants *Vermessor pergandei* and *Pogonomyrmex californicus*. As for orthopterons, Tinkham (personal communication) names the lubber grasshopper *Tytthotyle maculata*, the shield-backed cricket *Capnobotes fuliginosus*, *Arphia aberrans*, and *Trimerotropis p. pallidipennis* as occurring.

Driving sand is often arrested about a bush which then becomes the core of a growing sand dune. Sand verbena is characteristic of such incipient dune areas where there is much sand present. Favorable conditions for burrowing here attract the large kangaroo rat and other rodents. A characteristic assemblage of species results, which might be appropriately called the aeolian sand community (Grinnell 1914). It also supports tenebrionid beetles, solpugids, and scorpions.

Baja California

The coastal plain of the Gulfo de California south to a little beyond 30° N. Lat. resembles the area near Yuma and Calexico in California. Practically all the plants mentioned in the Colorado Desert account above are present (Shreve 1926, Grinnell 1914). Desert ironwood is present and the elephant tree appears to be more common here than in the United States. Slopes facing east are nearly devoid of vegetation. Some slopes bear a few small barrel cacti, and near the base of other slopes there may be a few creosote bushes and bur sages. In the bajadas, ocotillo reaches the greatest size that has been observed anywhere in its range, 18 to 26 feet (5.5 to 8 m) in height and sometimes with over one hundred branches. The Merriam's and desert kangaroo rats, pocket mouse, round-tailed ground squirrel, black-tailed jack rabbit, and coyote occur (Nelson 1921).

Sonora

The entire desert area in Sonora is an undulating plain broken frequently and irregularly by low hills and mountains. Aside from the Colorado River, no stream reaches the Gulf except for short periods during floods. The types of vegetation already described are found throughout the plains of southwestern Arizona and northwestern Sonora south to about 30° N. Lat. Here, less xeric dominants begin to take possession. In the lower drainage of the Magdalena (Concepción) River, creosote bush occurs in extensive stands usually with bur sage, either *Franseria dumosa* or *F. deltoidea.* Between the Magdalena and Sonora rivers the vegetation remains very open but increases slightly in hieght and in the number of abundant species. The dominance of creosote bush tends to become confined to the immediate coast and to certain restricted habitats. *F. dumosa* reaches its southern distribution limit. The species which are dominant here are chiefly ones that occur in streamways and fans north of the Gila River (Shreve 1934). This is a transition to the Arizona–Sonora type of desert discussed beyond.

VIZCAÍNO DESERT OF BAJA CALIFORNIA

There is no ecological description of this desert. The Vizcaíno Desert lies principally between 30° N. Lat. and Bahia de La Paz at 24° N. Lat. About 75 per cent of the surface is below 500 meters, and between 40 and 50 per cent of it is below 200 meters. The total of these two relatively low areas is around 25,000 square miles (63,750 km²). The greater part of the lowland is along the Pacific Coast. The annual rainfall is from 10 to 20 centimeters; from 1 to 30 days per year this desert has one millimeter or more of rain, most of it coming in August and September. Two to five years may sometimes elapse without any rain. Strong southwest winds sweep across the area from March to October, and there are occasional hurricanes. The mean annual temperature ranges from 20° to 25° C. The extreme maximum of 43.5° C., compared with 49° C. in Arizona, shows the modifying influence of the sea and gulf (Nelson 1921 and Government Records).

The important plants are creosote bush, ocotillo, elephant wood, and goat nut, the last species being used for food and drink by the natives. Cirio forms scattered polelike forests over the lower tablelands in the central district south of 28° 15′ N. Lat. There are many species of *Agave* and a great variety of cacti, including gigantic creeping and massive tree forms. There are many spectacular endemic plants (Nelson and Goldman 1926).

The antelope ground squirrel, white-footed mouse, common pocket mouse, black-tailed jack rabbit, Merriam's kangaroo rat, mountain lion, and coyote are present but are frequently represented by different subspecies than in other parts of the desert. Mule deer and probably pronghorns come down out of the mountains in winter. Noteworthy birds are the California quail, white-winged dove, three species of woodpeckers, and others of subspecific difference from those farther north. Nelson (1921) names 12 species of lizards for the area, and Mosauer (1936b) describes how some forms are highly adapted to sand.

ARIZONA–SONORA DESERT

Arizona

The *tall cactus–Gila woodpecker–ocotillo faciation* is a spectacular desert in the rainiest part of the western desert (Fig. 15-3). Creosote bush, brittle bush, and rock pocket mouse are typical constituents. The rock pocket mouse *Perognathus intermedius* outnumber all other mammals. *P. baileyi* is second in abundance and *P. amplus* also occurs. Other small mammals are rock squirrel, Harris' antelope squirrel, southern grasshopper mouse, and hispid cotton rat. Larger mammals are the ringtail, spotted skunk, kit fox, cottontail, white-throated wood rat, peccary, and white-tailed deer (Dice and Blossom 1937).

The Gila woodpecker is generally associated with the giant cactus. Tinkham (personal communication) states that the fruit of the tall cactus supplies much food for many species, although the foliage does not attract many insects. The Gila woodpecker, elf owl, ferruginous owl, screech owl, ash-throated flycatcher, and purple martin live in holes in the giant cactus. In the Organ Pipe Cactus National Monument, the roadrunner, great horned owl, red-tailed hawk, sparrow hawk, gilded flicker, and the Gila woodpecker are common permanent residents; the Gambel's quail is abundant (Hensley 1951). The cactus wren nests in the cholla cactus; the verdin in *Condalia*, palo verde, and cat claw; the black-tailed gnatcatcher in *Condalia*. The red-tailed hawk was found to take the lizards *Sceloporus magister* and *S. clarki*, round-tailed ground squirrels, and unidentified snakes. Nineteen per cent of the resident and wintering birds are carnivorous, 43 per cent are insectivorous, 8 per cent are miscellaneous feeders, and 30 per cent are plant feeders, notably on seeds. The dominant creosote bush and bur sage are not important as food for birds.

The tiger, black-tailed, and Mojave rattlesnakes, bullsnake, western patch-nosed snake, glossy snake, and Sonora lyre snake belong primarily in this community (Tinkham, personal communication). Eight species of lizards are present. The Gila monster (Fig. 15-5) belongs typically to the tall cactus community; it is crepuscular or nocturnal and buries its eggs in sandy spots. Its food is little known but apparently includes eggs and other reptiles. The western collared lizard is diurnal and dashes for holes in the ground when approached; its food is chiefly insects and other animals of comparable size. There is the regal horned lizard, an uta, a whiptail, a desert spiny lizard, the desert tortoise, and the Arizona zebra-tailed lizard (Gloyd 1937, Smith 1946).

Fig. 15-5. A. Gila monster (photo by H. K. Gloyd). B. Crevice spiny lizard (photo by Rozella B. Smith).

At the former desert laboratory of the Carnegie Institution near Tucson, Arizona, the crab spider *Misumenops celer* and the orthopterons *Trimerotropis p. pallidipennis* and *Platylactista azteca* were taken in July, 1944. Other insects commonly present are the bruchid *Acanthoscelides amicus,* the snout beetle *Apion ventricosum,* and the ants *Novomessor cockerelli* and *Pogonomyrmex barbatus nigrescens.* Several species of cicadas occur, and the grasshopper *Schistocerca v. vaga* (Fig. 15-2B) is known to feed on creosote bush (Tinkham, personal communication). *Zapata salutator* and *Lactista oslari* are found characteristically on tall cactus and palo verde but are not numerous.

Sonora

In central Sonora, creosote bush becomes localized in its occurrence. The desert vegetation is very open and its dominants are much larger plants than in the north, the principal ones being ironwood, palo verde, and mesquite. Other small trees which are abundant in this region are *Cercidium sonorae, Fouquieria macdougalii, Acacia occidentalis,* and *Guaiacum coulteri.* Large cacti are frequent but scarcely as abundant as in southern Arizona. They include *Cereus thurberi, C. schottii, C. almosensis, Opuntia thurberi,* and infrequent examples of most of the species that are abundant in Arizona. The ground among the small trees may be either bare or covered by heavy stands of *Franseria deltoidea* or *Encelia farinosa* (Shreve 1942).

Southward, near the coast, the vegetation becomes still taller. There is little change in the physiognomy between the Sonora and Yaqui rivers. The xerophytic species of the Colorado Desert drop out, and their places are taken by *Cercidium sonorae, Acacia occidentalis, Fouquieria macdougalii, Cereus pringlei, Opuntia thurberi,* and *Guaiacum coulteri.* The leguminous trees *Olneya, Prosopis,* and *Cercidium* and the shrub *Encelia* give character to the vegetation. As the creosote bush disappears, the brittle bush *Encelia* takes its place as the most abundant dominant. Cacti are conspicuous but form a very small percentage of the plant composition.

The characteristic mammals of Arizona are present although they may be represented by different subspecies (Burt 1938). Several grassland species appear here in the desert, such as the Allen or antelope jack rabbit, rock squirrel, valley gopher, white-throated wood rat, and desert cottontail. The Gila woodpecker and Gambel's quail are widely distributed.

COMMUNITY DEVELOPMENT IN THE WESTERN DESERT

Succession is evident on floodplains, both on well-drained areas between low and high water marks and in ponded areas (Fig. 15-6). The Salton Sea has been receding in recent years, and the relation of different types of vegetation to the height of the water table here is shown in Figure 15-7.

Willow–Cottonwood Community

Between Needles and the mouth of the Colorado River (Grinnell 1914) are two species of willow, *Salix nigra,* which also occurs along the Mississippi River, and *S. exigua,* the cottonwood *Populus fremontii,* and the undershrub batamote *Baccharis glutinosa.* Very large trees must have been present a century or more ago before the stream margins were stripped. At the present time, long stretches along the lower river are without trees of significant size, and the desert appears to come almost to the water's edge. Lower portions of the floodplain were originally inundated annually in early summer by from a few inches to as much as 12 feet (3.6 m) of water. The reed *Phragmites communis* grows in dense jungles on the more stable silt deposits resulting from this flooding.

Trees three to four feet (1+ m) in diameter occur in the floodplain forest of the Santa Cruz River in the St. Xavier Indian Reservation near Tucson, Arizona (Tinkham, personal communication). These trees are of great age. It is probable that as the ridges on which the trees occur reach a high level above the water, the trees die during drought periods and are followed by arrowweed or saltbush.

The desert pallid bat, Arizona myotis, and Mexican free-tailed bat divide the insect supply with the smaller birds. The only rodent of wide occurrence

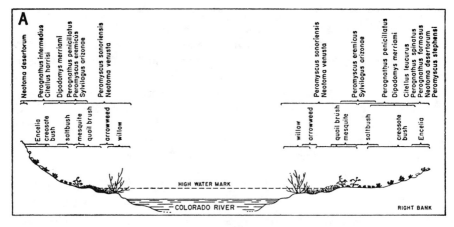

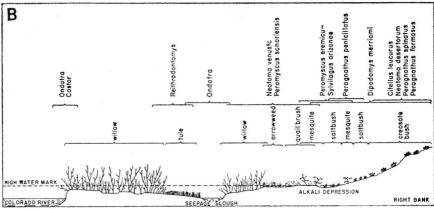

Fig. 15-6. Idealized cross-sectional profiles of the Colorado River looking downstream, showing the principal plant–animal communities and chief animal constituents. A. Cross-section at Needles where no floodplain sloughs are present. B. Cross-section east of Picaho, 20 miles north of Yuma (from Grinnell 1914).

is the deer mouse. Other mammals of this community are the mule deer, hispid cotton rat, western harvest mouse, desert cottontail, mountain lion, gray fox, striped skunk, and raccoon.

Since the willow–cottonwood community is the only forest in the region, it supports a concentration of birds, including the resident Cooper's hawk, screech owl, ladder-backed woodpecker, and redwinged blackbird. The most characteristic summer breeders are yellow warbler, Bell's vireo, summer tanager, black-chinned hummingbird, brown-headed cowbird, white-winged and mourning doves. The common ground dove occurs at St. Xavier. Snakes present are *Crotalus atrox, Pituophis catenifer affinis,* and *Lampropeltis getulus californiae;* lizards *Sceloporus clarki, S. magister* (Tinkham, personal communication), and others. Orthopterons in the area are *Platylactista azteca, Trimerotropis p. pallidipennis, Schistocerca v. vaga,* and *Conalcaea.*

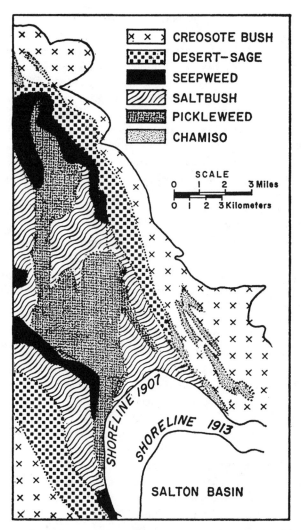

Fig. 15-7. Portion of the Salton Basin southeast of Thermal (from a figure by Shantz and Piemeisel 1924). Small cultivated areas have been restored to the original vegetation type in accord with topography. One or more seres dependent on soil moisture are suggested by the vegetation but are not discussed by Shantz and Piemeisel.

Arrowweed Community

Pluchea sericea grows densely over extensive areas slightly above the willows, commonly to the exclusion of everything else. Deer mice, wood rats, cottontails, and the two skunks of the region are present in low populations along with the lesser goldfinch and song sparrow.

On the east side of the Colorado near Blythe, California, the substratum is less sandy than in other adjacent areas. In July, 1944, arrowweed was found generally distributed with cat claw common and mesquite and creosote bush present. Approximately 12 arthropods per square meter were collected. The jumping spider *Habronattus elegans,* the leaf-eating beetle *Stenopodius vanduzeei,* and the weevil *Apion ventricosum* were relatively numerous. Tinkham (personal communication) names the following characteristic grasshoppers for the area: the alkali grasshoppers *Anconia integra, Aeoloplides californicus,* and *A. tenuipennis,* which feed on *Sarcobatus* and *Atriplex canescens,* and *Poecilotettix longipennis,* which feeds on floodplain composites and various shrubs. The extreme desert grasshopper *Derotmema delicatulum* was taken along with the lantern fly *Acanalonia saltonia* and the treehopper *Campylenchia curvata,* making a rather long list of insects for the desert. The principal ant was the black and yellow *Dorymyrmex pyramicus bicolor;* a large velvet ant was *Dasymutilla satanas.*

Quail Brush Community

On slightly higher ground, *Atriplex lentiformis* grows in dense clumps three to eight feet (1 to 2.4 m) high. The Gambel's quail, desert cottontail, desert kangaroo rat, round-tailed ground squirrel, and cactus mouse reach their maximum populations for the region in this community, and the badger and the white pocket gopher are not found elsewhere in the area.

Ponds and Tule Marsh Communities

As the first stage in the pond sere, a depression may hold water throughout most of the year and contain carp, brown bullhead, hump-backed sucker, and the large minnow. The bullheads and carp may be abundant and serve as food for numerous herons. Some 32 birds occur in these river habitats (Grinnell 1914).

The tule or marsh community usually contains some bulrushes and cat-tails, sedges, and considerable salt grass. The salt grass is commonly surrounded by arrowweed. Saltbush occurs in depressions that are only temporarily wet.

The amphibians *Bufo punctatus, B. alvarius* and *Hyla arenicolor* occur around permanent water in canyons. The western harvest mouse, muskrat, and pallid raccoon are present. The redwinged blackbird, song sparrow, and yellowthroat are resident birds.

Western Honey Mesquite Community

On the highest terraces above the tules and quail brush, the western honey mesquite takes possession. Originally it made an almost continuous narrow belt along the Colorado River. In 1910, trees were found up to eight inches (20 cm) in diameter. The community was largely depleted for fuel for steamers which plied between Yuma and Needles. The mesquite serves as a host of the parasitic mistletoe *Phoradendron californicum,* while willow and cottonwood support *P. flavescens.* Mistletoe, when in blossom, is visited by myriads of insects and produces an abundant and almost continuous crop of berries. Several of the winter and resident birds of the community depend almost wholly on these berries for their food.

Of the mammals, the mule deer, deer mouse, desert pocket mouse, desert cottontail, and striped skunk are present. The bobcat probably has its maximum abundance in the region here, and the white-throated wood rat is a characteristic species.

Cat Claw Community

Cat claw ranges in size from a large shrub to a small tree 20 feet (6 m) high and is prevalent in the drier washes. Associated with it is the abundant desert ironwood. Its leafy and thorny branches form ideal refuges and nest sites for small birds. The apparently leafless blue palo verde occurs along the smaller ravines leading into the hills. Some trees reach 48 inches (120 cm) in circumference of trunk 2 feet (0.6 m) above the ground and as much as 30 feet (9 m) in height.

Various birds forage here, including Gambel's quail, great horned owl, roadrunner, poor-will, house finch, lesser goldfinch, loggerhead shrike, and cactus wren. Mammals inhabiting the cat claw are large numbers of the mule deer and desert pocket mouse and cactus mouse, Merriam's kangaroo rat, long-tailed pocket mouse, and spiny pocket mouse in lesser abundance. The black-tailed jack rabbit and its enemies, the bobcat and gray fox, are also present. The western pipistrelle is a bat found in this stage (Grinnell 1914).

MOJAVE DESERT

The Mojave Desert is marked off physiographically by scattered mountain ranges which stretch from the southern tip of Nevada across southern California in an irregular line. The Mojave Desert is characterized by the usual absence of the tall cactus and palo verde, a reduced number of pocket mouse species, and the absence of the Gila monster except in the northeastern corner. It may be designated the *creosote bush–pocket mouse–yucca association.* Characteristic plants are essentially endemic. The most notable are the Joshua tree (Fig. 15-8), Parry saltbush, indigo bush, Mojave sage, and woolly bur sage. The characteristic mammals are the long-tailed and the little pocket mice. The antelope squirrel is more conspicuous in the Mojave than in the Sonoran Desert. The round-tailed ground squirrel is present in sandy areas. The kit fox, desert coyote, and spotted skunk, and the birds Gambel's quail, roadrunner, and mourning dove are present. There is segregation of some of these species according to altitude (Johnson *et al.* 1948, Benson and Darrow 1944, Tinkham, personal communication 1954).

Creosote Bush–Desert Kangaroo Rat Faciation

This faciation extends from the lowest parts of the area at 1800 feet to about 3300 feet (545 to 1000 m). The soil may be sandy, gravelly, pebbly, or stony, often forming a desert pavement. The terrain is mostly flat or gently sloping. Widely spaced creosote bushes are the most conspicuous plants, but with them are many other xerophilous plants such as burrobush and *Yucca schidigera.*

White bur sage is present throughout the Mojave, Colorado, and Arizona-Sonora deserts.

The desert iguana, sidewinder rattlesnake, desert tortoise, phainopepla bird, and Merriam's kangaroo rat are probably more abundant here than anywhere else (Johnson *et al.* 1948).

The desert side-blotched lizard is one of the daytime animals in evidence. Additional reptiles are *Callisaurus draconoides, Uta graciosus, Gopherus agassizi, Pituophis catenifer,* and *Cnemidophorus tessellatus* (Tinkham, personal communication).

Fig. 15-8. Mojave Desert. A Joshua tree is at the right, and the stub in the foreground contains a hole, probably made by a woodpecker, in which there was a screech owl nest (from Johnson *et al.* 1948, courtesy Alden Miller).

Insects living on the creosote bush are the grasshoppers *Bootettix punctatus* and *Ligurotettix coquilletti,* the katydids *Capnobotes fuliginosus* and *Insara covilleae,* and the very rare shield-bearing crickets *Anoplodusa arizonensis* and *Ateloplus splendidus.* Ground-frequenting species are *Cibolacris parviceps aridus, Derotmema delicatulum, Trimerotropis p. pallidipennis,* and *Poecilotettix sanguineus* (Tinkham, personal communication). The scorpion also occurs.

Yucca–Ladder-backed Woodpecker Faciation

Characteristic plant species in this faciation are three yuccas, the Joshua tree, Mojave dagger, and Whipple yucca. There is a wide overlapping of the creosote bush and yucca communities in most places. Joshua trees reach their maximum development on slopes above 3300 feet (1000 m). Whipple yucca also grows only at the upper elevation limits of the Mojave Desert, from 3000 to 4000 feet (900 to 1200 m). The Mojave dagger delimits the ecotone area between the Mojave and Arizona-Sonora deserts. Mojave dagger and Whipple yucca grow equally well on gently sloping mesas and in the rough terrain at the base of the mountains (Johnson *et al.* 1948). Several kinds of cacti, particularly *Opuntia basilaris, O. ramosissima, O. bigelovii,* and *Echinocereus troglochidiatus* are abundant in the yucca habitat and provide homes for animals.

The following vertebrates were apparently restricted to the yucca habitat in the Providence Mountain region: banded gecko, desert night lizard, Mojave rattlesnake, gilded flicker, cactus wren, and Bendire's thrasher (Johnson *et al.* 1948). Other species that occur in largest numbers in the yucca community are long-nosed leopard lizard, desert side-blotched lizard, checkered whiptail, prairie falcon, roadrunner, Costa's hummingbird, ladder-backed woodpecker, western kingbird, mockingbird, loggerhead shrike, Scott's oriole, badger, Botta's pocket gopher, southern grasshopper mouse, and desert wood rat. The desert spiny lizard is about equally abundant in the yucca and rock land habitats, and the panamint kangaroo rat occurs similarly in the yucca and sagebrush habitats. The following birds, not already noted, occur in both the Mojave and Colorado deserts but principally in the former: Gambel's quail, white-winged dove, burrowing owl, lesser nighthawk, LeConte's thrasher, California thrasher, black-tailed gnatcatcher, phainopepla, Lucy's warbler, crissal thrasher, poor-will, sage sparrow, and black-throated sparrow. Some of these species occur also in sagebrush or chaparral (Miller 1951).

An important lepidopteron in the community is the yucca moth. This moth cross-pollinates the flowers of the Joshua tree and deposits its eggs in the pistil of the flower. The larvae feed in the seed capsules, but they miss some that then develop normally. When the larvae are mature, they enter the soil where they may remain for more than a year. The time of pupation is uncertain, but their leaving the soil appears to depend upon rain. This is also true for various beetles which require rain to soften the soil. The moth pupae are provided with hooks which bring them to the surface with proper flexing movements of the body. The Joshua tree fails to bloom in some years (Riley 1892).

Orthopterons occurring in the association are *Cibolacris parviceps aridus, Trimerotropis p. pallidipennis,* and *Poecilotettix sanguineus,* which are also found in the creosote bush association, *Oedaleonotus b. borckii, Morsea c. californica, Idiostatus aequalis,* and *Aglaothorax segnis* (Tinkham, personal communication).

Sand Dunes

Large deposits of wind-blown sand are present in the creosote bush area southwest of Kelso. These deposits cover several square miles and in some instances rise over 400 feet (120 m) above the level of the surrounding flats. Small patches of wind-blown sand also occur on the floor of the valley situated west of Clark Mountain and in Ivanpah Valley. Grasses and creosote bushes grow sparingly on the dunes but more abundantly in the shallower sands near their edges.

The Merriam's kangaroo rat is common in the sand dunes, and the black-tailed jack rabbit and coyote are often present. Desert kangaroo rats in the Mojave Desert are largely restricted to the sand dune habitat, and round-tailed ground squirrels occur in greater numbers in the sand dune areas than elsewhere. The only family of burrowing owls found by Johnson *et al.* was on the edge of the dunes, probably because the sand furnished suitable burrowing sites. Orthopterons of the Kelso sand dunes are *Anoplodusa arizonensis, Coniana snowi, Xeracris minimus, Anconia integra, Eremiacris pallida,* a sand-treading camel cricket, and burrowing cockroaches (Tinkham, personal communication).

Desert washes occur in both the creosote bush and yucca communities. The desert washes typically have a sandy or gravelly floor in or on the margins of which cat claw, mesquite, and desert willow grow. Cliff rose occurs locally in the washes. In Cedar Canyon and in some other areas, rabbitbrush is a constituent. The long-tailed brush lizard, the verdin, and the hooded oriole are found in these washes (Johnson *et al.* 1948).

Rock Lands

Rock lands have a scattered vegetation, the composition of which depends on the adjacent vegetation. Cacti, Mojave dagger, Whipple yucca, and agave are characteristic plants at elevations below 5500 feet (1667 m). The characteristic vertebrates of the rock-land habitat are the ringtail, spotted skunk, rock squirrel, long-tailed pocket mouse, canyon mouse, western collared lizard, chuckwalla, and southwestern speckled rattlesnake. In canyons, the lizard *Sceloporus occidentalis,* broad-tailed hummingbird, scrub jay, Bewick's wren, cañon wren, rufous-sided towhee, and gray fox occur in their greatest numbers (Johnson *et al.* 1948).

Death Valley

Death Valley is a depression about 500 feet (150 m) below sea level, which influences its biota only slightly. Creosote bush, bladder stem, and small sage make up scattered vegetation over the valley floor. Only the leafhopper *Exitianus obscurinervis* was found numerous here.

On stable sand dunes, the large gummy composite arrowweed *Pluchea sericea* is abundant; also the salt grass *Distichlis stricta* is scattered between shrubs and hammocks of the grass *Sporobolus airoides.* This community was found to have more invertebrates than any others in the valley, including spiders of three genera, the leafhopper *Exitianus obscurinervis,* the snout beetle *Apion caricorne,* the grasshopper *Anconia integra,* a parasitic wasp, and an aphid.

The so-called devil's cornfield is occupied by arrowweed, shadscale, and iodine bush or pickerelweed. On a surface of flat, encrusted clay, the snout beetle *Apion vericorne* is abundant; the ant *Pogonomyrmex californicus,* the bug *Nysius strigosus,* and the horse fly *Tabanus punctifera* are present. On moving dunes, mesquite is present with a little creosote bush. In this area there are tracks of wood rats, also mice, kit fox, desert iguana, and probably the crested lizard. Thirty-nine desert mammals and 100 species of desert birds are known to occur in the valley.

EASTERN OR CHIHUAHUAN DESERTS

Although the eastern desert may be subdivided into two faciations, one with a predominance of yucca and the other with a predominance of succulents as secondary dominants, it is difficult to make such a separation with animals. The two faciations together, the *creosote bush–pocket mouse–tar bush association*, occupy west-central New Mexico, and Texas east to the Pecos River mixed with a great variety of other communities, the eastern one-fourth of Chihuahua, the western three-fourths of Coahuila, a narrow strip along the eastern part of Durango, the northeastern one-third of Zacatecos, and a large part of the plateau portion of San Luis Potosí and southwestern Nuevo León. The range in elevation is generally between 2000 and 6000 feet (600 and 1800 m). Various small ranges of low mountains are scattered over the area and commonly have mesquite grassland above the desert.

The rainfall varies from about 3 inches (7.5 cm) in the broad undrained basins of Coahuila to 12 to 16 inches (30 to 40 cm) in the more elevated stations at the southern and eastern edge of the desert. There are only 30 to 60 days per year with 0.1 inch (2.5 mm) or more of rain.

Sixty-five to 80 per cent of the rainfall comes from June through September. The other months are very dry, especially January through May. Accordingly, there is only one growing season. Frost is general over the area and severe above 5500 feet (1667 m) (Shreve 1942).

Of the 12 characteristic and common plants of the western and Chihuahuan deserts, there are only 3 species that occur in both: creosote bush, ocotillo, and mesquite. Other common and characteristic plants of the Chihuahuan Desert are white thorn, mortonia, and tar bush.

The animal dominants include Merriam's kangaroo rat and desert pocket mouse which are also important throughout the Sonoran Desert. Altogether, there are five species of pocket mice and four species of kangaroo rats in the Chihuahuan Desert. The other familiar groups of desert mammals are present, including the widely distributed carnivores, some with characteristic subspecies. The scaled quail and the lizard *Sceloperus poinsetti* have their maximum populations centering in this desert.

Near Alamogordo, New Mexico, creosote bush, mesquite, crucifixion thorn, sotol, ocotillo, and yuccas are dominant, and *Rhus trilobata, Atriplex canescens,* and *Yucca elata* occur on the sand dunes (Ruthven 1907). Kangaroo rats and pocket mice are numerous.

Northern Mexico is a gravel and sand-covered plain with scanty vegetation, mainly *Yucca elata,* creosote bush, and mesquite. There are three large drainage areas ending in temporary lakes or bolsons (sinks). There are hundreds of square miles of arid country in which creosote bush outnumbers all other plants. The white crucillo occurs in connection with sand dunes (Brand 1936).

A large series of succulents comes into the desert south of the Rio Conchos in the Big Bend area and in Coahuila. Two species of aborescent opuntias are widespread; two species of large barrel cacti, one of them doubtless *Echinocac-*

tus wislizeni, are more conspicuous than abundant. Cacti 6 to 12 inches (15 to 30 cm) high are common; as many as eight to ten species may be found on 100 square feet (9.3 m²). The semisucculents are yuccas, bear grass, and sotol; leaf succulents are species of *Agave* and *Hechtia.* Columnar cacti occur in the extreme southern part of the plateau desert (Shreve 1942). In addition there are *Opuntia imbricata, O. leptocaulis,* and *Mammillaria* (Muller 1939, 1940, 1947). Succulent desert occupies eastern Chihuahua south of the Rio Conchos, western Coahuila, eastern Durango, and southward. The remainder of the eastern desert is dominated by yucca.

Creosote Bush–Desert Cottontail–Yucca Faciation

This faciation occupies the south-central plain of New Mexico which surrounds the Rio Grande Valley and extends north in a narrow strip nearly to Albuquerque. Its eastern boundary is the San Andreas Mountains and, near El Paso, the Franklin Range. The faciation probably originally extended southeastward to a point opposite the mouth of the Rio Conchos near Presidio, Texas, and covered the northeast quarter of the state of Chihuahua. The remainder of the boundary ran slightly west of south up the Rio Conchos to the Rio San Pedro (Brand 1936).

Twenty-two species of mammals occur, but their distributions do not coincide well with the boundaries of the faciation (Dice 1930, Bailey 1931). Birds likewise do not differ greatly from those in the creosote bush desert elsewhere. At Alamogordo, the mourning dove, mockingbird, horned lark, Cassin's kingbird, black-throated sparrow, and common raven are seen. The roadrunner is also recorded for this portion of the desert. The principal lizards are checkered whiptail, desert spiny lizard, southern prairie lizard, leopard lizard, and the horned lizards *Phrynosoma coronatum* and *P. modestum* (Ruthven 1907).

In the white sands area southwest of Alamogordo, New Mexico, moving sand shows nothing but tracks of lizards, mice, and birds. On more stable sand, there is *Yucca elata,* the sumac *Rhus trilobata, Atriplex canescens,* and *Ephedra.* The bleached earless lizard is a common early dune inhabitant, and the fence lizard and six-lined racerunner also frequent the dunes. Tenebrionid beetles are numerous around yucca. The pinacote beetle is more abundant on the dunes than in the depressions. A pocket mouse *Perognathus apache* in this area is white (Dice and Blossom 1937). Orthopterons are *Trimerotropis texana* and *Ammobaenetes phrixocnemoides* (Tinkham 1948).

About 120 miles (192 km) north of Chihuahua City, the predominant plant is creosote bush, with tar bush, white crucillo, white thorn, and *Rhus microphylla* also present. Collections made in 1944 indicate that the following invertebrates are resident in the area: the millipede *Orthoporus chihuanus;* the spider *Misumenops coloradensis;* the grasshoppers *Eremiacris virgata, Bootettix argentatus,* and *Trimerotropis pallidipennis;* the neuropteron *Chrysopa carnea;* the beetles *Pachybrachis nigrofasciatus* and *P. xantholucens;* the lantern flies *Hysteropterum sepulchralia* and *H. unum;* the leafhopper *Exitianus obscurinervis;* the heteropterons *Nysius californicus* and *Phytocoris ramosus;* the ants

Dorymyrmex pyramicus and *Novomessor cockerelli;* and the termite *Amitermes perplexus.*

On the north side of the Rio Grande, 68 miles (109 km) below El Paso near McNary, Texas, the outstanding plants are *Larrea, Yucca, Acacia,* and *Fouquieria.* Invertebrates collected in 1944 were the spiders *Misumenops coloradensis* and *Sassacus papenhoei,* the hemipteron *Nysius ericiae,* and homopterons including *Norvellina flavida* and the treehopper *Centrodontus atlas.* The ants were represented by *Pogonomyrmex barbatus* and *Iridomyrmex;* a single creosote grasshopper was *Bootettix argentatus.* Several of these species occur in both desert and mesquite grassland. The white-throated wood rat may also be a desert-edge species (Bailey 1931).

Succulent Desert Faciation

Many small cacti and leaf succulents are present (Borell and Bryant 1942, Muller 1947). Cacti are especially associated with limestone and rocky areas. In northern Coahuila the species *Echinocactus, Echinocereus, Mammillaria,* and *Opuntia* occur; *Agave lecheguilla* and *Jatropha dioica* are semisucculents. Grassland occurs above the desert in the Chinati and Davis mountains (Tinkham 1948).

West of Saltillo, along the highway to Torreo, an area visited in late July, 1944, was dominated by creosote bush, tar bush, *Sericodes greggii,* and mariola. There was a large group of arthropods found. These included the spider *Misumenops dubius,* the orthopterons *Cycloptilum comprehendens fortior, Goniatron planum,* and *Diapheromera,* the beetles *Microrhopala rubrolineata* and *Pachybrachis xantholucens,* the heteropteron *Thyanta cursitor,* the homopteron *Hysteropterum auroreum,* and the ants *Myrmecocystus melliger* and *Crematogaster opaca punctalata.*

Orthopterons usually found on creosote bushes are *Clematodes larreae* and *Eremopedes scudderi;* on yuccas, *Rehnia cerberus* and *Diapheromera covilleae;* and on the ground, *Trimerotropis p. pallidipennis* and *T. texana* (Tinkham, personal communication).

Farther south, the vegetation at the eastern edges of the desert along the 150 kilometer contact between Nuevo León and San Luis Potosí at an elevation around 2000 meters contains widely spaced creosote bush, sotol, century plant, and ocotillo. There are thickets and scattered shrubs of *Prosopis juliflora, Celtis tala pallida, Mimosa biuncifera, Condalia spathulata,* and *Koeberlinia spinosa* (Muller 1939).

About 20 kilometers west of the Nuevo León border at points near Matehuala in San Luis Potosí, Dalquest (1953) secured Nelson's and lined pocket mice and Ord's, Nelson's, and Merriam's kangaroo rats. A bobcat had two Merriam's kangaroo rats in its stomach. Species of lesser influence were several mice (*Peromyscus maniculatus, P. difficilis, P. melanophrys* and *P. eremicus*).

In the Big Bend area of Texas, characteristic vertebrates include spotted ground squirrel, the pocket gopher *Thomomys umbrinus,* black-tailed jack rabbit, desert cottontail, peccary, lesser nighthawk, coachwhip, and western diamondback rattlesnake (Taylor *et al.* 1944).

Floodplain Development

The principal dominants in gravelly washes are *Acacia greggii, A. constricta, Chilopsis linearis, Condalia lycioides, Forestiera angustifolia, Diospyros texana,* and *Prosopis juliflora*. In washes at higher altitudes occur *Fallugia paradoxa, Juglans rupestris,* and *Ungnadia speciosa* (Taylor *et al.* 1944).

The Rio Grande floodplain and large sand washes and arroyos have as their principal dominants, mesquite, *Baccharis glutinosa,* and burrobush, either mixed or singly. On the banks of the Rio Grande, a strip is occupied by the willows *Salix interior* and other species, the common reed *Phragmites communis,* and occasionally the cottonwood *Populus palmeri. Baccharis glutinosa* is the pioneer shrub on the sandbars. Among animals found chiefly on the floodplain are yellow-faced pocket gopher, Ord's kangaroo rat, hispid cotton rat, black-crowned night heron, common ground dove, and checkered garter snake (Taylor *et al.* 1944).

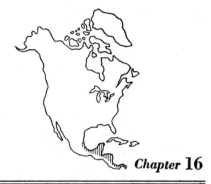

Chapter 16

The Tropical Rain Forest

Tropical biotic communities do not limit themselves to a zone bounded by 23 1/2° Lat. on each side of the equator (Figs. 16-1, 16-2, 16-3); tropical deciduous forest with coatis occurs north beyond 28° N. Lat. in Sonora and in Tamaulipas (Fig. 14-1). Thorn forest extends nearly to 25° N. Lat. (Gentry 1942, Leopold 1950b). On the other hand, grassland dominated largely by the grama grasses extends into Jalisco on the Mexican Plateau below 22° N. Lat. (Shreve 1942). The badger, a typical grassland animal, has been reported near Mexico City.

Snow and ice occur on the high mountains. Below the snow, there are meadows with many flowers. Elfinwood is poorly developed or absent; pine forests, except for occasional epiphytes, simulate the ponderosa or lodgepole pine forests of the north. Oak woodlands resemble those of California. In the *tierras calientes,* the rain forest and gallery forest are evergreen, but in locations with less rain the trees become almost completely bare in the dry season. This deciduous forest merges into thorn forest where there is still less rainfall or into tree or bush savanna or even into scrub vegetation (Fig. 16-4). We are here concerned chiefly with the rain forest.

Rainfall is usually less along the eastern coast than a short distance inland. Annual rainfall in the tropical rain forest as a whole may be as little as 120 or may exceed 350 centimeters (Table 16-1). It is usually lowest in February and March, March and April, or, rarely, January and February. The precipitation may be as little as two centimeters for the drier months. The greatest rainfall in most places comes in August and September, but it may be delayed until October and November. The highest rainfall is near the south coast of eastern Mexico and Guatemala, but the dry seasons here are less favorable than elsewhere.

395

The soils of the rain forest vary from sand or sandy loam to clays, but are apt to be heavy. In the flatter and lower lying sections, large portions of them are poorly aerated, underlaid with an impervious hardpan, and sodden or even miry. Such soils are usually acid. Although they are commonly considered to be rich, many of them actually are quite infertile (Barbour 1941).

Agriculture was the primary means of existence of the natives before the Spanish conquest of Mexico and Central America, except for part of Nayarit in west-central Mexico, the Isthmus of Panama, and the Atlantic side of Nicaragua and Honduras where it was relatively unimportant (Kroeber 1939, map 17). Tools were very primitive and vegetation was often removed by fire. At the time of Spanish settlement there were some great native empires, such as that of the Aztecs, but the Mayan empires of Yucatan and Palenque were in decline. It is difficult to estimate the extent to which pristine communities have been destroyed or the amount which they have been modified since these early empires were destroyed. The introduction of tools and weapons, the

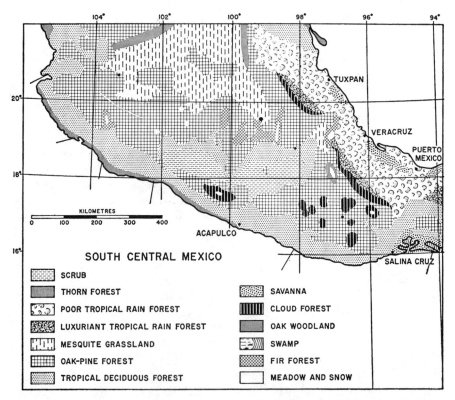

Fig. 16-1. Diagram of the biotic communities of middle Mexico. The boundaries of states are marked at the coast by projecting lines. A few of the well-known cities are marked by dots, the largest dot being Mexico City. Two kinds of swamp are recognized: freshwater and mangrove. The latter is indicated by curved vertical lines.

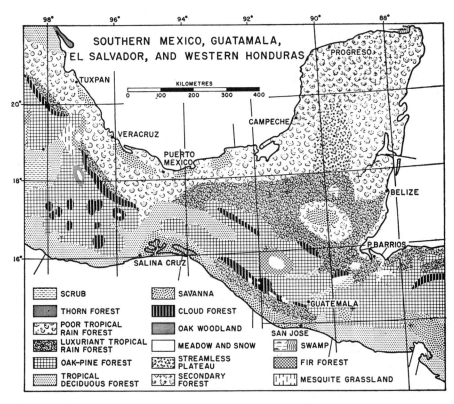

Fig. 16-2. Diagram of the biotic communities of southern Mexico and adjacent areas. The boundaries of the countries and states of Mexico are marked at the coast by projecting lines. Important boundary line corner and junction points are marked by +. Two kinds of swamp are recognized: freshwater and mangrove. The latter is indicated by curved vertical lines.

horse and burro for transportation, and the cow for food must have given great impetus to the destruction of natural communities, but apparently this was nowhere as extensive as what took place north of the Rio Grande (Cook 1909).

The rain forest probably least disturbed in A.D. 1500 was on the Atlantic side of Nicaragua and Honduras, in the area occupied by "uncivilized" tribes which did not build cities or subsist wholly by agriculture (Kroeber 1939). This area has more recently been considerably exploited (Carr 1940) for mahogany and locally cleared for bananas but still is probably most likely to hold pristine rain forest. Most of the Mayan civilization was in the Yucatan Peninsula, westward into Tabasco and Chiapas, and southward into Guatemala and British Honduras. Some 400 to 500 years have elapsed since this civilization disappeared, but the vegetation is still well marked by their agriculture of centuries ago.

The moist tropical forest has been called *selvas pluviales* by the natives, tropical evergreen forest, tropical rain forest, evergreen rain forest, broadleaved evergreen forest, and wet forest (Barbour 1941). Tropical rain forest

Table 16-1. Rainfall and Temperature at Various Localities in Central America and in States of Mexico

LOCALITY	ALTITUDE IN METERS	RAINFALL IN CENTIMETERS		MEAN ANNUAL TEMPERATURE	REFERENCE
		MEAN ANNUAL	DRIEST 2 MONTHS		
NORTH AND EAST COAST					
Colón, Panama.............	5	317	8	27° C.	Clayton 1927
La Ceiba, Honduras.........	2	316	8	..	Wise and Zacarias 1958
Lancetilla, Honduras.......	40	317	11	..	"
Payo Obispo, Quintana Roo...	4	125	5	26	Government
Valladolid, Yucatan.........	22	114	6	25	"
Campeche (Ciudad), Campeche...	25	89	2	26	"
Jonuta, Tabasco...........	14	170	6	..	"
Tlacotalpan, Veracruz.......	38	108	2	27	"
Puerto México, Veracruz.....	14	288	10	25	"
INLAND STATIONS					
Barro Colorado Island, Canal Zone...	100	225	8	27	Laboratory
Teapa, Tabasco.............	80	396	31	25	Government
Palenque, Chiapas..........	210	318	14	26	"
La Providencia, Chiapas.....	785	122	1	23	"
Cordova, Veracruz..........	871	227	10	20	"
Xilitla, San Luis Potosí.....	1035	254	13	21	"
SOUTH AND WEST COAST					
La Aurora, Chiapas.........	245	443	6	27	"
Estrella, Chiapas..........	625	334	5	..	"
Santa Cecilia, Guatemala.....	350	483	7	23	McBryde 1945
Tapachula, Chiapas.........	137	249	1	26	Government

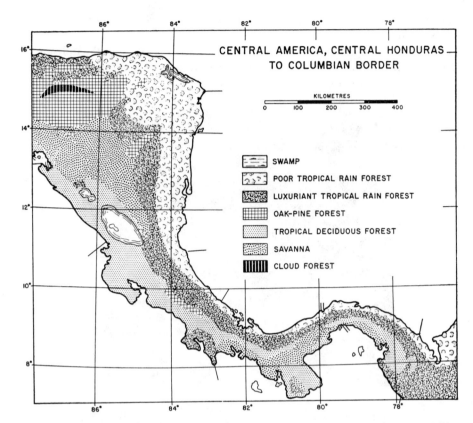

Fig. 16-3. Diagram of the biotic communities from Honduras to Colombia. The boundaries of the countries are marked at the coast by projecting lines. The Panama Canal is marked by a double line.

is the generally accepted term but is often applied erroneously or refers to several types of forest (Smith and Johnston 1945).

In addition to the mature tropical rain forest, an area may include hydroseral and xeroseral stages, some of which may still be pristine because they can not be used for agriculture. Hydroseral stages are likely to support evergreen vegetation. Miranda (1952) recognizes savannas containing evergreen gallery forest. Certain xeric areas appear unable to support forest (Fig. 16-4). Some secondary forests (Fig. 16-5) may be evergreen but they will not have the two or more tree layers typical of a rain forest. One should thus distinguish between evergreen forest (Leopold 1950b) or poor rain forest and the typical luxuriant rain forest.

Depending on the amount of leaf fall, one should also differentiate between tropical evergreen and tropical deciduous forest. The forests on Trinidad were separated by Beard (1946) on the percentage of trees without leaves during the driest part of the year: evergreen or rain forest, dominants about 6 per cent, understory negligible; semi-evergreen forest, dominants 16.6 per cent, understory 10 per cent; deciduous forest, dominants 64 per cent, understory 25 per cent.

In the tropical rain forest, ample rainfall occurs every month and only a few scattered trees are without leaves at any time. Flowering and fruiting of plants, as well as the reproductive activities of animals, are distributed through-

Fig. 16-4. A. Bush savanna seen on approaching the coastal plain west of Veracruz, east-northeast of Huatusco (Cia. Mexicana de Aviacon 1940), generally mapped as poor, tropical rain forest. There is apparently forest in the ravines and trees, bushes, and grass on the upland; the exact character of the vegetation appears not to have been studied. B. Close view in a scrubby evergreen forest near Papantla, Mexico, which is halfway between Tampico and Veracruz, north of the area of A (photo by C. J. Goodnight).

out the year. Light on the forest floor is dim and diffuse and varies but little in intensity from season to season. A tall overstory of scattered dominant trees 20 to 40 meters high is an outstanding feature (Schimper 1903, Richards 1952, Miranda 1952). Subordinate to these, are one, two, or more stories of trees 14 to 24 meters high, and large shrubs 3 to 6 meters high (Fig. 16-6). All these strata are heavily burdened with thick-stemmed lianas and epiphytes. The dominant trees have long, clean boles which extend one-half to two-thirds of the total height. The crowns are relatively restricted in size and tend to be spherical. Buttressing may be well or poorly developed. Stilt roots and pneumatophores are often evident but scattered in occurrence; thorns and spines

Fig. 16-5. A. Evergreen forest along the Pan-American Highway near Valles. B. *Sabal* on floodplain. C. An area 35 kilometers south of Cordova, Veracruz, that approaches rain forest. Note the three stories of trees. The tall individual has its trunk paralleled by three multiple strands of lianas. D. Valles River in flood showing a stand of *Sabal* on the floodplain.

are only rarely present. Palms may or may not be abundant; often they are scattered, immature, stemless plants. Compound leaves are often present in the upper strata but simple leaves predominate in the lower ones. The black howler monkey and a species of strangler fig are generally present (Schimper 1903, Barbour 1941, Richards 1952).

The greater part of the rain forest is on the eastern and northern coasts of Central America and southern Mexico and is generally most luxuriant away from the coast at altitudes of 100 to 200 meters above sea level. There are small areas of rain forest on the Pacific side near Gulfo Dulce in Costa Rica, perhaps northwest of Lake Managua in Nicaragua (Allen 1955), in Guatemala, and in Chiapas, Mexico.

The rain forest varies significantly in composition so that a number of regions or faciations may be distinguished.

(1)   The Panamanian region is characterized by the imitator capuchin monkey, the large trees being *Virola panamensis* and *V. warburgii*.

(2)   The Costa Rican region is characterized by the trees *Luehea seemanii*

Fig. 16-6. Profile of the *Orbignya–Dialium–Virola* association in British Honduras (after N. S. Stevenson 1942): 1, ironwood *Dialium guianense;* 2, banak *Virola koschnyi;* 3, cohune *Orbignya cohune;* 4, white copal; 5, wild coffee *Rinoria;* 6, waika plum *Rheedia edulis;* 7, mammee circuela *Lucuma;* 8, negrito *Simaruba glauca;* 9, red copal *Cupania;* 10, mountain trumpet *Pourouma aspera;* 11, cacao *Theobroma cacao;* 12, timbersweet Launaceae; 13, maya *Miconia;* 14, cacho vennado *Eugenia capuli;* 15, mata palo *Ficus;* 16, monkey tail *Chamaedorea;* 17, hone *Bactris;* 18, tree fern *Alsophila;* 19, pacuca *Geonoma;* 20, kerosene wood.

and *Pentaclethra*. Different subspecies of the spider monkey occur in this region and regions 4, 5, and 8 described below.

(3)   Important tall trees in the rain forest of the Gulfo Dulce region are *Tabebuia pentaphylla* and *Anacardium excelsum*. Probably the Costa Rican black squirrel is a characteristic animal.

(4)   The Caribbean lowland and eastern slope region in Nicaragua and northeastern Honduras is largely unexplored biologically (Griscom 1926). Perhaps some of the vegetation is second-growth (Carr 1950), although this is one of the areas where agriculture was not extensively practiced by the early natives (Kroeber 1939). Belt (1874) mentions the "cortess," probably *Tabebuia chrysantha,* as a tree overtopping the general forest canopy in the highlands of the Caribbean slope, but it and others noted are widely distributed. He also mentions *Marcgravia nepenthoides* which climbs to the tops of the tallest trees; it has a remarkable flower and appears to be characteristic of the high altitude rain forest (Standley 1931).

(5)   In the valleys between the mountains that rise along the shore of Honduras occur the best areas of rain forest, such as the one in the Lancetilla Valley (Standley 1931). The abundant trees are *Brosimum terrabanum* and *B. costaricanum*.

(6)   The characteristic trees of the evergreen forest on the Pacific slope of Guatemala and Chiapas, Mexico, are *Erblichia xylocarpa mollis, Sloanea ampla,* and *Dussia cuscatlanica* (Miranda 1952, Standley and Steyermark 1945).

(7)   The east slope rain forest in Alta Verapaz, Guatemala, contains *Vochysia hondurensis* and *Engelhardtia guatemalensis* as important trees (Standley and Steyermark 1945).

(8)   In the Yucatan Peninsula, the zapote tree *Colocarpum sapota* appears in climaxes throughout the area.

(9)   In the Veracruz and Tabasco region (Goldman and Moore 1946), the Mexican rubber tree *Castilla elastica* is probably abundant in the wetter parts. *Pachira aquatica* is a flourishing river-margin tree. There are few important trees in this area which do not also extend into Yucatan Peninsula or to the Pacific Coast. The great tinamou is a characteristic bird of the region; the little tinamou is in the lowland forest.

A HONDURAN VALLEY FOREST

Standley (1931) describes in a picturesque manner, here abbreviated, a forest in the Lancetilla Valley, 16 kilometers west of La Ceiba, Honduras. This area has an annual rainfall of 234 to 274 centimeters.

There is stillness; great blue butterflies float silently about. Flocks of parakeets chatter as they feed on the fruit of a high fig tree, or macaws or parrots scream as they alight on the branches. On the ground are turkeys and curassows. Wood-quail frequent the forest, but they are seldom seen. Rabbit-like agoutis nibble at palm seeds; suddenly a band of white-faced monkeys races across the tree tops, leaping recklessly from branch to branch. Just before a heavy rain, the air reverberates with the roaring of the howler monkeys. These animals live here in large numbers, but they are shy and very rarely seen. Occasionally a band of peccaries may be scented or heard racing away through the bushes. There are tapirs and probably an occasional tiger or jaguar. There are snakes; noteworthy is the deadly barba amarilla or fer-de-lance. However, in months spent walking through the forest, only a very few harmless snakes and one or two coral snakes are likely to be encountered.

The trunks of the trees, especially those of the fig and ceibas, assume massive proportions, and their enormous weight is supported by radiating, bracket-like buttresses that brace them against the wind, which is felt only by their tops. When a wind storm passes over, it is barely perceptible below on the ground, but a heavy branch overweighted with epiphytes may be torn loose by the force of the storm and come crashing down. Immediately after a heavy storm is a favorable time for collecting specimens of trees and epiphytes, for many such branches may be found lying fresh upon the ground.

The tree trunks are of somber color. Some of the trunks are smooth and quite free of epiphytes; others are rough and heavily laden with them. Some of the epiphytes, especially the aroids, cling tightly to the trunks, but more of them clasp the loftiest branches. Most of these trees, as well as their epiphytes, have small and inconspicuous flowers, but even when the flowers are showy, they can rarely be discerned from the ground, because they usually appear on the very outside of the foliage in the full sunlight.

A conspicuous constituent of this section of the forest consists of the lianas, great woody vines similar to the largest grapevines of northern forests. These vines commonly do not embrace the tree closely but dangle loosely about its trunk, several individuals finding support upon a single tree. A frequent plant of this habit in the forest above Lancetilla is *Marcgravia nepenthoides,* here called cachimba (tobacco pipe) in allusion to the curious form of the flowers, or rather the nectaries. The flowers bloom at the top of the forest and may be seen only after they have withered and fallen. The fresh flowers and brightly colored flowers of many other plants attract the hummingbirds. Some of the lianas and epiphytes develop long cordlike aerial roots that hang from the branches.

The understory consists of smaller trees that cannot exist in full sunlight. Regardless of the large number of species, certain trees occur in much greater abundance than others. A very large proportion of this high Honduran forest consists of *Brosimum terrabanum, B. costaricanum, Virola, Guarea excelsa, Calocarpum sapota, C. viride,* and *Dialium.* Other trees less common but noted more or less frequently are *Naucleopsis, Pourouma, Zollernia, Ormosia,* which has bright red, beanlike seeds, *Simaruba, Protium, Tetragastris, Cedrela, Vochysia,* which has yellow blossoms, *Astronium, Ceiba,* wild figs in great variety, *Calophyllum, Rheedia,* and *Lucuma izabalensis.* A rare tree is *Aspidosperma,* with broadly-winged seeds which sail like butterflies through the air. Occasionally there appears the smooth, copper-colored trunk of a giant *Bursera simaruba,* called indio desnudo (naked Indian).

At Lancetilla the understory is composed largely of palms. The most abundant and conspicuous is the cohune or corozo *Orbignya cohune* which attains its best development and is most plentiful on the lower hill slopes. They often stand closely together and shut out most of the light filtered through the trees above. Their huge leaves, frequently nine meters long, wither after falling to the ground and make a thick mulch over it. Other shorter palms grow with the cohune. The lancetilla *Astrocaryum mexicanum,* with its offensively armed stems, which has given the name to the valley, is noteworthy. There are also *Bactris* species, two Geonomas, and several species of *Chamaedorea.* One of the neatest of the palms found in such situations is *Reinhardtia gracilis.* It attracts attention because of its airy habit and especially on account of its cross-shaped leaves with rows of perforations or "windows" close to the midrib. A palm of less admirable characteristics is the balaire *Desmoncus.* It is a clambering vine and possesses pinnate leaves with thorny tips.

The forest understory includes a fair variety of trees other than palms. There is the yucca-like *Dracaena* with long grasslike leaves. *Guatteria amplifolia* is scattered in the forest, and *Mollinedia butleriana* is common. *Quararibea fieldii* is a frequent tree, noteworthy for its odor of slippery elm. It is interesting to find cacao here, apparently quite wild, in association with wild zapotes and avocados. There is the tree violet *Rinorea guatemalensis,* and there are small trees of *Xylosma sylvicola,* their trunks armed with branched thorns. *Carpotroche* attracts attention by its brightly colored, ball-like, winged fruits, set closely against the trunk. Another prominent tree is *Pentagonia,* with its ungainly branches and bunches of huge leaves, in contrast with the bright red flowers clustered at their bases.

There is likewise a good variety of inconspicuous shrubs; *Heisteria* is rather handsome because of the round, saucer-shaped, red calyx in which sits a contrasting black drupe. Several melastomes are grown in the deep shade, notably *Tococa grandifolia,* with large leaves of a particularly fine tint of green and small but fine flowers. The herbaceous plants of the heavy forest are characterized more for the number of individuals than for variety of species. Certain to receive notice is the bright *Aphelandra aurantiaca,* with green-bracketed spikes of fiery red flowers. *Begonia popenoei* grows on moist banks. Two broad-leaved grasses *Pharus glaber* and *Streptochaeta* are common. In low places in the forest, especially along streams, there are often great clumps of certain plants that have an important part in giving the undergrowth of the tropical forest its distinctive aspect. These are the heliconias, relatives of the banana, with canna-like leaves and brightly colored red, orange, and yellow inflorescences with long, narrow, spreading bracts.

Under the densest of the cohunes, where the ground is covered with rotting leaves, there are very few herbaceous plants. It is here that one may stumble upon several small and very inconspicuous saprophytic plants. The most plentiful is *Sciaphila,* which may be found almost anywhere if one looks closely. The wiry, reddish stems and the minute flowers blend so thoroughly with the background that the plants are scarcely discernible when one is standing.

In the hill forest the cohune palms form a belt of vegetation which has a rather definite limit. The cohunes end abruptly at an elevation of about 200 meters, and the forest above them assumes a slightly different aspect. Just below the uppermost cohunes there are small areas of silk-grass. Above the dense cohune belt the forest is much more open. The trees are generally of the same species, but they average considerably shorter in height. There is a much more conspicuous understory of dicotyledonous trees and a much greater variety of shrubs. *Trophis chorizantha* is a frequent small tree at this elevation, and another is *Lunania piperoides* with a whitish inflorescence. Slender shrubby pipers are frequent in these woods, and various bushy Rubiaceae, all much alike in appearance but remarkably diverse when inspected critically, also occur. The low, almost herbaceous *Psychotria uliginosa*, with large fleshy leaves, pale underneath, and stalked clusters of bright red berries, is present; also there is the shrub or slender tree *Myginda eucymosa*, with plumlike, cherry red fruits. The shrubs *Guarea bijuga* and *Siparuna tonduziana* are scattered.

The debris deposited at the upper limit of high tide along the coast near Tela, Honduras, contains various seeds. Palm seeds predominate; among them are representatives of certain species that do not grow nearby. There are quantities of *Pterocarpus* seeds, stones of *Spondias,* and usually the sea beans, the seeds of *Mucuna, Caesalpinia,* and *Entada.* A good many acorns may be found, and their source is puzzling for no oak trees grow close to the coast in this part of Central America. Probably these were brought from far inland by some river.

THE FOREST ON BARRO COLORADO ISLAND, PANAMA CANAL ZONE

Barro Colorado Island was formed by the damming of the Chágres River when the Panama Canal was built in 1914. It has an area of about six square miles (15 km²) (Table 16-2). The forest covering Barro Colorado Island is only partly virgin. There is a great abundance of small forest palms and tree ferns. The presence of these plants usually is proof that the area has not been under cultivation, at least for a very long time (Standley 1933).

A large number of species of trees, shrubs, herbs, ferns, as well as vertebrate and invertebrate animals of relatively large size characterize the area. However, few of the tree species listed by Standley above for the forest in Honduras are reported in the Canal Zone.

The tall tree dominants stand in the forest in an orchard-like arrangement above the shorter trees. This arrangement of overstory trees helps to distinguish the virgin from the second-growth vegetation (Enders 1935). Individuals of a given tree species are well separated one from another. It is difficult to designate any particular species as the main dominants. There are at least six times as many tree species as are found in the beech–maple–hemlock forests of northeastern United States (Kenoyer 1929). Noteworthy trees are *Iriartea*

Table 16-2. *Physical Conditions in the Barro Colorado Forest*[a]

	Relative Light Intensity	Wind Velocity, Kilometers per Day	Maximum Temperature Variation in a Month	Water Evaporated, Cubic Centimeters per Day
The main forest canopy.......	25	16	15.4° C.	2.0
Lower tree tops..............	6	....	9.5°	2.0
Small trees.................	5	....	....	...
Higher shrubs...............	..	1.6	....	1.9
Forest floor and low herbs.....	1	....	5.9°	...
Subterranean...............	0	0	Slight	Slight

[a] Above the tree tops the light intensity was 10,500 footcandles and the wind velocity 387 kilometers per day (Allee 1926)

exorrhiza, Ficus glabrata, Virola panamensis, Platypodium maxonianum, Zanthoxylum panamense, Hura crepitans, Spondias mombin, Bombacopsis fendleri, Sterculia apetala, Symphonia globulifera, Grias fendleri, Terminalia amazonia, Jacaranda copaia, Tabebuia pentaphylla, and *Tabebuia guayacan.* Perhaps the most remarkable tree in this list for size and tenacity of life is *Bombacopsis fendleri.* It comes in during early seral stages and persists well into the climax. The leaves fall in January and the flowers appear a few weeks later.

Buttresses (Fig. 16-7A, 16-9B) and stilt roots are prominent in this forest. Buttresses are found on *Bombacopsis, Ficus, Hieronyma alchorneoides, Hura crepitans,* and others; stilt roots on *Cecropia* and the stilt palm *Iriartea exorrhiza.* In general these features are observable only on large or tall-growing trees. Smooth bark is very prevalent; only a few trees, such as *Spondias mombin* and *Grias fendleri,* have the shaggy bark characteristic of temperate zone species. Thorny trunks characterize *Bombacopsis fendleri, Zanthoxylum, Hura crepitans, Erythrina,* and several of the palms. Cauliflory, or the production of flowers on the old wood, is exceptional as it occurs only in *Theobroma purpureum* (Kenoyer 1929).

The majority of the trees are evergreen. Some of them, as *Coumarouna panamensis, Tabebuia,* and *Hura crepitans,* may become leafless or nearly so during the dry season. *Cordia alliodora* may become leafless during July and August. A large tree of *Ficus involuta* began shedding its leaves August 17 and was completely bare five days later. After five days more the ground was littered with stipular bud-scales, and a new crop of leaves was out (Kenoyer 1929). Richards (1952) states that the length of life of leaves of tropical species is 13 to 14 months. Only a few renew their leaves at a constant rate throughout the year. Some trees shed their leaves at regular intervals, others very irregularly.

About 16 per cent of all plant species are lianas. A great variety of huge woody stems, sometimes reaching 15 centimeters in diameter, run up along-

Fig. 16-7. A. Base of a large tree, *Bombacopsis fendleri*, 46 meters in height, and with a crown diameter of 61 meters. The surrounding forest is second-growth (Kenoyer 1929). B. Spider monkey on a small island near the laboratory on Barro Colorado Island, Panama Canal Zone. C. Giant anteater at the edge of the laboratory clearing. D. Keel-billed toucan (from *Life in an Air Castle* by Frank J. Chapman, reproduced by permission of Appleton-Century-Crofts, Incorporated). E. Capuchin. F. Undergrowth in the forest on Barro Colorado Island.

side the trunks of the trees or festoon the branches. Vascular epiphytes constitute more than 10 per cent of the Barro Colorado plants; there are epiphytic species of ferns, 21; aroids, 20; orchids, 14; bromeliads, 8; *Peperomias,* 5; cacti, 2; gesneriads, 2; and figs, 1. Limbs of the big trees are commonly very heavily burdened. A single tree that had fallen with its load was found to have supported one hemi-epiphyte, six species of lianas, eight other species of vascular plants, and a great number of mosses, lichens, and liverworts (Kenoyer 1929). The strangler fig *Ficus* begins as an epiphyte, then sends a network of stems to take root in the ground. The stems anastomose and exert pressure around the trunk of the tree and may kill it, leaving the fig alone after the tree has decayed and disappeared.

Shrubs and herbs are relatively scarce on the dimly lighted forest floor. Various species of *Psychotria* and other Rubiaceae are the most prominent shrubs. Herbs include several species of ferns, *Selaginella,* and a number of monocotyledonous plants. One vascular parasite, a species of *Apodanthes,* grows within the trunk and branches of *Xylosma hemsleyana* and is apparent only when the clusters of white waxy flowers, seven millimeters long, erupt through the bark of the host.

Fungi are abundant on decaying wood and humus. *Peziza*-like forms are especially conspicuous. Flowering saprophytes occur. Other saprophytes are two Gentianaceae, *Leiphaimos alba* and *L. simplex,* and the rare orchid *Triphora cubensis* (Kenoyer 1929).

RAIN FOREST OF PETEN, GUATEMALA

The best rain forest in this region has in its uppermost tree layer at a height of 20 to 25 meters *Calophyllum brasiliense rekoi, Swietenia macrophylla, Rheedia edulis, Lucuma campechiana, Sideroxylon amygdalinum,* and *Ficus.* To the middle tree layer belong *Achras zapota, Vitex gaumeri, Ficus radula, Cecropia mexicana, Bursera simaruba, Spondias mombin, Aspidosperma megalocarpon, Brosimum alicastrum, Pseudolmedia spuria,* and species of Lauraceae and Leguminosae. The lower tree layer is unusually well developed in the somewhat open parts of the forest; many of the trees average eight to ten meters in height. The principal species are *Trichilia minutiflora, Sideroxylon meyeri, Sebastiania longicuspis, Parmentiera edulis, Lucuma durlandii, Laetia thamnia, Sabal, Protium copal, Ocotea lundellii* and *Zanthoxylum* (Lundell 1934, 1937).

Lianas abound; the Bignoniaceae appear to be widely represented. Shrubs are *Ruellia stemonacanthoides, Odontonema callistachyum, Psychotria flava, P. fruticetorum,* and *Beloperone aurea.* Palms, chiefly *Cryosophila argentea* and *Chamaedorea,* are often present in considerable numbers. Among the herbaceous plants, an occasional patch of the grass *Pharus parvifolius* occurs, and on the damp shaded forest floor, the fern *Dryopteris subtetragona* may be common.

The root-climbing aroids *Monstera* and *Philodendron smithii,* and the non-climbing *Anthurium tetragonum yucatanense* cover the trunks of many trees. The small epiphytic *Peperomia polochicana* and *P. glutinosa* grow on branches in the thin mantle of humus and moss.

In each large region there appear to be several groupings or faciations of dominant tree species, yet relatively few species are common to all parts of the region. The forest, just described, is called the zapotal faciation by Lundell (1937), even though zapote is in the subordinate middle layer.

The ramonal faciation, named by Bartlett (1935), is for the ramon *Brosimum alicastrum,* which also occurs in the middle tree layer. The vegetative parts of this species are used by deer. Lundell considered this faciation as one of the climaxes, although its occurrence in groves is probably the result of Mayan or other human manipulation. The height of the trees is 15 to 20 meters. The next lower tree layer includes the zapote, *Talisia olivaeformis,* and other species. The layer below this one contains *Celtis trinervia.* The shrub layer has species of *Piper, Guatteria,* and others. Herbaceous growth, woody vines, and tree trunk orchids are present. A third important faciation of this forest is the caobal, after the caoba *Swietenia macrophylla* listed above (Lundell 1937).

RAIN FOREST OF CHIAPAS, MEXICO

The rain forest near Pichucalco, at an elevation of 107 meters, is characterized by *Dialium guianense,* which grows to a height of 40 meters, *Sterculia mexicana, Gilbertia arborea, Poulsenia armata, Alchornea latifolia,* and *Cassia doylei.* The lower vegetation includes *Carludovicia palmata, Chamaedorea tepejilote, Geonoma magnifica,* and *Castilla elastica* (Miranda 1952).

Another forest type near the Palenque ruins at 210 meters is characterized by *Pithecellobium leucocalyx, Sweetia panamensis, Quercus skinneri,* and *Carapa guinanensis.* This upper tree layer is at 30 to 40 meters height. The understory is 10 to 25 meters in height and has *Simaruba glauca, Couepia polyandra, Poulsenia armata,* and *Xylopia frutescens.* The shrub layer is 3 to 4 meters in height and is composed of thatch palm *Cryosophila argentea* and small trees (Goodnight and Goodnight 1956).

Regardless of the taxonomy of the constituent plants in the various localities, the vegetation adheres to a characteristic structural plan. There is likewise a similar pattern of social organization among the animal constituents in all parts of the rain forest. The animal constituents appear not to differ taxonomically as sharply as do the plants, variations in different parts of the rain forest being chiefly at the subspecies rather than the species or genus level.

ANIMAL CONSTITUENTS

The number of animal species in the tropical rain forest is large and they occur in all strata. They vary in their adjustments depending on whether they are aerial, arboreal, terrestrial, or subterranean.

Influents of the Two Tree Top Strata

There are only about ten species of mammals about whose influence in the community is known, even in a slight way. The larger animals do not distinguish in their activities between the scattered large overstory trees and the subordinate trees of the next lower stratum. A behavior or mores pattern, representative of the canopy mammals, is that of the black howler monkeys. The monkeys go around in groups ranging from 4 to 35 individuals. Their density varies from about one per 4 to one per 2.6 hectares. However, the groups may travel over 120 hectares. They move through the trees and are remarkable for their climbing and acrobatic feats.

They feed upon foliage and fruit and spend about one-quarter of the daylight time in feeding. Carpenter (1934) lists 55 species of trees and vines that the species utilizes. These include nine of the trees above listed as dominant or important. The very large fruit of *Ficus glabrata* is most important. Like the fruit of other species, it is dry, sweet, and contains many seeds. It often also contains insect larvae which provide protein. The leaves (to some extent) and the inner hull or aril of the nuts of *Virola panamensis* are eaten. *Prioria copaifera* is a tall forest tree from which the howlers eat the leaves and the unique seed pods. *Platypodium maxonianum,* another tall tree, provides leaves, flowers, and fruit. *Zanthoxylum panamense* furnishes leaves and buds, and *Spondias mombin,* leaves. *Bombacopsis fendleri* has its buds, flowers, and leaves eaten, and *Tabebuia guayacan,* its flowers and possibly the leaves. *Tabebuia pentaphylla* is utilized by howlers for its flowers and leaf buds. The impact of this destruction of seeds and leaves is unevaluated (Carpenter 1934, Belt 1874).

The spider monkey occurs throughout the communities under consideration (Fig. 16-7). They generally move about in groups of 12 to 15 (Goldman 1920). Their climbing ability is extraordinary. They will chatter like a flock of shrill-voiced chickens at an animal on the ground and will break off branches of trees and let them fall on observers (Belt 1874, Dalquest 1953). Their food is largely fruit, but some insects and small vertebrates may be taken. There are three other species of monkeys in the Panamanian rain forest which do not reach north as far as Mexico: the insect- and vertebrate-eating capuchin (Fig. 16-7), the squirrel marmoset or titi monkey which feeds on seeds, fruits, and flowers, and the tree-cavity night monkey or douroucouli.

A prehensile-tailed porcupine, the puerco pinna, obtained in Panama, had its stomach distended with vegetable matter, a greenish part apparently leaves and a white mass which had the appearance of fruit pulp (Goldman 1920). In San Luis Potosí, a related species is nocturnal and sluggish, commonly remains in the same tree for a long time, and feeds on leaves and fruit. The kinkajou is frugivorous, strictly arboreal, and moves slowly through the trees, usually in groups or pairs. Its food includes star-apple, almendros, guavo, avocado, *Persea americana,* and espave beans *Anacardium.* It ranges through the evergreen forest into San Luis Potosí, keeping the same habits even though the species of trees change (Dalquest 1953, Enders 1935, Goldman 1920).

At least 15 species of bats occur on Barro Colorado Island (Enders 1935). Of these, 7 belong to the canopy. The Brazilian long-nosed bat lives in tree cavities; the greater white-lined bat hangs on the bark of trees, while the fish-eating bat fishes at night. A spear-nosed bat occurs in the trees. Gervais' fruit-eating bat cuts leaves for shelter and collects espave beans. The big fruit-eating bat is restricted to the tropical rain forest (Dalquest 1953). Seventeen bats occur in tropical San Luis Potosí, most of them are in the moist areas

In 1938, about 110 species of forest birds were known to occur regularly on Barro Colorado Island. The toucans are medium-sized fruit-eating birds. The keel-billed toucan (Fig. 16-7) is abundant in the tall forest, occurring in flocks as large as 22 individuals (Van Tyne 1935). They appear to prefer the forest margins when not breeding. Fruit-bearing trees important to this toucan are *Malmea depressa,* which is sometimes in the climax, and *Exothea paniculata* and *Trichilia moschata* which are evidently seral. The chestnut-mandibled toucan is absent north of Honduras, but the collared aracari extends north into Veracruz.

There are four species of trogons on Barro Colorado Island; the common ones are the slaty-tailed trogon and violaceous trogon. They seize fruits while on the wing by snatching or grabbing them and falling back to allow their body weight to break them off. They also feed on seeds and large insects. A nest of the slaty-tailed trogon was found excavated in an abandoned termite nest six meters above the ground on the black palm *Astrocaryum standleyanum.*

The birds in the upper layers of the forest at the Palenque ruins in Chiapas are king vulture, black vulture, roadside hawk, white hawk, white-crowned parrot, black and white owl, slaty-tailed trogon, keel-billed toucan, collared aracari, pale-billed woodpecker, and others. Species and subspecies are different from those in the Canal Zone, but in general their habits are similar (Goodnight and Goodnight 1956).

Throughout the rain forest there is a group of birds in the high layers which possesses a degree of ecological equivalence from place to place. For instance, frequenting the highest trees on Barro Colorado Island are blue-headed parrot, crimson-crested woodpecker, lesser kiskadee, and many others. In the Palenque forest of Chiapas, white-crowned parrot, pale-billed woodpecker, great kiskadee, and cinnamon becard have roughly similar habits.

Invertebrates also show segregation to different strata in their activities. The most abundant ant in the upper canopy is *Azteca trigona.* It has large carton nests which are meter-long inverted cones that hang like a paper stalactite from large limbs 24 meters or more above the ground. The tick *Amblyomma,* abundant near the ground, is markedly absent from the canopy. Carton-like nests of the termite *Nasutitermes* occur at all levels in the trees, but four species occur at this upper level (Allee 1926).

Lower Tree, Tree Trunk, and Shrub-Herb Strata

Animals, carrying out most of their activities in the shrub layer at three or four meters, may, however, go up to 10 or 15 meters in the trees. The rufous-breasted hermit and other hummingbirds frequent the shrub level. In Nicaragua, flycatchers, tanagers, woodcreepers, and woodpeckers traverse the forests in flocks (except when breeding) at lower levels where observation is possible (Belt 1874). On Barro Colorado the long-billed starthroat, a hummingbird, cross-pollinates flowers, the short-billed jacobin hummingbird catches flying insects, and the purple-crowned hummingbird plucks spiders from the undersides of leaves. In northern Peten, Guatemala, in 1931, the widely-distributed hook-billed kite was found feeding on snails of the genus *Pomacea*, a swallow-tailed kite had eaten beetles and lizards, a spectacled owl had eaten a small opossum, and a mottled wood-owl had eaten only insects (Van Tyne 1935).

The large nocturnal gecko *Thecadactylus rapicaudus* (Fig. 16-10B) is widespread in tropical America and has adhesive well-developed pads for climbing into trees (Park 1938). Another widely distributed arboreal lizard is the giant *Iguana iguana*. Both occur on Barro Colorado Island. In Peten, the same tree gecko and five species of anoles occur. All are insectivorous. The rough-coated *Sceloporus serrifer* also frequents the trees (Stuart 1935) and the tree frog *Centrolene fleischmanni* occurs on foliage in ravines (Park 1938).

Animals Using Both Trees and Ground

The abundant coati on Barro Colorado Island eats all the fruits, nuts, insects, and rodents common on the island. It also occurs in small numbers in the rain forest of northern Peten and San Luis Potosí. The ocelot (Fig. 16-8) makes use of low trees but hunts on the ground where it takes spiny rats and agoutis for the bulk of its food. Its use also of pacas, snakes, lizards, and occasionally peccaries and deer is well known. There is some social organization among individuals (Enders 1939). In San Luis Potosí, it is most abundant in the dense vegetation of the evergreen forest and the forest near the coast. The jaguar is less of a tree animal but uses trees at times. It feeds on much the same animals as the ocelot and is present throughout the evergreen forest (Belt 1873).

The opossum *Didelphis marsupialis* occurs everywhere in the Panamanian forest and feeds on guavo, papaya, figs, star-apple, mammals, birds and their eggs, and insects (Goldman 1920). Boa snakes sometimes swallow opossums. In San Luis Potosí, the species also occurs outside of the tropics.

In the Canal Zone, the brown opossum and four-eyed opossum are omnivorous and nocturnal, sleeping in burrows during the day. In San Luis Potosí, the brown opossum is absent (Enders 1939, Dalquest 1953).

The Canal Zone squirrel *Sciurus variegatoides* has much the same habits as the northern tree squirrels, with a preference for areas near openings in the forest. The red-bellied squirrel takes its place in the more northern evergreen forests. The Canal Zone pygmy squirrels appear to have a strong preference for dense, mature forest and divide their time between the trees and the ground. The little dull squirrel of San Luis Potosí is an ecological equivalent in the north. The three-toed sloth feeds on *Cecropia* leaves. The two-toed sloth is similar in general habits. Both swim well and leave the trees on occasion. The northern limit of sloths is in Nicaragua (Dalquest 1953, Enders 1939, Goldman 1920).

The crested guan or pavon of Panama and Mexico is a fairly large and common bird that shifts back and forth between the high trees and ground of the rain forest. The chachalaca *Ortalis garrula* of Panama also occurs both in the tall trees and on the ground; in Peten and Mexico, its place is taken by *O. vetula*. In northern Peten, the chachalaca was noted to feed on the fruit of *Rivina humilis*. The nests of these birds are near the ground (Van Tyne 1935). The plain xenops behaves like a nuthatch while the wedge-billed woodcreeper climbs tree trunks like a brown creeper. The spotted woodcreeper and the red-billed scythebill are also birds of the tree trunks in Panama. The latter uses its long curved bill to probe for insects. The northern royal-flycatcher frequents the lower branches and undergrowth of heavy forests.

Fig. 16-8. Night photographs on Barro Colorado Island (by F. M. Chapman); flashes were set off by the animals. A. Ocelot (courtesy the American Museum of Natural History and the *National Geographic Magazine*). B. Baird's tapir. C. White-lipped peccary (B and C from *Life in an Air Castle* by Frank J. Chapman, reproduced by permission of Appleton-Century-Crofts, Incorporated).

Of the smaller birds on Barro Colorado Island the black-faced ant-thrush spends much of its time close to the ground. The slaty antshrike is common. The white-bellied antbirds have been taken with their stomachs filled with ants, but they eat chiefly the prey of the foraging army ants (Johnson 1954). The black-faced ant-thrush and the ravine-frequenting spotted antbird are noteworthy (Sturgis 1928). Antbirds are common and several species occur together in roving flocks in the rain forest; there are nine species in southeastern Mexico. The golden-collared manakin makes courtship areas on the forest floor by clearing away leaves and debris with its bill. These areas vary in size and shape but are irregularly elliptical in outline and average 76 by 51 centimeters (Fig. 16-9). The courtship activities of the golden-collared manakin are spectacular (Chapman 1938).

Near Palenque in Chiapas, the white-whiskered puffbird, which feeds on beetles and grasshoppers, and the northern royal-flycatcher, which makes its nest from roots and moss, are shrub and ground species. Other species of significance are the white-breasted wood-wren and the sulphur-rumped flycatcher (Tashian 1952, Goodnight and Goodnight 1956).

The anole lizards *Anolis limifrons* and others on Barro Colorado Island are insectivorous. They are preyed upon by snakes, birds, and some mammals. Their chief protection lies in the color of the body which matches that of the background. The helmeted lizard is found in Peten. The gecko *Sphaerodactylus lineolatus* is active only during the day. The interactions of both these species are similar to those of *Anolis*. The horned palm viper hangs in tangles of vines and bushes and catches birds as they pass, hanging on to them until the poison that they inject does its work. The species has on occasion made fatal strikes into the faces of men. The harmless *Chironius carinatus*, an attractive inconspicuously colored snake, climbs anywhere through the lower strata. Netting (1936) took two *Imantodes* (Fig. 16-10) on bushes at night on Barro Colorado Island.

Invertebrates are numerous in both wet and dry seasons. In the dry season the ant *Camponotus sericeiventris* has been seen climbing in several trees. On one tree they were carrying bark and bits of wood into a hole some three meters above the ground where toucans were observed to fly, perch, and apparently feed on the ants. The ant *Paracryptocerus cordatus,* commonly found 25 to 35 meters high, has been seen running over large tree trunks near the ground (Allee 1926). *Cryptocerus cordatus* is found in small numbers near the ground in the middle of the forest. The leaf-cutting ant *Atta* has conspicuous nests on the ground from which roadways radiate in all directions to trees and shrubs. The ants take pieces of leaves back to their nests for use in the cultivation of the fungus that they use for food. The army ant *Eciton hamatum* makes upward excursions from the forest floor, as high as 15 meters, to reach the insects which feed on oozing sap.

At Palenque, Chiapas, sweep net collections of the lower vegetation yielded only about one-tenth as many animals as occur in the temperate deciduous forest at the same time of year. *Helicina oweniana* was the most common snail,

Fig. 16-9. A. Golden-collared manakin above its "court," which is the cleared space below the bird which is whirring (from *Life in an Air Castle* by Frank J. Chapman, reproduced by permission of Appleton-Century-Crofts, Incorporated). B. Common toad *Bufo marinus* near the Barro Colorado laboratory. C. Stilt palm. D. Fallen tree trunk with *Peripatus*. E. American cockroach. F. Common tarantula.

with *H. ghiesbreghti* and *H. zephyrina* less frequent. The notable ground spiders were *Eustala fuscovittata, Verrucosa arenata, Pseudometa alboguttata,* and 15 or more other genera in which species were either not described or not easily identified. The common phalangids were the long-legged *Geaya lineata* and *Prionostemma foveolatum* which feed on small insects. Among the insects there were also termites, notably the tree nest-maker *Nasutitermes ephrates* which also occurs in Panama. The ants included the stinging ant *Wasmannia auropunctata,* much despised by botanical collectors, *Ponera nitidula, Neoponera*

Fig. 16-10. A. *Imantodes,* total length nearly one meter, Piedras Negras, Guatemala. B. *Thecadactylus rapicaudus,* total length 15 centimeters, Piedras Negras, Guatemala, moderately common, rain forest only. C. *Ameiva festiva edwardsii,* total length 25 centimeters, tropical rain forest only (photos by Rozella B. Smith).

unidentata, and a number of others. The ground beetle *Onoto angulicolis* was found and five species of click beetles could be identified only to genus; this was true also of rove beetles. Butterflies were represented by 12 species including the colorful *Morpho, Papilio,* and *Anarita fatima venusta;* the last is recorded also at the northern extremity of the tropical deciduous forest near Victoria, Mexico. The larvae and food plants of these species are largely unknown. Hemipterons were represented by the stink bug *Euschistus spurculus* and others. The lygaeids included *Myodocha unispinosa.* Homopterons were less well represented. Among the leafhoppers, *Draeculacephala clypeata,* a common genus in the United States, and *Homalodisca atrata,* which occurs in Texas, and nearly a dozen others were present (Goodnight and Goodnight 1956).

Animals of the Forest Floor

Most of the mammals of the forest floor permeate through all seral stages, and some have a wide range of distribution. On Barro Colorado Island, Baird's tapir (Fig. 16-8) was observed to have eaten *Cecropia* leaves in an old clearing. A wild papaya tree had been straddled and then "walked down" until the stem broke; the fruit was then eaten and here and there a mouthful of leaves consumed. Tapirs bathe at the water's edge and frequent the stream courses to a considerable extent; they are fond of grasses which grow near rivers (Enders 1939). The tapir occurs throughout all types of evergreen forest in moist regions, from northwestern Colombia to Tamaulipas. The brocket deer in the dense rain forest occurs as far as Tamaulipas, Mexico. The deer in the poor rain forest of the Yucatan Peninsula is a different species.

White-lipped peccaries (Fig. 16-8) are known to feed upon the wild banana *Heliconia*, eating both the tender base of the stalk and the succulent leaf base as well as the thick, erect stem. The grass clearing around the laboratory on Barro Colorado Island supported three white-lipped peccaries for more than a month in 1932 (Enders 1939). The species does not extend north of southern Veracruz. The collared peccary occurs but is less abundant than in the tropical deciduous forest. The rabbit *Sylvilagus gabbi* occurs in the dense forests of Panama and north into San Luis Potosí and northern Veracruz.

The spiny rat inhabits dense forest on Barro Colorado Island, living under tree roots, fallen trunks, and in other cavities. Their food near Gatun Lake is the para grass *Panicum barbinode* and *P. grande;* inside the forest, almendro nuts are taken. The food of the agouti in the Canal Zone is herbs supplemented by fruit and nuts that fall to the ground. The husk of the palm nut is eaten; both the fleshy covering and the kernel of the almendro are consumed. The species is absent north of Chiapas. The paca is common in the forest on Barro Colorado Island where roots are dug and fallen fruit eaten. They make burrows for shelter.

At Palenque in Chiapas the population of spiny pocket mice is too low to estimate quantitatively, but there are about 20 Mexican deer mice per hectare and about 8 climbing rats *Ototylomys phyllotis* per hectare. Desmarest's spiny pocket mouse and the Talamancan rice rat were also abundant on the forest floor both here and in San Luis Potosí (Goodnight and Goodnight, personal communication).

The total population of various species of mammals on Barro Colorado Island was estimated in 1932 at tapirs, 20; forest deer, 8 to 10; two-toed anteaters, 41,000 (2.5/hectare); pumas, 8 or more; ocelots, 8; and white-lipped peccaries, several hundred (Enders 1935). The howler monkey was censused in 1932 at 398 individuals divided between 23 groups; in 1933 at 489 individuals in 28 groups (Carpenter 1934). By 1939 there was an enormous increase in terrestrial rodents and armadillos. Most monkeys, both species of deer, agouti, paca, and tayra had increased. The coati, kinkajou, squirrel marmoset, and opossum were about the same. The puma, ocelot, tapir, and the two peccaries had decreased.

The great tinamou, a ratite bird, nests on the ground and is a poor flier. The marbled wood-quail and the gray-necked wood-rail are ground birds but little is known of their habits. The great curassow in Panama is a ground bird but flies into the trees when in danger; it is also found in Mexico and Peten (Aldrich and Bole 1937, Belt 1874, Blake 1953, Sturgis 1928, Van Tyne 1935).

A turtle and the colubrid snakes *Drymobius boddaerti* and *Dendrophis dendrophis* are common. The egg-laying bushmaster is becoming less numerous while the viviparous fer-de-lance is perhaps increasing. *Bothrops atrox* remains plentiful and originally occurred as far north as Tamaulipas. The coral snake is a burrower.

The amphibians of Barro Colorado Island include three species of toads, two *Eleutherodactylus* and *Bufo marinus* (Fig. 16-9). Their influence is small or unknown. The frog *Engystomops pustulosus* is a ground inhabitant and breeds in temporary pools; another species, *Leptodactylus pentadactylus,* occupies burrows in the forest floor during the day but is active and vocal at night (Park 1938) (Fig. 16-11). At the Palenque ruins, Goodnight and Goodnight (1956) found the frogs *Eleutherodactylus rhodopsis* and *E. rugulosus* present in considerable numbers. The toad *Bufo valliceps* was abundant. The ground lizards *Sceloporus teapensis* and *Basiliscus vittatus* were common. The most conspicuous diurnal invertebrate of the area was *Pseudothelphusa,* a crab 13 centimeters wide.

The forest floor on Barro Colorado Island is thickly carpeted with dead leaves which in March make a loose layer some 7.5 centimeters thick. There is very little humus, however, because the leaves disintegrate rapidly or are washed away by the torrential rains of the wet season. Williams (1941) did not find many species in the leaf litter, but there were 9822 individuals per square meter. This is to be compared with only 2480 per square meter obtained in an oak–hickory forest in North Carolina (Pearse 1946). Millipedes are always present, making up 1.19 per cent of the total population. The frequency of occurrence of the commonest species is *Barroxenus panamanus,* 50 per cent; *Barrodesmus isolatus,* 40; *Eutynellus flavior,* 60; and *Stemmiulus canalis,* 50. Mites, collembolons, and ants are most numerous, both here and in the North Carolina area. In the wet season at the Palenque ruins, the millipedes *Rhysodesmus tabascensis* and *Amplinus flavicornis* are generally present. Centipedes appear less frequently. Spiders in the ground samples include *Verrucosa arenata.* Staphylinid beetles are common in the rain forest; *Thoracophorus brevicristatus* has been identified. A few snails, including *Oleacina guatemalensis,* were noted. Pseudoscorpions, mites, earthworms, sowbugs, and ants make up about one-third of the ground population (Goodnight and Goodnight 1956). In general, snails are few in the rain forest, both in the wet and the dry season, but the number of arthropods increases enormously during the wet season (Allee 1926, Park 1938, Williams 1941).

Beetles are perhaps the best known of the tropical insects. Williams recorded 21 species of that group, including the bark beetle *Xyleborus propinquus.*

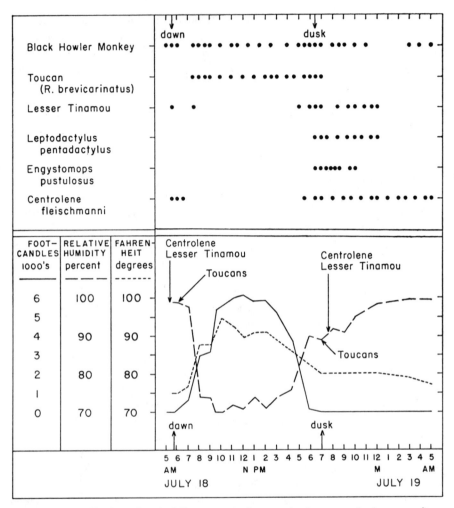

Fig. 16-11. Vocalization of typical Panama rain forest animals over a single twenty-four-hour cycle in July. Upper half, the black howler monkey, a toucan, the lesser tinamou, and three species of rain forest frogs, namely, the arboreal *Centrolene,* the floor dwelling *Engystomops,* and the semisubterranean *Leptodactylus.* Lower half, twenty-four-hour cycle in a clearing in the Panama rain forest of light intensity in footcandles (solid line), air temperature (short dashes), and relative humidity (long dashes). Dawn and dusk are indicated at two footcandles. The diurnal toucans begin calling shortly after dawn, continue through the day, and give their last call shortly after dusk; the nocturnal tree frog *Centrolene* and lesser tinamou stop vocalization just before dawn and begin calling shortly after dusk (after Park 1938).

Xyleborus affinis appeared in Allee's (1926) collections; it attacks broad-leaved trees and is recorded on dead chestnuts in New Jersey.

Army ants are conspicuous in the rain forest as they move about in military-like columns. The movements of *Eciton hamatum* were sufficient between August 5 and September 16, 1938, to suggest that they may traverse the entire length of Barro Colorado Island in a season. Between movements, the colony lives for periods of 1 to 14 days in bivouacs (Fig. 16-12). Here the workers remain clustered around the queen at night but make raids into the surrounding area during the daytime. These raids are for food and plunder, which they seize, kill, and carry back. Soft-bodied creatures are preferred. There are many ants of other habits, including the leaf-cutting ant *Atta*. Allee lists 10 species and Williams 22 additional ones. Two ponerine ants are present in both wet and dry seasons. *Paraponera clavata*, large ants about 2.5 centimeters

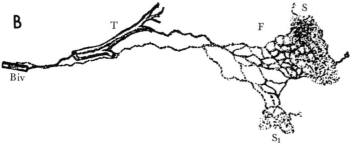

Fig. 16-12. A. Bivouac of the military ant *Eciton hamatum* in the nomadic phase. B. Diagram of a raid by *Eciton burchelli* which began at 6 A.M. At 11 A.M. the swarm (S) was approximately 55 meters from the bivouac (Biv) and about 12 meters wide; (F) network of columns; (T) large fallen tree through which swarm first raided; (S₁), auxiliary swarm produced by secondary division of main body shortly after 10:00 A.M. (after Schneirla 1944).

long, burrow in the ground. One colony of ants *Odontomachus haematoda opaeiventris* has been found occupying a part of an old log in close relations with the termite *Nasutitermes pilifrons,* which is unusual since the two do not normally come in contact (Allee 1926). The soft-bodied primitive arthropod *Peripatus braziliensis vagans* is found in rotting logs (Fig. 16-9). Additional log inhabitants are the American cockroach (Fig. 16-9) and the snake *Imantodes* (Fig. 16-10A). The land crab *Pseudothelpusa richmondi* is active during the daytime (Park 1938). The main mosquito flight takes place in the morning at less than one footcandle light intensity (Allee 1926).

POOR TROPICAL RAIN FOREST

The northern parts of the rain forest in Mexico and the area along the coastal plain of the Gulf of Mexico and the Caribbean Sea are less luxuriant than the forest areas just described. These forests are, however, more or less evergreen (Leopold 1950b, Miranda 1952) and should be considered in this biome (Fig. 16-13).

San Luis Potosí, Tamaulipas, and Veracruz Lowland Forest

The Pan-American Highway enters this forest about 20 kilometers south of Antiguo Morelos, Tamaulipas, and leaves it beyond Tamazunchale. There is little good rain forest on the coastal plain below 250 meters except here and there in local areas (Fig. 16-5). The poor rain forest at the present time is mainly second-growth or considerably disturbed. The mapping of "evergreen forest" by Smith and Johnston (1945) may possibly represent the potential climax area for this community. At Valles with an elevation of only 70 meters, the annual precipitation is only 115 centimeters. The forest is more mesic away from the highway at Xilitla, San Luis Potosí, at 1035 meters elevation, where the rainfall is 254 centimeters per year.

The three characteristic trees of the poor rain forest *Ceiba pentandra,* the strangler fig, and *Cecropia mexicana* are present. There are also representatives here of *Caesalpinia* and *Bursera* and the rubber tree *Castilla elastica,* but some of these could have been planted. The latter two are said to be deciduous.

The low growth is similar to that at Papantla, Veracruz, 200 kilometers to the southeast. Kenoyer (1941) names as a common undershrub the stinging spurge *Jatropha.* Epiphytes are present, only a few of which are partially or wholly parasitic. Among the epiphytes near Tamazunchale, San Luis Potosí, are no less than 60 species. Even more common epiphytes are the bromeliads (Smith 1941), aerial relatives of the pineapple. These are often coated with silvery scales which aid in the absorption of water or with leaf-bases so arranged as to form a reservoir which catches water. Climbing plants are also important, including an inedible grape, several trumpet creepers, a host of morning-glories, and cucurbits.

Among the tree-inhabiting animals, there is a nocturnal honey bear or kinkajou. Spider monkeys occur in bands up to 20 individuals. There are no recent records of the black howler monkey, but it was probably in the area

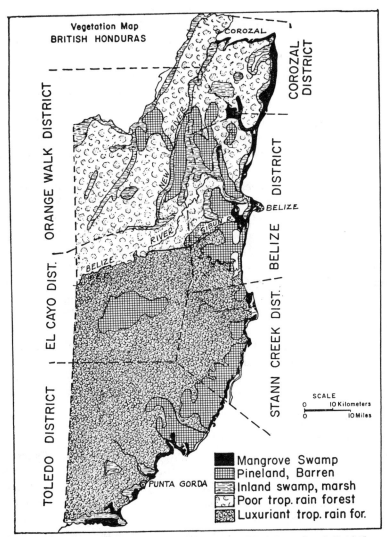

Fig. 16-13. Vegetation map of British Honduras (modified from Lundell 1945, courtesy Ronald Press Company).

in 1600. The red-bellied squirrel is common and widely distributed, and the prehensile-tailed porcupine is an important constituent. The Jamaica fruit-eating bat lives in hollow trees; it occurs also on Barro Colorado Island. The Brazilian brown bat is probably tree-inhabiting and insectivorous, and it also occurs in the Canal Zone (Dalquest 1953).

Some of the showy birds of the tropics find here their most northerly out-post, notably the scarlet macaw, olive-throated parakeet, and great curassow. The crimson-collared grosbeak is endemic to the area of this forest.

Most of the bats live in caves but utilize the forest for food. The Brazilian small-eared bat is numerous and frugivorous; Pallas' long-tongued bat feeds on nectar and fruit juices. The yellow-shouldered bat and Anthony's bat both feed on fruit. The little black bat occurs both here and on Barro Colorado Island. The hairy-legged vampire is a blood-sucker known to attack birds and mammals (Dalquest 1953).

The coati, four-toed anteater, black cat or tayra, grison, mountain lion, jaguar, ocelot, and margay cat use both the trees and the ground in the forest. The last four range into less moist tropical communities. The absence here of the bobcat of the north and sloths of the south is noteworthy. The collared peccary and both the white-tailed and brocket deer are present. According to Hershkovitz (1954), Baird's tapir occurs in this area earlier. A species of coyote is common.

The group of smaller tropical mammals includes the forest rabbit *Sylvilagus brasiliensis* which sleeps in dense vegetation during the day, the paca which burrows and feeds on succulent vegetation, the nine-banded armadillo which burrows as does the paca, black-eared rice rat, alfaro rice rat, Mexican harvest mouse, brush mouse, deer mouse, and Mexican wood rat. The cotton rat finds plenty of thick cover.

Common lizards are *Sceloporus variabilis, Cnemidophorus gularis, Ameiva undulata,* and *Leptodeira septentrionalis polysticta.* The common tree frog is *Hyla baudinii.*

The insect population is large, but only a few have been identified. A *Diabrotica* beetle is common. Moths found are *Anarita fatima, Phoebis agarithe,* and *Callosamia calleta.* The ant *Pachycondyla harpax* is common. Tree snails, especially *Practicolella berlandieriana* and *Guppya gundlachi,* are also common. The spiders *Euryopis scriptipes* and *Paraphidippus* are common, and the millipede *Rhysodesmus stemmiulus* is numerous.

Coastal Plain Semideciduous Tall Forest Along Streams

On the Gulf and Carribean coastal plain, as near Acayucan, Veracruz, the vegetation is mostly semideciduous forest (Miranda 1952), but a more luxuriant forest occurs along the streams, characterized by large trees and an abundance of lianas and epiphytes. *Pachira macrocarpa* and *P. aquatica* are important species. On the higher ground this forest fades into grassy areas dotted with groves of stunted trees. These groves vary from a few trees or bushes to tracts of woodland or thickets several hectares in extent. The Indians cut down and burn the woody vegetation to clear areas for cultivation, but the fires once started are allowed to spread wherever there is fuel to carry them. Once deforested, grass takes possession of uncultivated areas and a savanna becomes established (Cook 1909).

In this particular area, known animal constituents among the lizards are *Anolis biporcatus, A. sallaei, Ameiva undulata,* and *Bolitoglossa platydactyla;* and among the snakes *Constrictor c. imperator, Ctenosaura acanthura,* and *Drymarchon corais couperi.* Frogs are represented by *Hyla baudinii* (Ruthven 1912).

Behind Pearl Lagoon on the coastal plain of Nicaragua, much of the forest is second-growth, following cultivation or blowdowns by hurricanes. There are many local forest faciations with one or two species of smaller trees conspicuously dominant in each. Thickets are formed by various palms, notably *Attalea cohune,* and in the better-drained areas by bamboo. The "jungle" forest, while lacking the character of the tall rain forest, offers even better conditions for the larger animals than typical rain forest. Such browsers as deer and tapir require a varied understory such as the jungle affords, and tinamous, guans, and agoutis also reach their greatest abundance in the tangled woods. While the tree-top fauna in general prefers the typical rain forest, the toucan and white-faced monkey constituents occur more commonly in the smaller trees of the jungle. The great nomadic hordes of white-lipped peccary range widely in the climax forest and may reside there when some preferred fruit is falling, but their optimum permanent niche is in the jungle thickets where concealment is easier and sloughs and creeks are numerous (Carr 1950).

Pacific Coastal Plain

The forest is badly disturbed nearly everywhere on the Pacific coastal plain, but at a point near Tonala, Chiapas, Miranda (1952) lists the trees *Licania arboria, Poeppigia procera, Tabebuia palmeri, Calycophyllum candidissimum, Bursera simaruba, Couepia polyandra, Erythroxylon areolatum, Dalbergia granadillo, Cochlospermum vitifolium,* and *Swietenia humilis.* Miranda also mentions and maps *palmares,* some of which are dominated by *Sabal mexicana,* that have as subordinate vegetation *Pseudocalymma macrocarpum.* Rainfall in this area is small and the dry seasons are long.

A GUATEMALA–YUCATAN TRANSECT (FIG. 16-14)

Shore and Coastal Plain

The Pacific beach a short distance southeast of Champerico, Guatemala, is paralleled by a ridge with a lagoon behind it and with mangrove swamp sometimes choking the lagoon. On the drier ground there are thickets of low growing fan palms. Miranda's list (1952) for the shore zone includes *Pithecellobium dulce, Capparis flexuosa, Bursera excelsa, Achatocarpus nigricans,* and others. This zone is very narrow but on the lower coastal plain occurs the *palmares* described in the above paragraph. Many of the species mentioned by various authors are widespread types scattered along the Pacific Coast from Sinaloa, Mexico, to Panama and are often planted by man. These include *Tabebuia pentaphylla, Ceiba pentandra, Gliricidia sepium, Bursera simaruba, Cochlospermum vitifolium,* and *Sterculia apetala.* McBryde (1945) refers to palms as occurring on the coastal plain in scattered groves "relics of more extensive and luxuriant forest of former times." The gallery forest (Fig. 17-3A) is widespread along the numerous streams. A diverse forest exists, much of which is of a scrubby type, probably second-growth. The wide stream-skirting or gallery forests provide conditions suitable for forest animals. Of the mammals, the

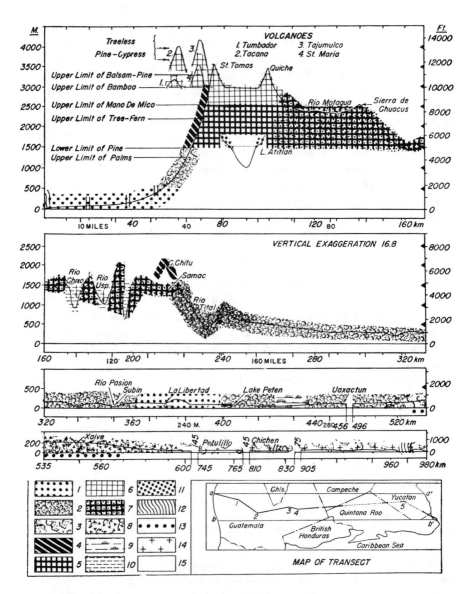

Fig. 16-14. The Guatemalan–Yucatan transect. On the map in the lower right corner, the Campeche–Quintana Roo boundary runs north and south. The transect strip begins at the Pacific shore with a width of 115 kilometers between lines a and b, and runs 22½° east of north (treated as north) ending between a¹ and b¹ on the northern coast of Yucatan with the sea level width throughout. The line of detailed description of vegetation, shown by the heavy line, has five changes of direction: the line from a to (1) C. Quiche is straight and the positions of the volcanic peaks St. Tomas and Quiche to the north, shown in the top diagram, are at right angles to this line; at C. Quiche the line turns east 15° and runs straight to (2) Samac which is about 7 kilometers west of Cabon; here the line turns back toward the north 22° to (3) Paso Subin, from here to (4) Flores,

white-tailed deer and the opossum *Philander opossum* are abundant. The squirrel *Sciurus socialis,* paca, and species of *Oryzomys* have sizable populations. Prominent resident birds are the chachalaca *Ortalis vetula* and the great curassow which is mainly terrestrial but roosts in trees. The tapir, originally, spider monkey, and yellow-headed parrot occur along the streams in the wide gallery forests. These three species and the coati also occur in the bocacosta forest (Griscom 1932, Villa 1948). Inland between elevations of 100 and 200 meters and in a distance of 10 kilometers rainfall increases from 200 to 300 centimeters a year. The coastal plain blends into the slope of the massive Guatemalan dome. Between 200 and 1450 meters altitude, rainfall increases from 300 to 450 centimeters per annum. The dry season comes in midwinter. The rainfall follows the pattern of the Atlantic Coast, but two wet seasons come in June and September.

Bocacosta Forest

The physiognomy of this forest has not been described. The area is now largely devoted to coffee. The forest begins at an altitude of 400 to 500 meters, although somewhat lower near the Mexico–Guatemalan border in the vicinity of Tapachula where the annual rainfall is 250 centimeters. In the transition to bocacosta near Tapachula, a net collection of 48 sweeps made in January, 1943, from herbs and shrubs yielded about 150 individuals including the spider *Wagneriana tauricornis;* the orthopteron *Conocephalus;* the fulgorids *Copicerus irroratus, Gaectulia plenipennis,* and *Monoflata palescens;* the hemipterons

thence to (5) Chichén-Itzá (410 km) where it turns east 30° and ends at b[1] at Holbox de Palomino on Isla Holbox. Because of lack of detailed study and apparent similarity of conditions, several long sections of the line in Quintana Roo and Yucatan, totaling 305 kilometers, were omitted from the 980 kilometers total length at sea level. The choice of the line and of the transect was determined by the location of areas that had been investigated and mapped.

The legend is put on as horizonal strips so as to indicate that land surfaces at those levels are covered in the main by the same communities throughout the width of transect belt. An exception is the thornbush–cactus vegetation in the valley of the Chixoy and Motagua which is strictly local. The different symbols are: 1, savanna (km 1 to 60 and 360 to 400) ; 2, tropical rain and evergreen forest (km 70 and 220 to 520) ; at km 70, the forest is a peculiar type called "bocacosta" (550 to 1400 m), the dark area of legend (km 220 to 250) is luxuriant; 3, poor evergreen or rain forest especially on the Yucatan Peninsula (km 70 and 720 to 980) ; 4, cloud forest (km 70 and 210) ; 5, oak–pine forest mainly cleared for agriculture (km 70 to 220) ; 6, pine forest with bunch grass and subalpine meadow and some areas of cypress (km 70 to 100) ; 7, oak–pine more varied than on the south slopes, including some pine savanna (km 100 to 220) ; 8, the streamless elevated area on the base of the Yucatan Peninsula (km 510 to 590) ; 9, freshwater swamp (km 420) ; 10, scrub or cactus–thornbush (at km 170, 185, and 200) ; 11, chaparral around Lake Atitlan and nearby; 12, mangrove swamp; 13, limestone sinks on dolines in southeastern Campeche (km 530 to 570) ; 14, secondary vegetation (km 810 to 980) ; 15, snow on Tacana and Tajumulca (km 50 to 70). Compare with Figure 16-2 (drawn from American Geographical Society maps; biological data from Stuart 1935 and personal communication 1954, McBryde 1945, Schmidt 1936, Lundell 1934).

Garganus albidivittis, Pycnoderes pallidirostris, and *Pachybrachis bilobatus;* and the dipterons *Leptopsilopa similis, Leucophenga varia,* which also occurs in central Illinois, and *Limonia (Rhiphidia) domestica.*

Some of the characteristic trees of the bocacosta forest in Mexico and Guatemala are *Sterculia mexicana, Dipholis minutiflora, Rheedia edulis, Pithecellobium arboreum,* and *Trophis chorizantha* (Miranda 1952) ; and additional species listed by Standley and Steyermark (1945) are *Dussia cuscatlantica, Billia colombiana, Heisteria macrophyllum,* and *Mollinedia guatemalensis. Bursera* is one of the principal shrubs. Some authors have named the herbaceous plants of the area, but whether they occurred under pristine conditions is in doubt. Palms disappear before the middle of this forest is reached.

The white-tailed deer is common and the brocket is present. The spider monkey once occurred. The poto or kinkajou, tayra, gray fox, the squirrels *Sciurus deppei* and *S. griseoflavus,* and three species of white-footed mice are or were present. The plain-breasted ground-dove, crested owl, and the blue-throated goldentail occur (Griscom 1932, Villa 1948). Schmidt (1936) found two species of the salamander *Oedipus* confined to this forest.

Altos

This is the land above 1450 meters. Mixed oak and pine woods are now scattered through these highlands where once there was continuous forest. Cloud forest occurs on the ocean front of the mountains. The brocket deer and the black chachalaca and quetzal birds occur. Oak–pine, alpine vegetation (Chap. 18), and scattered desert predominate across the mountain dome and down to an altitude of 1200 meters on the Gulf slope.

Rain Forest

Rain forest begins below Samac near Rio Sulba and continues for 120 kilometers to the swamps south of La Libertad, Peten. In this rain forest, a large number of palms occur. *Vochysia hondurensis, Engelhardtia guatemalensis, Calocarpum viride, Hymenaea courbaril,* and *Persea schiedeana* are common trees and shrubs. Isolated in these forests are also such northern genera of trees as *Magnolia, Berchemia, Gelsemium, Liquidambar, Carpinus,* and *Rhus.* There is an especial abundance and variety of Orchidaceae and Bromeliaceae. The moja blanca *Lycaste skinneri,* the national flower of Guatemala, is widespread in this portion of the country (Standley and Steyermark 1945). The great and little tinamous occur.

Communities of the Upland, Well-drained, Calcareous Land

The vegetation in the upland scarcely ever reaches a height of more than 20 meters and is known as the zapotal community by the natives. Zapotal covers about 85 per cent of the area. The zapote is the most abundant of the dominant plants. The gumbo limbo or chaca *Bursera simaruba* is second. Additional frequent trees are *Talisia diphylla* and *T. floresii, Sideroxylon meyeri, Lucuma campechiana,* and *Aspidosperma megalocarpon.* The mahogany is not prominent. The understory trees include *Spondias mombin, Eugenia capuli,* and *Trichilia minutiflora;* shrubs, *Piper gaumeri;* vines, *Serjania yucatanensis.* Groves of the ramon or breadfruit tree *Brosimum alicastrum* are widely scattered; the distinct ecological areas known as the ramonales are, therefore, not so prominent as in other divisions of the peninsula. Associated with the ramon tree is the persimmon *Diospyros anisandra* and the widely distributed zapote. The southern Campeche division may be negatively defined as the region where palms, mahogany, ferns, *Heliconia, Calathea,* and strangler figs are distinctly rare.

Throughout Campeche and in the states of Yucatan and Quintana Roo there are lakes, aguadas, and limestone sinks. The upland forest is poor, but evidence exists that even where the forest is almost completely obliterated by fire and cultivation, the zapotal still is the natural vegetation (Lundell 1934).

The animals in the zapotal area are in the main those of the tropical deciduous and poor rain forests of Central America and Mexico. At Tuxpeña, Campeche, 20 kilometers west of the edge of the transect but in the western edge of the zapote forest, the howler monkey, spider monkey, and kinkajou are present. At both Tuxpeña and Chichén-Itzá, Yucatan, within the transect, white-tailed deer, brown brocket deer, collared peccary, and most of the common tropical mammals, birds, and reptiles occur. A cottontail is an additional species at Chichén-Itzá.

Grasshopper outbreaks were reported in Yucatan from time to time between 1740 and 1871. These outbreaks appear to have centered in secondary vegetation south and west of Chichén-Itzá, and swarmed to the west, north, and south. The insects attack all kinds of plants, cultivated and native, and famines sometimes result. They spread into the dense forest of Peten and British Honduras and evidently attack the forest trees. They defoliate the shade trees of coffee plants and thus do extensive damage to the crop. There was another center for grasshopper outbreaks in San Salvador (Agacino 1952). There are bacteria (*Coccobacillus acridiorum* and others) in the bodies of grasshoppers which under unfavorable weather conditions increase and destroy the viscera (d'Herelle 1911). The Yucatan grasshopper is *Schistocerca paranensis* and has both a solitary and a gregarious phase. Their ravages reach to the coast where the transect includes freshwater marshes and mangrove swamps as discussed in Chapter 19.

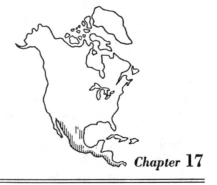

Chapter 17

Tropical Deciduous Forest and Related Communities with a Dry Season

In contrast with the evergreen rain forest, tropical forests that shed their leaves in the dry season are chiefly on the Pacific side of tropical America, although in Mexico they also occur elsewhere (Fig. 17-1). The vegetation varies from trees as tall as those in the rain forest to scrub only one or two meters high. Certain ground doves and cotton rats extend over nearly the entire deciduous forest and scrub vegetation. The chief character of the climate is the long dry season.

Much of the human population of tropical America lives within areas of the deciduous forests, as a result of which the forest has been more greatly modified, especially from "conuco" agriculture, than has any other type of tropical vegetation (Barbour 1941). There have been essentially no ecological studies in the area, although some collecting of plants and animals has been done. One must go back to Belt's 1874 account of the Nicaragua highlands for the only real biotic description of communities.

Tall tree forest communities occur on the Pacific slope from Panama to Chiapas, Mexico, in the lower parts of the interior valley of Chiapas, on part of the Chiapan Pacific coastal plain, and in the San Luis Potosí–Nuevo León–Tamaulipas region. The *lowland savanna communities* of Panama and northward into Oaxaca, Mexico, vary from the condition shown in Figure 17-6 to that in Figure 17-7. The savannas of southeastern Mexico are regarded as largely the result of clearing (Fig. 17-2B). *Upland savanna communities* are found in Panama, Nicaragua, Honduras, El Salvador, Guatemala, and in the states of Chiapas and Oaxaca, Mexico. The commonest pattern is for the trees to occur along the streams and valleys with the hilltops bare (Figs. 17-3, 17-4). The valley vegetation in dry climates may, however, be partly shrubs. *Short tree forest communities* extend along the Pacific slope behind the thorn forests

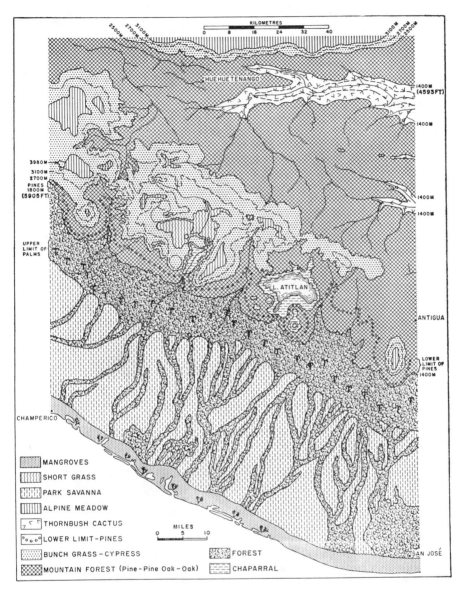

Fig. 17-1. The communities of a portion of southern Guatemala (from McBryde 1945), modified to conform with the interpretation of Miranda for the coast of Mexico.

from Sonora, Mexico, to Guatemala, across the escarpment of the Eje Volcanico, on both sides of the interior valley of Chiapas above the tall tree forest, and along the Pacific shore of Chiapas as a very narrow belt (Gentry 1942, Shreve 1937). *Thorn forest communities* occupy a narrow strip of the Pacific Coast between the shore and the short tree forest southward from southern Sonora, and an area in northeastern Mexico in Sierra de Tamaulipas. The type is poorly defined and doubtfully mapped, and there is almost no information on the animals of this forest type. The largest area of *arid tropical scrub communities* stretches across northern Guerrero along the Rio Balsas into Puebla, Mexico (Fig. 17-12), and in other restricted valleys of Mexico and Guatemala (Fig. 17-13). Leopold (1950b) reports upland thorny scrub on poor soil in Honduras (not mapped).

Fig. 17-2. A. Semideciduous gallery forest north of San José, Guatemala, with wild papaya underneath. B. Savanna just outside of Veracruz.

All of these vegetation types are climax or late seral. The coastal plain of Chiapas and Guatemala has been regarded as savanna (McBryde 1945), but this is probably due to clearing. McBryde (1945) calls the bocacosta forest, tall tree tropical deciduous (monsoon) forest, although other plant students consider it essentially tropical rain forest. Tall tree deciduous forest covers much of Cuba; there is a little potential tall tree forest in extreme southern Florida and on some of the Bahama Islands. The island areas are treated separately in Chapter 19.

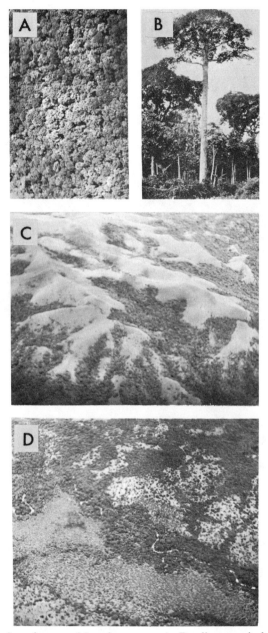

Fig. 17-3. Vegetation of areas with a dry season. A. Excellent tropical semideciduous or rain forest northwest of David (Pan-American Airways, 1941). B. Cuipo tree *Cavanillesia platanifolia,* a giant of the semiarid forest. C. Grassy ridges and forested ravines near Villanueva and northwest of Lake Managua, Nicaragua (Pan-American Airways 1941). D. Bush savanna valley of a branch of the Rio Chalutega, south of Tegucigalpa, Honduras. The darker upper areas and along the stream probably contain oaks. The smaller lighter trees are probably *Crescentia alata.* The light open spaces are grass and sedge (Pan-American Airways 1941) .

Fig. 17-4. Vegetation of areas with a dry season. A. Pacific-facing slope of the mountains opposite the Mar Muerto; height of mountains about 1500 meters; a savanna-like area in lower right; ridges covered with grass (Pan-American Airways 1941). B. Mountain ridges in the Rio Tehuantepec drainage (Pan-American Airways 1941). Tropical decidu-ous forest extends up the ravines and valleys, with grass on the ridges, and a few fair-sized conifers at the top; altitude of mountains about 2100 meters.

TALL TREE TROPICAL DECIDUOUS FOREST

Besides tropical deciduous forest, this vegetation has been called monsoon forest, trade wind forest, and *selvas de hojas deciduas*. There is less uniformity of nomenclature for this type of forest than for the rain forest into which it grades. With decreasing rainfall, the deciduous forest blends into savanna.

Differences in plant communities are correlated with total annual rainfall, its seasonal distribution, and soil conditions (Barbour 1941). The life form, the stature of the plant constituents, the number and height of the strata, the abundance of epiphytes, the spacing of the plants, the frequency of thorns and the mechanical interdependence of individuals are more important for classifying the vegetation than either the physical conditions or taxonomic composition. The species composition of both the plant and the animal con-stituents varies greatly from area to area (Barbour 1941, Smith and Johnston 1945).

Soil and Climate

Much of the soil is derived from volcanic deposits. Differences in soil and in the amount of erosion influence the type of vegetation, but rainfall is of outstanding importance. In most parts of the tall tree forest, the rainy season is from May through October and the dry season from December through April (Table 17-1).

Mean temperatures during the dry season are: Copainalá, 21.4° C.; San Salvador, 23.2° C.; and Balboa, 26.0° C. The mean temperatures for the rainy months are: Copainalá, 23.8° C.; San Salvador, 23.3° C.; and Balboa, 25.7° C.

Table 17-1. Rainfall at Various Stations in the Tall Tree Tropical Deciduous Forest

LOCALITY	DEC.-APR.	MAY-OCT.	NOV.	YEARLY TOTAL	DRIEST MONTH	WETTEST MONTH
Tampico, Mexico.........	14.0 cm	105.5 cm	5.6 cm	125 cm	Apr.	Sept.
Copainal Chiapas.......	9.0	96.0	2.6	105	Feb.	Sept.
San Salvador, El Salvador.	8.3	176.4	3.9	189	Mar.	June
Rivas, Nicaragua.........	5.2	157.8	11.7	175	Jan.	Oct.
Miravalles, Costa Rica....	8.4	182.0	19.0	209	Mar.	Oct.
Balboa, Canal Zone.......	18.2	124.6	25.0	169	Mar.	Oct.

The total amount of rainfall varies greatly from year to year; at Mojica, Costa Rica, it was 102 centimeters in 1930, and 300 centimeters in 1933 (Becker 1943). At San Salvador the record for 1886 was 386 centimeters; for 1903, 117 centimeters (Clayton 1927).

Physiognomy

The tall tree tropical deciduous forest ranges from open to moderately dense stands, and the trees are one- or, at most, two-storied. There may be a thick underbrush, which is sometimes thorny. Vines and creepers are not abundant. There is usually a horizontal open view three meters or so above the ground. The ground cover is sparse, but sometimes dense mats of wild pineapples are found. Grass is often abundant. The litter and humus layers are usually fairly deep (Barbour 1941).

The trees vary in size with the conditions. They often attain very large diameters; white boles and rounded crowns are common. The root systems are well developed and usually deep. The boles are not heavily buttressed and are of moderate taper. The limbs are large and fairly wide spreading and form rounded crowns. The twigs often bear great quantities of epiphytic growths such as bromeliads. The bark is variable but usually thick and deeply furrowed. The leaves are as a rule medium-sized and abundant. Neither the flowers nor the fruits have especially distinguishing characteristics. The wood is extremely variable; in many species there are conspicuous concentric growth rings due to the complete or partial interruption of growth during the dry season (Barbour 1941).

According to Griscom (1935) there are two types of tall tree deciduous forests in Panama: a type, also noted by Schimper (1903), which is an open forest with little undergrowth and with the tree trunks resembling columns in a large hall; and a type that has an understory of shrubs and low trees. The boundary between the pillared forest and rain forest in Darien, Panama, is sharp.

Much of the forest, except near streams, loses its leaves in the dry season. A number of trees make their vegetative growth in the wet season and flower and mature fruit in the dry season. The extent and time of leaf fall varies from year to year, depending on rainfall (Goldman 1920). On the Azuero Peninsula, Panama, only about 50 per cent of the trees lose their leaves (Aldrich and Bole 1937). The shrub stratum in this forest is occupied almost entirely by vines which hang from the branches of the trees. The forest floor is destitute of ground cover. The Azuero forest is probably climax. Floodplain and swamp forests resemble rain forests and furnish niches within the deciduous forest region for animals from the tropical rain forest.

Trees, Shrubs, and Herbs

The large over-topping trees are the dominants. One important tree is the gumbo limbo *Bursera simaruba* which is prominent in relict areas and occurs from Panama to Tamaulipas. The list below is primarily for Panama, but several species or related species occur northward into Mexico. The two largest trees of tropical America are *Enterolobium cyclocarpum* and *Ceiba pentandra,* which reach 40 meters in height and have large crowns. *Cavanillesia platanifolia* is also a very large tree, but otherwise most of the species are smaller than those of the Atlantic lowlands. Aside from these species the larger trees of the dense deciduous forests are *Anacardium occidentale, A. excelsum, Licania platypus,* three species of the wild fig *Ficus, Platymiscium polystachyum, Pithecellobium oblongum,* and *Sweetia panamensis. Hura crepitans* is a medium-sized tree. Other species include *Apeiba tibourbou, Cordia alliodora, Cassia grandis, Cedrela mexicana, Andira inermis, Dalbergia retusa,* and *Cecropia mexicana. Roupala darienensis, Triplaris americana, Gliricidia sepium,* and *Cavanillesia platanifolia* are characteristic of the forest in southern Central America but do not go north into Mexico. About half of the trees listed for the Chiapas tall tree forest occur in Panama. The palmetto *Sabal mexicana* occurs regularly in this forest, sometimes in fairly thick pure stands and sometimes mixed with other trees.

The shrubs and herbs *Croton cortesianus, Malvaviscus arboreus, Eugenia capuli, Adelia barbinervis, Sida acuta,* and *Agave* were collected from 1942 to 1944 in Tamaulipas and are of significance here because of the invertebrates taken from them.

Animals Using the Trees

Principal sources of information on animals are Goldman (1920) for Panama, Aldrich and Bole (1937) for the Azuero Peninsula, Panama, and Dalquest (1953) for Tamaulipas and Veracruz.

When the animals of the deciduous forest on the Azuero Peninsula are arranged in their order of abundance, the capuchin and black howler monkeys are last in the list of common mammals, instead of first as in the rain forest. The howler monkey originally went north in the deciduous forest as far as Tamaulipas.

The white-faced capuchin is recorded in some other localities in the deciduous forest area which follows the stream-skirting gallery forests into relatively dry areas. The capuchins travel in large bands; they give assistance to fallen young and break off dead or weak limbs on their routes of travel through the tree tops. A troop of capuchins causes the withdrawal of most birds in the vicinity and great alarm among the chachalacas. On the other hand, flocks of jays sometimes annoy the monkeys by dashing at them. The monkey's appetite for birds' eggs and young is probably the basis for this behavior (Aldrich and Bole 1937).

Spider monkeys are generally present where the forest is continuous. They have remarkable climbing powers, making considerable use of their tail, and go in parties of 12 to 15. They are absent from the short tree deciduous forest and the west coast of Mexico from Oaxaca northward. Their food consists of fruit and occasional insects. The behavior of the monkey in trees without leaves has not been described.

The kinkajou is confined to the trees and is nocturnal. The squirrel *Sciurus variegatoides* lives in trees, feeding on nuts and fruits; it is not recorded in the rain forest. Its ecological equivalents in San Luis Potosí are the fire-bellied squirrel and the little dull squirrel. The former also occurs in the rain forest. There is no data as to their food; probably they do not enter shrubby areas. The gray three-toed sloths are usually seen curled in a ball, high in the trees. They do not occur north of Nicaragua. Some of the same species of bats that occur in the Panama forests are also found in Tamaulipas and northern Veracruz.

Among the birds the blue-crowned motmot is common in the deciduous forest and uncommon in the rain forest; it occurs also in Chiapas. The orange-chinned parakeet occurs in large numbers on the Azuero Peninsula, feeding on fruits (Aldrich and Bole 1937); the olive-throated parakeet is characteristic of the northern Veracruz forest. The red-lored parrots of the Azuero Peninsula feed on fruit; the species is also found in northern Veracruz and Tamaulipas and in gallery forests north of the deciduous forest. The slaty-tailed trogon occurs in groves of trees on Azuero Peninsula; its place is taken by the bar-tailed trogon in northern Veracruz. The keel-billed toucan occurs widely.

The chestnut-headed oropendola (Fig. 17-5) occurs throughout Panama. In building its hanging nests during the dry season, it wraps a long green tendril, used like twine, about a limb many times and forms the foundation of the nest bag. Air rootlets, fine strips of bark, filamentous blossoms, plant fibres, and stringlike strips from the stem of *Monstera*, torn off by the strong bill of the bird, form the rest of the nest. The only enemies of the oropendolas are other birds; a cowbird, a flycatcher, and a phoebe harass the builder. In response to alarm calls from the guarding males, the entire colony hastily dives

Fig. 17-5. A. Nests of the chestnut-headed oropendola on a tree west of Balboa, Panama. B. Two coatis on Barro Colorado Island (courtesy of the American Museum of Natural History). C. Two collared peccaries on Barro Colorado Island (from *Life in an Air Castle* by Frank J. Chapman, reproduced by permission of Appleton-Century-Crofts, Incorporated).

into the lower forest growth. An owl *Pulsatrix* makes an opening in the lower part of the nest bag to get at the young; defense against this predator is almost nil (Chapman 1928). The Montezuma oropendola of northern Veracruz has similar habits. The cowbird, a flycatcher of a related species, and the same phoebe are present here.

Animals Using the Trees and Ground

Inhabiting both the ground and trees on the Azuero Peninsula and in the Tamaulipas and Veracruz areas are the coatis (Fig. 17-5), which travel in parties of females and young males. The old males are solitary. The animals are mostly nocturnal and arboreal but often are quite active on the ground. They are nearly omnivorous, climbing the trees to secure fruit or to catch arboreal lizards. This is the same species that occurs in the rain forest. The three-toed anteater in Panama is also present in both deciduous and evergreen forest. It spends considerable time on the ground feeding on ants. A collected specimen had a half kilogram of ants in its stomach, including the species *Camponotus a. abdominalis, Dolichoderus bispinosus, Pseudomyrmex pallida, Aphaenogaster,* and *Crematogaster.* The anteater is also present in the deciduous forest area at Ebana in San Luis Potosí. The tayra and Mexican mouse-opossum are common in the deciduous forest but less so elsewhere. The ring-tailed cat, jaguarundi, and ocelot are present.

On the Azuero Peninsula, the crested guan and the great curassow birds fly into trees when in danger. They are less abundant in the deciduous forest than in the rain forest. The following common species in the deciduous forest utilize both the ground and trees: blue-crowned motmot, white-whiskered puffbird, the woodcreepers *Xiphorhyncus guttatus* and *X. flavigaster,* xenops, lance-tailed manakin, dusky-capped flycatcher, black-crested jay, gray-headed greenlet, yellow-billed cacique, and others. A few of these species are also common in the rain forest, the rest are uncommon or absent there.

Major Permeant Influents of the Ground

Ground animals are a little more important than in the rain forest. The collared peccary (Fig. 17-5) extends from the tropical rain forest through the deciduous forest into open bush-covered country (Enders 1935). It travels in single file from feeding grounds to resting places and wallows. Where the trails go through thickets, they are too narrow and too low for a man to pass except by crawling. The trails are, however, important pathways for smaller animals, such as the coati, octodonts, and several genera of marsupials. In feeding on the forest floor, the peccaries leave their trails and spread out, rooting here and there as they go. They feed much like hogs, taking roots, fruits, nuts and other vegetable products, reptiles, and any other available animals. They are numerous where wild figs, nut palms, and other fruit-bearing trees are present, and in regions where they are little disturbed, they have regular feeding stations, called *comederos*. Their food varies with the season, and the peccaries shift around with the changing supply. When almendro nuts are ripe, nuts cracked by them may be found under every tree, and trails of the peccary lead from resting place to tree and from tree to tree (Enders 1935).

The white-lipped peccary is also present in the deciduous forest of Panama and north into southern Tamaulipas. The tapir at one time was in southern Tamaulipas as well as in Panama and undoubtedly frequented the gallery forest in both areas (Hershkovitz 1950). The white-tailed deer occurs.

Small Mammals

Rodents of mouse size are few in species and individuals. Talamancan rice rats are common on the Azuero Peninsula, numerous in the deciduous forest, and infrequent in the rain forest. Other small mammals in this locality are the spiny rat *Proechimys semispinosus* and the pale-bellied agouti. This latter species frequents thickets and is said to live in burrows. In southern Tamaulipas and San Luis Potosí, the hispid cotton rat is present in both the evergreen and deciduous forests.

Birds

On Azuero Peninsula, the white-whiskered puffbird, keel-billed toucan, chestnut-backed antbird, buff-throated woodcreeper, and several other species were common in both the deciduous and rain forest from February through April, 1932. The great tinamou and crested guan, which are common in the rain forest, are uncommon in the deciduous forest (Aldrich and Bole 1937).

In the Tamaulipas–Veracruz area the rufescent tinamou is a ground bird, the blue-crowned motmot is a bush inhabitant, and the ivory-billed woodcreeper uses tree trunks. The birds observed on two hectares at 651 kilometers on the Pan-American Highway, August 7, 1942, were white-winged dove,

2; white-tipped dove, 1; bicolored hawk, 1; turkey vulture, 2; black vulture, 6; and white-eyed vireo, 3.

Amphibians and Reptiles

Occurring in the deciduous forest are a variety of amphibians and reptiles, including the widely distributed toad *Bufo valliceps,* the frog *Rana pipiens,* the tree snake *Leptodeira septentrionalis,* the speckled ground snake *Drymobius margaritiferus,* the ribbon snake *Thamnophis sauritus,* the rat snake *Elaphe flavirufa,* and *Constrictor c. imperator* (Hobart Smith, personal communication).

Invertebrates

In the August, 1942, study various butterflies were found, including *Phoebis agarithe* and the zebra butterfly *Heliconius charithonius,* which also occurs in the Antilles. The wedge-shaped beetle *Heterispa westwoodi,* the langurid beetle *Camptocarpus longicollis,* the tenebrionid *Pyanisia tristis,* and *Tarpela* were common. The millipede *Rhinorcicus potosianus* was abundant and *Orthoporus* present. The scorpion *Vejovis bilineata* and the centipede *Mayobuys vistoriae* were secured. Spiders and mites were *Eustala devia, Cyrene delecta,* and *Amblyomma cajennense.* The ants found were *Pachycondyla harpax, Pheidole* (*flavens* group), *Ecetatoma turberculatum,* and *Pseudomyrmex,* which lives in *Acacia* thorns. There were grasshopper nymphs and the termite *Tenuerotritermes.* The snails included *Praticolella berlandieriana* and *P. griseola* (also occurring farther north and in Cuba), *Helicina zyphyrina* (also at Orizaba), *H. vanattae, Streptostila, Bulimulus dealbatus, Euglandina corneola, E. texasiana* (also in Texas and Veracruz), and *Polygyra implicata.*

LOWLAND SAVANNA COMMUNITIES

Lowland savannas are of various types: scattered trees in grassland, groves of trees or bushes, or gallery forests. The coarse tropical grasses may be short or of mid-height. Tall grasses occur only in moister habitats or are introduced species (Fig. 17-2). The grasses are accompanied by short-stemmed forbs. The taller plants are palms, conifers, or dicotyledonous trees or shrubs. The great amount of gallery forest provides suitable conditions for animals and subordinate plants. These species were doubtless also favored by the more extensive forest before white settlement. The variable character of the vegetation in areas with great differences in rainfall (Beard 1953) and the lack of knowledge of climax and developmental stages make natural classification impracticable.

Savanna Communities of the Pacific Slope of Panama

There is a gradient in Panama from tall deciduous forest to savanna and grassland. The amount of grassland increases in the north with the greater areas of south-facing slopes on the mountains. Clearing and cultivation prevent the development of deciduous forest in many areas. Savanna tends to occur near the coast on the Azuero Peninsula and between there and the Canal Zone, with forest in the highland (Figs. 17-6, 17-7). The western savannas are crossed by streams with wooded borders. Some are leafless in the dry season. There is little scrub.

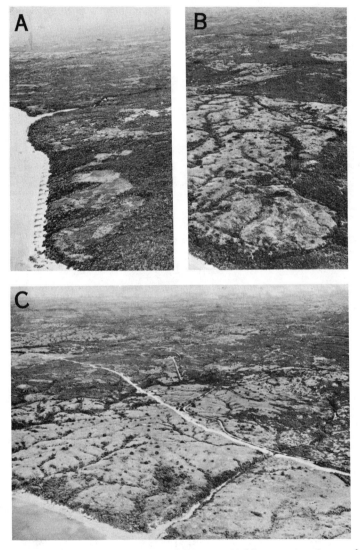

Fig. 17-6. Vegetation in the Chame area of Panama, 60 kilometers southwest of Balboa, the Pacific Ocean on the left. A. Forest and areas cleared by man. B. The hills are grass covered; the woody vegetation is largely bushes. C. A large portion of the land is in grass. A wide ancient trail is paralleled by the Pan-American Highway in the foreground. Only a few clearings are evident (U.S. Army photo, September 9, 1941, courtesy of Major Hugh J. Denney, and James Zetek).

The following grasses occur in the savannas: *Bouteloua americana, B. fili-formis,* and *B. pilosa; Andropogon tener* which is a bunch grass; *Eragrostis acutiflora; Sporobolus ciliatus; Digitaria panicea;* and *D. singularis. Thrasya hitchcockii* is in the open grassland at La Chorrera; it is the only genus not found in the United States. In the rainy season the grassland vegetation becomes a bright green sward with dwarfed herbs, *Crotalaria, Mimosa, Sty-losanthes, Zornia,* and other species. Two species of *Indigofera, Sida linifolia,* and *Meibomia barbata* are herbs occurring in the grassland.

Fig. 17-7. Pure grassland with trees along streams near Chorrera, Panama, 25 kilometers west of Balboa.

Some shrubs, such as *Miconia pteropoda* and *Lantana camara,* occur in thickets. Other shrubs are *Indigofera suffruticosa* and *I. panamensis.* The grasses *Lasiacis sorghoidea, Setaria vulpiseta,* and others grow in the thickets and at forest edges. Of the small trees, *Guazuma ulmifolia, Roupala complicata, Curatella americana, Cornutia grandifolia,* and *Duranta repens* are important. The stream-skirting forest trees include *Enterolobium cyclocarpum* and *Pithecel-lobium vahlianum* (Standley 1928).

Some mammals and birds find their homes along the gallery or other forest borders, but secure food in the open grass-covered areas. The savannas during the dry season are often swept by fire, which destroys much of the population of small vertebrates and arthropods. Some of the hawks are said to have learned to hunt along the fire lines, pouncing upon small rodents and other creatures that attempt to escape (Goldman 1920).

Arboreal and semiarboreal mammals are represented by the savanna marmosa *Marmosa mexicana,* one of which was found in a nest of leaves placed in tangled vines on a small tree. The pale woolly opossum *Caluromys derbianus* and the squirrel *Sciurus variegatoides* occur (Goldman 1920). Aldrich and Bole (1937) found both *S. variegatoides* and *S. granatensis* in bushy second-growth and scrubby woods.

White-tailed deer appear to be most common in Panama in the partly open savanna region, forest borders, and mixed growth of small trees and shrubby vegetation which springs up wherever the original forest is cut.

The rabbit *Sylvilagus brasiliensis* ranges at low elevations in the savanna regions of southern Panama. Cherrie's cane rat is nocturnal and makes its nests of grass. Three rice rats, *Oryzomys tectus, O. fulvescens,* and *O. caliginosus,* and the spiny pocket mouse *Liomys adspersus* appear to be grass inhabitants (Aldrich and Bole 1937). The Chiriqui pocket gopher *Macrogeomys cavator* is known to avoid heavily forested areas (Goldman 1920).

The birds characteristic of the savanna are the yellowish pipit, red-breasted blackbird, which frequents open grassland, yellow-rumped cacique associated with the chestnut-headed oropendola, yellow-faced grassquit, and fork-tailed flycatcher. In the little islands of woodland scattered over the savanna, the smaller tyrant flycatchers, ant-thrushes, wrens, and other brush and forest-loving species are found in small numbers (Bangs *et al.* 1906). Other birds are the roadside hawk, groove-billed ani, rufous-breasted wren, eastern meadowlark, grasshopper sparrow, and ruddy-breasted seedeater (Goldman 1920, Sturgis 1928). Aldrich and Bole (1937) list 26 species common in the shrubby, grassy, small secondary tree-covered areas.

The reptiles and amphibians of the Panamanian savanna have been little studied. Barbour (1906) suggests that *Ameiva surinamensis* is the most common lizard and the oblique-lidded skink *Mabuya agilis* is present along with the toad *Bufo spinulosus* and tree frog *Hyla ebraccata*. The invertebrates have received almost no attention.

Peten Savanna

South of La Libertad, Guatemala, at altitudes below 200 meters, occurs savanna vegetation that is probably produced by dry season fires. The open country is covered with the grasses *Trachypogon montufari, Leptocoryphium lanatum, Paspalum plicatulum, Andropogon condensatus,* and other species of *Andropogon*. Perennial forbs are chiefly legumes. The area is dotted with wooded islands. In passing from the open grassland into the islands, the first stage encountered is shrubby, usually dominated by *Conostegia xalapensis, Acacia,* and bromeliads, but with some palms and other species. Inside these shrubs are the taller species *Miconia argentea, Metopium brownei, Cecropia, Acacia,* and the woody vines *Davilla kunthii* and *Cnestidium*. This stage intergrades into the high forest in the center of the islands. Here *Xylopia frutescens, Bursera simaruba,* species of the Araliaceae *Simaruba glauca, Spondias mombin, Cecropia,* and *Ficus* occur. Older seral ones may contain the mahogany *Swietenia macrophylla*. The undergrowth is characterized by the shrubs *Mouriria parvifolia* and *Alibertia edulis* and a lower layer of *Piper* and *Psychotria* (Stuart 1935).

Among the constituent mammals are the white-tailed deer, ocelot, mountain lion, and collared peccary. Less common are the armadillo *Dasypus novemcinctus* and the squirrel *Sciurus yucatanensis;* the howler monkey is heard occasionally in the larger patches of forest. On the open savannas, the eastern meadowlark, bobwhite, and sparrows are the most common birds, while hummingbirds, doves, and the parrot *Amazona* occur to a lesser extent. Black vultures are distributed over the savannas. The amphibians and reptiles of the forest include two common frogs, four anoles, one tree gecko, and eight other reptiles (Stuart 1935).

Among the invertebrates, ants, especially the leaf-cutting ant *Atta,* the acacia ants *Pseudomyrmex,* and the army ants *Eciton,* are very numerous. Termites are common throughout the bush areas and around habitations. Following the rains a great variety of dragonflies, leafhoppers, and grasshoppers emerge. Flies and mosquitoes are surprisingly rare on the savannas (Stuart 1935). Goodrich and Van der Schalie (1937) comment on how bleached snail shells reveal the prevalence of fire. Only three species were found: *Practicolella griseola, Euglandina cumingi,* and *Bulimulus unicolor.*

UPLAND SAVANNA COMMUNITIES

In most of Central America the inland and upland portions of the Pacific slope are also occupied by savanna communities. East of Lake Nicaragua (Belt 1874) there are rock knolls covered with spiny cacti, low feathery-leaved trees, the slender, spiny, thicket-forming palm *Bactris minor,* prickly acacias, and thorny bromelias. An important tree is *Crescentia cujete* (Fig. 17-3D). Open spaces, especially on hills, are grass and sedge covered. Carr (1950) states that in Honduras, thorn scrub, notably that formed by *Mimosa tenuifolia,* is widespread and associated with arid savanna.

Numerous armadillos, which fed on ants and other insects, occurred in Nicaragua in the period of Belt's travels, 1868 to 1872. Doves of various sizes were numerous, including the inca dove, common ground dove, and plain-breasted ground dove. Belt sighted a small pack of coyotes, and their howling could often be heard in the early mornings. A troop of capuchin monkeys was observed on the ground among low scattered trees. He reports late June flights of the butterfly *Marpesia chiron* each year toward the forests in the southeast. The flights were in columns about 50 meters in width and a few hundred meters apart. Hundreds of individuals were always in sight; no return migrations were noted. Many beetles were present which he does not name, and the yellow-banded wasp *Rubrica surinamensis* hunted flies on men and mules. He mentions lizards, trogons, and antlike spiders. Buprestid beetles were numerous on the shrubs.

Locally there are dry gravelly hills covered with low scrubby bushes and trees, principally prickly acacias, nancitos, *Psidium guajava,* and *Crescentia cujete.* The last species was the most abundant and in a size and distribution pattern resembling an orchard. There are occasional *Curatella americana* with their thick coriaceous leaves that are used by the natives instead of sandpaper. Belt makes one reference to deer in the state of Nueva Segovia which appear now to be exterminated. A common bird is the insectivorous grackle

Cassidi mexicanus, which includes grasshoppers in its diet. The tick *Ixodes bovis* is abundant during the dry season and infests lizards, snakes, birds, and mammals.

At an altitude of about 200 meters in Nicaragua along the Jigalpa and Acaoyapa rivers, the banks have many high, thick-foliaged trees with lianas hanging from them and bromelias, orchids, ferns, and many other epiphytes on their branches. The spot-breasted oriole and *Trogon citreolus* are the most conspicuous birds. They catch insects on the wing and sometimes open nests of termites. The parrot *Amazona* flies past in screaming flocks and alights in the trees. The flycatcher *Myiarchus tyrannulus* sits on small branches and darts off every now and then after passing insects. A couple of motmots make short flights after the larger insects. The swallow *Petrochelidon* skims past in circling flights. The hawk *Buteogallus anthracinus* searches for land and freshwater crabs. The long-tailed manakin, a bird of the gallery forest, is present in the deep shade. In the bushes are warbling orange-and-black sisitotis (Belt 1874).

SHORT TREE TROPICAL DECIDUOUS FOREST

The description of this community must center around an extreme northern area because the more extensive and centrally located areas have not been well studied. Various areas of this short tree deciduous forest have little taxonomic unity except in genera, but they all conform to the particular vegetation type. A small but somewhat broken area of this forest has its northern extremity in the Rio Mayo Basin of southeastern Sonora, Mexico (Gentry 1942) (Fig. 17-8). It extends southward inside an area of thorn forest or savanna to the coast of Guerrero, then continues directly adjacent to the coast to eastern Oaxaca, where it turns inland behind the first mountain chain. Another strip in the upper southern foothills of the Eje Volcanico on the escarpment of the Mexican Plateau reaches three-fourths of the way across the country in the states of Michoacan, Mexico, Guerrero, Morelos, and Puebla (Miranda 1942, 1947, Leopold and Hermandes 1944, Leavenworth 1946, Hall and Bernardo 1949, Leopold 1950b). There are also areas in Chiapas (Figs. 16-1, 16-2).

Climate

The short tree forest is almost free of frost even in Sonora. Damage to plants by frost occurs only at intervals of several years. There are no official stations for temperature in the Sonoran forest, but Gentry (1942) gives the summer maximum as 39° C. and the minimum as 17° C. (Table 17-2). The driest months are February, March, April, and November. The rainiest month is July with 17 centimeters. Compared with the rain forest, the decline in moisture in this forest has more effect either because of its lesser annual total or because of the longer dry season and brings a decrease in the height of the vegetation, an increase in thorns and/or cacti, and a decrease in volume of plant material.

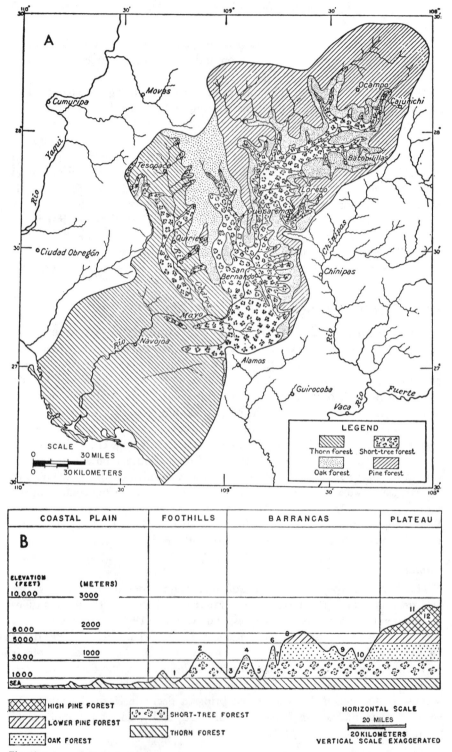

Fig. 17-8 legend on facing page.

Table 17-2. Rainfall and Temperature at Localities in the Short Tree Tropical Deciduous Forest°

LOCALITY	ALTITUDE	ANNUAL RAINFALL	SUMMER RAINFALL	LENGTH OF RAINY SEASON	TEMPERATURES DURING SUMMER GROWING SEASON
Sonora (average).....	300-900 m	50 cm	38 cm	12 weeks	27.8° C.
Colima, Colima......	500	88	71	16	28
Janecatepec, Morelos..	1300	85	72	16	23
Tuxtla, Chiapas......	536	94	77	16	25

ª Sonora data from Gentry 1947; data for other localities from government records.

Sonora and Sinaloa Forest

Sonora and Sinaloa contain the largest area of short tree tropical deciduous forest in Mexico. In Sonora, the forest ranges from an altitude of 300 meters to 1070 meters. Its northern boundary lies in the Rio Mayo Basin 50 to 60 kilometers inland from the coast. The forest occurs principally on the steep slopes of canyons and is nearly confined to the barranca region. It is bordered on the east by oak forest.

The forest stature is highly variable. The lesser trees, including *Bursera confusa,* rarely exceed 8 meters in height. The tallest trees, such as *Ceiba acuminata,* are 12 to 18 meters high. Along the arroyos, *Taxodium mucronatum, Platanus racemosa,* and several species of *Ficus* rise to heights of 18 to 24 meters. The larger trees dominate lesser trees. On the whole, the average height of the forest is about 12 meters. The short tree forest lacks the broken canopy characteristic of much of the thorn forest.

Miranda (1942, 1947) names 34 species of trees in the short tree deciduous forest; Gentry (1942) names 32. Only three species are common to the two lists, and only one of the three is in both lists of important species. The heterogeneous nature of the short tree deciduous forest is evident, and there is a considerable number of tropical species.

The most important plants in one climax stand, with the maximum height of the species and the number of individuals per hectare, follow: *Bursera inopinnata,* 15 meters, 60 individuals; *Lysiloma divaricata,* 15, 40; *Ceiba acuminata,* 15, 10; *Cassia emarginata,* 10, 50; *Bursera grandifolia,* 15, 10; *Jatropha cordata,* 8, 340; *Jatropha platanifolia,* 3.6, 690; *Willardia mexicana,*

Fig. 17-8. A. Map of the vegetation of the Rio Mayo Basin, Sonora, Mexico. B. Cross-section along axis of Rio Mayo, on line drawn through Navojoa and Nemelichi: 1, Rio Cedros; 2, Sierra Sutucame (south tip); 3, Arroyo Guajaray; 4, Sierra de la Ventana (south tip); 5, Rio Mayo; 6, Sierra Charuco; 7, Arroyo de Loreto; 8, Sierra Canelo; 9, Rio Batopilillas; 10, Arroyo de Santisimo; 11, Sierra Cajurichi; 13, Sierra Madre (after Gentry 1942).

10, 35; *Guaiacum coulteri*, 8, 85; *Coursetia glandulosa*, 6, 95; *Karwinskia humboldtiana*, 6, 95; *Randia echinocarpa*, 4.5, 55; *Cereus pectenaboriginum*, 11, 305; *Cereus thurberi*, 8, 70; and *Cereus alensis*, 4.5, 10. *Ceiba acuminata* (15, 18) was present in some areas.

The tropical element is represented in the short tree forest by the hydromorphic *Ficus*, the lianas *Arrabidaea littoralis*, *Marsdenia edulis*, and *Gouania mexicana*, and the climbing *Pisonia capitata*. Epiphytes are represented by the orchid *Oncidium cebolleta* and bromeliads by *Tillandsia inflata* and *Hechtia*. Other prominent plants of tropical distribution are *Guazuma ulmifolia*, *Cestrum lanatum*, *Bursera grandifolia*, *Cassia emarginata*, *Trichilia hirta*, *Vitex mollis* (Gentry 1942). Most of these plants have leaves or leaflets of large size (up to 32 x 45 cm), of a mesomorphic character, and without hirsute covering, epidermal thickening, or other features commonly found in arid environments.

Periods of growth during the rainy season alternate with periods of domancy during the dry season. Many plants have underground storage systems which enable them to make a quick response to summer rains and give them maximum growth during the growing season. In this group are *Ceiba acuminata* and *Ipomoea arborescens*, trees with storage roots; *Hymenocallis sonorensis*, a lily with a bulb; *Phaseolus caracala*, a vine with a thickened root; *Vincetoxicum caudatum*, with a tuber; *Salpianthus macrodontus*, with a tuberous root; and *Tigridia pringlei* with a bulb. Their manner of response is represented by *Jarilla chocola*. This is a dioecious, tolerant, minor species of the forest. With the coming of summer rains, turgid stems spring up from a crown of tubers and in a few weeks attain heights of 50 or 75 centimeters. With the last dwindling rains, it dies, leaving its fruit on the ground. In its niche among the shrubbery, it has relatively constant soil moisture, and transpiration is reduced by shade (Gentry 1942).

Prominent mammals of the short tree forest are the deer, pronghorn, jack rabbit, raccoon, coyote, bobcat, and ocelot. The Sonoran wood rat *Neotoma phenax* makes nests in low trees or thick bushes. Persimmons are an important part of its diet. Coatis go in troops of 30 to 40 individuals and feed on fruits, nuts, tender vegetation, insects, and carrion. Merriam's kangaroo rat, Goldman's pocket mouse, and three other species of pocket mice are present. There are a number of records of white-footed mice and cotton rats (Burt 1938).

The blue-rumped parrotlet occurs in the short tree tropical deciduous forest, principally in flocks at altitudes around 360 meters. The lilac-crowned parrot occurs in the same area and also in the oak zone. The great horned owl, ferruginous owl, Gambel's quail, elegant quail, berylline and black-chinned hummingbirds, coppery-tailed trogon, russet-crowned motmot, Gila woodpecker, and a host of songbirds of the southwestern United States are present (van Rossem 1945). Almost no biotic notes are available regarding them.

In the Sierra Suratato, inland from Guamúchil, Sinaloa, some 200 kilometers south of the Rio Mayo Basin, there is a short tree tropical deciduous forest near 300 meters elevation. Near Varomena, a good mixed deciduous forest occurs with mature trees 9 to 12 meters high. The relatively large mesic leaf, averaging about 75 centimeters, is more common than the smaller pinnatifid type represented by the Mimosaceae and Burseraceae (Gentry 1946b). In the Sierra Taonichoma, 200 kilometers farther south, the great bulk of the mountains is covered with a climax virgin short tree forest, persisting through 915 meters or more elevation. The general height of the forest dominants is 10 to 12 meters or more elevation. The plain around the west base of the mountain is covered with a uniform short tree forest about 9 meters high. The trees are evenly distributed. Animal constituents of this plant-rich wilderness are mountain lions and jaguars, which live in the great canyons and kill deer. Peccaries occupy the underbrush; coatis climb cliffs and trees, securing a living from roots and fruits. Among the many indigenous birds are parrots, chachalacas and game birds (Gentry 1942).

Michoacan–Guerrero–Puebla Forest

The annual rainfall along the Pacific Coast increases toward the southeast; at Manzanillo, Nayarit, it is 105 centimeters; at La Unión, Guerrero, 112 centimeters; at Acapulco, Guerrero, 150 centimeters; at Jamiltepec, Oaxaca, 205 centimeters. Beyond Jamiltepec, however, rainfall decreases. This probably explains the dropping out of the arid thorn forest near Acapulco and the change back to savanna in the Salina Cruz area in southeastern Oaxaca.

The Hoogstral Expedition of the Chicago Natural History Museum in 1940 and 1941 provided information for a transect between the Rio Tepalcatepec and Corro de Tacintaro in northern Michoacan. Leavenworth (1946) describes the tropical deciduous forest portion of this transect. Among the trees which he mentions are *Ficus padifolia* and *Haematoxylum brasiletto*. Miranda (1947) states the greater part of Leavenworth's deciduous forest is gallery forest. His list also includes trees on the uplands of the genera *Bursera* and *Lysiloma*, which are important dominants in Sonora. The tropical deciduous forest, thorn forest, and scrub are not fully differentiated in the bird studies of Blake and Hanson (1942) or by Leopold and Hermandes (1944).

When the highway from Mexico City to Acapulco was surveyed under the direction of Professors G. Gandara and I. M. M. Lumbier, a record was made of the plants growing adjacent to the highway route. On the road from Cuernavaca to Acapulco one leaves the oak–pine forest and enters the short tree forest. In the stretch between Cuernavaca and Iguala, the following short trees and shrubs, which also occur in Sonora, were recorded: *Pithecellobium dulce, Zanthoxylum fagara, Exogonium bracteatum, Acacia farnesiana,* and *Taxodium mucronatum,* together with 12 species of other genera. After crossing a ridge in the last few kilometers before reaching the coast, they added *Alvaradoa amorphoides, Croton ciliato-glandulosus, Lysiloma acapulcensis, Bauhinia latifolia, B. pes-caprae,* and *Nectandra sanguinea* (Fig. 17-9). The forest near Acapulco may be secondary.

The short tree deciduous forest mammals have received less close observation as to habits and choice of vegetation than the tall tree forest mammals. Areas have become modified by human activity and fire has done much damage (Miranda 1942). The white-tailed deer occurs from Colima to the Isthmus of Tehuantepec, mainly in the deciduous forest and lower edge of oak–pine bordering on the desert (Leopold and Hermandes 1944). The mountain lion, jaguar, ocelot, wolf, and coyote occur throughout most of Guerrero, probably frequenting the deciduous forest. The peccary enters the forest along the Pacific Coast. The coati occurs abundantly in Colima, has been recorded at Apatzingán, Michoacan, and occurs in Guerrero. The murine opossum *Mar-*

Fig. 17-9. Communities in Guerrero, Mexico. A. Scrub with white pillar-like cacti, shrubs, and trees. B. Common lizard *Ctenosaura pectinata* of the short tree forest (photos by Rozella B. Smith).

mosa canescens is present (Fig. 17-11C). The gray fox occurs in the upper deciduous forest (Leopold and Hermandes 1944, Hall and Bernardo 1949). Occurring statewide in Michoacan are opossum, raccoon, hooded skunk, gray fox, coyote, seven species of bats, armadillo, and deer. All find suitable places in the deciduous forest (Hall and Bernardo 1949). Two rice rats, the tropical pygmy mouse, one harvest mouse, and one spiny pocket mouse are confined to the lower parts of Michoacan. Most of these species were taken in Morelos by Davis and Russell (1953).

As to the birds, the banded quail occurs throughout the tropical deciduous forest area and into arid llanos of the Rio Balsas, which suggests they use secondary vegetation (Fig. 17-10). The tropical deciduous forest in Guerrero supports the magpie-jay, squirrel cuckoo, orange-fronted parakeet, white-

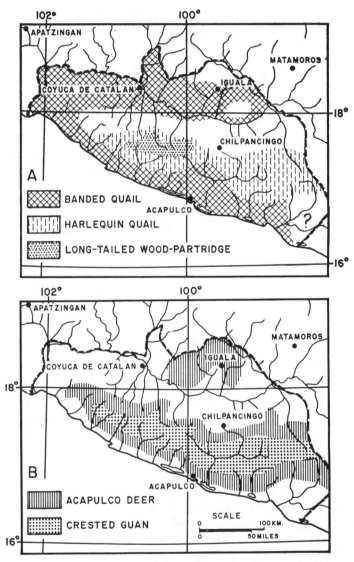

Fig. 17-10. Distribution of birds and mammals in Guerrero, Mexico. A. Banded quail in a very diversified area with mattoral (deciduous forest) and *llanos aridus;* harlequin quail in the oak–pine; long-tailed wood partridge above 8000 feet (2400 m); the singing quail has a similar distribution to the partridge and frequents the oak forest. B. Acapulco deer and the crested guan in both cloud forest and oak–pine forest (from Leopold and Hermandes 1944).

tipped dove, and turkey. Davis and Russell (1953) list a large series of birds and mammals in Morelos, but these are difficult to relate to types of vegetation. Mammals and birds that probably originally utilized deciduous forest in Morelos include coati, hispid cotton rat, fork-tailed emerald hummingbird, dusky hummingbird, russet-crowned motmot, rufous-backed robin, streak-backed oriole, and squirrel cuckoo. Predatory birds which are widely distributed and enter the deciduous forest are gray hawk, common black hawk, crested caracara, sparrow hawk, and great horned owl (Jean W. Graber, personal communication).

Hobart M. Smith (personal communication) took the lizards *Anolis gadovi*, *A. taylori,* and *Phyllodactylus delcampi* in this tropical deciduous forest and nowhere else in Mexico. He also found *Sceloporus stejnegeri, Ctenosaura pectinata* (Fig. 17-9), the anurans *Bufo gemmifer* and *Acrodytes inflata,* and the caecilian *Gymnopis multiplicata.*

Farther east Miranda (1942) diagrams a reconstruction of the vegetation of the Eje Volcanico in the upper Balsas area (Fig. 17-11). A forest of fig, willow, and baldcypress makes up the gallery forest. Miranda believes a huamuchiles–mesquite forest *Pithecellobium dulce–Prosopis juliflora* once occupied the flats at the base of the slopes, while the slopes were covered with short tree tropical deciduous forest of *Bursera. B. odorata* is important and may be either a shrub or a tree, as is the case also with *B. jorullensis. Bursera*

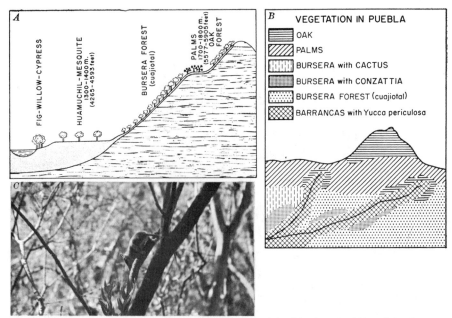

Fig. 17-11. Reconstruction of the vegetation of Puebla, Mexico. A. Miranda's diagram of the vegetation of the escarpment of the Mexican Plateau. B. Distribution of vegetation in central Puebla which has been called tropical deciduous forest (after Miranda 1942). C. Mouse opossum *Marmosa* near Tehuantepec (photo by Rozella B. Smith).

vegetation occurs generally on the Eje Volcanico in the less moist areas and is associated with the often numerous white pillar cactus (Fig. 17-9A). On slopes with much moisture, *Hauya rusbyi*, and *Coccoloba acapulcensis*, the latter species sometimes a shrub, become dominants.

Succession in a lake, Laguna de Epatlan, at an altitude about 1350 meters, is quite similar to that in the United States. The emergent plants are the well-known reed *Phragmites communis*, bullrush *Scirpus californicus*, and cat-tail *Typha latifolia*. A later stage is dominated by *Acacia farnesiana* which occurs in Texas (Cantu 1942). As the habitat becomes drier, the proportion of northern species decreases. The predominant mature forest of the area is dominated by species of *Bursera*.

Oaxaca–Chiapas Forest

The Oaxaca forest of the coast extends inland in a northeasterly direction in the Tehuantepec area to connect with that in the central valley of Chiapas (Fig. 16-2) and crosses the Guatemalan border. Miranda (1952) maps part of the valley between 700 and 1000 meters or higher, as short tree deciduous forest. The Guatemalan forest trees are 8 to 12 meters in height. The most frequent tree in the forest is *Alvaradoa amorphoides*. Others occurring are *Heliocarpus reticulatus*, *Fraxinus purpusii*, *Lysiloma acapulcensis*, *Haematoxylum brasiletto*, *Ceiba acuminata*, *Cochlospermum vitifolium*, *Bursera bipinnata*, *Bumelia celastrina*, *Pistacia mexicana*, *Gyrocarpus americanus*, *Piscidia piscipula*, *Bursera excelsa*, *Swietenia humilis*, *Ficus cookii*, and *Zuelania guidonia*. *Haematoxylum brasiletto*, *Ceiba acuminata*, and *Cochlospermum vitifolium* also occur in Sonora. In the barrancas the forest trees are 15 to 20 meters in height. Miranda (1952) lists 90 plants of importance. There are no biotic studies in this forest.

THORN FOREST COMMUNITIES

Thorn forest accompanies declining rainfall and is generally an intermediate stage between the short tree deciduous forest and scrub. The vegetation has been called xerophytic forest, acid forest, semidesert forest and *selvas secas*. It becomes deciduous in the dry season. The physical nature of the soil is such that only a fraction of the rainfall becomes available to the trees because of rapid runoff. The soil is often impregnated with saline substances which favor thorn forest. The soils of dry forest regions are usually porous and often thin and rocky. They are seldom acid in reaction but often alkaline. The forest forms open, parklike stands, with occasional dense thickets of cactus and thorny shrubs. Ground cover is sparse and composed of creeping cactus, agaves, and other species (Barbour 1941).

The largest areas of thorn forest are in southern Sonora and in Sinaloa, and most of the following description concerns these areas. There is a narrow strip of thorn forest along the Pacific Coast south of Guerrero and on the Gulf Coast in southern Tamaulipas (Leopold 1950b). There is also an area in the

Sierra de Tamaulipas (Martin *et al.* 1954). Leopold maps some thorn forest
in extreme northern Yucatan, and small areas occur in various other localities;
many of them may be secondary (Figs. 14-1, 16-1).

The climate of the thorn forest area is represented by only a few meteorologi-
cal stations (Table 17-3). At Mazatlán, 225 kilometers south of Culiacán, the
mean annual rainfall (1889-1936) is 77 centimeters, with 54 centimeters coming from July through September. From November to May inclusive, the mean
rainfall is only 8 centimeters. Near Aldama, 40 kilometers from the coast in
southern Tamaulipas, an annual rainfall of 72 centimeters, with 56 centimeters
coming from June through September, is characteristic.

Table 17-3. Rainfall and Temperature at Localities in the Thorn Forest

Locality	Altitude	Annual Rainfall	Summer Rainfall	Length of Rainy Season	Temperature During Growing Season
Sonora (average).....	100-300 m	30 cm	20 cm	10 weeks	32° C.
Navojoa, Sonora.....	38	29	40	16	32
Culiacán, Sinaloa.....	53	60	43	16	28
Punta Jerez, Tamaulipas........	1	122	101	16	27

ᵃ Sonora data from Gentry 1942; data for other localities from government records.

Shreve (1937) regards this forest as more closely related to desert vegetation
than to the short tree forest. Its pronounced xeric features are the scanty and
open canopy, the strong predominance of small leaves or very small leaflets,
the occurrence of cacti, and the great difference between the vegetation of the
uplands and the luxuriant floodplains. In the Rio Mayo region of Sonora the
thorn forest has an average height of about 6.5 meters, with rather spindling
shrubs and forest undergrowth. The total density of perennials in typical
stands is 10,000 to 12,500 stems per hectare. The forest occupies the low
basaltic hills and mesas and, to a lesser extent, the lowland valleys. In elevation
it ranges from sea level to 600 meters. The thorn forest lies adjacent to the
sea on the west and to the desert on the north, while on the east it intergrades
with the short tree deciduous forest (Gentry 1942). The thorn forest on the
coastal plain of the northern half of Sinaloa exhibits considerable uniformity,
but there is gradual change farther south through the lower end of the state.
The belt of forest near the coast is more arid and open than in the interior.

Stem growth may be erect and single but is more often angularly ascending
and cespitose (*Acacia cymbispina, Bursera, Cereus thurberi*), woody and
dense (*Acacia*), soft, thick, and semisucculent (*Bursera, Fouquieria macdougalii*), or strictly succulent (*Cactaceae, Pedilanthus macrocarpus*) (Gentry
1942). There is a preponderance of trees with compound leaves which have

leaflets less than 25 square millimeters in area (*Acacia cymbispina, Lysiloma divaricata, L. microphylla, Prosopis juliflora, Pithecellobium sonorae, Eysenhardtia orthocarpa, Acacia farnesiana, A. pennatula*). Trees which have compound leaves with leaflets larger than 2500 square millimeters in area (*Cassia atomaria, Lonchocarpus megalanthus, Caesalpinia platyloba, Guaiacum coulteri, Diphysa suberosa*) are also frequent. The most common trees are the evergreen *Zizyphus sonorensis* and the deciduous *Ipomoea arborescens*. Other species are *Jatropha cinerea, Bunchosia palmeri, Cordia sonorae, Ficus petiolaris,* and *Sapium lateriflorum* (Shreve 1937).

In Sonora, there are two types of vegetation, determined chiefly by the nature of the terrain: the thorn forest itself (mesas and slopes) and mixed associations (arroyo margins and valleys). The former is uniform in growth, with close regular spacing of individuals and the plants elongated and crown-branched; the latter is varied and irregular in formation and individual spacing. The valleys are often dominated by very large mesquite (Gentry 1942).

The height of the perennial species in the thorn forest is 5.5 to 8 meters. The numbers per hectare of individuals of dominant species in the thorn forest of Sonora are *Fouquieria macdougalii,* 310; *Jatropha cordata,* 300; *Coursetia glandulosa,* 225; *Mimosa palmeri,* 90; *Bursera,* 85; *Acacia cymbispina,* 80; *Lysiloma divaricata,* 75; and *Karwinskia humboldtiana,* 10. *Acacia cymbispina* is the most important single dominant. The following shrubs are 1.2 to 2.3 meters in height: *Randia obcordata,* 45 per hectare; *Mimosa,* 35; *Lycium berlandieri longistylum,* 15; *Caesalpinia palmeri,* 15; *Piscidia mollis,* 5; *Cercidium torreyanum,* 5 and *Cereus thurberi,* 20. The most abundant and conspicuous species of cactus is *Cereus pectenaboriginum*. The thorn forest is rich in both annual and perennial herbaceous plants. It is not always possible to distinguish readily between ephemerals and root perennials. The greatest development of herbaceous plants comes during the summer months (Gentry 1942).

The major permeant influent is the white-tailed deer, which apparently is not numerous and restricted to higher elevations in the thorn forest. Present also are the coyote and opossum. Four species of bats occur in the area (Burt 1938).

Minor influents are the antelope jack rabbit, desert cottontail, bobcat, hog-nosed skunk, and badger. The white-throated wood rat nests in the cactus. Burrowing rodents include the painted spiny pocket mouse and the southern pocket gopher, which appears to be largely confined to the thorn forest. The pocket mouse *Perognathus goldmani* occurs in both the thorn forest and short tree forest. Its food consists chiefly of grass seeds. *Perognathus pernix* is widely distributed in the thorn forest. It is most numerous in thick cactus groves, and the seeds of *Opuntia* are an important article in its diet. Merriam's kangaroo rats inhabit bushy areas and their food is seed of grasses and forbs. The southern grasshopper mouse prefers open cactus-covered areas, grass, and bush-covered areas. Its food is insects and other arthropods.

Gambel's quail is generally distributed in the thorn forest and short tree forest along the Rio Mayo. The inca dove evidently occurs in both the thorn and short tree forest and also in the desert but with a declining population toward the north. The white-fronted parrot is a regular resident of the thorn forest. The groove-billed ani arrives in the Mayo Valley in May and June to breed. It and the screech owl are found in both the thorn and short tree forests. The elegant and Gambel's quails occur in the thorn forest area of Sonora. The military macaw, the blue-rumped parrotlet, and the white-fronted parrot are to be expected in both the thorn and short tree forests. The Gila and ladder-backed woodpeckers are common in the thorn forest and short tree forest. There is a long list of small resident and migrant birds, including the social flycatcher, various sparrows, swallows, and warblers (van Rossem 1945, Blake 1953).

The bush-climbing caleyea snake *Leptodeira punctata* was taken crawling along the bank of a temporary stream and *L. maculata* under a log. Two *Masticophis lineatulus* were captured, one near a pool and another in a tree trying to swallow a *Hyla baudinii*. *Masticophis bilineatus* was observed crawling in sparse shrubs. The ubiquitous *Constrictor constrictor imperator* was recorded (Taylor 1936).

At an altitude of 450 to 550 meters near Colima, Kannowski (1954) reports a two-layered forest at the contact of the short tree and thorn forests. He lists four overstory trees: *Acacia cymbispina,* other *Acacia, Caesalpinia pulcherrima,* and *Calsearia pringlei.* Hooper (1955) reports the southern pygmy mouse, spiny mouse, hispid cotton rat, and white-tailed deer. Kannowski (1954) found the soil-burrowing ant *Novomessor manni* abundant in the upper thorn forest.

ARID TROPICAL SCRUB COMMUNITIES

Iguala, Guerrero, at 735 meters, is near the mean elevation of the *basque espinosa* of Leopold and Hermandes (1944) and has an annual rainfall of 114 centimeters. Of this, 95 centimeters fall between June and September and only about 2 centimeters between December and March. The mean annual temperature is 27.0° C., with the temperature from June to September averaging 25.6° C. May is the hottest month with 30.7° C. Farther east at Huehustlan in Puebla the climatic conditions are quite similar to those at Iguala, but the dry season is a little more severe.

The arid areas near the mid-slope of Eje Volcanico north of Apatzingán, Michoacan, noted by Blake and Hanson (1942) and the area shown by Leopold and Hermandes around Teloloapan, Mexico, appear to be largely scrubby *Bursera* vegetation with the cactus *Cephalocereus* (Fig. 17-9A), *Acacia,* and *Caesalpinia.* According to the map of Leopold and Hermandes, this vegetation is duplicated in the rolling hills south of the Rio Balsas. Figure 17-12 is a diagrammatic representation of these facts and an attempt to suggest that the primitive forest was largely short tree deciduous forest.

A second type of scrub occupies the Rio Balsas Basin with *Acacia, Lysiloma, Pithecellobium,* and *Prosopis.* On slightly more humid uplands near Coyuca de Catalan, large spreading shrub *Bauhinia chlorantha,* other shrubs *Zizyphus sonorensis* and *Desmanthus virgatus,* along with the leguminous trees and blue-stem bunchgrass *Andropogon fastigiatus* are found. *Ficus mexicana, Salix humboldtiana,* and *Taxodium mucronatum* (10 to 30 m high) are typical stream-side plants.

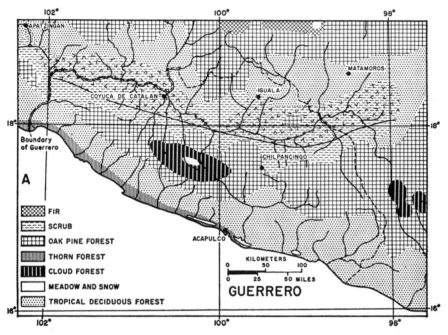

Fig. 17-12. Diagrammatic reconstruction of the vegetation in and adjacent to the state of Guerrero, Mexico. The **V**'s in the northern half of the scrub legend indicate lesser density of chaparral scrub and the presence of the grass *Andropogon fastigiatus* that is called *llanos aridus* (Leopold and Hermandes 1944, Leopold 1950b, Blake and Hanson 1942, Davis and Russell 1953, Miranda 1942).

Large mammals are generally absent in the arid tropical scrub. The peccary, however, is found throughout Guerrero, which includes much scrub. Characteristic birds are the plain chachalaca, gray hawk, and ruddy ground-dove. The banded quail occurs both here and along the coast. The sharp-shinned hawk is reported from the Balsas Valley. Gadow (1908) visited the Balsas scrub and found the peculiar wormlike lizard *Bipes canaliculatus.* It burrows in sandy alluvial soil along the Rio Balsas above the reach of ordinary floods.

In protected Guatemalan valleys, the scrub type of vegetation has been called a cactus–thornbush community by McBryde (1945) and "desert chaparral region" by Standley and Steyermark (1945) (Fig. 17-13). The birds here have been listed by Tashian (1953a). Somewhat similar communities occur near Tehuacán, Mexico, and in the upper part of the Salada Valley in Puebla. There are also areas near the eastern edge of the Mexican Plateau north of Mexico City.

Fig. 17-13. Scrub in dry mountain pockets. A. A ridge in the upper Rio Salada Valley north of Tehuacan (1951) along the highway to Orizaba with small shrubs and yuccas. B. The valley floor near the highway in the same area with yuccas and two types of cacti and shrubs. C. The Motagua Valley near Uzumatlan at an altitude of 200 meters looking in a northerly direction toward the Sierra de las Minas (1951). The cacti with the white apexes are *Cereus maxonii,* the predominant shrubs are *Mimosa* and *Calliandra.* D. The cactus without the white tip is *Cereus eichlamii.* The low trees are *Karwinskia calderoni* and *Jacquinia aurantiaca* (photos by R. E. Tashian).

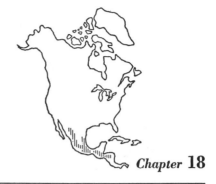

Chapter **18**

The Oak and Pine Forests, Cloud Forest, and Other Mountain Communities

The communities considered in this chapter are confined to mountains and foothills. *Pinus pseudostrobus* and white-tailed deer are distributed throughout, the deer especially through the oak communities. At the time of the Spanish conquest, the native population was relatively dense in the oak areas. The higher altitude pine areas were only scatteredly occupied.

Oak forest forms a narrow belt extending from Sonora and Nuevo León probably to Panama. The altitude of the center of the belts ranges from 1000 to 2300 meters. Several species of oaks may occur at a particular point but be almost entirely different at another point only a couple hundred kilometers distant. On the Pacific Coast, oak woodland as known in the United States ends in Jalisco, Mexico; in the interior the limit comes a little farther north at about 22° N. Lat. In the Rio Mayo area of Sonora, oak forest comes in above the short tree tropical deciduous forest.

Pine forest occurs at elevations of 1300 to 3900 meters from 22° N. Lat. in southern Mexico through Nicaragua. A large gap in the pine belt occurs at Tehuantepec and in eastern Guatemala. North temperate species and genera of insects, birds, and mammals extend south as constituents of the pine communities. Only one genus of small rodents is found that is not also present in the north temperate mountains. In Guatemala much of the forest has been destroyed by thousands of years of occupation by the Mayas.

Forest communities with two or three species of fir mixed with pine usually occur at 2800 to 3600 meters elevation on the high mountains of Mexico and Guatemala. Dense stands of cypress occur in the pine zone below 3000 meters, especially in Guatemala.

Cloud forest occurs at 900 to 2700 meters in isolated areas from Tamaulipas into Panama.

459

Alpine meadow with decumbent juniper occurs at 3800 meters and higher on Cerro Potosí, Orizaba, and other places.

OAK AND PINE FORESTS

It is impracticable to separate the description of pine and oak communities when animals are considered, because of lack of detailed studies, disturbance to the vegetation, and uncertain ecological position of the oaks. The best biotic study of the oak–pine community was conducted by Gadow (1908) on Mount Orizaba. Other accounts are very limited in scope with almost nothing con-

Fig. 18-1. A, C, and D. Oak–pine in Michoacan. B. Oak–pine in Honduras.

cerning insects, which are usually important influents among coniferous **trees.**
The close resemblance to northern communities has made the study of **the**
pine forest unattractive to investigators (Figs. 18-1, 18-2).

Soil and Climate

The soil is derived mainly fro.n rock, broken pieces of which are usually
present, although in many areas the soil is of volcanic origin. There is a **little**
limestone, and the mesas are often covered with a layer of calcareous clay and,

Fig. 18-2. A. Pine forest northwest of Tegucigalpa, Honduras, at an altitude near 1500
meters, probably second-growth following fire: *Pinus oocarpa,* bracken fern *Pteridium,*
and *Calea, Conostegia,* and *Rubus* (plant identifications by Paul H. Allen). B. Pine **at**
lower altitude and not burned as recently as A; epiphytes on some of the tree **trunks**
indicate the tropics. C. The palm *Paurotis cookii,* shown at the right and back, **in the**
upper oak and pine zones (identification by Paul H. Allen).

pebbles. Runoff is commonly rapid, so that only a small percentage of the rainfall is retained, although in Nuevo León there is a litter of leaves and a humus layer over the ground. There is a gradient of rainfall from the north southward. Oak woodlands in Arizona at 1300 to 1800 meters have an annual rainfall of 25 to 40 centimeters, while in the south it increases to over 100 centimeters (Table 18-1). In Arizona the oak woodland slopes rise from an arid plain at 1200 meters rather than the 0 to 200 meter coastal plain in the Rio Mayo region, Sonora.

Table 18-1. Precipitation and Temperature at Various Localities in Oak and Pine Forests[a]

LOCALITY	VEGE-TATION	ALTI-TUDE	ANNUAL PRECIP-ITATION	RAINY SEASON	SUMMER RAINFALL	MEAN ANNUAL TEM-PERATURE
Rio Mayo, Sonora......oak		1054 m	63 cm	June-Sept.	48 cm	23.3° C.
Rio Mayo, Sonora......pine		1830	97	May-Sept.	71	20.0
Galeana, Nuevo León...oak		1654	47	June-Sept.	25	18.0
Mexico, D. F..........oak		2280	59	June-Sept.	45	11.3
Desierto de los Leones..pine		3220	128	May-Sept.	108	15.0
Comitán, Chiapas......oak		1635	100	June-Sept.	65	23.3
Las Casas, Chiapas.....oak & pine		2128	117	May-Sept.	92	14.5
Tegucigalpa, Honduras..oak		936	...	June-Sept.	...	....

[a] Sonora data from Gentry 1947; data for other localities from government records.

Oaks

At Rio Mayo the principal oaks are *Quercus chihuahuensis, Q. albocincta, Q. tuberculata, Q. gentryi*. The pine zone at Rio Mayo, where the elevation of the mountains is low, is poorly developed. None of the oak or pine dominants is known to occur north of the Mexico–United States boundary. There are four species of oaks scattered in the lower part of the pine zone: *Q. arizonica* occurs in Arizona but is not a dominant in the oak forest; *Q. chihuahuensis* is the most important species of oak. It is a small tree, irregularly branched, with a trunk 30 centimeters or more in diameter, and grows to a height of 4.6 to 7.6 meters. It forms an open, evenly-spaced cover, varying according to conditions. On the arid open slopes and mesas, the trees are commonly 9.1 to 15.2 meters apart; where sufficient water is available, they are close enough together for their crowns to touch.

The average density of perennial plants in the oak forest is between 750 and 10,000 individuals per hectare of which 50 to 350 are trees with an average height of about 9.1 meters. The trees are usually leafless during the spring dry season from late February through most of June. Only in the moist canyon bottoms are the trees evergreen. The palm *Sabal uresana* also occurs near water courses. *Yucca, Nolina matapensis,* and *Agave* are abundant on arid slopes. Tropical plants in the understory of the oak forest are *Cornus disciflora, Hoffmannia rosei,* and *Solanum madrense.*

At San Angel, D. F., Mexico, oaks occur at elevations from 2350 to 2900 meters where rainfall is 85 to 135 centimeters. Pines come in from 2800 to 3000 meters with rainfall 125 to 145 centimeters. The important oaks here are *Quercus centralia, Q. pulchella,* and *Q. rugosa.* An *Agave* was the only Rio Mayo species noted at San Angel (Rzedowski 1954). There *Indigofera densifolia, Cosmos bipinnatus, Cupressus lindleyii, Calea pedunculatus, Archibaccharis hirtella,* and *Geranium latum* are mentioned as important. The coarse grasses are mostly different species in the oak and pine communities. No species cf *Muhlenbergia* occurs in both communities although there is a good series of species present in each community.

Relations of species are the same in the Sierra Madre Oriental. At 1900 meters on Cerro Grande, the dominant oaks are *Quercus coccolobaefolia* and *Q. crassifolia* with *Arctostaphylos pungens* and *Pinus teocote* near Ciudad, San Luis Potosí. The annual rainfall here is 36 centimeters. At El Tablon near Zaragoza at an altitude of 2350 meters, the oak dominants are the same as at Cerro Grande with *Q. affinis* and *Q. castanea* added. In the understory, *Q. rugulosa* and *Q. hartwegii* are present. None of the oaks in San Luis Potosí is at San Angel. However, *Pinus teocote* is important at San Angel and *Arbutus xalapensis* is present in both places. On the east slope of the Sierra Madre Oriental in Nuevo León, around 1500 meters elevation, *Quercus clivicola, Q. monterreyensis, Q. cupreata, Q. canbyi,* and *Q. polymorpha* are most notable. Apparently none of oaks at Nuevo León occurs in the other localities under consideration. In exposed situations, branching occurs at two meters above the ground and the canopy is open (Muller 1939). *Pinus teocote* is occasionally mixed with the oak. This pine occurs at 1500 to 2800 meters on east slopes and from 2500 to 3000 meters on the western slopes of the plateau. Relict oak–pine areas occur at 2100 meters in Costa Rica (Calvert and Calvert 1917).

Pines

Near the Rio Mayo, *Pinus flexilis reflexa* is the principal dominant in the dry tablelands, while *P. ayacahuite* is the dominant in moist situations; *P. arizonica* is important locally. *P. ayacahuite* extends south into Chiapas and Guatemala while *P. f. reflexa* extends into Canada. *P. lumholtzii* is limited to the plateau area. Snow is of annual occurrence near the Rio Mayo. The density of trees and shrubs varies from 200 to 1200 per hectare. *Sabal* and *Agave* are present among the pines. *Alnus* occurs near streams.

In the San Angel, D. F., area, *Pinus hartwegii* and *P. teocote* are important, with *P. montezumae* in lesser abundance; *Abies religiosa* is present.

Pinus ayacahuite, P. occarpa, P. montezumae, and *P. teocote* are the most widely distributed species. *P. hartwegii* and *P. rudis* occur under cold climates from 2800 to 4000 meters altitude. The firs *Abies religiosa* and *A. guate-*

malensis are often associates of these two pines. *P. strobus* and *P. tenuifolia* occur in rainy areas. Below the pine belt, the oak forest constitutes an ecotone with the tropical or desert communities at lower elevations. The oak forest takes the place of pinyon–juniper woodland and oak–bush of the United States, with which it is in contact at its northern extremity.

Animal Influents and Dominants

Entering the oak–pine in northern Mexico are three grizzly bears, black bear, and wolf. The wolves and bears drop out between 19° and 22° N. Lat. in central Mexico. The gray fox, raccoon, and mountain lion continue into South America. The coyote is found in Nicaragua. The white-tailed deer is represented by various subspecies south into Panama. The peccary, coati, gray fox, and raccoon enter the oak–pine at its lower levels. Near the Rio Mayo and southward, peccaries, coatis, jaguars, parrots, and other animals enter mixtures of pines and oaks. The mountains here are too low for pure pine communities (Gentry 1946a, 1946b). The tree squirrel *Sciurus truei* is common in the oak near the Rio Mayo. The eastern cottontail covers the highland south to Costa Rica. *Sylvilagus cunicularius* is in the high chain of volcanoes along the southern escarpment of the Mexican Plateau, the Eje Volcanico. The rock squirrel is widely distributed in the oak–pine forest from Nayarit to Morelos and Nuevo León.

In the large oak areas of Michoacan, Mexico, spiny pocket mice, cotton rats, and piñon mice occur in abundance. The southern pocket gopher and the llano pocket gopher are found generally distributed. Harvest mice, deer mice, ferruginous wood rats occur along with subspecies of the other mammals noted north of Michoacan. As to the mammals of the pine forest in Michoacan, the volcano mouse, Mexican vole, and wood mouse are present. The white-eared hummingbird, gray-sided chickadee, pigmy nuthatch, Mexican junco, Sierra Madre sparrow, and others occur in the Morelos pine and extend mainly northward (Davis and Russell 1953).

The turkey is clearly an oak–pine inhabitant, occurring from Michoacan to Veracruz and Oaxaca. The band-tailed pigeon is in oak forest near Rio Mayo and extends in oak–pine forest south into Nicaragua; it feeds on acorns. The green violet-ear hummingbird, red-shafted flicker, and hairy woodpecker are in the oak forest throughout the southern half of the plateau, including Morelos. Most of those named are also in the pine forest. In the oak–pine forest of Guatemala, there are a number of races of northern Rocky Mountain birds including among others the hairy woodpecker, golden-crowned kinglet, Steller's jay, and brown creeper. The Mexican junco is one of the most abundant birds in the pine forest from southern Arizona to Guatemala.

In the pine forest of central Guerrero, the common lizards are *Sceloporus formosus scitulus, Anolis liogaster, Barisia g. gadovii,* and *Abronia;* snakes, *Geophis omiltemana, Rhadinaea omiltemana, Bothrops barbouri,* and *Crotalus o. omiltemanus;* frogs, *Tomodactylus amulae* and two *Eleutherodactylus* (Hobart M. Smith, personal communication).

Near Samac, Guatemala, at an altitude of 1300 meters, in an original forest of almost pure pine with undergrowth, the snails *Streptostila lattrei, Salasiella margaritacea, Epirobia polygyrella, Poteria bisinuatus,* *Chondropoma rubricundum,* and *Dryaemus lattrei,* were reported by Van der Schalie (1940).

Bark beetles feed on the cambium of pine and other trees throughout the hemisphere. Hopkins (1909) noted that the bark beetle *Dendroctonus parallelocollis* destroyed forests of *Pinus teocote, P. leiophylla,* and *P. ayacahuite* in Morelos. *D. mexicana* attacked the same species in Michoacan. Kinsey (1929) has pointed out a number of species of oak–apple gall-forming cynips which occur on the highlands of Mexico and Guatemala.

CLOUD FORESTS

Cloud forest has also been called temperate rain forest, upland forest, and *selvas nubladas.* Cloud forests occupy a mountain zone beginning between 1000 to 2000 meters depending on where moisture in the ascending aerial masses condenses. Rain rarely occurs, but the forests are dripping wet throughout the year from the mist and from direct moisture condensation on the vegetation. This moisture must enter the soil to be absorbed by the trees. The climate is usually cool, but freezing temperatures are uncommon. The mountain slopes below the cloud forest may be grassy and even arid. Near their upper limits, the cloud forest becomes low and scrubby and may merge into bleak and cold treeless regions. Their soils are usually thin and rocky but fertile. Areas of cloud forest occur in scattered and isolated sections of the mountains of Mexico, Central America, and on higher peaks of the West Indies islands and Andes Mountains. The forests are unusually rich in orchids and other epiphytic plants. Tree flowers and fruits are neither abundant nor conspicuous (Barbour 1941).

Cloud forests do not appear to be a distinct vegetation type in the sense of being dominated by certain kinds of trees but are forests of various taxonomic composition that are locally fog-ridden. Their character is determined vegetatively more by the heavy tropical understory than by the overstory. These subordinate plants crowd the lower strata and occupy the trunks and branches of the dominant trees. The reproduction of trees goes on successfully, however, in spite of the seedlings being subject to unusual competition. Tropical cloud forest may be recognized at elevations of 900 to 2500 meters in Panama and 1500 to 3000 meters in Guatemala (Fig. 18-3), and temperate cloud forest usually above 1200 to 2300 meters in Chiapas and 900 to 1700 meters in Tamaulipas (Carr 1950, Martin *et al.* 1954, Miranda 1952, Schmidt 1936).

The tropical cloud forest is very dense. There are several strata composed of shade tolerant species and often a heavy nonthorny undergrowth of tree ferns. Vines and creepers are not abundant. The dense ground cover is composed of ferns, cryptogams, and herbaceous plants. The trees range from medium to fairly large size, are thin, and have their root systems often partially exposed. The tree boles, only slightly buttressed, are long and the crowns are large and rounded. The bark is variable but is usually thin and smooth. Foliage is very abundant, with large multipored leaves (Barbour 1941).

In Panama, the upper tropical zone has an irregular lower margin, depending largely on variations in moisture with change in slope exposure. This forest is without conifers; three species of oak and one of *Prunus* are present (Goldman 1920). The cloud forest in Honduras contains oaks and a dense undergrowth dominated by melastomes. Tree ferns are present, and epiphytes are numerous and of great variety. The many small peaks through the area are usually crowned with pine (Carr 1950). The cloud forest in southern Guatemala is characterized by epiphytes and a greater luxuriance of plants, especially tree ferns, *Matudaea* being one of them (Miranda 1952). The shrub *Miconia,* which also occurs in Panama, is abundant. Above 2300 meters tree ferns de-

Fig. 18-3. Cloud forest on Mount Ovando near Escuintla, Chiapas, Mexico: A, *Quercus boqueronae* (about 20 m tall); B, *Billia hippocastanum;* C, the abundant shrub *Miconia* (grows to 2.5 m tall); D, ascending stem of *Bignonia;* E, the epiphyte *Tillandsia fasciculata* (photo by Rozella B. Smith; provisional identification of plants by Ezi Matuda of Mexico City).

crease and the curious tree *Cheiranthodendron pentadactylon* enters, ranging up to 2900 meters. Above 2600 meters there is an increase of bamboo in the undergrowth to form dense thickets at the upper border of the cloud forest. The last 300 meters of the cloud forest may be characterized by birch and exhibits a transition to an oak–pine, pine, or other coniferous forest (Schmidt 1936).

Large areas of primeval cloud forest formerly existed in the high country of Guatemala, in some places even at 3500 meters. The pines and oaks averaged nearly two meters in diameter, and the trees were lofty and loaded with epiphytes. Eruptions of lava often destroyed areas of the cloud forest, resulting subsequently in a series of successional stages. If the eruption covered the ground with ash, it took about 25 years for a thick vegetation of grasses and bushes to take possession. Pine forest appeared next, followed after a very long time by the cloud forest again. Erosion is excessive on volcanic ashes, sands, and tufa, such as cover most of this highland. The upper slopes of the volcanoes are consequently seamed with ravines, with brooks at their bottoms, and knifelike ridges between. Dense subtropical growth often occurs in the ravines, with pines on the sterile ridges (Griscom, 1932).

In northern Chiapas, Mexico, the trees listed by Miranda (1952) include *Clethra suaveolens, Saurauia villosa, Turpinia paniculata, Cedrela pacayana, Prunus lundelliana, Persea schiedeana,* and *Orepanax xalapense.* On Mount Ovando, near Escuintla, at 2000 meters (Fig. 18-3), trees of the same genera are present, but the species are different. Herbs include the large-leaved *Gunnera insignis* and some ten other genera. The most northerly area of cloud forest is in Tamaulipas. This is an unusual forest that includes such temperate deciduous forest trees as sweetgum, blackgum, American hornbeam, and dogwood.

Most of the vertebrates (Carr 1950) and invertebrates (opilionids: Goodnight and Goodnight 1953) found in cloud forests belong to tropical genera. On the Azuero Peninsula, Panama, the vesper rat is the common mammal; the keel-billed toucan, spot-crowned barbet, ochre-bellied flycatcher, and white-throated robin are common birds. The tropical species of peccaries, deer, large cats, opossums, anteaters, gray fox, and armadillo range into this cloud forest (Aldrich and Bole 1937). Griscom (1932, 1935) states that the quetzal occurs in much of the cloud forest from Costa Rica to central Mexico. Hairy woodpecker, acorn woodpecker, and broad-tailed hummingbird occur in the cloud forest from Costa Rica to central Mexico. The little green tanager *Chlorospingus* is by far the commonest bird in this cloud forest, nesting in tall tree ferns. Without distinguishing the different types of cloud forest, the quetzal and numerous other birds, several mice, two species of shrews of the genus *Cryptotis,* and a large number of other animals, both vertebrate and invertebrate, recognize the lower limits of the hardwood cloud forest and rarely cross them. A few animals from the semiarid lowland habitats range well into the higher altitude pine cloud forest. On the other hand, the black chachalaca and the ocellated quail may not discriminate between hardwoods and high pine. The toad *Bufo coccifer,* the lizard *Sceloporus malachiticus,* and the snake *Imantodes* occur in all faciations.

In Chiapas, Hobart M. Smith (personal communication) found the lizard *Sceloporus malachiticus taeniocnemis;* the snakes *Rhadinaea lachrymans* and *Bothrops bicolor;* two species of the frog *Ptychohyla;* and the salamanders *Chripterotriton xolocalcae* and *Magnadigita nigroflavescens.* Schmidt (1936) found five species of the salamander *Oedipus* in the tropical cloud forest of southern Guatemala.

Near Samac, Guatemala, in an original hardwood cloud forest at an elevation of 1300 meters, Van der Schalie (1940) reports the snails of the forest litter as *Schasicheila pannucea, S. minuscula, Brachypodella morini, Euglandina monilifera, Poteria texturatus,* and *Lysinoe ghiesbreghti. Schasicheila pannucea* also occurred in pine forest. In the vegetation was the snail *Helicina tenuis.*

TIMBERLINE FOREST, ALPINE MEADOW, PARAMO, AND SNOW

Mountains reaching 3400 to 3700 meters in altitude occur in the Sierra Madre Occidental and Oriental and the Eje Volcanico of Mexico and in Guatemala and Costa Rica. Volcan Chirique of Panama is 3350 meters high, but no mountain in Nicaragua, Honduras, or San Salvador approaches this height. Accordingly, the communities described here are limited very largely to the northern areas named.

The subalpine and timberline forest on Cerro Potosí, Nuevo León, at 3800 meters is composed of *Pinus montezumae* and fir. The stunted alpine form of *P. montezumae,* var. *hartwegii* persists to the limit of tree growth. The timberline area is an open forest of stunted and gnarled, randomly spaced trees which finally gives way entirely to open meadow. In its upper few meters (and occasionally at lower altitudes) *Pinus flexilis mexicana* and *Juniperus mexicana* come in as shrubs that are several times broader than they are high. In extreme form they are mats only 30 centimeters in height (Muller 1939). Above approximately 3600 meters the vegetation changes entirely to herbs which form a well-developed meadow. This extends over the 100 to 200 meters of the broad crown of the Potosí peak. With the exception of *Pentstemon gentianoides* and *Lupinus cacuminis,* the species all assume the typical low form of alpine plants, most of them producing rosettes. Herbs found in the alpine meadow, named approximately in the order of decreasing importance, are *Poa annua, Phleum alpinum, Bidens muelleri, Potentilla leonina, Castilleja bella, Draba neomexicana, Phacelia platycarpa, Senecio scalaris, Androsace subulifera,* and *Gnaphalium sprenglei.* Some of these species also occur in the subalpine forest. One of the common animals in the alpine meadow is the Mexican vole. It feeds on the seeds of *Lupinus, Senecio scalaris,* and *Pinus montezumae* and destroys the snail *Humboldtiana.* Clark's nutcracker was observed catching a vole and carrying it away (Koestner 1941).

Alpine meadow occurs on the Eje Volcanico peaks at 3700 to 4700 meters. Species listed for Cerro Potosí are not common here. The pine trees near the treeline are upright and small; matlike junipers take the place of elfinwood. Plants of the genus *Castilleja* and *Draba* usually go highest; *Senecio* and *Lupinus* with their showy flowers are characteristic. The older naturalists barely mention grasses but stress the luxuriance of the alpine meadow plants (Heilprin 1892). Goldman (1951) states that the sacaton grass extends to very high altitudes.

On the crest of the Cerro de la Muerto in the Talamanca Range of Costa Rica, there are paramos or paramillos such as are characteristic of arid portions of the high Andes. Paramos are made up of thick-rooted short-stemmed or stemless plants which often form mats, and some plants of this type are in these high areas of Guatemala. Alpine scrub communities in Guatemala have predominantly boreal genera (Griscom 1932).

MOUNT ORIZABA TRANSECT

Mount Orizaba, El Pico, or Volcan de Orizaba is an extinct volcano. It is the highest mountain in North America south of Alaska, rising to 18,314 feet (5575 m). To illustrate the relation of communities, a transect is here considered, beginning at the northern outskirts of Veracruz, extending through Cordova to the top of the mountain, then to the northern outskirts of Puebla (Fig. 18-4).

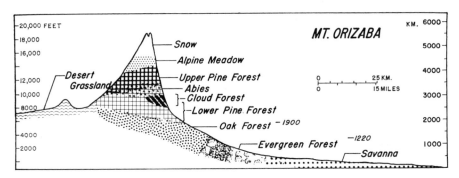

Fig. 18-4. Mount Orizaba transect (vertical exaggeration about 19.5 times).

Savanna Zone

Starting from the coast, two or three kilometers north of the city of Veracruz, the area is somewhat swampy (Fig. 17-3). In the west southwest direction to Cordova, the first 30 kilometers bring a rise to 100 meters, the second 30 kilometers to 600 meters, and the third 30 kilometers to about 900 meters. There are 17 streams. The savanna vegetation that occurs here extends to 450 meters and is probably original, rather than the result of clearing, although introduced grasses, such as Guinea grass, Bermuda grass, and Egyptian grass, may have brought some modification of the flora and fauna (Scovell 1893). In traveling across this area, one gets the impression of a grassy plain with very few trees or shrubs. Groves vary in size from a few trees or bushes to tracts of several hectares. Along the streams there is luxuriant forest characterized by large trees and an abundance of lianas and epiphytes.

In the grassy, scrubby fields are bobwhites, ground doves, tropical kingbirds, white-collared seedeaters, and blue-black grassquits; where the fields are marshy several sorts of herons are to be seen and often a mangrove swallow flying overhead. In the scattered groves of big trees and along the river's edge, the most regularly observed birds are the red-billed pigeon, ruddy ground-dove, squirrel cuckoo, golden-fronted woodpecker, sulphur-bellied flycatcher, great kiskadee flycatcher, brown jay, boat-tailed grackle, melodious blackbird, and black-throated oriole (Edwards 1955). A little farther south are found a number of characteristic reptiles, such as the snakes *Drymarchon corais* and *Coniophanes* (Ruthven 1912).

Tropical Rain Forest

At San Alejo, 87 kilometers from Veracruz, there is an abrupt change to hills covered with second-growth tropical trees. The region between 450 and 1220 meters altitude is tropical rain forest. The annual rainfall is generally 230 to 260 centimeters; at Cordova it is 227 centimeters, at Orizaba 212 centimeters. The soil at Cordova is derived from volcanic and limestone rocks and is exceptionally fertile. The forest trees are seldom large but they exist in great variety, bearing ferns, orchids, bromeliads, and other plants in great profusion on their trunk and branches.

Originally the animals occurring here were those of the rain forest. Around the south base of Orizaba, Baker (1922) found two varieties of the snails *Oligyra flavida* and *Helicina zephyrina* on leaves of shrub and trees up to 4.5 meters in height. Many *Guppya gundlachi* occurred on the ground among the leaves and humus and *Guppya trochulina* on the leaves of palms and other trees.

Oak Zone

There are scattered oaks as low as 500 meters, but according to Heilprin (1892), oaks form a dense forest from 1220 to 1830 meters and a poorer forest up to 2440 meters and are mixed with pines up to 2750 meters. The alders *Alnus jorullensis* and *A. castanaefolia* become mixed with oaks up to 3550 meters, higher on the eastern than on the western slope. Deer, coyotes, mountain lions, opossums, coatis, tree squirrels, and a mixture of tropical and northern birds were originally present, as they now are higher up on the mountains and in oak forests farther south.

Pine Zone

Pinus leiophylla, without fir, extends from 1800 to 2750 meters. Above 2100 meters, species of plants and animals are more like those in the north, although there are no forests similar to those in the temperate zone. There are oaks, elders, mustards, plantains, chickweeds, dock, violets, and familiar ferns. Sparrows, meadowlarks, blackbirds, crows, woodpeckers, and hummingbirds are common along with many less familiar species. At about 2450 meters, Blatchley (1892) took less than a dozen species of butterflies and beetles, and other animals appeared to diminish in numbers as rapidly as the Lepidoptera. Birds, however, were as numerous in species and individuals at 2450 meters as below. At this altitude in Mexico, only the wolf, coyote, bobcat, and mountain lion are to be expected among the larger mammals; the latter is not rare near Orizaba. The eastern cottontail occurs at this level.

Gadow (1908) ascended the southeast face of the mountain up to snow line, via Perla, Veracruz, and the west slope from Chilahicomula. On the southeast slope from 2450 to 2750 meters, he found openings in the forest of *Pinus leiophylla* and *P. montezumae* bordered with various deciduous and evergreen oaks, arbutus trees, and the alder *Alnus jorullensis*. In the open patches were dense shrubby masses of small-leaved *Arbutus spinulosa*. *Fuchsia microphylla* formed shrubs four meters high. *Tillandsia punctulata* made clusters on the pine trees. Orchids were well represented by the broad-leaved, white-flowering *Catasetum,* which grew in mold-filled recesses of hollow oak trees. Ferns in the forest and openings were scanty, except for bracken, although a few grew on the mossy stems of rough-barked trees such as the alder. Tropical conditions were still indicated by the presence of gigantic lianas such as *Bignonia*. Salvias, dahlias, begonias, geraniums, oxalis, fuchsias, tradescantias, irises, and *Ipomoea purga* were the commonest herbs and shrubs.

There were five kinds of land salamanders that prowled about at dusk and during the night in search of insects and small centipedes. Some of them hid in the daytime under moss and scraps of pinebark. *Pseudoeurycea leprosa* led a partly arboral life, hunting and hiding in places among the clusters of epiphytic plants, such as tillandsias, orchids, and the climbing philodendron. *P. orizabensis* was present up to at least 3820 meters; *Bolitoglossa platydactyla* occurred from 2750 meters down to the lowlands of the coast. Most of the salamanders are viviparous and independent of water. Their respiration is carried on through their permanently moist skin and throat.

Rattlesnakes do not ordinarily occur in moist forests, but *Crotalus triseriatus,* a small species with a poor rattle, was found from 2750 to nearly 3900 meters. The garter snake *Thamnophis scalaris* was also present. Lizards were relatively few. *Sceloporus formosus* occurred on rocks below 2450 meters. *S. grammicus microlepidotus* ranged from the hot low country of southern Oaxaca to the treeline on Citlaltepetl at about 4120 meters. It is truly arboreal, ascending into the tree tops at sunrise in search of insects. *S. aeneus* occurred in grass-covered ground, in moist pine forests, and on more barren, grassy, lava-strewn slopes almost up to the snow line. There were four species of the viviparous *Abronia; A. taeniata gramineus* ascends the highest trees in search of insects and makes its nests in hollow oaks, pines, and arbutus.

Birds were not abundant. No owl, eagle, hawk, or falcon was seen or heard by the Gadow party. Some tits, a tree-creeper, woodpeckers, and jays were noticed.

Mammals included *Peromyscus, Heterogeomys hispidus* and rock squirrels. Armadillos were found near Contreras at 2450 meters and burrows at 2750 meters.

Pine–Fir Zone

The fir *Abies religiosa* begins at about 2750 meters. Forests of *Pinus monte-zumae* and fir thin out near 3600 meters and disappear in most parts of the mountain at about 3900 meters, although scattered trees occur up to 4200 meters. Arbutus and alder are important. The undergrowth is composed of broom, elder bushes, and various shrub and treelike heaths. *Senecio chrysactis* extends from about 3650 meters to the limit of the pines. The yellow "asters" and *Lupinus vaginatus* form a compact undergrowth, especially where the pines thin into groves.

At 3800 meters there are mice of species not encountered lower down, some squirrels, and deer. Bluebirds, warblers, and tits feed in the lichen-covered trees. A few hummingbirds visit the lupine. Rattlesnakes are seen. Of the reptiles, the highest altitude is reached by the lizard *Sceloporus aeneus.* The salamanders *Barisia viridiflava* on the grass tussocks and *Pseudoeurycea leprosa* occur nearly up to the limit of trees (Gadow 1908).

Scattered Tree Juniper Belt

Above 3900 almost to 4370 meters, juniper spreads out over the rocks; herbaceous plants occur from 3810 to 4020 meters to the tree line and include mustards, composites, *Castillejas,* and a few other plants, but only two grasses. There are no Ranunculaceae, claytonias, willows, or other hydric plants. At

Fig. 18-5. Yucca grassland east of Puebla, Mexico, characteristic of the west end of the Mount Orizaba transect.

high elevations, *Pinus montezumae* becomes reduced to only three to four meters and is weatherbeaten. It occurs in scattered groups on the southern and western slopes, leaving the northern and eastern slopes bare. Near the lower limit of snow, the beetle *Phellopsis* and myriapods are under stones. The orthopterons *Pezotettix* and *Geuthophilus* and white lepidopterons are present in small numbers. The mouse *Peromyscus melanotis* goes up to 4620 meters (Blatchley 1892, Heilprin 1892, Gadow 1908).

On the west side of the mountain, the Montezuma pine belt reaches from timber line down to 3900 or 3350 meters. Below this, fir, alder, and sacaton grass go down to 2900 meters. Other pines descend to 2400 meters. Around 3500 meters occur *Pinus leiophylla, P. montezumae,* oaks, arbutus, and alders. A little lower there was a zone of the cherry *Prunus capuli* mixed with pine and oak. *Tillandsia* was much less luxuriant at 2600 meters than on the southeast side. The oak and cherry gives way at about 2300 meters to desert grassland (Fig. 18-5). Large yuccas are characteristic there. Little is known of the animals.

Chapter 19

The Communities of Southern Florida, Cuba, and the Shores of the Mainland

There are similarities in the topography of areas in Cuba, Florida, coastal British Honduras, Guatemala, and Yucatan. Sandy low wet land and dry ridge land restrict the development of climax vegetation to local areas. There are a number of plants which occur near the shore in Cuba as well as in Florida, the Yucatan Peninsula, and British Honduras. Mangrove swamps, gumbo limbo, slash pine, paradise tree, strangler fig, palmetto palm, and a number of other closely related species with similar habits and habitat relations occur in all areas (Lundell 1934, Bartlett 1935, Beard 1953). Vertebrate species are few and peculiar. The taxonomic relations of the community constituents are chiefly with South America except where endemism occurs.

SOUTHERN FLORIDA

The southern boundary of the warm temperate magnolia forest climax is near Palm Beach on the Florida east coast and in a line that curves southward to Lemon Bay near Naples on the west coast. This line is established by the occurrence of tropical species in the late subclimax hammocks; however, the transition between magnolia forest and subtropical communities is some 75 to 150 miles (120 to 240 km) wide. The vegetation is very varied. Three climaxes appear probable, depending on the substratum: (a) subtropical hammocks with some northern plants, (b) tropical hammocks with no northern plants, and (c) dry scrubby vegetation on the Keys. There is rain in every month, and frosts occur only occasionally. The natives prior to 1600 did not disturb the vegetation greatly; they were beachcombers and used sea food to a considerable extent. The greatest changes have been introducd in recent time by the installation of canals to drain the marshes and swamps.

474

The tropical and subtropical areas of Florida are at or below 23 feet (7 m) above sea level. The only elevations of any consequence are the old dunes along the east and west coasts. The dunes, especially along the east coast, usually have white sand at the surface, but at a depth of a foot or two the soil is rusty yellowish and slightly indurated sand (Harper 1927). Various soils are present on the southern coast and islands; a deficiency in potash may be general. Peat soil is present near the south shore (Davis 1943). There are large areas of partially exposed limestone with very little soil, both on the Keys and mainland. The Everglades area, which is in the ecotone, lies on a plain of limestone and is partially surrounded by a rock rim. Originally the drainage of the Everglades was through the Caloosahatchee River, several small rivers, and general overflow.

The area has 50 or more inches (125 cm) of rain from February through October and a mean temperature of 65° F. (18° C.), October through February (Fig. 19-1). The dry season is not intense but falls from November through February with a total rainfall of 10 inches (25 cm), an average of about 2.5 inches (6.2 cm) per month. The interiors of hammocks southwest of Miami are frost free, apparently because of the protection provided by the surrounding vegetation (Davis 1943).

Tropical Moist Hardwood Forest

Although not necessarily characteristic of the species, the tropical moist hardwood forests are made up of trees having crooked trunks and hard, heavy wood, resulting from slow growth (Harper 1927). Stiff evergreen leaves are characteristic in spite of limestone and abundant humus. The trees make a dense shade, and there are very few herbs but often a great profusion of airplants on limbs and leaning trunks of the rough-barked trees. In essence, this is a poor rain forest.

Large trees occurring on the Keys and southern coast of Florida are the gumbo limbo *Bursera simaruba, Coccoloba diversifolia, Sideroxylon foetidissimum, Eugenia confusa, Lysiloma bahamensis,* the Florida strangler fig *Ficus aurea,* the West Indies mahogany *Swietenia mahogoni, Sapindus saponaria,* the royal palm *Roystonea elata,* which is often not present, *Krugiodendron ferreum, Chrysophyllum oliviforme,* the lignumvitae *Guaiacum sanctum,* the thatch palms *Thrinax parviflora* and *T. microcarpa,* and *Hippomane mancinicella.* Nine of the 13 species are evergreen.

Understory trees and shrubs are *Duranta repens, Simaruba glauca, Nectandra coriacea, Zanthoxylum fagara, Z. flavum, Metopium toxiferum, Exothea paniculata, Citharexylum fruticosum, Prunus myrtifolia, Psychotria ligustrifolia, Gymnanthes lucida, Picramia pentandra, Rapanea guianensis,* and *Piscidia piscipula,* which is particularly common on the Florida Keys. Other shrubs and small trees may be more common locally than these (Harper 1927, Davis 1943, Little 1953). Twenty-three of the 27 species listed by Harper are evergreen. Most of these trees and shrubs occur in Cuba and the West Indies, several

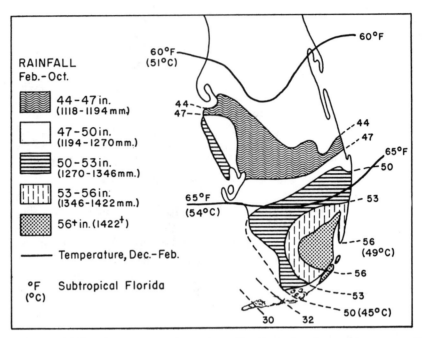

Fig. 19-1. Rainfall and temperature in southern Florida during the growing season; the subtropical portion is roughly south of the December-February isotherm of 65° F. (18.3° C.).

also in Yucatan and Guatemala, indicating the wide tropical relationship of the communities of the Gulf and Caribbean shores.

Some of the more common vines are *Berchemia scandens, Vitis munsoniana* and *V. sicyoides, Ampelopsis arborea, Chiococca racemosa, Pisonia aculeata, Hippocratea volubilis, Rhabdadenia biflora,* and a number of herbaceous vines such as *Ipomoea* (Davis 1943, Standley 1920-26). Some ferns are stiff and coarse and may be 10 feet (3 m) tall, as, for example, the giant bracken *Pteris tripartita.* Some are delicate and small, such as the filmy fern *Trichomanes punctatum.* Other species are epiphytic and some are vines, such as *Polypodium plumlula.* The more common orchids are of the genus *Epidendrum,* particularly *E. tampense, E. nocturnum,* and *E. chochleatum.* Other orchids are *Polystachya luteola, Cyrtopodium punctatum, Polyrrhiza,* and a number of *Oncidium.* The air-plants *Tillandsia fasciculata, T. utriculata,* and *T. aloifolia* are numerous (Davis 1943).

Among the major permeant influent animals is the white-tailed deer, which is now largely absent. There have been a few reports of the wolf *Canis niger* in the past 25 years. The black bear occurs all through southern Florida, including the Keys. They often plunder the buried eggs of marine turtles (De-Pourtales 1877). Bears prepare a bed on the ground in dense thickets, pulling together grass and moss which they tread down. Their food is chiefly of vegetable origin: nuts, cabbage palmetto berries, and buds. A large bear will climb to the top of a cabbage palmetto, put its arms around the top, and sway back and forth until the big bud loosens, then fall to the ground on its back with the bud securely grasped in its paws. The mountain lion still exists in small numbers in southern Florida, particularly in Collier, Lee, and Hendry counties. They feed on deer, rabbits, turkeys and other animals. In July, 1940, a collaborator of Hamilton (1941) observed mountain lion signs in the magnolia–tropical ecotone area about 17 miles (27 km) east of Fort Myers. A full grown sow had been pulled back into the palmettos with several branches covering it. A gravid doe had been killed and the foetus removed and eaten. The bobcat is common near Cape Sable and Fort Myers, while the red-shouldered hawk preys upon the smaller mammals. The raccoon is common.

The marsh rabbit, a minor permeant influent, occurs in the Cape Sable hammocks and farther north. Near the southern limits of the city of Fort Myers, Hamilton reported that on March 23, 1941, workmen captured two young marsh rabbits in a shallow nest; they found another nest with five young beneath a cabbage palm. In this immediate area, several large diamondback rattlesnakes were killed. It is suspected that swamp rabbits provide a substantial part of the food of these big snakes. The spotted skunk appears at dusk to play or search for food. Stomachs of trapped individuals contained remains of the skink *Eumeces inexpectatus* and a few fragmentary insects. The fox squirrel has been observed in thickets bordering small cypress swamps. The cotton rat is generally distributed, but the cotton mouse is largely restricted to the tropical hammocks (Opsahl 1951).

Although the vegetation is largely of tropical species, only 7 of 58 species of breeding land birds are of tropical origin. In the tropical hammock forest of Lignumvitae Key, Robertson (1955) found more than 18 pairs of nesting birds belonging to five species, per 40 hectares. Only the black-whiskered vireo and great crested flycatcher nest in the closed, probably nearly climax forest. The vireo occurs in Cuba and other islands.

The characteristic lizards of the tropical hummocks are the Carolina anole and the brown red-tailed skink, and the snakes are the rough green snake, Key rat snake, and the southern Florida coral snake. Frequent species are the toad *Bufo terrestris,* the frogs *Hyla squirella* and *Eleutherodactylus ricordi;* the lizard *Eumeces inexpectatus;* the snakes, black racer *Coluber c. constrictor,* eastern indigo snake *Drymarchon corias couperi,* and eastern diamondback rattlesnake *Crotalus adamanteus* (Carr 1940).

Subtropical Hammocks

About 80 per cent of the plant species in the hammocks near Miami (Fig. 19-2) also occur in the West Indies. All lack some of the truly tropical trees; all have some northern trees, such as live oak, pignut hickory, and seral species of the magnolia forest. *Sabal palmetto* and the redbay *Persea borbonia* are among the common seral trees (Phillips 1940, Harper 1927).

Both tropical and subtropical hammocks originally supported numbers of tree snails of the genus *Liguus,* chiefly varieties of *L. fasciatus* (Fig. 19-5C)

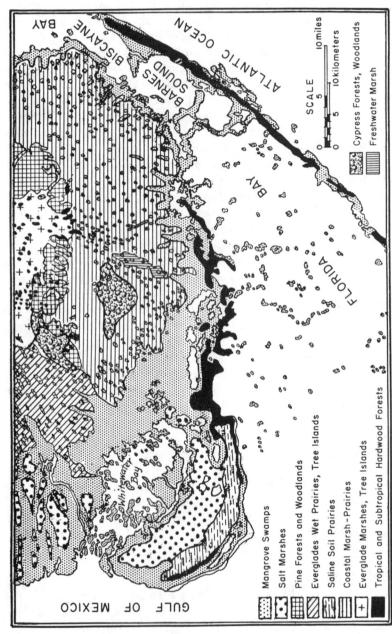

Fig. 19-2. Prominent community types in southern Florida, all are early seral stages except the subclimax tropical forest. The so-called "prairies" are in part swamps; the small circles in the "prairies" are subtropical hammocks (after Davis 1943).

(Young 1951). They are large, with brightly colored or variegated rings (Simpson 1929). There are about two dozen species of which three or more occur in Cuba and Haiti. The snails leave the trees only to lay eggs or make migrations across the country. During the dry season, from November to May, the snails become dormant after secreting a gummy substance which hardens and attaches the shells firmly to the trees. Before the beginning of the rain season, sometimes as early as April, the snails become active and mate. They live for the most part on smooth-barked trees, but occasionally even on rough-barked live oaks. Although almost wholly confined to hammocks, they occur occasionally in the hydrosere. Their food is chiefly algae that they obtain from the bark. Other snails in the tropical and subtropical hammocks are *Dryaemus dormani* and *Helicina orbiculata,* genera which are well represented in the tropics. Simpson (1929) names 13 snails found both in Florida and the West Indies.

Ten species of spiders taken on Paradise Key are known to be in Georgia. Butterflies and moths are numerous; of those listed by Watson (1926), *Anaena floridalis* occurs in the tropical keys and is restricted to Florida and the West Indies. Most of the others have no definite relation to hammocks and occur through the Gulf states and into Mexico. Very few species are common to Florida and the adjacent tropical islands. As for ants, *Odontomachus haematoda insularis* is frequent and occurs in the West Indies and Central America; *Myrmothrix abdominalis floridanus* has close relatives in these areas.

The flies *Leucophenga varia* and *Neogriphoneura sordida* are common in the summer; both are widely distributed. The crane flies *Tipula* (*Limonia*) *duplex, T.* (*Lunatpula*) *umbrosa,* and *Dicranomyia liberta* are present and are not recorded in northern Florida. Orthopterons are the same as those in the magnolia forest. Thus very few species of insects and other arthropods in Florida are found also in the tropics farther south.

Dry Tropical Communities of the Westerly Keys

Rainfall decreases rapidly westward on the Florida Keys (Fig. 19-1). The vegetation is thicket-like; the principal plants are two species of the thatch palm *Thrinax, Piscidia piscicola, Coccoloba uvifera,* the Florida silver palm *Coccothrinax argentata,* cabbage palmetto, and slash pine. Harper (1927) lists sixteen species of shrubs, including saw palmetto but none of the species of the tropical hardwood forest. The vegetation is probably not related ecologically to the mainland climax. Some areas in the Bahamas have a similar vegetation. The Key deer, a small variety of the white-tailed deer, occurs. Reptiles and amphibians include *Eumeces laticeps, Cnemidophorus sexlineatus,* and the snake *Natrix sipedon;* the salt water terrapin is reported at the Marquesas Key, between Key West and the Tortugas (DePourtales 1877).

Mangrove Halosere

The mangroves are important on most tropical and subtropical coasts of all parts of the world (Fig. 19-3). Due to the effect of the sea, they occur about

50 miles (80 km) farther north along the coasts of Florida than do most other subtropical species. Mangrove communities occur in three zones related to the level and salinity of the surface and soil water. These zones occur within a successional trend from off-shore pioneer communities to upland, freshwater, nonhalophytic communities inland from the mangrove swamps (Davis 1940, 1943).

The red mangrove associes of most swamps includes an outer pioneer belt and an inner mature and full grown one. The plants have an extensive bracing prop root system and produce viviparous seedlings that germinate on the parent tree (Fig. 19-4). Animal constituents on the roots below low tide are clusters of the snail *Melina alata*. A little higher in the intertidal level, the oyster *Ostrea floridensis* forms dense masses that may completely cover the mangrove roots. The snail *Strobilops hubbardi stevensoni* occurs under the bark of dead limbs (Bartsch, personal communication). Many red mangroves are killed by larvae of the beetle *Chrysobothrus tranquebanca* (Snyder 1919). Accumulation of debris and sediment under the red mangrove makes the habitat unsuited to the red mangrove, and the black mangrove invades.

In the black mangrove associes the halophytes *Batis maritima* and *Salicornia perennis* are the most constant salt-marsh associates, but the grasses *Sporobolus*

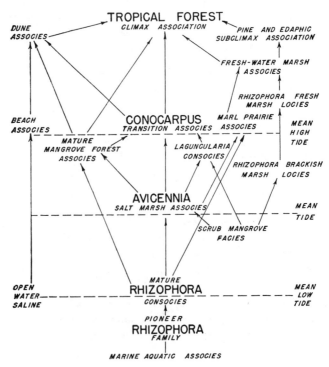

Fig. 19-3. The successional relations of mangrove and other communities (after Davis 1940).

virginicus and *Spartina alterniflora* are locally abundant. The black mangrove trees usually form open stands with the other plants growing between them. They usually produce many erect roots known as pneumatophores which project upward through the mud and water and have the appearance of asparagus tips. The community is inundated over its outer part by every average tide, but not over much of its inner part. The white mangrove grows as a secondary species with either of the other mangroves but occasionally forms thickets inland from the black mangrove zone. The white mangrove seldom has pneumatophores. Mangroves go some distance inland along tidal estuaries but are finally crowded out by other vegetation.

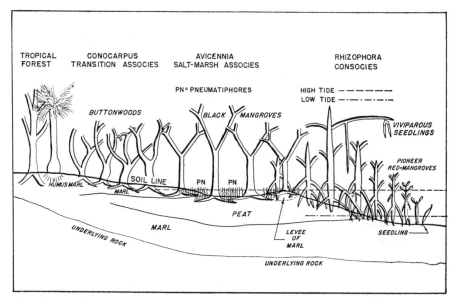

Fig. 19-4. A diagrammatic transect through mangrove communities from pioneer *Rhizophora* to tropical hammock forest (slightly modified from Davis 1940).

Button-mangrove, which is not a true mangrove, forms a transition community between the main mangrove communities and an inland, nonhalophytic type of vegetation (Fig. 19-4). Associated with it is a variety of small trees and shrubs, many of them tropical hardwoods, which together resemble a hammock forest. Invading trees are *Drypetes diversifolia*, which also occurs in the West Indies, and among others, species of *Sideroxylon, Bursera, Swietenia, Krugiodendron,* and *Simaruba.* Some of the mature tropical or subtropical hammock forests adjacent to the mangrove swamps are on soils of mangrove origin and represent the subclimax or climax of the halosere. Other nonhalophytic vegetation inland from the saline or brackish-water swamps are freshwater marshes and marl prairies. Some of these are also on soil of mangrove origin. These communities are transformed into the climax forest through seral stages similar to those of salt and freshwater marshes of different origin.

Fox squirrels and raccoons occur in the mangrove swamps (W. B. Robertson, personal communication). Black-whiskered vireos, mangrove cuckoos, double-crested cormorants, and various herons and egrets nest in the mangrove. A katydid, the crickets *Anaxipha scia, Orocharis gryllodes, Nemobius cubensis,* and *Hygronemobius alleni,* and the earwig *Euborellia ambigua* occur here. The butterfly *Phocides batabano* occurs also in Cuba; the larvae feed on the red mangrove (Watson 1926).

Salt and Freshwater Marshes

The grass *Spartina alterniflora* grows best in the outer deep-water sandy soil of swamps and may be effective in holding soil materials, thus aiding in establishment of the pioneer red mangrove (Fig. 19-2) (Chap. 3) (Davis 1940). In the mangrove sere, the gradual change from mangrove swamps toward freshwater marshes is in zones; the soils of each zone being progressively higher in the landward direction. The plant dominants are, first *Batis,* then a mixture of *Batis* and *Salicornia,* then *Distichlis* with patches of *Spartina* and occasionally the sedge *Fimbristylis castanea,* then more *Spartina* or perhaps patches of *Juncus,* and finally the nearly fresh-water marshes, such as the cat-tail.

There are many inland freshwater ponds and swamps in the ecotone between magnolia forest and the tropical woodland of the southern coast and the northern keys. The developmental stages of hammocks which contain tropical species are very largely those of the magnolia forest (Laessle 1942). A swamp in a small depression may develop into a low hammock by the invasion of sweetbay, along with live oak, *Ilex cassine, Myrica cerifera,* and others. The low hammock may become a high hammock, such as the Paradise Key (Royal Palm Park) (Fig. 19-5) with the addition of tropical trees. These hammocks, of which there are many, frequently have live oak persisting on their outer margins.

In the wide expanses of the southern Everglades occur the spike-rush *Eleocharis cellulosa* which stands in shallow water that frequently disappears during the dry season, the tall saw-grass *Cladium jamaicensis* (a sedge) which stands in shallow water that usually persists throughout the year, bush-covered islands, palmetto islands, marl and rock prairies, and cypress ponds and woodlands. This series may constitute a sere, but this has not been demonstrated.

At Ochopee, Florida, in July 1942, the spike-rush extends above the surface of water that was 5 to 12 inches (12.5 to 30 cm) deep. These pond areas had been dry in December. The golden top minnow, least killifish, bluefin killifish, and mosquitofish occur in combined numbers estimated at 80 per square meter. The tree frog *Hyla cinerea* and its eggs are present. Amphipods and the snail *Physa cubensis* are in small numbers, about 5 per square meter. Immature insects, including Ceratopogonidae (Diptera), Belostomidae (Hemiptera), dragonfly nymphs, and water tigers total 25 per square meter. In a tall saw-grass marsh, the plants grow in 2 to 3 inches (5 to 7.5 cm) of water. Aquatic animals are few. The least and bluefin killifish and most of the other aquatic species occur in smaller numbers than in the deeper water of the spike-rush area. These marshes are frequented by the raccoon, bobcat, and marsh rabbit (Blair 1935). The cotton rat is often very abundant in grass or sedge-covered areas when they are not submerged. Before the water level over much of the area was lowered by drainage, it furnished a home for thousands of mottled ducks, herons, egrets, limpkins, and Everglade kites, but most species now occur in reduced populations.

Bush-covered Island is a dense tangle of shrubs with black moist soil on a high spot in the marshes that were studied in 1942. The common plants in such habitats are Florida elder, pond-apple, buttonbush, and such herbs as the vanilla-plant. The birds in the vicinity are those common in the magnolia forest area. The lizard *Anolis carolinensis* is present; it is common throughout

Fig. 19-5. A. Strangler fig in the former Royal Palm State Park, Florida. B. Young royal palm in the same area. C. Tree snails in a late stage of subtropical forest (National Park Service photo by Victor Cahalane). D. *Sphaerodactylus notatus* which occurs both in Florida and in Cuba and the Bahamas (courtesy Comstock Publishing Company, Inc.).

Fig. 19-6. A. The saw-grass *Mariscus jamaicensis.* B. Palmetto islands in the Everglades. C. Inner edge of the mangrove, blending with a "grass"-covered marsh. D. Freshwater coastal marsh "prairie," south of Homestead, Florida. E. A hammock too young to contain palmettos.

the southern states. Various insects and spiders, including the ant *Camponotus abdominalis floridanus,* and the spider *Hentzia palmarum,* are found. The cabbage palmetto probably invades such bush-covered areas. There are many cabbage palmetto islands in the area of Ochopee, and below and surrounding the palmettos there is usually a dense mass of shrubs (Fig. 19-6). The snails *Liguus* and *Chondropoma dentatum* occur in young palmetto hammocks (Bartsch, personal communication).

Portions of semidry prairies, on Miami limestone, intersect the Miami pinelands. Besides the saw-grass and spike-rush, the switch grass *Spartina bakeri* occurs. The sedge *Schoenus nigricans* and a few saw palmettos are present. Harper (1927) lists more than twenty herbaceous plants in the area of study, but only the grasslike plants are conspicuous. There is evidence of fire. The substratum is limestone with a thin layer of black soil. The limestone substratum is very irregular, and some of the depressions are damp. The animal constituents include the toad *Bufo quercicus* and the usual grassland arthropods. An unusual feature in the summer of 1942 was the large number of the land snails *Polygyra septemvola volvoxis* and *Polygyra uvulifera*. Present were the long-bodied spider *Tetragnatha pallescens* and the orb weaver *Acanthepeira moesta,* both of which belong to Florida and the Gulf Coast. Orthopterons were represented by *Neotettix femoratus* and nymphs of *Conocephalus* and *Orchelimum* in considerable numbers. The leafhopper *Chlorotettix viridius* was present. In December, 1942, collections obtained the long-horned grasshopper *Conocephalus gracillimus,* a Florida species, the bug *Protenor australis,* the clerid beetle *Isohydnocera aegra,* and several spiders. This community is probably a late stage of the tall saw-grass marsh.

CUBA AND THE ISLE OF PINES

Cuba covers 110,480 square kilometers (Fig. 19-7), which is about the area of Florida east of the Suwannee River. The lowland plant and animal communities are more complicated than those of Florida because of the variations in substratum and altitude. Large areas of Cuba are below seven meters above sea level; the Colon Plain has been flooded across Matanzas and western Santa Clara from coast to coast. More than half of Cuba is less than 92 meters above sea level. Another one-third is between 92 and 300 meters. There are moun-

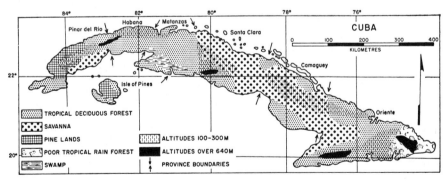

Fig. 19-7. The probable original distribution of vegetation in Cuba. This diagram was based in part on a map by Brother Leon (personal communication 1924) and in consultation with Carabia (1945); Leon's map shows 19 relict areas of tropical forest; these were connected and intervening blank areas filled in. It was assumed that considerable portions of western Cuba that are now savanna have become so through human modification of the so-called tropical deciduous forest. Mountains have been largely ignored in making the diagram.

tainous areas with peaks above 610 meters in northeastern Pinar del Río, in southeastern Santa Clara, and in southern and eastern Oriente. Pico Turquino has an altitude of 2000 meters.

Throughout the island, rainfall occurs mainly from May through October and ranges from 110 to 140 centimeters in the lowlands. The dry season runs from December through March. The greatest rainfall is in the most westerly Pinar del Río Province, where locally it is 175 centimeters, although in general, rainfall increases toward the east. The greater rainfall in the eastern part of Cuba is apparently offset by greater water loss through the substratum (Seifriz 1943), as the vegetation becomes less and less luxuriant. Frost does not occur except rarely in the highest mountains. The mean temperature for the wet season is 27.1° C. and for the dry season, 22.7° C.

The island has a foundation of ancient metamorphic and igneous rocks, chiefly serpentines, diorite, and schists. These were originally covered in part by limestone of Tertiary origin. Where the limestone has been eroded away, the igneous rock is exposed. Rich soil derived from the limestone has been washed into the valleys and plains of the northern and southern coastal regions and into many central districts as well. The black soil may be 30 meters deep locally and apparently was originally covered with a mahogany forest. Where only a thin covering of earth lies over a flat smooth limy deposit, the forest is less luxuriant (Barbour 1941). Serpentine rock is abundant and irregularly exposed by erosion. It is hydrous magnesium silicate, and soils resulting from it are low in calcium and high in magnesium, nickel, and chromium, with marked effect on plants. There are two areas of serpentine soil, one in central Camagüey and another in central Santa Clara, each about 25 kilometers in diameter, and many other small ones. Seifriz's (1943) soil map shows a mosaic of 23 generalized soil types occurring on the island.

The vegetation may be classified as tropical deciduous forest, savanna, rain forest, pineland, and swamps (Leon 1926 and personal communication). Data on the leaf fall of Cuban trees is wanting. Considering the large number of evergreen tropical tree species common to both tropical Florida and Cuba, a largely evergreen rain forest would be expected in Cuba were it not for the reduced rainfall. While the seasonal distribution of rainfall is similar in extreme southern Florida and Cuba, rainfall is about 20 per cent greater in tropical Florida than in the greater part of the Cuban lowlands. The Cuban biotic communities have been greatly modified by human occupation. They have generally been described in terms of natural regions based on the geological substratum.

Tropical Semideciduous Forest

Although primarily below 100 meters, some tropical semideciduous forest is found up to 300 meters where there is more moisture. In the Havana–Matanzas area (Fig. 19-8), the natural vegetation contains *Bursera simaruba, Coccoloba diversifolia, Sideroxylon foetidissimum, Eugenia confusa, Lysiloma bahamensis, Swietenia mahogoni, Sapindus saponaria, Krugiodendron ferreum, Guaiacum sanctum, Thrinax parviflora, T. microcarpa, Hippomane mancinicella, Dipholis salicifolia,* and *Cupania glabrae* which also occur in Florida, and in addition *Eugenia buxifolia, E. axillaris, Guettarda calyptrata, Catesbaea spinosa, Malpighia cubensis.* Among the most abundant vines are snowberry, the showy-flowered *Bauhinia heterophylla,* and *Bignonia sagraeana.* The Cuban royal palm *Roystonea regia* is replacing *Sabal florida* to a large degree, which is more abundant toward the west. Individuals occurring in meadows are *Gliricidia sepium, Cecropia peltata, Samanea saman, Crescentia cujete, Chrysophyllum oliviforme, Cedrela odorata,* mahogany, lignumvitae, *Jatropha curcas, Urera baccifera,* and others (Leon 1926, Seifriz 1943).

Fig. 19-8. General view of palms along the Havana–Batabano Road in 1940.

Animal constituents include several species of bats. Columbus noted among the mammals of the island, the "perro mudo" (dumb dog), which the Cuban naturalist, Felipe Poey y Aloy, thought was the raccoon of Florida. Birds common in the wooded hills near Havana include as winter residents numerous marsh hawks, sparrow hawks, and an occasional painted bunting. The resident Cuban birds include the earless owl, Cuban blackbird, Cuban trogon, Cuban lizard cuckoo, bobito, Cuban tody, and smooth-billed ani. Caves in the limestone hills throughout the island are utilized by the bare-legged owl (Barbour 1923, Leon 1926).

Some of the most numerous trees that occur in the forest at higher elevations, locally up to 1060 meters, are *Calophyllum antillanum,* almacigo gumbo limbo, *Clusia rosea,* and the strangler fig. Present in the moister forest of the hills in central Cuba and on Pico Turquino are *Dipholis jubilla, Oxandra lanceolata, Cecropia peltata, Prunus occidentalis, Dendropanax arborea, Prockia crucis, Juglans insularis, Celtis trinervia, Cedrela odorata, Hibiscus tiliaceus, Zan-*

thoxylum spinifex, Terminalia chicarronia, mahogany, and others. On Turquino, there are a few vines, such as *Vitis tilifolia,* known as water lianas, which bleed when cut. Another liana is the legume *Entada scandens.* Epiphytic plants are not abundant, owing to insufficient moisture, but the narrow-leaved *Tillandsia* and broad-leaved *Guzmania monostachya* are fairly common (Leon 1926, Seifriz 1943, Carabia 1945).

The cat-sized South American octodonts *Capromys melanurus, C. pilorides,* and *C. prehensilis* are the principal mammals. They are chiefly plant feeders. Flocks of parrots and the Cuban parakeet are in this forest (Leon 1926), as well as the two crows *Corvus minutus* and *C. nasicus* (Barbour 1923). Reptiles are represented by the large iguana *Cyclura carinata* and many others of the same genera as found on the mainland. *Sphaerodactylus notatus* (Fig. 19-5D) enters houses in Cuba, and, in fact, many of the 40 species of geckos and lizards that occur have learned to live in or about houses and gardens. The same is true of amphibians and snakes.

Several species can be added from the forest in the Sagua Baracoa Mountains (Leon 1926). Noteworthy is the rare solenodon which is insect eating and nocturnal and has its nest underground. It was formerly more widely distributed. *Capromys melanurus,* although not so rare, seems also to be confined in recent decades to the Oriente province. The Cuban ivory-billed woodpecker was at one time abundant. The elusive butterfly *Chlosyne perezi* and the rare *Clothilda cubana* and *Cydimon poeyi* may be found here.

The biotic communities of Pico Turquino are divisible into several zones, differing sharply with regard to sun and wind exposure. The mountains are arid through the first 100 meters. From 100 to 800 or 900 meters is the tropical semideciduous forest which is the vegetation type most broadly distributed over Cuba. From 800 to about 1100 meters is an open highland forest composed of *Hymenaea courbaril, Brunellia comocladifolia, Tabebuia pentaphylla, Magnolia cubensis,* and *Coccoloba monticola,* all of which species or genera are also found in Mexico. *Pinus cubensis* and *Juniperus barbadensis* occur on isolated hilltops. From 1100 to 1800 meters are subalpine herbaceous fields. From 1800 to 2000 meters there is alpine thicket. Cloud forests are poorly developed (Seifriz 1943, Carabia 1945).

Savannas

The savannas of Pinar del Río are on sandy siliceous soils. The palm *Sabal florida* is scattered and interspersed with grass and forbs, and the bottle palm *Colpothrinax wrighti* occurs. The generally typical trees of the savanna are *Roystonea regia, Samanea saman, Guazuma tomentosa, Ceiba pentandra, Crescentia cujete, Chlorophora tinctoria, Cecropia peltata, Gliricidia sepium,* and *Bursera simaruba.* In some cases they represent a remnant of lowland forest. Pinelands also occur. Where soil is very good, grasses of the genera *Panicum, Paspalum, Arundinella,* and *Arthrostylidium* are common. *Andropogon virgatus, A. gracilis, Sporobolus indicus,* and *Leptocoryphium* are important. In older savannas, a group of grasses and sedges are replaced by large groves of *Sabal florida, Copernicia torreana, C. hospita, C. baileyana,* and *Coccothrinax miraguama.*

In Santa Clara Province and farther east, notably in central Camagüey, compact serpentine soil occurs. The plants here are *Belairia mucronata, Randia spinifex, Rheedia aristata, Guettarda holocarpa,* and others. Snails are supposedly wanting in the serpentine area. The serpentine areas reach a climatic edaphic climax which is different from the climax common to half dozen other soils over large areas. Serpentine vegetation is often scrubby; on uplands in Northern Oriente it is largely herbaceous, endemics are numerous, and pines are rare.

Due to human activity most of Cuba below 100 meters appears to be savanna. In these dryer parts, the mockingbird is native along with the palm swift and bobwhite (Carabia 1943, Leon 1926, Seifriz 1943).

Seral Communities

Under pristine conditions, the permanent lakes of Cuba supported the familiar zones of aquatic vegetation. At present the lakes in western Cuba are often wholly covered with water lettuce *Pistia stratiotes,* water hyacinth, and floating islands of willows and custard apples. The lake surface has the appearance of a green plain with scattered clumps of trees. High winds move the islands about, crowd them together, or blow them ashore to expose large areas of open water that may persist for weeks. The amount of water in the lakes varies greatly between wet and dry seasons (Barbour 1923).

In the largest swamps, much of the surface may be composed of semi-floating areas of bulrushes or saw-grass. Where conditions are favorable water-lilies, the cat-tail *Typha domingensis,* and buttonbush occur in their proper seral positions along with *Salix occidentalis* and more rarely *Fraxinus cubensis.* The Everglade kite lives in lagoons and swamps, feeding on snails especially of the genus *Ampullaria;* there are also several species of hawks and other birds (Barbour 1923).

Mangrove swamps occur along half to three-fourths of the coast in a similar manner as along the Florida coast. Zapata swamp lies northeast of the Isle of Pines and is brackish near the coast. In some parts, most of the vegetation is herbaceous, principally *Typha angustifolia, Cladium jamaicensis,* and many sedges. Patches of mangrove trees are scattered here and there. In other areas, *Lysiloma bahamensis, Pithecellobium tortum, Davilla rugosa, Rheedia brevipes, Banisteriopsis pauciflora,* and patches of the palma corojo *Acrocomia crispa* are found. Remnants persist of *Pithecellobium discolor* and *Pisonia rotundata.* The mammal *Capromys nana* was recently discovered in these woods. It is much smaller than the other hutias present, *C. prehensilis poeyi* and *C. p. pilorides.* The mangrove warbler and clapper rail are common. Two crocodiles occur in the Zapata swamp, the caiman *Crocodilus acutus* and the crocodrilo *C. rhombifer* (Leon 1926).

Pinelands

Pinelands occur within the rain forest regions of eastern Central America, notably in Nicaragua and Honduras, southern Florida, and Cuba, particularly on sandy or rocky and gravelly, usually acid, sterile soils. Few trees except *Pinus caribaea* can exist until plant and animal reactions have improved the soil. The forests are open and scattered with little ground cover except grasses. Protected from fire, the vegetation gradually develops into tropical rain or deciduous forest.

In Pinar del Río Province and on the Isle of Pines are extensive low sandy areas of grassland containing widely scattered pine trees and here and there thin woods or park woodland. In addition to *Pinus caribaea,* there is *P. tropicalis* which is endemic to western Cuba. The live oak *Quercus virginiana* which occurs in and around hammocks in Florida grows singly or in thin woods. As the habitat develops with time *Myrica cerifera* and such ericaceous shrubs as *Vaccinium ramonii* and *Befaria cubensis* invade, the parkland becomes more and more heterogenous, and the savanna-like arrangement of plants becomes lost (Seifriz 1943). Pineland vegetation affords the last refuge for the rare little pine crow, pine warbler, a few woodpeckers, and eastern meadowlark; sandhill cranes may be present in early seral stages (Barbour 1923).

MEXICO, BRITISH HONDURAS, AND GUATEMALA

Mangrove Communities

Mangrove usually occurs in narrow strips along the coast of the mainland and in the estuaries of rivers. It sometimes broadens out to several miles in width while growing in peat, as described for Florida. In the Yucatan Peninsula, the mangrove communities are generally followed on higher land by a scrubby tree growth characterized by *Zanthoxylum caribaeum, Conocarpus erectus,* and *Citharexylum caudatum.* The scrubby growth then intergrades into either broad-leaved or coniferous forest. Where beaches occur, they support an herbaceous growth which includes *Cakile edentula, Tournefortia gnaphalodes, Philoxerus vermicularis,* various sedges, and the scrubby species *Coccoloba uvifera, Scaevola plumierii, Chrysobalanus icaco, Batis maritima,* and *Ernodea littoralis.* Behind the beach, mangrove may come in (Lundell 1934).

Sand Dunes and Sand Ridges

On the beaches and shore dunes of British Honduras, the most abundant and conspicuous trees are the coconut palms (Fig. 19-9). The goatfoot morning-glory *Ipomoea pes-caprae* and *Canavalia maritima* both have coarse, ropelike, creeping stems several meters long. *Remierea guianensis* grows in small tufts in loose sand along with *Croton maritimus* and various other plants. Another morning-glory *Ipomoea stolonifera* is often plentiful. Common beach grasses are the sandbur *Cenchrus, Chloris petraea,* and *Stenotaphrum.* Other plants thriving in the sand are *Cakile, Vigna repens,* the spurge *Euphorbia ammannioides,* the mallow *Sida cordifolia, Waltheria americana, Lippia nodiflora, Stachytarpheta jamaicensis, Diodia maritima,* and *Melanthera nivea* (Standley 1931). Several of these species occur also in southern Florida.

Pinelands occupy the flat areas between rivers where the soil is sandy and frequently submerged (Fig. 19-10). The moister depressions are occupied by sedges and the primenta palm *Acoelarraphe pinetorum. Pinus caribaea* comes onto higher ground along with grasses. In older areas the shrub *Curatella americana* occurs. Still later the oak *Quercus oleoides australis* appears (Bartlett 1935). This vegetation is typically savanna, such as is common farther south (Austin 1929). The next stage that follows contains the cohune palm

Fig. 19-9. Coconut palms on the beach at Ceiba, Honduras.

Orbignya cohune, which persists into the evergreen forest where mahogany and ironwood are important dominants (Charter 1940). Bartlett (1935) considers this forest subclimax to tropical rain forest, and he calls it zapatal, of which there are several faciations, depending on the soil (Chap. 16).

The red-lored parrot occurs regularly on pine ridges, the common night-hawk breeds there, and the acorn woodpecker, boat-billed flycatcher, gray-crowned yellowthroat, golden-crowned warbler, rufous-capped warbler, and rusty sparrow are typical inhabitants. Of 62 species of birds found in the rain forest climax, only one is represented in the pine ridge country (Austin 1929). On a low pineland several deer tracks, a number of runways in the sedges probably made by rice rats, and tracks of raccoons were observed. Rodents of the genus *Sigmodon* are present; squirrels enter the area for acorns. Gray foxes, white-lipped peccaries, coatis, young opossums, and the eyras *Felis yagouaroundi* are visitors from the adjacent rain forest (Murie 1935, Sanderson 1941).

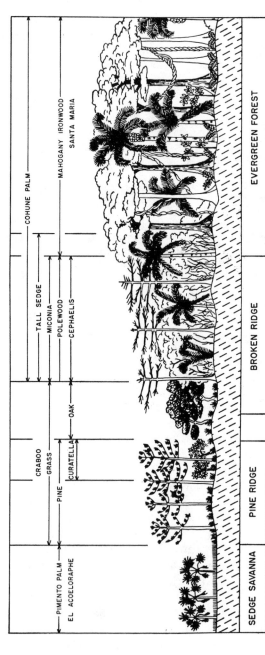

Fig. 19-10. Tropical forest development in the ridge area of British Honduras (redrawn from Charter 1940).

Hydroseral Stages on the Yucatan Peninsula

There are only minor differences between ponds here and in the United States, and many of the same species and genera of plants occur. Scattered throughout the uplands are depressions or aguadas which are the chief source of water for animals. They are often covered with a floating mass of *Pistia stratiotes*. In some basins, sedges thrive in the shallow water, *Eleocharis retroflexa* being common; on the saturated mud banks of some aguadas *Eleocharis retroflexa, Fuirena simplex, Rynchospora cephalotes*, and *R. cyperoides* occur. These fresh waters support the snails *Physa* and *Ferrissia*, and the electric light bug *Belostoma* (Lundell 1937, Sanderson 1941).

The aguadas commonly become so completely silted that during dry seasons only occasional small pools remain in their centers. These silted lagoon beds support a mixed forest, strikingly different from the upland mesophytic forest, both in floristic composition and physiognomy. The water which floods the lagoons and aguadas during the rainy season is coffee-colored and contains a high percentage of decomposing organic matter.

The silted basins are fringed with the tree *Pachira aquatica* and the pools by the palo tinta tree *Haematoxylum campechianum* from which the tintal community derives its name (Lundell 1937). In general the height of the tintal varies from 5 to 11 meters, increasing away from the center. A few large erect trees reach much higher, standing out above the mass of the swamp forest. *Talisia floresii* reaches a height of 20 meters. The main stratum of low trees and shrubs contains *Haematoxylum campechianum, H. brasiletto, Diospyros bumelioides, Guettarda gaumeri,* and various species of *Eugenia, Mimosa, Croton, Caesalpinia, Phyllanthus,* and others. The Leguminosae and Euphorbiaceae appear to dominate. The trees and shrubs are gnarled, twisted, and thorny and have interlocked knotty branches. Some, such as *Haematoxylum campechianum,* are shaggy barked and deeply grooved (Lundell 1937).

Mammals move to high ground during the rainy season and return to the submerged portions as the dry season advances. For this reason deer are hunted more successfully during the rainy season. Some of the more permanent aguadas may be populated with several species of fishes which indicate that they have connections with a river during the rainy season. When aguadas dry up, the fish die. Cats and opossum frequent these ponds to feed on the fish and to drink (Murie 1935). Only two birds are recorded from the aguadas, a little blue heron was observed being killed in a tree at the edge of the pond by a hawk-eagle. However, two other herons, a limpkin, a tiger-heron, and a few waders are mentioned as occurring at Uaxactun where there is a large aguada and other water-holding depressions (Van Tyne 1935).

Very little is known about the development of forest on floodplains within the American tropics. Standley (1931) mentions the Central American willow *Salix chilensis,* which is conspicuous on account of its flexible, drooping branches and pale foliage, and the native bamboo *Guadua aculeata,* which towers above most of the trees, for northern Honduras. Prominent in the herbaceous vegetation is *Calathea lutea,* a coarse plant two meters high or more, that has huge, thin, stiff, vertical leaves that are white on the underside.

EPILOGUE

While it has on the whole been a pleasure to bring together this information on the bioecology of North America, it seems fitting to say that the basic defects in our fund of knowledge are (1) the lack of studies of both plants and animals in the same community, (2) the lack of quantitative data on the populations of animals and their food habits and on the density of plants, and (3) the lack of consideration of the interrelation of animals and the vegetation. These deficiencies need to be remedied. Let us hope that every effort is made to learn the structure, composition, and dynamics of the original communities of North America before they finally succumb before the advance of civilization.

Literature Cited

ADAMS, C. C. 1905. The postglacial dispersal of North American biota. Biol. Bull. 9:53-71.

——. 1909. An ecological survey of Isle Royale, Lake Superior. Rept. Univ. Michigan Mus.; Rept. Bd. Geol. Surv. for 1908. 468 pp.

——. 1915. An ecological study of prairie and forest invertebrates. Bull. Illinois St. Lab. Nat. Hist. 11:33-119.

ADAMS, C. C., G. P. BURNS, T. L. HANKINSON, B. MOORE, and N. TAYLOR. 1920. Plants and animals of Mount Marcy, New York. Ecology 1:71-94, 204-233, 274-288.

ADAMS, R. H. 1941. Stratification, diurnal and seasonal and migration in a deciduous forest. Ecol. Monogr. 11:189-227.

AGACINO, E. M. 1952. The locust problem in Central America and Mexico. FAO, Plant Prot. Bull. 1:18-20.

AIKMAN, J. M. 1926. Distribution and structure of the forests of eastern Nebraska. Univ. Nebraska Stud. 26:1-75.

——. 1928. Competition studies in the ecotone between prairie and woodland. Iowa Acad. Sci. Proc. 35:99-103.

——. 1935. Native vegetation of the region. Possibilities of shelterbelt planting in the Plains region. U.S. Govt. Printing Office, 155-174.

ALBERTSON, F. W. 1937. Ecology of mixed prairie in west central Kansas. Ecol. Monogr. 7:483-546.

——. 1941. Prairie studies in west central Kansas. 1940 Trans. Kansas Acad. Sci. 44:4-13.

ALBERTSON, F. W., and J. E. WEAVER. 1945. Injury and death or recovery of trees in prairie climate. Ecol. Monogr. 15:393-433.

ALBRECHT, W. A. 1946. Soil as the basis of wildlife management. 8th Midwest Wildl. Conf. 8:1-8.

——. 1957. Soil fertility and biotic geography. Geogr. Rev. 47:87-105.

ALDOUS, A. E. 1942. The white-necked raven in relation to agriculture. U.S. Dept. Int., Fish and Wildl. Serv. Rept. 5:1-54.

ALDOUS, A. E., and H. L. SHANTZ. 1924. Types of vegetation in the semiarid portion of the U.S. and their economic significance. J. Agr. Res. 28:99-127.

495

ALDRICH, J. M. 1905. Catalogue of North American Diptera. Smithsonian Misc. Coll. 1444:1-680.

ALDRICH, J. W. 1943. Biological survey of the bogs and swamps in northeastern Ohio. Amer. Midl. Nat. 30:346-402.

ALDRICH, J. W., and B. P. BOLE, JR. 1937. The birds and mammals of the western slope of the Azuero Peninsula. Cleveland Mus. Nat. Hist. Sci. Publ. 7:1-96.

ALDRICH, J. W., and J. DUVALL. 1955. Distribution of American game birds. U.S. Dept. Int., Fish and Wildl. Serv. Circ. 34:1-42.

ALDRICH, J. W., and P. GOODRUM. 1946. Virgin hardwood forest, tenth breeding bird census. Audubon Field Notes 10:144-145.

ALEXANDER, G. 1951. Occurrence of Orthoptera at high altitudes with special reference to Colorado Acrididae. Ecology 32:104-112.

ALLEE, W. C. 1926a. Measurement of environmental factors in the tropical rain-forest of Panama. Ecology 7:273-302.

————. 1926b. Distribution of animals in a tropical rain-forest with relation to environmental factors. *Ibid.* 7:445-468.

ALLEE, W. C., A. E. EMERSON, O. PARK, T. PARK, and K. P. SCHMIDT. 1949. Principles of animal ecology. W. B. Saunders Company, Philadelphia. xii + 837 pp.

ALLEN, P. H. 1955. The conquest of Cerro Santa Barbara Honduras. Ceiba 4:253-270.

ALVORD, C. W., and L. BIDGOOD. 1912. The first exploration of the trans-Allegheny region by the Virginians 1650-1674. Arthur H. Clark Co., Cleveland, Ohio. 275 pp.

AMERICAN ORNITHOLOGISTS' UNION. 1957. Check-list of North American birds. 5th ed. American Ornithologists' Union, Baltimore, Maryland. ix + 691 pp.

ANDERSON, R. M. 1946. Catalogue of Canadian recent mammals. Nat. Mus. Canada Biol. Ser. 31, Bull. 102:1-237.

ANDREWS, H. J., *et al.* 1936. Forest map of Oregon; forest map of Washington. U.S. Dept. Agr., For. Serv.

Anonymous. 1912. Game animals, birds and fishes of British Columbia, Canada. Bur. Prov. Inf. Bull. 17:1-82.

ANTHONY, H. E. 1928. Field book of North American mammals. G. P. Putnam's Sons, New York. xxvi + 674 pp.

AREND, J. L., and O. JULANDER. 1948. Oak sites in the Arkansas Ozarks. Univ. Arkansas Coll. Agr., Agr. Expt. Sta. Bull. 484:3-42.

ARIAS, A. C. 1942. Mapa de las provincias climatológicas de la República Mexicana. Secretaria de Agricultura y Fomento, Inst. Geogr., Tacubaya, Mexico.

ARNASON, A. P. 1941. Arthropod populations of the vegetation of wheatland and native grassland at Saskatoon, Saskatchewan. Thesis abstract, Univ. Illinois, 3-21.

ASHBY, E. 1932. Transpiration organs of *Larrea tridentata* and their ecological significance. Ecology 13:182-188.

AUSTIN, O. L., JR. 1929. Birds of the Cayo District, British Honduras. Bull. Mus. Comp. Zool. [Harvard] 69:363-393.

BAILEY, F. 1950. Ultraviolet boosts egg production. Poultry Trib. 56:34.

BAILEY, R. W., H. G. UHLIG, and G. BREIDING. 1951. Wild turkey management in West Virginia. Conserv. Comm., West Virginia Div. Game Mgmt. Bull. 2:1-47.

BAILEY, V. 1905. Biological survey of Texas. N. Amer. Fauna 25:1-222.

————. 1926. Biological survey of North Dakota. *Ibid.* 49:1-2.

————. 1930. Animal life of Yellowstone National Park. Charles C. Thomas, Springfield, Illinois. 241 pp.

————. 1931. Mammals of New Mexico. N. Amer. Fauna 53:1-412.

————. 1936. The mammals and life zones of Oregon. *Ibid.* 55:1-416.

BAKER, F. C. 1939-48. Field book of land snails. Illinois Nat. Hist. Surv. Manual 2:i-xi, 1-166.

BAKER, F. S. 1949. A revised tolerance table. J. For. 47:179-181.

BAKER, H. B. 1922. Mollusca collected by the University of Michigan-Walker Expedition in southern Vera Cruz, Mexico 1. Occ. Pap. Mus. Zool., Univ. Michigan 106:1-94.

BAKER, R. H. 1951. Mammals from Tamaulipas, Mexico. Publ. Mus. Nat. Hist., Univ. Kansas 5:207-218.

————. 1956. Mammals of Coahuila, Mexico. *Ibid.* 9:125-335.

BALL, E. D. 1932. The food plants of the leafhoppers. Ann. Ent. Soc. Amer. 25:497-502.

BALL, E. D., E. R. TINKHAM, R. FLOCK, and C. T. VORHIES. 1942. The grasshoppers and other Orthoptera of Arizona. Univ. Arizona Tech. Bull. 93:257-373.

BANGS, O., J. E. THAYER, T. BARBOUR, and S. GARMAN. 1906. Vertebrates from the savanna of Panama. II. Mammalia. III. Aves. IV. Reptilia and Amphibia. Bull. Mus. Comp. Zool. 46:211-230.

BANKS, N. 1910a. Catalogue of Nearctic Hemiptera-Heteroptera. American Entomological Society, Philadelphia. 111 pp.

————. 1910b. Catalogue of Nearctic spiders. U.S. Natl. Mus. Bull. 72:1-80.

BARBOUR, T. 1906. Reptilia and Amphibia. In Bangs *et al.* 1906, 224-229.

————. 1923. The birds of Cuba. Mem. Nuttall Orn. Cl., Cambridge, Massachusetts. 141 pp.

BARBOUR, T., and C. T. RAMSDEN. 1919. Herpetology of Cuba. Mem. Mus. Comp. Zool. [Harvard] 47:73-212.

BARBOUR, W. R. 1941. Forest types of tropical America. Caribbean For. 3:137-150.

BARNARD, E. C. 1900. Roseburg Quadrangle, Oregon. U.S. Geol. Surv., 21st Ann. Rept. 5:577; plate CXXVIII.

BARNES, R. D. 1953. Ecological distribution of spiders in non-forest maritime communities at Beaufort, North Carolina. Ecol. Monogr. 23:315-337.

BARTLETT, H. H. 1935. Botany of the Maya area: miscellaneous papers. I. A method of procedure for field work in tropical American phytogeography based upon a botanical reconnaissance in parts of British Honduras and the Peten Forest of Guatemala. Carnegie Inst. Washington Publ. 461:1-25.

————. 1937. Unpublished manuscript on Tamaulipas grassland.

BARTRAM, W. 1944. Diary of a journey through the Carolinas, Georgia, and Florida, from July 1, 1765 to April 10, 1766. Amer. Philos. Soc., n.s. 33 (part 1):1-243.

BAXTER, D. V., and F. H. WADSWORTH. 1939. Forest and fungus succession in the lower Yukon Valley. Univ. Michigan Sch. For. and Conserv. Bull. 9:1-52.

BEARD, J. S. 1944. Climax vegetation in tropical America. Ecology 25:127-158.

————. 1946. Natural vegetation of Trinidad. Oxford Univ. Sch. For., Oxford For. Mem. 20:1-152.

————. 1953. Savanna vegetation of northern tropical America. Ecol. Monogr. 23:149-215.

————. 1955. Classification of tropical American vegetation. Ecology 36:89-99.

BECKER, H. F. 1943. Land utilization in Guanacaste Province, Costa Rica. Geogr. Rev. 33:74-85.

BEED, W. E. 1936. A preliminary study of the animal ecology of the Niobrara Game Preserve. Conserv. and Surv. Div., Univ. Nebraska Bull. 10:1-33.

BELT, T. 1874. The naturalist in Nicaragua. A narrative of a residence at the gold mines of Chontales — journeys in the savannas and forests. J. Murray, London. xvi + 403 pp.

BENNETT, G. W. 1947. Fish management — a substitute for natural predation. Trans. 12th N. Am. Wildl. Conf. 12:276-283.

BENNITT, R. and W. O. NAGEL. 1937. A survey of the resident game and fur bearers of Missouri. Univ. Missouri Stud. 12:1-215.

BENOIT, J., F. X. WALTER, and I. ASSENMACHER. 1950. Nouvelles recherches relative a l'action de lumieres de differentes longueurs d'onde sur la gonadostimulation du canard male impubere. C.R. Soc. Biol. 144:1206.

BENOIT, J., I. ASSENMACHER, and F. X. WALTER. 1950. Reponses du mecanisme gonado-stimulant a l'eclairement artificiel et de la prehypophyse aux castrations bilaterale et unilaterale, chez le canard domestique male, au cours de la periode de regression gesticulaire saisonniere. C.R. Soc. Biol. 144:573.

BENSON, L., and R. DARROW. 1944. Manual of southwestern trees and shrubs. Univ. Arizona Biol. Sci. Bull. 6:440-443.

BENT, A. C. 1937. Life histories of North American birds of prey. Part 1. U.S. Natl. Mus. Bull. 167:i-viii, 1-409.

——. 1938. *Ibid.* Part 2. 170:i-viii, 1-482.

——. 1939. Life histories of North American woodpeckers. *Ibid.* 174:i-viii, 1-334.

BEQUAERT, J. C. 1933. Botanical notes from Peninsula of Yucatan. (In Shattuck, G. C. The Peninsula of Yucatan.) Carnegie Inst. Washington Publ. 431:505-524 (chap. XXVII).

BILLINGS, W. D. 1938. The structure and development of old field shortleaf pine stands and certain associated physical properties of the soil. Ecol. Monogr. 8:437-499.

——. 1945. The plant associations of the Carson Desert region, western Nevada. Butler Univ. Botan. Stud. 7:89-123.

——. 1949. The shadscale vegetation zone of Nevada and eastern California in relation to climate and soils. Amer. Midl. Nat. 42:87-109.

——. 1951. Vegetational zonation in the Great Basin of western North America. In Les bases ecologiques de la regeneration de la vegetation des zones arides. Ser. B [U.I.S.B., Paris] 9:101-122.

BIRD, R. D. 1930. Biotic communities of the aspen parkland of central Canada. Ecology 11:356-442.

BISHOP, S. C. 1943. Handbook of salamanders. Comstock Publishing Company, Inc., Ithaca, New York. 555 pp.

BLAIR, W. 1935. Some mammals of southern Florida. Amer. Midl. Nat. 16:801-804.

——. 1938. Ecological relationships of the mammals of the bird creek region, northeastern Oklahoma. *Ibid.* 20:473-526.

——. 1940. A contribution to the ecology and faunal relationships of the mammals of the Davis Mountain region, southwestern Texas. Mus. Zool., Univ. Michigan 46:7-39.

——. 1941. Annotated list of mammals of the Tularosa Basin, New Mexico. Amer. Midl. Nat. 26:218-229.

——. 1943. Activities of the Chihuahua deermouse in relation to light intensity. J. Wildl. Mgmt. 7:92-97.

BLAKE, E. R. 1953. Birds of Mexico. University of Chicago Press, Chicago. 644 pp.

BLAKE, E. R., and H. C. HANSON. 1942. Notes on a collection of birds from Michoacan. Fld. Mus. Nat. Hist. Zool. Ser. 22:513-551.

BLAKE, I. H. 1926. A comparison of the animal communities of coniferous and deciduous forest. Illinois Biol. Monogr. 10:371-520.

——. 1931. Biotic succession on Katahdin. Appalachia 18:409-424.

——. 1945. An ecological reconnoissance in the Medicine Bow Mountains. Ecol. Monogr. 15:207-243.

BLATCHLEY, W. S. 1892. An entomologist in Mexico. Ent. News 3:131-136.

——. 1910. On the Coleoptera known to occur in Indiana. Indiana Dept. Geol. Nat. Res. Bull. 1:1-1386.

——. 1920. Orthoptera of eastern North America. Indianapolis, Indiana. 784 pp.

——. 1926. Heteroptera or true bugs of eastern North America. Nature Publishing Co., Indianapolis, Indiana. 1116 pp.

BOGERT, C. M., and R. B. COWLES. 1947. Results of the Archbold expeditions. No. 58. Moisture loss in relation to habitat selection in some Floridian reptiles. Amer. Mus. Novit. 1358:1-34.

BORELL, A. E., and D. BRYANT. 1942. Mammals of the Big Bend area of Texas. Univ. California Publ. Zool. 48:1-63.

BOYCE, S. G. 1954. The salt spray community. Ecol. Monogr. 24:29-67.

BRAGG, A. N. 1939. Geographical distribution of Acridae in northern Oklahoma. Amer. Midl. Nat. 22:660-675.

BRAND, D. B. 1936. Notes to accompany a vegetation map of northwest Mexico. Univ. New Mexico Bull., Biol. Ser. 4:5-27.

BRAUN, E. LUCY. 1940. An ecological transect of Black Mountain, Kentucky. Ecol. Monogr. 10:193-241.

———. 1941. The differentiation of the deciduous forest of the eastern United States. Ohio J. Sci. 41:235-241.

———. 1942. Forests of the Cumberland Mountains. Ecol. Monogr. 12:414-447.

———. 1950. The deciduous forests of eastern North America. Blakiston Company, Inc., Philadelphia. 596 pp.

BRAY, W. L. 1905. The vegetation of the sotol country in Texas. Univ. Texas Bull. [Sci. Ser. 6] 60:1-244.

———. 1910. Distribution and adaptation of vegetation of Texas. *Ibid.* [Sci. Ser. 10] 82:1-108.

BREHM, A. E. 1895. From North Pole to Equator. Studies of wildlife and scenes in many lands. Trans. by Margaret Thomson. Blackie, London. 592 pp.

BRETT, C. H. 1947. Interrelated effects of food, temperature and humidity on the development of the lesser migratory grasshopper, *Melanoplus mexicanus mexicanus* (Saussure) (Orthoptera). Oklahoma Agr. Expt. Sta. Tech. Bull. 26:5-50.

BROOKS, A. H. 1906. The geology and geography of Alaska, U.S. Geol. Surv. Prof. Paper 45:1-327.

BROOKS, M. G. 1930. Notes on the birds of Cranberry Glades, Pocahontas County, West Virginia. Wilson Bull. 42:245-252.

———. 1943. Birds of the Cheat Mountains. Cardinal 6:25-48.

BROWN, CLAIR A. 1944. Historical commentary on the distribution of vege.ation in Louisiana and some recent observation. Proc. Louisiana Acad. Sci. 8:35-47.

———. 1945. Louisiana trees and shrubs. Louisiana For. Comm. Bull. 1:1-262.

BROWN, D. 1941. The vegetation of Roan Mountain; a phytosociological and successional study. Ecol. Monogr. 11:61-97.

BROWN, H. L. 1946. Rodent activity in a mixed prairie near Hays, Kansas. Trans. Kansas Acad. Sci. 48:448-456.

———. 1947. Coaction of jack rabbit, cottontail and vegetation in a mixed prairie. *Ibid.* 50:28-44.

BROWN, L. G., and L. E. YEAGER. 1945. Fox squirrels in Illinois. Illinois Nat. Hist. Surv. 23:449-536.

BROWN, MARY JANE. 1931. Comparative studies of the animal communities of oak-hickory forests in Missouri and Oklahoma. Publ. Univ. Oklahoma, Biol. Surv. 2:231-261.

BROWN, W. H. 1912. The plant life of Ellis, Great, Little and Long lakes in North Carolina. Contr. U.S. Natl. Herb. 13:323-343.

BRUMWELL, M. 1951. An ecological survey of the Fort Leavenworth Military Reservation. Amer. Midl. Nat. 45:187-231.

BRUNER, L. 1891. Destructive locusts of North America, together with notes on the occurrence in 1891. U.S. Dept. Agr. Insect Life 4:18-24.

BRUNER, W. E. 1931. The vegetation of Oklahoma. Ecol. Monogr. 1:99-188.

BRYANT, H. C. 1911. The relation of birds to an insect outbreak in northern California during the spring and summer of 1911. Condor 13:195-208.

BUECHNER, H. K. 1946. Birds of Kerr County, Texas. Trans. Kansas Acad. Sci. 49:357-364.

BUELL, M. F. 1946. Jerome Bog, a peat filled Carolina Bay. Torreya 73:24-33.

BUELL, M. F., and HELEN FOOT BUELL. 1941. Surface level fluctuation in Cedar Creek Bog, Minnesota. Ecology 22:317-321.

BUELL, M. F., and R. L. CAIN. 1943. The successional role of southern white cedar, *Chamaecyparis thyoides,* in southeastern North Carolina. Ecology 24:85-93.

BUGBEE, R. E., and A. REIGEL. 1945. The cactus moth, *Melitara dentata* (Grote), and its effect on *Opuntia macrorhiza* in western Kansas. Amer. Midl. Nat. 33:117-127.

BUKOVSKY, B. 1935. To the criticism of the basic problems and concepts of biocenology (English summary). Probl. Ecol. Biocen., 95-99.

BULLOUGH, W. S. 1951. Vertebrate sexual cycles. John Wiley & Sons, Inc., New York. 117 pp.

BUNKER, C. D. 1940. The kit fox. Science 92:35-36.

BURGER, W. 1949. Review of experimental investigations on seasonal reproduction in birds. Wilson Bull. 61:221-230.

BURLISON, W. L., C. L. STEWART, R. C. Ross, and O. L. WHALIN. 1934. Production and marketing of red-top. Bull. Illinois Agr. Expt. Sta. 404.

BURT, W. H. 1938. Faunal relationships and geographic distribution of mammals in Sonora, Mexico. Univ. Michigan Misc. Pap. Mus. Zool. 39:7-75.

——. 1940. Territorial behavior and populations of some small mammals in southern Michigan. *Ibid.* 45:1-58.

——. 1952. Field guide to mammals. Houghton Mifflin Company, Cambridge, Massachusetts. 200 pp.

CAHALANE, V. H. 1928. A preliminary wildlife and forest survey of southwestern Cattaraugus County, New York. Roosevelt Wild Life Bull. 5:9-144.

CAIN, S. A. 1930. An ecological study of the heath balds of the Great Smoky Mountains. Butler Univ. Botan. Stud. 1:177-208.

——. 1935. Studies on virgin hardwood forest. III. Warren's Woods, a beech-maple climax forest in Berrien County, Michigan. Ecology 16:500-513.

CALHOUN, J. B. 1941. Distribution and food habits of the mammals in the vicinity of Reelfoot Lake Biological Station. J. Tennessee Acad. Sci. 16:177-185, 207-225.

CALVERT, A. C. (SMITH), and P. P. CALVERT. 1917. A year of Costa Rican natural history. Macmillan Company, New York. 577 pp.

CANTU, D. R. 1942. Observations ecologicas sobre la vegetation franocarnica de la laguna de *Epatlan* (Puebla). Ann. Inst. Biol. Univ. Mexico 13:405-415.

CARABIA, J. P. 1945. A brief review of the Cuban flora. In Verdoorn 1945, 68-70.

——. 1945a. Vegetation of Sierra de Nipe, Cuba. Ecol. Monogr. 15:321-334.

CARPENTER, C. R. 1934. A field study of the behavior and social relations of howling monkeys. Comp. Psychol. Monogr. 10:1-168.

CARPENTER, J. R. 1935. Fluctuation of biotic communities. I. Prairie-forest ecotone of central Illinois. Ecology 16:203-212.

——. 1939. Fluctuations of biotic communities. V. Aspection in a mixed grass prairie in central Oklahoma. Amer. Midl. Nat. 22:420-435.

CARR, A. F. 1940. A contribution to the herpetology of Florida. Univ. Florida, Publ. Biol. Ser. 3:1-114.

——. 1950. Outline of a classification of animal habitats in Honduras. Bull. Amer. Mus. Nat. Hist. 94:567-594.

CARY, M. 1911. A biological survey of Colorado. N. Amer. Fauna 33:9-256.

——. 1917. Life zone investigations in Wyoming. *Ibid.* 42:1-95.

CAVERHILL, P. Z. 1926. British Columbia. In Shelford 1926, 153, 154, 159.

CHAMBERLIN, R. V., and W. IVIE. 1944. Spiders of the Georgia region of North America. Bull. Univ. Utah 35:1-267.

CHAPMAN, F. M. 1928. Nesting habits of Wagler's oropendola (*Zarhynchus wagleri*) on Barro Colorado Island. Bull. Amer. Mus. Nat. Hist. 58:123-166.

——. 1935. The courtship of Gould's manakin on Barro Colorado Island, Canal Zone. *Ibid.* 68:471.

CHARTER, C. F. 1940. Reconnaissance survey of the soils of British Honduras. U.S. Govt. Printing Office. Belize B. H. Separate, 1-31.

CHITTY, HELEN. 1950. Canadian arctic wild life enquiry, 1945-49, with a summary of results since 1933. J. Animal Ecol. 19:180-193.

CHRYSLER, M. A. 1910. The ecological plant geography of Maryland — coastal zone West Shore district. Maryland Weather Ser., Spec. Publ. 3:149-197.

CLARK, R. F. 1949. Snakes of the hill parishes of Louisiana. J. Tennessee Acad. Sci. 24:244-261.

CLARKE, C. H. D. 1939. Wildlife investigation in Banff National Park. Bur. Natl. Parks, 1-26 [mimeographed].

——. 1940. A biological investigation of the Thelon Game Sanctuary. Nat. Mus. Canada Bull., Biol. Ser. 96:1-135.

CLARKE, C. H. D., and I. M. COWAN. 1945. Birds of Banff National Park, Alberta. Canad. Fld. Nat. 59:83-103.

CLAYCOMB, G. B. 1919. Notes on the habits of *Heterocerus* beetles. Canad. Ent. 51:25.

CLAYTON, R. C. 1927. World weather records. Smithsonian Misc. Coll. 29:1099.

CLEMENTS, F. E. 1905. Research methods in ecology. University Publishing Co., Lincoln, Nebraska. 334 pp.

——. 1920. Plant indicators. Carnegie Inst. Washington Publ. 290:1-388.

CLEMENTS, F. E., and V. E. SHELFORD. 1939. Bio-ecology. John Wiley & Sons, Inc., New York. vi + 425 pp.

CLEMENTS, F. E., J. E. WEAVER, and H. C. HANSON. 1929. Plant competition. Carnegie Inst. Washington Publ. 398:1-340.

CLIFFORD, C. S., and G. A. HARDY. 1942. Report on a collecting trip to the Lac la Hache area, British Columbia. Rept. Prov. Mus. 1942:25-49.

CLOVER, E. U. 1937. Vegetational survey of the lower Rio Grande Valley, Texas. Madroño 4:41-66, 77-100.

COCKRUM, E. L. 1952. Mammals of Kansas. Publ. Mus. Nat. Hist., Univ. Kansas 7:1-303.

COGSWELL, H. S. 1947. The chaparral country. Audubon Mag. 49:75-81.

COMSTOCK, J. H. 1948. The spider book. Comstock Publishing Company, Inc., Ithaca, New York. 729 pp.

CONANT, R. 1958. A field guide to reptiles and amphibians of eastern North America. Houghton Mifflin Company, Boston. xv + 366 pp.

COOK, O. F. 1909. Vegetation affected by agriculture in Central America. U.S. Dept. Agr., Bur. Pl. Ind. Bull. 145:1-30.

COOK, W. C. 1924. The distribution of the pale western cutworm, *Porosagrotis orthogonia* Morr: a study in physical ecology. Ecology 5:60-68.

COOLEY, R. A. 1922. Grasshoppers, cutworms and other insect pests of 1921-1922. Montana Agr. Expt. Sta. Bull. 150:1-31.

COOPER, W. S. 1912. Ecological succession of mosses as illustrated upon Isle Royale, Lake Superior. Plant World 15:197-213.

——. 1922. The broad-sclerophyll vegetation of California. An ecological study of the chaparral and its related communities. Carnegie Inst. Washington Publ. 319:1-124.

——. 1923. The recent ecological history of Glacier Bay, Alaska. Ecology 4:93-128, 223-246, 355-365.

——. 1926. Vegetational development upon alluvial fans in the vicinity of Palo Alto, California. *Ibid.* 7:1-30.

——. 1928. Seventeen years of successional change upon Isle Royale, Lake Superior. *Ibid.* 9:1-5.

——. 1936. The strand and dune flora of the Pacific Coast of North America. Essays in geobotany in honor of William Albert Setchell, 141-187.

——. 1939. A fourth expedition to Glacier Bay, Alaska. Ecology 20:130-155.

——. 1942. Vegetation of the Prince William Sound region, Alaska; with a brief excursion into post-Pleistocene climatic history. Ecol. Monogr. 12:1-22.

CORE, E. L. 1929. Plant ecology of Spruce Mountain, West Virginia. Ecology 10:1-13.

Corps of Engineers, War Department. 1932, 1937. Quadrangles (maps) Hickman, Kentucky to Portageville, Missouri.

——. 1937. Levee and River charts (maps). Island no. 8 to Island no. 14.

CORY, C. B. 1912. The mammals of Illinois and Wisconsin. Fld. Mus. Nat. Hist., Zool. Series Publ. 51, 11:1-505.

COTTAM, C. 1939. Food habits of North American diving ducks. U.S. Dept. Agr. Tech. Bull. 643:1-140.

COTTLE, H. J. 1931. Studies on the vegetation of southwestern Texas. Ecology 12:105-155.

——. 1932. Vegetation on north and south slopes of mountains in southwestern Texas. *Ibid.* 13:121-134.

COUES, E. 1871. Notes on the natural history of Fort Macon, North Carolina and vicinity. Proc. Acad. Nat. Sci. Philadelphia, 1870, 1:12-49; 2:120-153.

COUES, E., and H. C. YARROW. 1878. Notes on the natural history of Fort Macon, North Carolina and vicinity. Proc. Acad. Nat. Sci. Philadelphia 4:21-28; 5:297-316.

COUPLAND, R. T. 1950. Ecology of mixed prairie in Canada. Ecol. Monogr. 20:271-315.

COWAN, I. M. 1936. Distribution and variation in deer (*Odocoileus*) of the Pacific Coast region of North America. California Fish and Game 22:155-247; maps pp. 204, 234.

——. 1945. The ecological relations of the food of the Columbian black-tailed deer, *Odocoileus hemionus columbianus* (Richardson), in the coast forest region of southern Vancouver Island, British Columbia. Ecol. Monogr. 15:109-139.

——. 1947. Timber wolf in the Rocky Mountain National Parks of Canada. Canad. J. Res. D 25:139-174.

COWAN, I. M., and J. A. MUNRO. 1944-46. Birds and mammals of Revelstoke National Park. Canad. Alp. J. 29:100-121, 237-256.

COWLES, H. C. 1899. The ecological relations of the vegetation on the sand dunes of Lake Michigan. Botan. Gaz. 27:95-117, 167-202, 281-308, 361-391.

——. 1901. Plant societies of Chicago and vicinity. *Ibid.* 31:73-108, 145-182.

——. 1928. Persistence of prairies. Ecology 9:380-382.

COWLES, R. B., and C. M. BOGART. 1944. Preliminary study of the thermal requirements of desert reptiles. Amer. Mus. Nat. Hist. 83:265-296.

Cox, C. F. 1933. Alpine plant succession in James Peak, Colorado. Ecol. Monogr. 3:1-372.

CRAIG, W. 1908. North Dakota life-plant, animal and human. Bull. Amer. Geogr. Soc. 40:321-332, 401-415.

CRIDDLE, N. 1925. The habits and economic importance of wolves in Canada. Canad. Dept. Agr. Bull., n.s. 13:1-24.

——. 1933. Studies in the biology of North American Acrididae development and habits. Proc. World's Grain Exhib. and Conf., Canada, 474-494.

CROCKETT, D. 1834. Narrative of the life of David Crockett of the State of Tennessee. Philadelphia. 211 pp.

CRONEMILLER, F. P. 1942. Chaparral. Madroño 6:199.

CRONEMILLER, F. P., and P. S. BARTHOLOMEW. 1950. The California mule deer in chaparral forests. California Fish and Game 36:243-265.

CROOK, C. 1935. Birds of late summer on Reelfoot Lake. J. Tennessee Acad. Sci. 10:1-18.

———. 1938. Food of some water birds of Reelfoot Lake. *Ibid.* 13:33-43.

CURRAN, C. H. 1934. Family and genera of North American Diptera. Published by author, New York City.

DALKE, P. D., W. R. CLARK, JR., and L. J. KORSCHGEN. 1942. Food habit trends of the wild turkey in Missouri as determined by dropping analysis. J. Wildl. Mgmt. 6:237-243.

DALQUEST, W. W. 1948. Mammals of Washington. Univ. Kansas Publ. 2:1-444.

———. 1953. Mammals of the Mexican State of San Luis Potosí. Louisiana St. Univ. Stud., Biol. Sci. Ser. 1:212.

DANSEREAU, P. 1948. Botanical excursions in Quebec Province: Montreal, Quebec, Gaspé Peninsula. Bull. Serv. Geogr. 2:1-20.

———. 1955. Biogeography of the land and inland waters. In Geography of the north lands, by G. H. T. Kimble and Dorothy Good, Amer. Geogr. Soc. and John Wiley & Sons, Inc., New York.

DANSEREAU, P., and F. SEGADAS-VIANNA. 1952. Ecological study of the peat bogs of eastern North America. I. Structure and evolution of vegetation. Canad. J. Botan. 30:490-520.

DARROW, R. A. 1944. Arizona range resources and their utilization. I. Cochise County. Univ. Arizona Agr. Expt. Sta. Tech. Bull. 103:311-366.

DASMANN, W. P. 1950. Basic deer management. California Fish and Game 36:251-284.

DAUBENMIRE, R. F. 1936. The "Big Woods" of Minnesota: its structure, and relation to climate, fire, and soils. Ecol. Monogr. 6:235-268.

———. 1942. An ecological study of the vegetation of southern Washington and adjacent Idaho. *Ibid.* 12:53-79.

DAVIS, D. H. S. 1936. A reconnaissance of the fauna of Akpatok Island, Ungava Bay. J. Animal Ecol. 5:319-332.

DAVIS, J. H., JR. 1937. Aquatic plant communities of Reelfoot Lake. J. Tennessee Acad. Sci. 12:96-103.

———. 1940. The ecology and geologic role of mangroves in Florida. Carnegie Inst. Washington Publ. 517:303-412.

———. 1943. The natural features of southern Florida, especially the vegetation and the Everglades. Florida Geol. Surv. Bull. 25:5-311.

DAVIS, W. B., and R. J. RUSSELL. 1953. Aves y mamiferos del estado de Morelos. Rev. Soc. Mexico Hist. Nat. 14:77-147.

DAY, G. M. 1953. The Indian as an ecological factor in the northeastern forest. Ecology 34:329-346.

DAYTON, W. A., *et al.* 1937. Range plant handbook. U.S. Dept. Agr. For. Serv. In four sections: grasses, 1-125; grasslike plants, 1-17; range weeds, 1-213; browse, 1-157. 355 pp.; index.

DEFOREST, H. 1921. The plant ecology of the Rock River woodlands of Ogle County, Illinois. Trans. Illinois St. Acad. Sci. 14:152-196.

DEFRIESE, L. H. 1884. See Kentucky Geological Survey.

D'HERELLE, F. 1911. Sur une épizootie de nature bactérienne sévissant sur les sauterelles au Mexique. C.R. Acad. Sci. 152:1413-1415.

DELONG, D. M., and D. J. KNULL. 1945. Check list of the Cicadellidae (Homoptera) of America north of Mexico. Ohio St. Univ. Grad. Sch., Biol. Ser. 1:1-102.

DEMAREE, D. 1932. Submergence experiments with *Taxodium.* Ecology 13:258-262.

De Nio, R. M. 1938. Stomach analysis of elk in northern Idaho and Montana. Univ. Idaho Sch. For. Bull. 8:34-36.

DePourtales, L. F. 1877. Hints on the origin of the flora and fauna of the Florida Keys. Amer. Nat. 11:137-144.

Dice, L. R. 1916. Distribution of the land vertebrates of southeastern Washington. Univ. California Publ. Zool. 16:293-348.

———. 1920. The land vertebrate associations of interior Alaska. Occ. Pap. Mus. Zool., Univ. Michigan 85:1-24.

———. 1922. Some factors affecting the distribution of the prairie vole, forest deer-mouse, and prairie deermouse. Ecology 3:29-47.

———. 1930. Mammal distribution in the Alamogordo region, New Mexico. Occ. Pap. Mus. Zool., Univ. Michigan 213:1-32.

———. 1937. Mammals of the San Carlos Mountains and vicinity. Geology and Biology of the San Carlos Mountains, Univ. Michigan Sci. Ser. 12:245-268.

———. 1940. Relationship between the wood-mouse and the cotton-mouse in eastern Virginia. J. Mammal. 21:14-23.

———. 1943. The biotic provinces of North America. University of Michigan Press, Ann Arbor. viii + 78 pp.

———. 1952. Natural communities. University of Michigan Press, Ann Arbor. 547 pp.

Dice, L. R., and P. M. Blossom. 1937. Studies of mammalian ecology in southwestern North America with special attention to the colors of desert mammals. Carnegie Inst. Washington Publ. 485:1-129.

Diettert, R. A. 1933. The morphology of *Artemisia tridentata*. Lloydia [Cincinnati, Ohio] 1:3-74.

Dirks-Edmunds, Jane C. 1947. A comparison of biotic communities of the cedar-hemlock and oak-hickory associations. Ecol. Monogr. 17:235-260.

Doane, R. W., E. C. Van Dyke, W. J. Chamberlin, and H. E. Burke. 1936. Forest insects. McGraw-Hill Book Company, Inc., New York. 463 pp.

Doering, Kathleen. 1942. Host plant records of Cercopidae in North America, north of Mexico (Homoptera). J. Kansas Ent. Soc. 15:65-72, 73-92.

Donaldson, R. C. 1947. Lake County bygones (with map of Mississippi River left bank 1885-1942). Lake County Banner, Tiptonville, Tennessee, 24 (9).

Dorman, C. C. 1926. The Kisatchie wold. In Shelford 1926, 463-464.

Dowding, Eleanor S. 1929. The vegetation of Alberta. III. The sandhill areas of central Alberta with particular reference to the ecology of *Arceuthobium americanum* Nutt. J. Ecol. 17:82-105.

Dozier, H. L. 1920. An ecological study of hammock and piney woods insects in Florida. Ann. Ent. Soc. Amer. 13:25-380.

Drew, W. B. 1942. The revegetation of abandoned cropland in the Cedar Creek area, Boone and Callaway counties, Missouri. Univ. Missouri Agr. Expt. Sta. Res. Bull. 344:3-52.

Dufresne, F. 1942. Mammals and birds of Alaska. U.S. Dept. Int. Circ. 3:1-37.

Dunn, E. R. 1931. Amphibians of Barro Colorado Island. Occ. Pap. Boston Soc. Nat. Hist. 5:403-421.

Dyar, H. G. 1902. List of North American Lepidoptera and key to the literature of this order of insects. Bull. U.S. Natl. Mus. 52:1-723.

Ecke, D. H. 1948. Reproduction in Illinois cottontails. Presented to 10th Midwest Wildlife Conference, Ann Arbor, Michigan, Dec. 9-11, 1948.

Ecke, D. H., and C. Johnson. 1952. Plague in Colorado. Public Health Monogr. 6(1):1-36.

Eddy, S. 1934. A study of freshwater plankton communities. Illinois Biol. Monogr. 12:6-93.

EDDY, S., and P. H. SHIMER. 1929. Notes on the food of the paddlefish and the plankton of its habitat. Trans. Illinois St. Acad. Sci. 21:59-68.

EDWARDS, E. P. 1955. Finding birds in Mexico. E. P. Edwards Co., Amherst, Virginia. 101 pp.

EGGLER, W. A. 1938. The maple-basswood forest type in Washburn County, Wisconsin. Ecology 19:243-263.

EGLER, F. E. 1942. Checklist of the ferns and flowering plants of the Seashore State Park, Cape Henry, Virginia. New York St. Coll. For., Syracuse, New York, 60 pp. [mimeographed].

EISENMAN, E. 1952. An annotated list of birds of Barro Colorado Island, Panama Canal Zone. Smithsonian Misc. Coll. 117:1-62.

———. 1955. The species of middle American birds. Trans. Linn. Soc. New York 7:1-128.

EKBLAW, W. E. 1926. Northwest Greenland. In Shelford 1926, 87-90. Danish Greenland. *Ibid.,* 90-98.

ELLIOT, D. G. 1903. A list of mammals collected by Edmund Heller, in the San Pedro Martir and Hanson Laguna Mountains and the accompanying regions of lower California with descriptions of apparently new species. Fld. Columbian Mus. Publ. 70, Zool. Ser. 3:199-232.

———. 1904. The land and sea mammals of middle America and the West Indies. Publ. Fld. Mus., Zool. Ser. 4:1-850.

ELROD, M. J. 1902. Biological reconnaissance in the vicinity of Flathead Lake. Univ. Montana 10, Biol. Ser. 3:1-182.

EMERSON, F. W. 1932. The tension zone between the grama grass and pinon juniper associations in northeastern New Mexico. Ecology 13:347-358.

ENDERS, R. K. 1935. Mammalian life histories from Barro Colorado Island, Panama. Bull. Mus. Comp. Zool. 78:386-502.

———. 1939. Changes observed in the mammalian fauna of Barro Colorado Island 1929-1937. Ecology 20:104-106.

ENGELS, W. L. 1942. Vertebrate fauna of North Carolina coastal islands, a study in the dynamics of animal distribution. I. Ocracoke Island. Amer. Midl. Nat. 28:273-305.

ENGLISH, L. L. 1954. Effects of D.D.T. sprays on mite and aphid populations on elms. J. Econ. Ent. 47:658-660.

ERRINGTON, P. L. 1945. Some contributions of a fifteen year local study of the northern bob-white to a knowledge of population phenomena. Ecol. Monogr. 15:1-34.

ESSIG, E. O. 1934. Insects of western North America. Macmillan Company, New York. 1035 pp.

EVANS, M. W. 1949. Vegetative growth, development and reproduction in Kentucky bluegrass. Ohio Agr. Expt. Sta. Res. Bull. 681:1-39.

EWING, J. 1924. Plant successions of the brush prairie in northwestern Minnesota. J. Ecol. 12:238-266.

EYLES, M. S. 1942. Plants of Reelfoot Lake with special reference to use as duck foods. J. Tennessee Acad. Sci. 17:14-21.

FAUTIN, R. W. 1946. Biotic communities of the northern desert shrub biome in western Utah. Ecol. Monogr. 16:251-310.

FAUVER, B. J. 1949. A study of the ecological distribution of breeding bird populations in eastern North America. Ph.D. Thesis, Univ. Illinois.

FERNALD, M. L. 1950. Gray's manual of botany. 8th ed. American Book Company, New York. 1632 pp.

FERREL, C. M., and H. R. LEACH. 1950. Food habits of the pronghorn antelope of California. California Fish and Game 36:21-26.

———. 1952. The pronghorn antelope of California with special reference to food habits. California Fish and Game 38:285-293.

FICHTER, E. 1939. An ecological study of Wyoming spruce-fir forest arthropods with special reference to stratification. Ecol. Monogr. 9:183-215.

———. 1954. An ecological study of invertebrates of grassland and deciduous shrub savanna in eastern Nebraska. Amer. Midl. Nat. 51:321-339.

FISK, H. N. 1944. Geological investigation of the Alluvial Valley of the lower Mississippi River. Mississippi River Commission, Vicksburg, Mississippi. 68 pp.; 80 figs.; 33 pls.

FITCH, H. S. 1948. Ecology of the California ground squirrel on grazing lands. Amer. Midl. Nat. 39:513-596.

FITCH, H. S., and R. L. McGREGOR. 1956. The forest habitat of the University of Kansas Natural History Reservation. Publ. Mus. Nat. Hist., Univ. Kansas 10:77-127.

FITCH, H. S., F. SWENSON, and D. D. TILLOTSON. 1946. Behavior and food habits of the red-tailed hawk. Condor 48:205-237.

FORBES, E. B., and S. I. BECHDEL. 1931. Mountain laurel and rhododendron as foods for the white-tailed deer. Ecology 12:323-334.

FORBES, S. A. 1894. Laboratory studies of the chinch bug by W. G. Johnson. Rept. Illinois Ent. 19:178-189.

FRASIER, W. P., and R. C. RUSSELL. 1954. Annotated list of the plants of Saskatchewan. University of Saskatchewan, Saskatoon.

FREEMAN, C. P. 1933. Ecology of the cedar glade vegetation near Nashville, Tenn. J. Tennessee Acad. Sci. 8:143-228.

FREUCHEN, P. 1915. Report of the first Thule expedition. Scientific work. Medd. Grønland 51:387-411.

FREY, D. G. 1951. Pollen succession in the sediments of Singletary Lake, North Carolina. Ecology 32:518-523.

FRIAUF, J. J. 1953. An ecological study of the Dermoptera and Orthoptera. Ecol. Monogr. 23:79-126.

FRICK, T. A. 1939. Slope vegetation near Nashville, Tenn. J. Tennessee Acad. Sci. 14:344-420.

FROHNE, W. C. 1957. Reconnaissance of mountain mosquitoes in the McKinley Park region, Alaska. Mosquito News 17(1):17-22.

FULLER, G. D. 1935. Postglacial vegetation of the Lake Michigan region. Ecology 16:473-487.

FULLER, M. L. 1912. The new Madrid earthquake. U.S. Dept. Int., Geol. Surv. Bull. 494:7-115.

GABRIELSON, I. N., and S. G. JEWETT. 1940. Birds of Oregon. Oregon State College, Corvallis. 650 pp.

GADOW, H. 1908. Through southern Mexico. Witherby, London. 527 pp.

GAMS, H. 1918. Prinzipienfragen der Vegetationsforschung. Zurich.

GANIER, A. F. 1933. Water birds of Reelfoot Lake. J. Tennessee Acad. Sci. 8:65-83.

GARDNER, J. L. 1950. Effects of thirty years of protection from grazing in desert grassland. Ecology 31:44-50.

GARMAN, P., and W. T. MATHIS. 1956. Mineral balance as related to occurrence of Baldwin spot in Connecticut. Connecticut Agr. Expt. Sta. Bull. 601:1-19.

GATES, F. C. 1912. The vegetation of the beach area in northeastern Illinois and southeastern Wisconsin. Bull. Illinois St. Lab. Nat. Hist. 9:255-372.

GENTRY, H. S. 1942. Rio Mayo plants: a study of the flora and vegetation of the valley of the Rio Mayo Sonora. Carnegie Inst. Washington Publ. 527:1-328.

———. 1946a. Sierra Tacuichamona—a Sinaloan plant locale. Bull. Torrey Bot. Cl. 73:356-362.

———. 1946b. Notes on the vegetation of the Sierra Surotato in northern Sinaloa. *Ibid.* 73:451-462.

GERSBACHER, E. O. 1939. The heronries at Reelfoot Lake. J. Tennessee Acad. Sci. 14:162-179.

GERSBACHER, E. O., and E. M. NORTON. 1939. Typical plant succession at Reelfoot. J. Tennessee Acad. Sci. 14:230-238.

GILLETTE, C. P. 1905. The western cricket. Colorado Agr. Coll. Expt. Sta. Bull. 101:1-10.

GLEASON, H. A. 1909. Ecological relation of the invertebrate fauna of Isle Royale, Michigan. In Adams 1909, 57-80.

———. 1912. An isolated prairie grove and its phytogeographical significance. Botan. Gaz. 53:38-49.

GLOYD, H. K. 1937. Herpetological considerations of faunal areas in southern Arizona. Bull. Chicago Acad. Sci. 5:79-136.

GLOYD, H. K., and H. M. SMITH. 1942. Amphibians and reptiles from the Carmen Mountains, Coahuila. Bull. Chicago Acad. Sci. 6:231-235.

GODDEN, F. W., and L. T. GUTZMAN. 1938. Bighorn sheep and mountain goats. Univ. Idaho Sch. For. Bull. 8:43-49.

GODFREY, W. E. 1949. Birds of Lake Mistassini and Lake Albanel, Quebec. Nat. Mus. Canada Biol. Ser. 38, Bull. 114:1-43.

GOFF, C. C. 1931. Flood-plain animal communities. M.S. Thesis, Univ. Illinois.

———. 1952. Flood-plain animal communities. Amer. Midl. Nat. 47:478-486.

GOLDMAN, E. A. 1920. Mammals of Panama. Smithsonian Misc. Publ. 68:1-267.

———. 1951. Biological investigations in Mexico. *Ibid*. 115:1-476.

GOLDMAN, E. A., and R. T. MOORE. 1946. Biotic provinces of Mexico. J. Mammal. 26:347-360.

GOODNIGHT, C. J., and MARIE L. GOODNIGHT. 1953. The opilionid fauna of Chiapas, Mexico and adjacent areas. Amer. Mus. Novit. 1610:1-81.

———. 1956. Some observations on a tropical rain forest in Chiapas, Mexico. Ecology 37:139-150.

GOODRICH, C., and H. VAN DER SCHALIE. 1937. Mollusca of Peten and North Alta Vera Paz, Guatemala. Univ. Michigan Misc. Publ. Mus. Zool. 34:7-50.

GOODRUM, P. D. 1940. A population study of the gray squirrel in eastern Texas. Texas Agr. Expt. Sta. Bull. 591:1-34.

GOODRUM, P. D., W. P. BALDWIN, and J. W. ALDRICH. 1949. Effect of DDT on animal life of Bull's Island, South Carolina. J. Wildl. Mgmt. 13:1-11.

GOODWIN, G. G. 1934. Mammals collected by A. W. Anthony in Guatemala, 1924-1928. Bull. Amer. Mus. Nat. Hist. 68:1-60.

GORDON, R. B. 1940. The primeval forest types of southwestern New York. New York St. Mus. Bull. 321:1-102.

GRAHAM, S. A. 1925. Two dangerous defoliators of the jack pine. J. Econ. Ent. 18:337-343.

———. 1937. The walking stick, a forest defoliator. Univ. Michigan, Sch. For. and Conserv. Circ. 3:1-28.

GRANGE, W. B. 1949. The way to game abundance. Charles Scribner's Sons, New York. 365 pp.

Great Plains Committee. 1936. The future of the Great Plains. U.S. Govt. Printing Office. 194 pp.

GREEN, CHARLOTTE HILTON. 1939. Trees of the south. University of North Carolina Press, Chapel Hill. xiv + 551 pp.

GREEN, E E. 1928. Animal succession in a forest series. M.S. Thesis, George Peabody Coll. for Teachers, Nashville, Tennessee.

GREEN, H. L. L. 1946. The elk of the Banff National Park. Natl. Parks Bur. Canada, 1-32 [mimeographed].

GREENE, R. A., and G. H. MURPHY. 1932. The influence of two burrowing rodents, *Diplomya spectabilis spectabilis* (kangaroo rat) and *Neotoma albigula albigula* (pack rat) on desert soils in Arizona. Ecology 13:359-363.

GREENE, R. A. and C. REYNARD. 1932. The influence of two burrowing rodents, *Dipodomys spectabilis* (kangaroo rat) and *Neotoma albigula albigula* (pack rat) on desert soils in Arizona. Ecology 13:73-80.

GRINNELL, J. 1909. Birds and mammals of the 1907 Alexander Expedition to southeastern Alaska. The birds. Univ. California Publ. Zool. 5:181-224.

———. 1914. An account of the mammals and birds of the Lower Colorado Valley with especial reference to the distributional problems presented. Univ. California Publ. Zool. 12:51-294.

———. 1915. Vertebrate fauna of the Pacific Coast. See Nature and science on the Pacific Coast, P. Elder and Co., San Francisco, California, 104-114.

———. 1923a. Observations upon the bird life of Death Valley. Proc. California Acad. Sci. 13:43-109.

———. 1923b. Burrowing rodents of California as agents in soil formation. J. Mammal. 4:137-149.

———. 1933. Review of the recent mammal fauna of California. Univ. California Publ. Zool. 40:71-234.

———. 1937. Mammals of Death Valley. Proc. California Acad. Sci. 23(9):115-169.

———. 1938. California's grizzly bears. Sierra Cl. Bull. 1938:71-81.

GRINNELL, J., and J. DIXON. 1918. Natural history of the ground squirrels of California. Missouri St. Comm. Hort. Bull. 7:597-708.

GRINNELL, J., J. DIXON, and J. M. LINSDALE. 1930. Vertebrate natural history of a section of northern California through the Lassen Peak region. Univ. California Publ. Zool. 35:594.

GRINNELL, J., and T. I. STORER. 1924. Animal life in the Yosemite. University of California Press, Berkeley. xviii + 752 pp.

GRINNELL, J., and H. S. SWARTH. 1913. An account of the birds and mammals of the San Jacinto area of southern California with remarks upon the behavior of geographic races on the margins of their habitats. Univ. California Publ. Zool. 10:197-406.

GRISCOM, L. 1926. In Shelford 1926, 604.

———. 1932. The distribution of bird-life in Guatemala. A contribution to a study of the origin of Central American bird-life. Bull. Amer. Mus. Nat. Hist. 64:3-425.

———. 1935. The ornithology of the Republic of Panama. Bull. Mus. Comp. Zool. 78:261-383.

GRISWOLD, S. B. 1942. Woody plants along the streams which cross Silis County, Kansas. Trans. Kansas Acad. Sci. 45:98-99, 102-103.

GROBMAN, A. B. 1944. The distribution of salamanders of the genus Plethodon in eastern United States and Canada. Ann. New York Acad. Sci. 45:261.

GROSS, A. O. 1931. Snowy owl migration 1930-31. Auk 48:501-511.

———. 1944. Food of the snowy owl. *Ibid.* 61:1-19.

HAECKEL, E. H. 1869. Entwickelungsgang und Aufgaben der zoologie. Jena. Leitsch. 5:353.

HALL, ADA, 1925. Effect of oxygen and CO_2 on the development of the white fish. Ecology 6:104-116.

HALL, E. R. 1951a. American weasels. Univ. Kansas Publ. Mus. Nat. Hist. 4:1-466.

———. 1951b. Synopsis of North American Lagomorpha. *Ibid.* 5:119-202.

HALL, E. R., and V. R. BERNARDO. 1949. An annotated check list of the mammals of Michoacan, Mexico. Univ. Kansas Publ. Mus. Nat. Hist. 1:431-472.

HALLIDAY, W. E. D. 1937. A forest classification for Canada. Canada Dept. Mines Resour., For. Serv. Bull. 89:1-189.

HAMILTON, A. G. 1936. Relative humidity and temperature in the development of three species of African locusts. Trans. Roy. Ent. Soc. 85:1-60.

HAMILTON, W. J., JR. 1941. Notes on some mammals of Lee County, Fla. Amer. Midl. Nat. 25:686-691.

————. 1943. The mammals of eastern United States. Comstock Publishing Company, Inc., Ithaca, New York. 432 pp.

HANSEN, H. P. 1949. Post glacial succession, climate and chronology in the Pacific Northwest. Trans. Amer. Philos. Soc. 37:1-188.

HANSON, H. C. 1924. A study of the vegetation of northeastern Arizona. Univ. Nebraska Stud. 24:1-94.

————. 1950. Vegetation and soil profiles in some solifluction and mound areas in Alaska. Ecology 31:606-630.

————. 1951. Characteristics of some grassland, marsh, and other plant communities in western Alaska. Ecol. Monogr. 21:317-378.

————. 1953. Vegetation types in northwestern Alaska and comparison with communities in other Arctic regions. Ecology 34:111-140.

HANSON, H. C., and C. T. VORHIES. 1938. Need for research on grasslands. Sci. Mon. 46:230-241.

HARDY, R. 1945. Breeding birds of the pigmy conifers in the Book Cliff region of eastern Utah. Auk 62:523-542.

HARE, F. K. 1950. Climatic and zonal divisions of the boreal forest formation of eastern Canada. Geogr. Rev. 40:615-635.

HARKNESS, W. J. K., and E. L. PIERCE. 1941. The limnology of Lake Mize, Florida. Proc. Florida Acad. Sci., 95-116.

HARPER, MARGUERITE. 1938. The ecological distribution of earthworms as found in the developmental stages of the floodplain. M.S. Thesis, Univ. Illinois, 1938.

HARPER, F. 1955. The barren ground caribou of Keewatin. Univ. Kansas Mus. Nat. Hist. Misc. Publ. 6:1-164.

HARPER, R. M. 1913. Economic botany of Alabama. Part 1. Geol. Surv. Alabama Monogr. 8:1-228.

————. 1914. Geography and vegetation of northern Florida. Florida St. Geol. Surv. Ann. Rept. 6:163-437.

————. 1921. Geography of central Florida. *Ibid.* 13:75-307.

————. 1927. Natural resources of southern Florida. *Ibid.* 18:27-206.

————. 1939. The Alabama pocosin. Amer. Bot. 45:53-58.

————. 1930. The natural resources of Georgia. Bull. Univ. Georgia 30. 105 pp.

HARSHBERGER, J. W. 1916. The vegetation of the New Jersey pine-barrens. Christopher Sower Co., Philadelphia. 329 pp.

HARTZ, N., and C. KRUUSE. 1911. The vegetation of northeast Greenland. Medd. Grønland 30:333-431.

HARVEY, L. H. 1903. A study of the physiographic ecology of Mt. Katahdin, Maine. Univ. Maine Stud. 5:1-50.

————. 1908. Floral succession in the prairie grass formation of southeastern South Dakota. Botan. Gaz. 46:81-108, 277-298.

HATT, R. T. 1930. The relation of mammals to the Harvard Forest. Roosevelt Wild Life Bull. 5:625-671.

HAYWARD, C. L. 1942. Biotic communities of Mt. Timpanogos and the western Uinta Mountains. Ph.D. Thesis, Univ. Illinois.

————. 1945. Biotic communities of the southern Wasatch and Uinta Mountains. Gr. Basin Nat. 4:1-124.

————. 1948. Biotic communtiies of the Wasatch chaparral, Utah. Ecol. Monogr. 18:473-506.

————. 1952. Alpine biotic communities of the Uinta Mountains of Utah. *Ibid.* 22:93-120.

HEARNE, S. 1911. A journey from Prince of Wales Fort in Hudson's Bay to the northern ocean. (New edition with notes and illustrations by J. B. Tyrrell.) Champlain Society, Toronto, Canada. xv + 437 pp.

HEFLEY, H. M. 1937. Ecological studies of the Canadian River floodplain in Cleveland Co. Okla. Ecol. Monogr. 7:345-402.

HEFLEY, H. M., and R. SIDWELL. 1945. Geological and ecological observations of some high plains dunes. Amer. J. Sci. 243:361-376.

HEILPRIN, A. 1892. The temperate and alpine floras of the giant volcanoes of Mexico. (A report from the Committee on the Michaux Legacy.) Proc. Amer. Philos. Soc. 30:4-22.

HEIMBURGER, H. V. 1924. Reactions of earthworms to temperature and atmospheric humidity. Ecology 5:276-282.

HELLER, E. 1909. Birds and mammals of the 1907 Alexander expedition to southeastern Alaska. The mammals. Univ. California Publ. Zool. 5:245-267.

————. 1910. Mammals of the 1908 Alexander Alaska Expedition. Univ. California Publ. Zool. 5:321-360.

HENDRICKSON, G. O. 1930. Studies on the insect fauna of Iowa prairies. Iowa St. Coll. J. Sci. 4:49-179.

————. 1943. Mearns cottontail investigations in Iowa. Ames For. 21:59-74.

HENSLEY, M. M. 1954. Ecological relations of the breeding bird population of the desert biome of Arizona. Ecol. Monogr. 24:185-207.

HERSHKOVITZ, P. 1954. Mammals of northern Columbia. Preliminary report 7: Tapirs (genus *Tapirus*). Proc. U.S. Natl. Mus. 103:465-496.

HEYWARD, F., and R. M. BARNETTE. 1934. Effect of frequent fires on chemical composition of forest soils in the longleaf pine region. Agr. Expt. Sta. Florida Bull. 265:1-39.

HITCHCOCK, A. S. 1935. Manual of grasses of the United States. U.S. Dept. Agr. Misc. Publ. 200:1-1040.

HODGKISS, H. E. 1930. Maple bladder gall mite *Varates quadripedes*. New York Agr. Expt. Sta. Tech. Bull. 163:5-45.

HOFMANN, J. V. 1920. The establishment of a Douglas fir forest. Ecology 1:49-53.

HOLT, W. P. 1909. Notes on vegetation of Isle Royale, Michigan. In Adams 1909, 217-248.

HOLTTUM, R. E. 1922. The vegetation of west Greenland. J. Ecol. 10:87-108.

HOOPER, T. 1955. Notes on mammals of western Mexico. Occ. Pap. Mus. Zool., Univ. Michigan 565:1-26.

————. 1957. Records of Mexican mammals. *Ibid.* 586:1-9.

HOPKINS, A. D. 1899. Report on investigations to determine the cause of unhealthy conditions of spruce and pine from 1880-1893. West Virginia Agr. Expt. Sta. Bull. 56:197-461.

————. 1909. I. The genus *Dendroctonus*. U.S. Dept. Agr. Ent. Tech. Ser. 17(1):1-64.

HOPKINS, D. M., and R. S. SIGAFOOS. 1951. Frost action and vegetation patterns on Seward Peninsula, Alaska. Geol. Surv. Bull. 974-C:51-100.

HOPPING, G. R. 1921. The control of bark beetle outbreaks in British Columbia. Dom. Dept. Agr. Ent. Branch Circ. 15:1-15.

————. 1928. The western cedar borer. Canad. Dept. Agr. Pamph. 94, n.s.:1-17.

HORN, E. E., and H. S. FITCH. 1942. Interrelations of rodents and other wildlife of the range. Univ. California, Bull. 663:96-129.

HORTON, J. S. 1941. The sample plot as a method of quantitative analysis of chaparral vegetation in southern California. Ecology 22:457-468.

HOTCHKISS, N., and R. E. STEWART. 1947. Vegetation of the Patuxent Research Refuge, Maryland. Amer. Midl. Nat. 38:1-75.

HOWELL, A. B., and I. GERSH. 1935. Conservation of water by the rodent *Dipodomys*. J. Mammal. 16:1-9.

HOWELL, A. H. 1918. A revision of the flying squirrels. N. Amer. Fauna 44:1-64.

———. 1921. A biological survey of Alabama. I. Physiography and life zones. II. The mammals. *Ibid.* 45:1-88.

———. 1932. Florida bird life. Florida Dept. Game and Fresh Water Fish, Coward-McCann, Inc., New York. 579 pp.

HUBBARD, D. M. 1941. The vertebrate animals of Friant Reservoir Basin with special reference to the possible effects upon them of the Friant Dam. California Fish and Game 27:198-215.

HUGHES, B. O., and D. DUNNING. 1949. The pine forest of California. Ybk. U.S. Dept. Agr. 1949:352-358.

HUMPHREY, H. H. 1935. A study of *Idria columnaris* and *Fouquieria splendeus*. Ecology 22:184-206.

HUNTINGTON, E. 1919. World power and evolution. Yale University Press, New Haven, Connecticut. 287 pp.

INGLES, L. G. 1929. The seasonal and associational distribution of the fauna of the upper Santa Ana River. Washington J. Ent. Zool. 21:1-48, 57-96.

———. 1947. Ecology and life history of the California gray squirrel. California Fish and Game 33:139-158.

International Wildlife Protection, American Committee. 1934. The present status of the muskox in Arctic North America and Greenland. Spec. Publ. 5:1-87.

Interstate Deer Herd Commission. 1951. The Devil's Garden deer herd. California Fish and Game 37:233-272.

ISELY, F. B. 1935. Acridian researches within northeastern Texas (Orthoptera). Ent. News 46:37-43, 69-75.

———. 1937. Seasonal succession, soil relations, numbers, and regional distribution of northeastern Texas acridians. Ecol. Monogr. 7:317-344.

———. 1938. Relations of Texas Acridians to plants and soil. *Ibid.* 8:551-604.

JACKSON, H. H. T. 1914. The land vertebrates of Ridgeway Bog, Wisconsin: their ecological succession and source of ingression. Bull. Wisconsin Nat. Hist. Soc. 12:4-54.

JACOT, A. P. 1935. Wildlife of the forest carpet. Sci. Mon. 40:425-430.

———. 1936. Soil structure and soil biology. Ecology 17:359-379.

JANES, M. J. 1937. Studies of certain phases of the biology of the chinch bug *Blissus leucopterus* (Say) under conditions of constant temperature and relative humidity. Iowa St. Coll. J. Sci. 12:132-133.

JANES, M. J., and ANNA HAGER. 1936. Studies of the incubation of the chinch bug egg. Iowa St. Coll. J. Sci. 10:395-402.

JENSEN, H. A. 1947. A system for classifying vegetation in California. California Fish and Game 33:199-266.

JENSEN, S. 1909. Mammals observed on Adrup's journeys to east Greenland, 1898-1900. Medd. Grønland 29:1-62.

JEPSON, W. L. 1925. A manual of the flowering plants of California. University of California Press, Berkeley. 1238 pp.

JOHANSEN, F. 1921. Insect life on the western arctic coast of America. Canad. Arctic Exped. 1913-18, 3:1K-61K.

———. 1924. General observations on the vegetation. *Ibid.* 1913-18, 5:1C-85C.

JOHNSON, D. H., M. D. BRYANT, and A. H. MILLER. 1948. Vertebrate animals of the Providence Mountains area of California. Univ. California Publ. Zool. 48:221-376.

JOHNSON, M. 1926. Activity and distribution of certain wild mice. J. Mammal. 7:245-277.

JOHNSON, R. A. 1954. The behavior of birds attending army ant raids on Barro Colorado Island, Panama Canal Zone. Proc. Linn. Soc. New York 63-65:4170.

JOHNSON, W. G. 1894. In Forbes 1894.

JONES, G. N. 1947. A botanical survey of the Olympic Peninsula, Washington. University of Washington Press, Seattle. 286 pp.

JONES, G. N., and G. D. FULLER. 1955. Vascular plants of Illinois. University of Illinois Press, Urbana. 593 pp.

JONES, G. T. 1938. Dormancy and leafing in white elm (*Ulmus americana*). University of Chicago Press, Chicago. 77 pp.

JONES, SARAH E. 1941. Influence of temperature and humidity on the life history of the spider *Agelena naevia walckenaer*. Ann. Ent. Soc. Amer. 34:557-571.

JORGENSEN, M. 1934. A quantitative investigation of the microfauna communities of the soil in East Greenland. Medd. Grønland 100:1-39.

KANNOWSKI, P. B. 1954. Notes on the ant *Novomessor manni* Wheeler and Creighton. Occ. Pap. Mus. Zool., Univ. Michigan 556:1-6.

KASTON, B. J. 1939. The native elm bark beetle *Hylurgopinus rufipes* (Eichhoff) in Connecticut. Connecticut Agr. Expt. Sta. Bull. 420:1-38.

KEARNEY, T. H. 1901a. Report on a botanical survey of the Dismal Swamp region. Contr. U.S. Natl. Herb. 5:321-500.

———. 1901b. The plant covering of Ocracoke Island; a study in the ecology of the North Carolina strand vegetation. *Ibid*. 5:261-321.

KEEN, F. P. 1938. Insect enemies of western forests. U.S. Dept. Agr. Misc. Publ. 273:1-209.

KEEVER, CATHERINE. 1950. Causes of succession on old fields of the Piedmont, North Carolina. Ecol. Monogr. 20:229-270.

———. 1953. Present composition of some stands of the former oak chestnut forest in the Blue Ridge Mountains. Ecology 34:44-54.

KELLOGG, R. 1937. West Virginia mammals. Proc. U.S. Natl. Mus. 84:443-479.

———. 1939. Annotated list of Tennessee mammals. *Ibid*. 86:245-303.

KENDEIGH, S. C. 1944. Measurement of bird populations. Ecol. Monogr. 14:67-106.

———. 1946. Breeding birds of the beech-maple-hemlock community. Ecology 27:226-244.

———. 1947. Bird population studies in the coniferous forest biome during a spruce budworm outbreak. Dept. Lands and For., Ontario, Canada, Div. Res. Biol. Bull. 1:1-174.

———. 1948. Twelfth breeding bird census. Audubon Field Notes 2:232-233.

———. 1954. History and evaluation of various concepts of plant and animal communities of North America. Ecology 35:152-171.

KENNEDY, C. H. 1917. Notes on the life histories and ecology of dragon flies (of central California and Nevada). Proc. U.S. Natl. Mus. 52:483-635.

KENOYER, L. A. 1929. General and successional ecology of the lower tropical rain forest at Barro Colorado Island, Panama. Ecology 10:201-222.

———. 1941. Plant life of the Pan American Highway. Power's guide to Mexico 1941-42. Pan American Tourist Bureau, Laredo, Texas, p. 69.

Kentucky Geological Survey. 1884. Timber and botany. B:1-177. (Reprinted). Reports by groups of counties; quantitative observations, 1875-1879; chief author, L. H. DeFriese, others, A. R. Crandall and I. Hussey.

KIMBALL, H. H. 1924. Records of total solar radiation intensity and their relation to daylight intensity. Mon. Weather Rev. 52:473-479.

KINCAID, T., *et al.* 1910. Harriman Alaska Expedition. Vol. 8 (1) (1904, Doubleday, Page, & Company, New York). Smithsonian Institution, Washington, D.C. ix + 238 pp.

KING, W. 1939. A survey of the herpetology of Great Smoky Mountain National Park. Amer. Midl. Nat. 21:531-582.

KINSEY, A. C. 1929. The gall wasp genus *Cynips*. Indiana Univ. Stud. 16:1-577.

KLYVER, F. D. 1931. Major plant communities in a transect of the Sierra Nevada Mountains of California. Ecology 12:1-17.

KOELZ, S. E. 1936. An ecological survey of the Sangamon floodplain, and adjacent areas, with special reference to vertebrates. M.S. Thesis, Univ. Illinois.

KOESTNER, E. J. 1938. Mammal population and territories in an elm-maple forest. M.S. Thesis, Univ. Illinois.

———. 1941. An annotated list of mammals collected in Nuevo Leon, Mexico, in 1938. Gr. Basin Nat. 2:9-15.

KOMAREK, E. V. 1937. Mammal relationships to upland game and other wildlife. Trans. N. Amer. Wildl. Conf. 2:561-569.

———. 1939. A progress report on southeastern mammal studies. J. Mammal. 20:292-299.

KOMAREK, E. V., and R. KOMAREK. 1938. Mammals of the Great Smoky Mountains. Bull. Chicago Acad. Sci. 5:137-161.

KORSTIAN, C. F. 1927. Factors controlling germination and early survival in oaks. Yale Univ. Sch. For. Bull. 19. 115 pp.

KORSTIAN, C. F., and T. S. COILE. 1938. Plant competition in forest stands. Duke Univ. Sch. For. Bull. 3:1-125.

KREFTING, L. W., and E. I. ROE. 1949. The role of some birds and mammals in seed germination. Ecol. Monogr. 19:269-286.

KROEBER, A. L. 1939. Cultural and natural areas of native North America. University of California Press, Berkeley. 242 pp.

KRUCKEBERG, A. R. 1954. Ecology of serpentine soils. Plant species in relation to serpentine soils. Ecology 35:267-274.

KUCHLER, A. W. 1946. The broadleaf deciduous forests of the Pacific northwest. Ass. Amer. Geogr. 36:122-147.

KUNS, M. L., and R. E. TASHIAN. 1954. Notes on the mammals of northern Chiapas, Mexico. J. Mammal. 35:100-103.

KURZ, H. 1942. Florida dunes, scrub and geology. Florida St. Bd. Conserv. Geol. 23:1-54.

LAESSLE, A. M. 1942. The plant communities of the Welaka area with special reference to correlations between soils and vegetational succession. Univ. Florida Publ., Biol. Ser. 4:1-143.

LARSEN, E. B. 1943a. Importance of master factors for the activity of Noctuids. Saertr. Ent. Medd. 23:352-374.

———. 1943b. Influence of humidity on the life and development of insects. Vidensk. Medd. Dansk Naturh. Foren. Kbh. 107:128-184.

LARSEN, E. C. 1947. Photoperiodic responses of geographical strains of *Andropogon scoparius*. Botan. Gaz. 109:132-149.

LARSEN, J. A. 1930. Forest types of the northern Rocky Mountains and their climatic control. Ecology 11:631-672.

LARSON, F. 1940. The role of the bison in maintaining the short grass plains. Ecology 21:113-121.

LATHAM, R. M. 1941. The history of the wild turkey in Pennsylvania. Pennsylvania Games News 12:6-7, 32.

LAWRENCE, W. E. 1926. In Shelford 1926, 181.

LEAVENWORTH, W. C. 1946. A preliminary study of the vegetation of the region between Cerro Tancitaro and the Rio Tepalcatepec Michoacan, Mexico. Amer. Midl. Nat. 36:137-206.

LEDERER, J. 1672. The discoveries of John Lederer in three marches from Virginia to the west of Carolina and other parts of the continent; begun March 1669 and ended in September, 1672. London. 30 pp.

LEHMAN, V. W. 1941. Atwater's prairie chicken: its life and management. N. Amer. Fauna 57:1-63.

LENG, C. W. *et al.* 1920-39. Catalogue of Coleoptera of America north of Mexico, including fossils by H. F. Wickham. 470 pp. Supp. 1, 2, 3, with A. J. Mutchler; 1, 1927, 78 pp.; 2-3, 1933, 112 pp.; Supp. 4 with R. E. Blackwelder, 1939, 146 pp. J. D. Sherman, Mount Vernon, New York.

LEON, BROTHER. 1926. Cuba. In Shelford 1926, 682-694.

LEOPOLD, A. S. 1950a. Deer in relation to plant succession. Trans. N. Amer. Wildl. Conf. 15:571-580.

————. 1950b. Vegetation zones of Mexico. Ecology 31:507-518.

LEOPOLD, A. S., and L. HERMANDES. 1944. Los recursos biologicos de Guerrero con referencia especial a los mamiferos y aves de casa. Com. Imp. y Coor. Inv. Cien. 1944:361-390.

LEOPOLD, A. S., T. RINEY, R. McCAIN, and L. TEVIS. 1951. The jaw bone deer herd. California Dept. Nat. Res. Game Bull. 4:1-139.

LeSUEUR, H. 1945. Ecology of the vegetation of Chihuahua, Mexico, north of parallel twenty-eight. Univ. Texas Publ. 4521:1-92.

LEWIS, F. J., ELEANOR S. DOWDING, and E. H. MOSS. 1928. The vegetation of Alberta. II. The swamp, moor and bog forest vegetation of central Alberta. J. Ecol. 16:19-70.

LEWIS, I. F. 1917. The vegetation of Shackleford Bank. North Carolina Geol. Econ. Surv. Econ. Pap. 46:1-32.

LEWIS, M., and W. CLARK. 1904-05. Original journals of the Lewis and Clark Expedition, 1804-1806. Ed. by R. G. Thwaites. Dodd, Mead & Company, New York. 8 vol. illus.

LIEBERG, J. B. 1900. Bitterroot Forest Preverse. Part 5. U.S. Dept. Int., Geol. Surv. 20th Rept. For Res., 317-410.

Life (L. BARNETT). 1954. The world we live in. Part X. The Arctic barrens. Editorial staff of *Life* and L. Barnett. Simon and Schuster, Inc., New York, 197-218.

LINDEBORG, R. G. 1941. Fluctuations in the abundance of small mammals in east-central Illinois, 1936-1939. Ecology 22:96-99.

LINDEMAN, R. L. 1941. The developmental history of Cedar Creek Bog, Minnesota. Amer. Midl. Nat. 25:101-112.

LINSDALE, JEAN M. 1938. Environmental responses of vertebrates in the Great Basin. Amer. Midl. Nat. 19:1-206.

————. 1946. The California ground squirrel. University of California Press, Berkeley. 475 pp.

LITTLE, E. L. 1953. Checklist of native and naturalized trees of the United States. Agr. Handb. 41:1-472.

LIVINGSTON, B. E., and F. SHREVE. 1921. The distribution of vegetation in the United States as related to climatic conditions. Carnegie Inst. Washington Publ. 284:1-284.

LONGHURST, W. M. 1940. The mammals of Napa County, California. California Fish and Game 26:240-270.

LONGHURST, W. M., A. S. LEOPOLD, and R. F. DASMAN. 1952. A survey of California deer herds, their ranges and management problems. California Dept. Fish and Game, Game Bull. 6:1-136.

Longstaff, T. G. 1932. An ecological reconnaissance in west Greenland. J. Animal Ecol. 1:119-142.

Longyear, B. O. 1927. Trees and shrubs of the Rocky Mountain region. G. P. Putnam's Sons, New York. 244 pp.

Loomis, H. F. 1944. Millipeds principally collected by Prof. V. E. Shelford in the eastern and southeastern states. Psyche 5:166-177.

Lowery, G. H., Jr., *et al.* 1952. Fifty-second Christmas bird count. Audubon Field Notes 6:138.

——. 1955. Louisiana birds. Louisiana State University Press, Baton Rouge. 356 pp.

Lundell, C. L. 1934. A preliminary sketch of the phytogeography of the Yucatan Peninsula. Carnegie Inst. Washington Publ. 436:258-355.

——. 1937. The vegetation of Peten. *Ibid.* 478:1-244.

——. 1940. Botany of the Maya area. Misc. Pap. XIV-XXI Carnegie Inst. Washington. 58 pp.

——. 1945. Vegetation and natural resources of British Honduras. In Verdoorn 1945, 270-273.

Lupo, P. H. 1923. A tundra trip in Alaska. Trans. Illinois St. Acad. Sci. 16:54-63.

Luttig, J. C. 1920. Journal of a fur trading expedition on the upper Missouri, 1812-1813. Ed. by Stella M. Drumm, Missouri Historical Society, St. Louis. 192 pp.

Lutz, H. J. 1930. Original forest composition in northwestern Pennsylvania as indicated by early land survey notes. J. For. 28:1098-1103.

Lyman, C. P. 1943. Control of the coat color in the varying hare. Bull. Mus. Comp. Zool. 93:393-461.

Lyon, M. W. 1936. Mammals of Indiana. Amer. Midl. Nat. 17:1-373.

McBryde, F. W. 1945. Cultural and historical geography of southwest Guatemala. Smithsonian Inst., Inst. Social Anthrop. Publ. 4:1-184.

McBryde, J. B. 1933. The vegetation and habitat factors of the Carrizo Sands. Ecol. Monogr. 3:249-297.

McClure, H. E. 1943a. Aspection in the biotic communities of the Churchill area. Ecol. Monogr. 13:1-35.

——. 1943b. Ecology and management of the mourning dove, *Zenaidura macroura* (Linn.), in Cass County. Iowa Res. Bull. 310:357-415.

McDougall, W. B. 1922. Symbiosis in a deciduous forest. Botan. Gaz. 73:200-212.

McDougall, W. B., and Charlotte Liebtag. 1928. Symbiosis in a deciduous forest. III. Micorhizal relations. Botan. Gaz. 86:226-234.

McDougall, W. B., and W. T. Penfound. 1928. Ecological anatomy of some deciduous forest plants. Ecology 9:349-353.

McDunnough, T. 1938-39. Checklist of the Lepidoptera of Canada and the United States; microlepidoptera. Mem. S. California Acad. Sci. 1:1-272; macrolepidoptera. *Ibid.* 2:1-171.

McLean, D. D. 1940. The deer of California, with particular reference to the Rocky Mountain mule deer. California Fish and Game 26:139-166.

——. 1944. The prong-horned antelope in California. *Ibid.* 30:221, 241.

MacLulich, D. A. 1937. Fluctuations in the numbers of the varying hare (*Lepus americanus*). Univ. Toronto Stud., Biol. Ser. 43:1-136.

McMinn, H. E. 1939. An illustrated manual of California shrubs. Stacey, San Francisco, California. 689 pp.

Macnab, J. A. 1945. Faunal aspection in the coast range mountains of northwestern Oregon. Ph.D. Thesis, Univ. Nebraska.

——. 1958. Biotic aspection in the coast range mountains of northwestern Oregon. Ecol. Monogr. 28:21-54.

Macnab, J. A., and Jane Claire Dirks. 1941. California redbacked mouse in the Oregon coast range. J. Mammal. 22:174-180.

MACNAB, J. A., and D. M. FENDER. 1947. An introduction to Oregon earthworms with additions to the Washington list. Northwest Sci. 21:69-75.

MALIN, J. C. 1947. The grassland of North America. Prolegomena to its history. Edwards Bros., Ann Arbor, Michigan. 398 pp.

MAJOR, J. L. 1931. Animals of the oak-chestnut and beech-maple forest. Thesis, George Peabody Coll. for Teachers, Nashville, Tennessee.

MANNICHE, A. L. V. 1910. The terrestrial mammals and birds of northeast Greenland. Medd. Grønland 45:1-200.

MANNING, T. H. 1943. Notes on the mammals of south, central and west Baffin Island. J. Mammal. 24:475.

———. 1946. Bird and mammal notes from the east side of Hudson Bay. Canad. Fld. Nat. 60:71-85.

MARR, J. W. 1948. Ecology of the forest-tundra ecotone on the east coast of Hudson Bay. Ecol. Monogr. 18:117-144.

MARSHALL, F. H. A., and F. P. BOWDEN. 1934. The effect of irradiation of different wave lengths on the oestrous cycle of the ferret with remarks on the factors controlling sexual periodicity. J. Exptl. Biol. 11:409-422.

MARTIN, A. C., and F. M. UHLER. 1939. Food of game ducks in the U.S. and Canada. U.S. Dept. Agr. Tech. Bull. 634:1-147.

MARTIN, P. S. 1955. Zonal distribution of vertebrates in a Mexican cloud forest. Amer. Nat. 89:347-361.

MARTIN, P. S., C. R. ROBINS, and W B. HEED. 1954. Birds and biogeography of the Sierra de Tamaulipas, an isolated oak-pine habitat. Wilson Bull. 66:38-57.

MASS, F. H. 1938. The deer situation in northern Idaho. Univ. Idaho Sch. For. Bull. 8:30-34.

MEAD, A. R. 1943. Revision of the giant west coast land slugs of the genus *Ariolimax morch* (Pumlonata: Arionidae). Amer. Midl. Nat. 30:675-717.

MEINICKE, E. P. 1936. Changes in California wild life since the white man. Commonwealth 12:253-275.

MELANDER, A. U., and M. A. YOTHERS. 1917. The coulee cricket. St. Coll. Washington Agr. Expt. Sta. Bull. 137:1-34.

MELIN, E. 1930. Biological decomposition of some types of litter from North American forests. Ecology 11:72-101.

MERRIAM, C. H. 1890. Results of a biological survey of the San Francisco Mountain region and the desert of the Little Colorado, Arizona. N. Amer. Fauna 3:1-113.

———. 1899. Results of a biological survey of Mount Shasta, California. *Ibid.* 16:9-179.

———. 1901. Prairie dogs of the Great Plains. Ybk. U.S. Dept. Agr. 1901:257-270.

MERRIAM, C. H., and L. STEJNEGER. 1891. Results of a biological reconnaissance of south-central Idaho. N. Amer. Fauna 5:1-113.

MERRIAM, J. C. 1899. Report on the expedition to the John Day fossil fields. Univ. California Chron. 2:217-225.

METCALF, Z. P. 1923. A key to the Fulgoridae of eastern North America with descriptions of new species. J. Elisha Mitchell Sci. Soc. 38:139-230.

METCALF, Z. P., and H. OSBORN. 1920. Some observations on insects of the between-tide zones of the North Carolina coast. Ann. Ent. Soc. Amer. 13:108-121.

MEYER, A. M. 1937. An ecological study of cedar glade invertebrates near Nashville, Tennessee. Ecol. Monogr. 7:403-443.

MILLER, A. H. 1951. An analysis of the distribution of the birds of California. Univ. California Publ. 50:531-644.

MILLER, F. H. 1941. Forest protection requirements report, Great Smoky Mountains National Park (abstract of the 1939 report in 1941). Files of the Natl. Park Serv. [typewritten].

MILLER, G. S., and R. KELLOGG. 1955. List of North American mammals. U.S. Natl. Mus. Bull. 205:1-954.

MIRANDA, F. 1942. Notas generales sobre la vegetacion des s. o. del Estado de Puebla especialmente de la zona de Itzocan de Matamoras. Ann. Inst. Biol. 13:417-438.

———. 1947. Estudios sobre la vegetacion de Mexico. V. Rasgos de la vegetacion en la cuenca del Rio de la Balsas. Rev. Soc. Mexico Hist. Nat. 8:95-114.

———. 1952. La vegetacion de Chiapas. Primer Parte. Dept. Prensa y Turismo Tuxla Gutierrez, Chiapas, Mexico. 334 pp.

Mississippi River Commission. 1938, 1947. Early stream channels (maps).

MÖBIUS, K. 1877. Die Auster und die Austernwirtschaft. Berlin (English transl.: The oyster and oyster culture. Rept. U.S. Fish Comm. 1880:683-751).

MOHR, C. O. 1940. Comparative populations of game, fur and other mammals. Amer. Midl. Nat. 24:581-584.

———. 1943. Cattle droppings as ecological units. Ecol. Monogr. 13:275-298.

———. 1947. Table of equivalent populations of North American small mammals. Amer. Midl. Nat. 37:223-249.

MOORE, A. W. 1940. Wild animal damage to seed and seedlings on cutover of Douglas fir lands of Oregon and Washington. U.S. Dept. Agr. Tech. Bull. 706:1-28.

MOORE, J. C. 1946. Mammals from Welaka, Putnam County, Fla. J. Mammal. 27:49-59.

MOSAUER, W. 1936a. The toleration of solar heat in desert reptiles. Ecology 17:56-66.

———. 1936b. The reptilian fauna of sand dune areas of the Vizcaino Desert and of northwestern lower California. Occ. Pap. Mus. Zool., Univ. Michigan 329:1-21.

MOSBY, H. S., and C. O. HANDLEY. 1943. The wild turkey in Virginia. Its status, life history, and management. Div. Game and Inland Fish., Richmond, Virginia. xx + 281 pp.

MOSS, E. H. 1932. The vegetation of Alberta. IV. The poplar association and related vegetation of central Alberta. J. Ecol. 20:380-415.

MUESEBECK, C. F. W., K. V. KROMBEIN, and H. K. TOWNES. 1951. Hymenoptera of America north of Mexico; synoptic catalogue. U.S. Dept. Agr. Monogr. 2:1-1420.

MULLER, C. H. 1939. Relation of the vegetation and climatic types in Nuevo León, Mexico. Amer. Midl. Nat. 21:687-729.

———. 1940. Plant succession in the *Larrea-Flourensia* climax. Ecology 21:206-212.

———. 1947. Vegetation and climate of Coahuila, Mexico. Madroño 9:33-57.

MUMA, M. H., and K. E. MUMA. 1949. Studies on a population of prairie spiders. Ecology 30:485-503.

MUNGER, T. T. 1940. The cycle from Douglas fir to hemlock. Ecology 21:451-459.

MUNGER, T. T., W. E. LAWRENCE, and H. M. WIGHT. Oregon. In Shelford 1926, 181-193.

MURIE, A. 1935. Mammals from Guatemala and British Honduras. Univ. Michigan Misc. Publ. Mus. Zool. 26:1-30.

———. 1941. Ecology of the coyote in Yellowstone Park. Fauna Natl. Parks U.S. Bull. 4:1-206.

———. 1944. The wolves of Mt. McKinley. Fauna Natl. Parks U.S., Fauna Ser. 5:1-238.

MURIE, O. J. 1935. Alaska-Yukon caribou. N. Amer. Fauna 54:1-93.

MURRAY, J. J. 1933. Summer birds of Mountain Lake, Va. Raven 4:1-2.

MYERS, R. M., and P. G. WRIGHT. 1948. Initial report on the vegetation of McDonough Co. Trans. Illinois St. Acad. Sci. 41:43-48.

NECKER, W L. 1934. Contribution to the herpetology of the Smoky Mountains of Tennessee. Bull. Chicago Acad. Sci. 5:1-4.

NEEDHAM, J. G. 1909. Neuropteroid insects (Isle Royale). In Adams 1909, 305.

NELSON, E. W. 1921. Lower California and its natural resources. Natl. Acad. Sci. 16:1-194 [first memoir].

———. 1925. Status of the pronghorned antelope, 1922-1924. U.S. Dept. Agr. Bull. 1346:1-64.

———. 1930. The wild animals of North America. National Geographic Society, Washington, D.C. 525 pp.

NELSON, E. W., and A. E. GOLDMAN. 1926. Mexico. In Shelford 1926, 574-596.

NETTING, M. G. 1936. Notes on a collection of reptiles from Barro Colorado Island, Panama Canal Zone. Ann. Carnegie Mus. 25:113-120.

NEWCOMBE, C. L. 1930. An ecological study of the Allegheny cliff rat (*Neotoma pennsylvanica* Stone). J. Mammal. 11:204-210.

NICE, M. M. 1931. The birds of Oklahoma. Univ. Oklahoma Biol. Surv. Oklahoma 3:1-222.

NICHOL, A. A. 1937. The natural vegetation of Arizona. Univ. Arizona Tech. Bull. 68:181-222.

NICHOLS, G. E. 1913. The vegetation of Connecticut. II. Virgin forests. Torreya 13:199-215.

———. 1918. The vegetation of northern Cape Breton Island, Nova Scotia. Trans. Connecticut Acad. Arts and Sci. 22:249-467.

———. 1935. The hemlock-white pine-northern hardwood region of eastern North America. Ecology 16:403-422.

NIELSEN, J. C. 1907. The insects of east Greenland. Medd. Grønland 29:363-414.

ODUM, E. P. 1943. The vegetation of the Edmund Niles Huyck Preserve, New York. Amer. Midl. Nat. 29:72-88.

———. 1944. Notes on the small mammal population at Mountain Lake, Virginia. J. Mammal. 25:408-410.

———. 1955. Fundamentals of ecology. W. B. Saunders Company, Philadelphia. 384 pp.

OLMSTED, C. E. 1943. Growth and development in range grasses. III. Photoperiodic responses in the genus *Bouteloua*. Botan. Gaz. 105:165-181.

———. 1945. Growth and development in range grasses. V. Photoperiodic responses of clonal divisions of three latitudinal strains of side-oats grama. *Ibid.* 106:382-401.

OLSON, S. F. 1938. Organization and range of the (wolf) pack. Ecology 19:168-170.

OOSTING, H. J. 1942. An ecological analysis of the plant communities of Piedmont, North Carolina. Amer. Midl. Nat. 28:1-126.

———. 1948. Ecological notes on the flora of east Greenland and Jan Mayen. Amer. Geogr. Soc. Spec. Publ. 30:225-269.

OOSTING, H. J., and W. D. BILLINGS. 1942. Factors effecting vegetational zonation on coastal dunes. Ecology 23:131-142.

———. 1943. The red fir forest of the Sierra Nevada. Ecol. Monogr. 13:259-274.

———. 1951. Comparison of virgin spruce-fir forest in the northern and southern Appalachian systems. Ecology 32:84-103.

OPSAHL, J. F. 1951. Small mammal populations in tropical Florida. M.S. Thesis, Univ. Illinois.

OSBORN, B., and P. F. ALLAN. 1949. Vegetation of an abandoned prairie-dog town in tall grass prairie. Ecology 30:322-332.

OSBORN, H., and Z. P. METCALF. 1920. Notes on the life history of the salt marshes cicada (*Tibicen viridifascia* Walk.). Ent. News 21:248-252.

OSGOOD, W. H. 1904. A biological reconnaissance of the base of the Alaska Peninsula. N. Amer. Fauna 24:1-86.

———. 1909. Biological investigations in Alaska and Yukon Territory. *Ibid.* 30:1-96.

———. 1926. In Shelford 1926, 141-146.

PALMER, T. S. 1897. The jackrabbits of the U.S. U.S. Dept. Agr., Div. Biol. Surv. Bull. 6:1-88.

PARK, O. 1938. Studies in nocturnal ecology. VII. Preliminary observations on Panama rain forest animals. Ecology 19:208-223.

PARK, O., J. A. LOCKETT, and D. J. MYERS. 1931. Studies in nocturnal ecology with special reference to climax forest. Ecology 12:709-727.

PARKER, J. R. 1930. Some effects of temperature and moisture upon *Melanoplus mexicanus mexicanus* Saussure and *Camnula pellucida* Scudder (Orthoptera). Univ. Montana, Agr. Expt. Sta. Bull. 233:5-132.

PEARSE, A. S. 1926. Wisconsin. In Shelford 1926, 284-287.

———. 1939. Animal ecology. McGraw-Hill Book Company, Inc., New York. xii + 642 pp.

———. 1946. Observations of the microfauna of the Duke forest. Ecol. Monogr. 16:127-150.

PEARSE, A. S., H. J. HUMM, and G. W. WHARTON. 1942. Ecology of sand beaches of Beaufort, North Carolina. Ecol. Monogr. 12:135-190.

PEARSON, G. A. 1931. Forest types in the southwest as determined by climate and soil. U.S. Dept. Agr. Tech. Bull. 247:1-133.

PELTON, J. 1953. Ecological life cycle of seed plants. Ecology 34:619-628.

PENDERGRAFT, A. 1946. Sidelights on the pronghorn antelope. Texas Game and Fish 4:22.

PENFOUND, W., and E. S. HATHAWAY. 1938. Plant communities in the marshland of southeastern Louisiana. Ecology 8:1-56.

PENFOUND, W., and M. E. O'NEILL. 1934. The vegetation of Cat Island, Mississippi. Ecology 15:1-17.

PESSIN, L. J. 1933. Forest associations in the uplands of the lower Gulf Coastal Plain (longleaf pine belt). Ecology 14:1-15.

PETTIT, E. 1932. Measurement of ultraviolet solar radiation. Astro-Physical J. 75:185-221. (Also Mt. Wilson Obs. 445:1-37, 369-405; Records 1924-38 in Int. Astr. U. Bull.)

———. 1924-38. Character figures of solar phenomena. Int. Astr. U. Quart. Bull. [Ultra-violet].

PHILLIPS, W. S. 1940. A tropical hammock on the Miami (Florida) limestone. Ecology 21:166-175.

PHIPPS, C. R. 1930. Blueberry and huckleberry insects. Maine Agr. Expt. Sta. Bull. 356:107-232.

PIEMEISEL, R. L., and F. R. LAWSON. 1937. Types of vegetation in the San Joaquin Valley of California and their relation to the beet leafhopper. U.S. Dept. Agr. Bull. 557:1-28.

PILSBRY, H. A. 1939-48. Land Mollusca of North America. Acad. Nat. Sci. Philadelphia 3. 4 parts.

PITELKA, F. A. 1941. Distribution of birds in relation to major biotic communities. Amer. Midl. Nat. 25:113-137.

PITELKA, F. A., P. Q. TOMICH, and G. W. TENCH. 1955. Ecological relations of jaegers and owls as lemming predators near Point Barrow, Alaska. Ecol. Monogr. 25:85-117.

POLUNIN, N. 1934-35. The vegetation of Apotok Island. J. Ecol. 22:337-395; 23:161-209.

———. 1948. Botany of the Canadian eastern Arctic, Part III. Nat. Mus. Canada 104:1-304.

POOL, E. L. 1940. A life history sketch of the Allegheny woodrat. J. Mammal. 21:249-270.

POOL, R. J., J. E. WEAVER, and F. C. JEAN. 1918. Further studies in the ecotone between prairie and woodland. Botan. Surv. Nebraska 18:1-47.

PORSILD, A. E. 1951. Botany of southeastern Yukon adjacent to the Canol Road. Nat. Mus. Canada, Bull. 121:1-400.

POTZGER, J. E. 1946. Phytosociology of the primeval forest in central-northern Wisconsin and upper Michigan, and a brief post-glacial history of the lake forest formation. Ecol. Monogr. 16:211-250.

——. 1951. The fossil record near the glacial border. Ohio J. Sci. 51:126-133.

POTZGER, J. E., and C. O. KELLER. 1952. The beech line in northwestern Indiana. Butler Univ. Botan. Stud. 10:108-113.

POTZGER, J. E., M. E. POTZGER, and J. McCORMICK. 1956. The forest primeval of Indiana as recorded in the original U.S. land surveys and an evaluation of previous interpretations of Indiana vegetation. Butler Univ. Botan. Stud. 13:95-111.

PREBLE, E. A. 1908. A biological investigation of the Athabaska-Mackenzie region. N. Amer. Fauna 27:1-574.

PRESTON, R. J. 1948. North American trees. Iowa State College Press, Ames, Iowa. 751 pp.

PRICE, W. A., and G. GUNTER. 1942. Certain recent biological changes in south Texas with consideration of probable causes. Proc. Texas Acad. Sci. 1942:3-21.

PUTNAM, J. A., and H. BULL. 1932. The trees of the bottomlands of the Mississippi River delta region. Southern For. Expt. Sta. Occ. Pap. 27:1-207 [mimeographed].

RAND, A. L. 1944. The southern half of the Alaska highway and its mammals. Nat. Mus. Canada Bull. 98:1-50.

——. 1945. The mammals of Yukon. *Ibid.* 100:1-93.

——. 1946. List of Yukon birds and those of the Canol Road. *Ibid.* 105:1-76.

——. 1947. The 1945 status of the pronghorn antelope. *Antilocapra americana* (Ord), in Canada. *Ibid.* 106:1-34.

RASMUSSEN, D. I. 1941. Biotic communities of Kaibab Plateau, Arizona. Ecol. Monogr. 11:229-275.

RAUP, H. M. 1951. Vegetation and cryptoplanation. Ohio J. Sci. 51:105-116.

RAUSCH, R. 1951. Notes on the Nunamiut Eskimo and and mammals of the Anaktuvuk Pass region, Brooks Range, Alaska. J. Arctic Inst. N. Amer. 4:147-195.

RHOADES, S. N. 1896. Contribution to the zoology of Tennessee. No. 3. Mammals. Proc. Acad. Nat. Sci. Philadelphia 1896:175-205.

RICE, L. A. 1946. Studies on deciduous forest animal populations during a two year period with differences in rainfall. Amer. Midl. Nat. 135:153-171.

RICHARDS, P. W. 1952. Tropical rain forest. Cambridge University Press, Cambridge. 450 pp.

RIGG, G. B. 1919. Early stages of bog succession. Publ. Puget Sound Biol. Sta., 195-209.

——. 1925. Some sphagnum bogs of the north Pacific Coast of America. Ecology 6:260-278.

RIGG, G. B., and C. T. RICHARDSON. 1938. Profiles of some sphagnum bogs of the Pacific coast of North America. Ecology 19:408-434.

RILEY, C. V. 1892. Yucca moth and yucca pollination. Missouri Botan. Gdn. Rept. 3:99-158; 10 plates.

RILEY, C. V., A. S. PACKARD, and C. THOMAS. 1877. Ravages of locusts in other countries. Ann. Rept. U.S. Ent. Comm. I(chap. XIX), 460-477.

RIVES, W. C. 1886. The birds of Salt Pond Mountain, Virginia. Auk 3:156-161.

ROBERTSON, W. B., JR. 1955. An analysis of breeding bird populations of tropical Florida in relation to the vegetation. Ph.D. Thesis. Univ. Illinois.

ROBERTSON, W. B., JR., and E. L. TYSON. 1950. Herpetological notes from eastern North Carolina. J. Elisha Mitchell Sci. Soc. 66:130-147.

Roe, F. G. 1951. The American buffalo: a critical study of the species in its wild state. University of Toronto Press, Toronto. 957 pp.

Rogers, J. S. 1930. The summer cranefly fauna of the Cumberland Plateau in Tennessee. Occ. Pap. Mus. Zool., Univ. Michigan 215:1-50.

——. 1933. Ecological distribution of the craneflies of northern Florida. Ecol. Monogr. 3:1-74.

Rosendahl, C. O. 1926. Minnesota. In Shelford 1926, 267-284.

Rowan, W. 1925. Relation of light to bird migration and developmental changes. Nature 115:494-495.

——. 1938. Light and seasonal reproduction in animals. Biol. Rev. 13:374-502.

Ruff, F. J. 1938. The white-tailed deer of the Pisgah National Game Preserve. U.S. Dept. Agr. For. Serv. Southern Region. 247 pp. [mimeographed].

Rumsey, W. E. 1926. West Virginia. In Shelford 1926, 341-347.

Ruthven, A. G. 1907. A collection of reptiles and amphibians from southern New Mexico and Arizona. Amer. Mus. Nat. Hist. 23:483-603.

——. 1912. The amphibians and reptiles collected by the University of Michigan. Walker Expedition in southern Vera Cruz, Mexico. Zool. Jahrb. Abt. F. Syst. 32:295-332.

Rydberg, P. A. 1954. Flora of the Rocky Mountains and adjacent plains. Hafner Publishing Co., Inc., New York. 1144 pp.

Rzedowski, J. 1954. Vegetacion del Pedregal de San Angel, Mexico, D.F. Ana. Esc. Nac. Cien. Biol. 8 (1-2):59-129.

——. 1955. Notas sobre la flora y la vegetacion del Estado de San Luis Potosí, Mexico, D.F. Ciencia 15:141-158.

——. 1956. Notas sobre la flora y vegetacion del San Luis Potosí. III. Vegetacion de la region de Guadalcazar. Ana. Inst. Biol. 27:169-228.

Sabrosky, C. W., I. Larson, and R. K. Nabours. 1933. Experiments with light upon reproduction, growth and diapause in grouse locusts. Trans. Kansas Acad. Sci. 36:298-300.

Sampson, A. W. 1944a. Plant succession on burned chaparral lands in northern California. Univ. California Agr. Expt. Sta. Bull. 685:1-144.

——. 1944b. Effect of chaparral burning on soil erosion and on soil moisture relations. Ecology 25:171-191.

Sanders, Nell Jackson, and V. E. Shelford. 1922. A quantitative and seasonal study of a pine-dune animal community. Ecology 3:306-320.

Sanderson, I. T. 1941. Living treasure. Viking Press, New York. 290 pp.

Savely, H. E., Jr. 1939. Ecological relations of certain animals in dead pine and oak logs. Ecol. Monogr. 9:321-385.

Schimper, A. F. W. 1903. Plant geography on a physiological basis (translation). Clarendon Press, Oxford. 839 pp.

Schmidt, K. P. 1926. Honduras. In Shelford 1926, 601-602; British Honduras, *Ibid.,* 600.

——. 1936. Guatemala salamanders of the genus *Oedipus*. Fld. Mus. Nat. Hist., Zool. Ser. 20:135-166.

——. 1950. Check list of North American amphibians and reptiles. 6th ed. University of Chicago Press, Chicago. viii + 280 pp.

Schmidt, K. P., and D. D. Davis. 1941. Field book of snakes of North America and Canada. G. P. Putnam's Sons, New York. 365 pp.

Schmidt, K. P., and L. C. Stuart. 1941. The herpetological fauna of the Salama Basin, Baja Verapaz, Guatemala. Fld. Mus. Nat. Hist., Zool. Ser. 24:233-247.

Schneirla, T. C. 1944. Studies of army ant behavior pattern. Nomadism in the swarm-raider (*Eciton burchelli*). Proc. Amer. Philos. Soc. 87:438-457.

SCHNEIRLA, T. C., and G. PIEL. 1948. The army ant. Sci. Amer. 178:16-25.

SCHWARTZ, C. W. 1941. Home range of the cottontail in Missouri. J. Mammal. 22:386-392.

SCOTT, J. W. 1942. Mating behavior of the sage grouse. Auk 59:477-498.

SCOTT, T. G. 1943. Food coactions of the northern plains fox. Ecol. Monogr. 13:427-479.

———. 1947. Comparative analysis of red fox feeding trends on two central Iowa areas. Iowa St. Coll. Agr. Expt. Sta. Res. Bull. 353:427-487.

SCOVELL, J. T. 1893. Mount Orizaba or Citlaltepetl. Science 21:253-258.

SEAMANS, H. L. 1926. The pale western cutworm. Dom. Canada Dept. Agr. Pamph. 71:1-8.

SEAMANS, H. L., and C. W. FARSTAD. 1938. *Agropyron smithii* Rydb. and *Cephus cinctus* Nort. Ecology 19:350.

SEIFRIZ, W. 1943. Plant life of Cuba. Ecol. Monogr. 13:375-426.

SELLARDS, E. H., R. M. HARPER, C. N. MOONEY, W. J. LATIMER, H. GUNTER, and E. GUNTER. 1915. The natural resources of an area in central Florida. Florida St. Geol. Surv. Ann. Rept., 117-188.

SETON, E. T. 1909. Life histories of northern animals. Charles Scribner's Sons, New York. 2 vols.

———. 1911. The arctic prairies. Charles Scribner's Sons, New York. 415 pp.

———. 1925-28. Lives of game animals. Doubleday-Doran & Co., New York. 4 vols.

SHACKLEFORD, MARTHA W. 1929. Animal communities of an Illinois prairie. Ecology 10:126-154.

———. 1939. New methods of reporting ecological collections of prairie arthropods. Amer. Midl. Nat. 22:676-683.

SHANTZ, H. L. 1924. The natural vegetation of the Great Plains region. Ann. Ass. Amer. Geogr. 13:81-105.

SHANTZ, H. L., and R. L. PIEMEISEL. 1924. Indication of the significance of the natural vegetation of the southwestern United States. J. Agr. Res. 28:72-801.

SHANTZ, H. L., and R. ZON. 1924. Natural vegetation. U.S. Dept. Agr., Atlas of Amer. Agr., Sec. E, 1-29.

SHAVER, J. M. 1933. A preliminary report on the plant ecology of Reelfoot Lake. J. Tennessee Acad. Sci. 8:48-54.

SHAVER, J. M., and M. DENNISON. 1928. Plant succession along Mill Creek. J. Tennessee Acad. Sci. 3:5-13.

SHELFORD, MABEL B. 1913. Decline in the primeval communities at the head of Lake Michigan. In Shelford 1913, 13-15.

SHELFORD, V. E. 1907. Preliminary notes on the distribution of the tiger beetles (*Cicindela*) and its relation to plant succession. Biol. Bull. 14:9-14, 47.

———. 1911a. Ecological succession. I. Stream fishes and the method of physiographic analysis. *Ibid.* 21:9-34.

———. 1911b. Ecological succession. II. Pond fishes. *Ibid.* 21:9-34.

———. 1911c. Physiological animal geography. J. Morph. 22:551-618.

———. 1911d. Ecological succession. III. A reconnaissance of its causes in ponds with particular reference to fish. Biol. Bull. 22:1-38.

———. 1912a. Ecological succession. IV. Vegetation and the control of land animal communities. *Ibid.* 23:59-99.

———. 1912b. Ecological succession. V. Aspects of physiological classification. *Ibid.* 23:331-370.

———. 1913. Animal communities in temperate America. University of Chicago Press, Chicago. viii + 368 pp.

———. 1926. Naturalist's guide to the Americas. (Articles by 106 additional authors, referred to this item in citing.) Williams & Wilkins Company, Baltimore. 761 pp.

———. 1932. An experimental and observational study of the chinch bug in relation to climate and weather. Bull. Illinois Nat. Hist. Surv. 19(6):487-547.

———. 1943. Abundance of the collared lemming in the Churchill area, 1929-1940. Ecology 24:472-484.

———. 1945. The relation of the snowy owl migration to the abundance of the collared lemming. Auk 62:592-596.

———. 1951a. Fluctuation of non-forest animal populations in the upper Mississippi basin. Ecol. Monogr. 21:149-181.

———. 1951b. Fluctuation of forest animal populations in east central Illinois. *Ibid.* 21:183-214.

———. 1952a. Paired factors and master factors in environmental relations. Trans. Illinois Acad. Sci. 45:155-160.

———. 1952b. Does ultraviolet light influence the quality of cow's milk? Guernsey Breeders J. August, 1132-1133.

———. 1954a. Some lower Mississippi Valley floodplain biotic communities, their age and elevation. Ecology 35:126-142.

———. 1954b. Antelope population and solar radiation. J. Mammal. 35:533-538.

SHELFORD, V. E., and M. W. BOESEL. 1942. Bottom animal communities of the island area of western Lake Erie in the summer of 1937. Ohio J. Sci. 42:179-190.

SHELFORD, V. E., and W. P. FLINT. 1943. Populations of chinch bugs in the upper Mississippi Valley from 1823-1940. Ecology 24:435-455.

SHELFORD, V. E., and S. OLSON. 1935. Sere, climax and influent animals with special reference to the transcontinental coniferous forest of North America. Ecology 16:375-402.

SHELFORD, V. E., and A. C. TWOMEY. 1941. Animal communities of the vicinity of Churchill, Manitoba. Ecology 22:47-69.

SHELFORD, V. E., and R. E. YEATTER. 1955. Some suggested relations of prairie chicken abundance to physical factors especially rainfall and solar radiation. J. Wildl. Mgmt. 19:233-242.

SHIMER, H. 1867. Notes on *Micropus* (*Lyarus*) *leucopterus* Say (the chinch bug) with an account of the great epidemic disease of 1865 among insects. Proc. Acad. Nat. Sci. Philadelphia 19:75-80.

SHORT, C. W. 1845. Observations on the botany of Illinois with special reference to the autumnal flora of the prairies. West. J. Med. Surg., n.s. 3:185-198.

SHREVE, E. 1916. An analysis of the causes of variations in the transpiring power of cacti. Physiol. Res. 2(3, Serial 13):73-127.

———. 1923. Seasonal changes in the water relations of desert plants. Ecology 4:266-292.

SHREVE, F. 1914. The direct effects of rainfall on hygrophilous vegetation. J. Ecol. 11:82-98.

———. 1915. Vegetation of a desert mountain range as conditioned by climatic factors. Carnegie Inst. Washington Publ. 217.

———. 1917. A map of the vegetation of the U.S. Geogr. Rev. 3:119-125.

———. 1924. Soil temperature as influenced by altitude and slope exposure. Ecology 6:128-136.

———. 1925. Ecological aspects of the deserts of California. *Ibid.* 6:93-103.

———. 1926. The desert of northern Baja California. Bull. Torrey Bot. Cl. 53:129-136.

———. 1928. Soil temperatures in redwood and hemlock forests. *Ibid.* 54:649-656.

———. 1929. Changes in desert vegetation. Ecology 10:364-373.

———. 1931. Physical conditions in sun and shade. *Ibid.* 12:96-104.

———. 1934. Vegetation of the northwestern coast of Mexico. Bull. Torrey Bot. Cl. 61:373-380.

———. 1936. The plant life of the Sonoran Desert. Sci. Mon. 62:195-213.

————. 1937. The vegetation of Sinaloa. Bull. Torrey Bot. Cl. 64:605-613.

————. 1942. Grassland and related vegetation in northern Mexico. Madroño 6:190-198.

————. 1944. Rainfall of northern Mexico. Ecology 25:105-111.

SHREVE, F., M. A. CHRYSLER, F. H. BLODGET, and F. W. BESLEY. 1910. The plant life of Maryland. Maryland Weather Serv. Spec. Publ. 3:1-533.

SHREVE, F., and A. L. HINCKLEY. 1937. Thirty years of change in desert vegetation. Ecology 18:463-478.

SHULL, A. F. 1907. Habits of the short-tailed shrew. Amer. Nat. 41:495-522.

SIMPSON, C. T. 1929. The Florida tree snails of the genus *Liguus*. Proc. U.S. Natl. Mus. 73:1-44.

SIMPSON, T. L. 1939. Feeding habits of the coot, Florida gallinule and least bittern on Reelfoot Lake. J. Tennessee Acad. Sci. 14:110-115.

SKINNER, M. P. 1927. The predatory and fur-bearing animals of Yellowstone National Park. Roosevelt Wild Life Bull. 4:163-281.

SMALL, J. K. 1933. Manual of the southeastern flora. Published by the author, New York. 1554 pp.

SMITH, A. C., and I. M. JOHNSTON. 1945. Phytogeographic sketch of Latin America. In Verdoorn 1945, 11-18.

SMITH, H. M. 1940. An analysis of the biotic provinces of Mexico as indicated by the distribution of lizards of the genus *Sceloporus*. Ana. Esc. Nac. Cien. Biol. 2:96-110.

————. 1941. Snakes, frogs and bromelias. Chicago Nat. 4:35-43.

————. 1946. Handbook of lizards. Comstock Publishing Company, Inc., Ithaca, New York. 557 pp.

SMITH, H. M., and E. H. TAYLOR. 1945. Annotated check list and key to the snakes of Mexico. U.S. Natl. Mus. Bull. 187:1-239.

————. 1948. Amphibians of Mexico. *Ibid.* 194:1-118.

————. 1950. Reptiles exclusive of snakes of Mexico. *Ibid.* 199:1-253.

SMITH, J. B. 1909. The insects of New Jersey. Ann. Rept. New Jersey St. Mus. 880 pp.

SMITH, R. C., *et al.* 1943. Common insects of Kansas. Rept. Kansas St. Bd. Agr. 17:3-440.

SMITH, VERA G. 1928. Animal communities of deciduous forest succession. Ecology 9:479-500.

SMITH-DAVIDSON, VERA G. 1930. The tree layer society of the maple-red oak climax forest. Ecology 11:601-606.

SMITH-DAVIDSON, VERA G., and MARTHA W. SHACKLEFORD. 1928. Aestival ecology of a moist coniferous forest. Manuscript in the library of Oklahoma Coll. of Women, Chickasha.

SNEAD, E., and G. O. HENDRICKSON. 1942. The food of the badger in Iowa. J. Mammal. 23:380-391.

SNELL, NEVA. 1933. A study of humus fauna. J. Ent. Zool. 25:33-40.

SNYDER, T. E. 1919. Injury to casuarina trees in southern Florida by the mangrove borer. J. Agr. Res. 16:155-164.

Society of American Foresters. 1940. Forest cover types of eastern U.S. Rept. Comm. For. Types. September, 1-39.

SPARKMAN, DREW. 1943. The distribution of ant species in relation to some North American biomes. M.S. Thesis, Univ. Illinois.

STANDLEY, P. C. 1920-26. The trees and shrubs of Mexico. Contr. U.S. Natl. Herb. 23:1-1721 [5 parts].

————. 1928. Flora of the Canal Zone. Contr. U.S. Natl. Herb. 27:1-416.

————. 1931. Flora of the Lancetilla Valley, Honduras. Fld. Mus. Nat. Hist. Publ. 283, Bot. Ser. 10:7-418.

————. 1933. The flora of Barro Colorado Island, Panama. Contr. Arnold Arbor. 5:1-230.

————. 1945. A brief survey of the vegetation of Costa Rica. In Verdoorn 1945, 64-66.

STANDLEY, P. C., and J. A. STEYERMARK. 1945. The vegetation of Guatemala. In Verdoorn 1945, 275-278.

STANWELL-FLETCHER, J. F., and T. C. STANWELL-FLETCHER. 1940. Naturalists in the wilds of British Columbia. Sci. Mon. 50:17-32, 125-137, 210-224.

STARLING, J. H. 1944. An ecological study of the Pauropoda of the Duke forest. Ecol. Monogr. 14:291-310.

STEENIS, J. H. 1947. Recent changes in the marsh and aquatic plant status at Reelfoot Lake. J. Tennessee Acad. Sci. 22:22-27.

STEFANSSON, V. 1921. The friendly Arctic. Macmillan Company, New York. 784 pp.

STENLUND, M. H. 1955. A field study of the timber wolf (*Canus lupa*) on the Superior National Forest, Minnesota. Minnesota Dept. Conserv. Tech. Bull. 4:1-54.

STEPHENS, F. 1906. California mammals. West Coast Publishing Co., San Diego. 351 pp.

STEVENSON, N. S. 1942. The forest associations of British Honduras. Caribbean For. 3:164-172.

STEWART, G., W. P. COTTAM, and S. S. HUTCHINS. 1940. Influence of unrestricted grazing on northern salt desert plant associations in western Utah. J. Agr. Res. 60:289-316.

STEWART, R. E., and J. W. ALDRICH. 1949. Breeding bird populations in the spruce region of the central Appalachians. Ecology 30:75-82.

STODDARD, H. L. 1939. The use of controlled fire in southeastern game management. Cooperative Quail Study Ass., Sherwood Plantation, Thomasville, Georgia, 1-21.

STONE, W. 1937. Bird studies at old Cape May. An ornithology of coastal New Jersey. Delaware Valley Orn. Cl., Acad. Nat. Sci. Philadelphia. 520 pp.

STORER, T. I. 1930. Summer and autumn breeding of the California ground squirrel. J. Mammal. 11:235-237.

————. 1932. Factors influencing wild life in California, past and present. Ecology 13:315-327.

STORER, T. I., F. P. CRONEMILLER, E. E. HORN, and B. GLADING. 1942. Studies of the valley quail. Univ. California Agr. Expt. Sta. Bull. 663:130-142.

STRODE, D. D. 1954. The Ocala deer herd. Florida Game and Freshwater Fish Comm. Publ. 1:1-42.

STUART, L. C. 1935. A contribution to a knowledge of the herpetology of a portion of the savanna region of central Peten, Guatemala. Univ. Michigan Misc. Publ. Mus. Zool. 29:7-56.

STURGIS, B. B. 1928. Field book of the birds of the Panama Canal Zone. G. P. Putnam's Sons, New York. 466 pp.

SUMMERHAYES, V. S., and C. S. ELTON. 1923. Contributions to the ecology of Spitsbergen and Bear Island. J. Ecol. 11:214-286.

————. 1928. Further contributions to the ecology of Spitsbergen. *Ibid.* 16:193-268.

SUMNER, F. B. 1925. Some biological problems of our southwestern deserts. Ecology 6:352-371.

SURFACE, H. A. 1906. The serpents of Pennsylvania. Pennsylvania Dept. Agr. Mon. Bull. 4:113-208.

SUTTON, G. M. 1932. Birds of Southampton Island. Mem. Carnegie Mus. 12:1-267.

SUTTON, G. M., and T. D. BURLEIGH. 1939. A list of birds observed on the 1938 Semple Expedition to northeastern Mexico. Occ. Pap. Mus. Zool., Louisiana St. Univ. 3:15-46.

SUTTON, G. M., and W. J. HAMILTON, JR. 1932. The mammals of Southampton Island. Mem. Carnegie Mus. 12:9-111.

Sutton, G. M., and O. S. Pettingill, Jr. 1943. Birds of Linares and Galleana, Nuevo León, Mexico. Occ. Pap. Mus. Zool. 16:273-291.

Swaine, J. M. 1918. Canadian bark beetles. Part II. Preliminary classification with an account of the habits and means of control. Dom. Canada, Dept. Agr., Ent. Branch Tech. Bull. 14:1-143.

Swaine, J. M., *et al.* 1931-32. Forest insects. [Two or three page brochures, each with a colored plate covering: the hemlock looper; the spruce budworm; the eastern spruce bark-beetle; the jack pine sawfly (with M. B. Dunn); the black-headed budworm (with R. E. Balch).] Canada Dept. Agr., Div. For. Insects.

Swarth, H. S. 1911. Birds and mammals of the 1909 Alexander Alaska Expedition. Univ. California Publ. Zool. 7:9-172.

———. 1922. Birds and mammals of the Stikine River region of northern British Columbia and southeast Alaska. Univ. California Publ. 24:125-314.

Sweet, Helen E. 1930. An ecological study of the animal life associated with *Artemisia californica*, Less., at Claremont, California. J. Ent. and Zool. 22:57-70.

Sweetman, H. L. 1931. Preliminary report on the physical ecology of certain Phyllophaga (Sarabaeidae, Coleoptera). Ecology 12:401-422.

Swift, E. 1946. A history of Wisconsin deer. Wisconsin Conserv. Dept. Publ. 323:1-96.

Taber, R. D. 1953. Studies of black-tailed deer reproduction in three chaparral cover types. California Fish and Game 39:177-186.

Talbot, M. W., and H. H. Biswell. 1942. The forage crop and its management (San Joaquin Experimental Range). Univ. California Agr. Expt. Sta. Bull. 663:13-49.

Talbot, M. W., J. W. Nelson, and R. E. Storie. 1942. The experimental area. Univ. California Agr. Expt. Sta. Bull. 633:7-12; soil map, p. 9.

Tansley, A. G. 1935. The use and abuse of vegetational concepts and terms. Ecology 16:284-307.

Tashian, R. E. 1951. Ecology of the dry season avifauna of southeastern Guatemala. Ph.D. Thesis, Purdue Univ.

———. 1952. Some birds from the Palenque region of northeastern Chiapas, Mexico. Auk 69:60-66.

———. 1953. The birds of southeastern Guatemala. Condor 55:198-210.

Taylor, E. H. 1936. Notes on the herpetological fauna of the Mexican state of Sinaloa. Univ. Kansas Sci. Bull. 24:505-536.

Taylor, W. P. 1922. A distributional and ecological study of Mount Rainier, Washington. Ecology 3:214-246.

———. 1935. Some relations to soil. *Ibid.* 16:127-136.

Taylor, W. P., and H. K. Buechner. 1943. Relationship of game and livestock to range vegetation in Kerr County, Texas. Cattleman, March. 3 pp.

Taylor, W. P., W. B. McDougall, and W. B. Davis. 1944. Plant and animal communities, etc. Preliminary report of an ecological survey of Big Bend National Park, 12-28 [mimeographed, in National Park files only].

Taylor, W. P., W. B. McDougall, C. C. Presnall, and K. P. Schmidt. 1946. The Sierra del Carmen in northern Coahuila, Texas. Geogr. Mag., Spring.

Taylor, W. P., and W. T. Shaw. 1929. Provisional list of land mammals of the State of Washington. Occ. Pap., Charles R. Conner Mus. [State College of Washington] 2:1-31.

Taylor, W. P., C. T. Vorhies, and P. B. Lister. 1935a. The relation of jackrabbits to grazing in southern Arizona. J. For. 33:490-493.

Tedrow, J. C. F. 1955. Soils and soil formation in the permanently frozen arctic tundra. Bull. Ecol. Soc. Amer. 36:96-97.

Tehon, L. R. 1928. Methods and principles for the interpreting of the phenology of crop pests. Bull. Illinois Nat. Hist. Surv. 17:321-346.

a forest survey of Illinois. Div. Illinois Nat. Hist. Surv. 16:1-102.

THARP, B. C. 1926. Structure of Texas vegetation east of the 98th meridian. Univ. Texas Bull. 2606:1-96.

———. 1939. The vegetation of Texas. Texas Acad. Sci., nontechnical series, xvi + 74 pp.

THOMAS, C. 1879. The chinch bug. U.S. Dept. Int., Ent. Comm. Bull. 5:1-44.

THOMPSON, D. H. 1941. Symposium on hydrobiology. University of Wisconsin Press, Madison, 214-216.

TIDESTROM, K. 1925. Flora of Utah and Nevada. Contr. U.S. Natl. Mus. 25:1-665.

TINKHAM, E. R. 1948. Faunistic and ecological studies on the Orthoptera of the Big Bend region of the Trans-Pecos Texas. Amer. Midl. Nat. 40:521-663.

TISDALE, E. W. 1947. The grasslands of the southern interior of British Columbia. Ecology 28:346-382.

TRANSEAU, E. N. 1903. On the geographic distribution and ecological relations of the bog plant societies of North America. Botan. Gaz. 36:401-420.

———. 1905. Forest centers of eastern America. Amer. Nat. 39:875-889.

———. 1935. The prairie peninsula. Ecology 16:423-437.

TRAPNELL, C. G. 1933. Vegetation types in Goethaab Fjord in relation to those in other parts of West Greenland, and with special reference to Isersiutilik. J. Ecol. 21:294-334.

TRIPPENSEE, R. E. 1948. Wildlife management, upland game and general principles. McGraw-Hill Book Company, Inc., New York. 477 pp.

TWOMEY, A. C. 1945. The bird population of an elm-maple forest with special reference to aspection, territorialism, and coactions. Ecol. Monogr. 15:173-205 [full account, 1937, Ph.D. Thesis, Univ. Illinois].

TUCKER, MARIE. 1943. Ecological and geographical relations of the Enchytraeidae. M.S. Thesis, Univ. Illinois.

VAN DER SCHALIE, H. 1940. Notes on mollusca from Alta Vera Paz, Guatemala. Occ. Pap. Mus. Zool., Univ. Michigan 413:1-11.

VAN DERSEL, W. R. 1938. Native woody plants of the U.S., their erosion-control and wild life values. U.S. Dept. Agr. Misc. Publ. 303:1-362.

VAN ROSSEM, A. J. 1945. A distributional survey of the birds of Sonora, Mexico. Occ. Pap. Mus. Zool., Louisiana St. Univ. 21:3-379.

VAN TYNE, J. 1935. The birds of northern Peten, Guatemala. Univ. Michigan Misc. Publ. 27:5-46.

VERDOORN, F. 1945. Plants and plant science in Latin America. (Articles by 40 additional authors, referred to this item in citing.) Chronica Botanica Co., Waltham, Massachusetts.

VEST, D. 1955. Biotic communities as epizootic highways. Symp. Ecol. Disease Trans. in native animals. Univ. Utah for Army Chemical Corp., Dugway, Utah. 112 pp.

VESTAL, A. G. 1913. An associational study of Illinois sand prairie. Bull. Illinois Nat. Hist. Surv. 10:1-94.

———. 1914a. A Black soil prairie station in northeastern Illinois. Bull. Torrey Bot. Cl. 41:351-363.

———. 1914b. Internal relations of terrestrial associations. Amer. Nat. 48:413-444.

———. 1917. Foothills vegetation in the Colorado front range. Botan. Gaz. 64:353-385.

———. 1938. Problems of the Garrigue-like bush of California. Bull. Ecol. Soc. 19:12-13.

VESTAL, A. G., and M. F. HEERMANS. 1945. Size requirements for reference areas in mixed forest. Ecology 26:122-134.

VILLA, B. 1948. Mamiferos del Soconusco, Chiapas. Ann. Inst. Biol. [Mexico Univ.] 19:485-531.

Viosca, P., Jr. 1932. Field excursions, (AAAS) New Orleans, La. to Pearl River, La., Dec. 31, 1931, and Jan. 1, 1932 [mimeographed].

———. 1933. Louisiana out-of-doors. A handbook and guide. Published by the author, New Orleans, Louisiana. 187 pp.

———. 1944. Distribution of certain cold blooded animals in Louisiana in relation to the geology and physiography of the state. Proc. Louisiana Acad. Sci. 8:47-62.

Visher, S. S. 1916. The biogeography of the northern Great Plains. Geogr. Rev. 2:89-115.

———. 1946. Evaporation regions of the United States. Sci. Mon. 62:453-457.

Vorhies, C. T. 1945. Water requirements of desert animals in the southwest. Univ. Arizona Agr. Expt. Sta. Tech. Bull. 107:487-525.

Vorhies, C. T., and W. P. Taylor. 1922. Life history of the kangaroo rat, *Dipodomys s. spectabilis* Merriam. U.S. Dept. Agr. Bull. 1091:1-40.

———. 1933. The life histories and ecology of jack rabbits. *Lepus alleni* and *Lepus californicus* sp. in relation to grazing in Arizona. Univ. Ariz. Agr. Expt. Sta. Tech. Bull. 49:467-587.

———. 1940. Life history and ecology of the white-throated wood rat, *Neotoma albigula albigula* Hartley, in relation to grazing in Arizona. Univ. Arizona Tech. Bull. 86:455-529.

Wailes, B. L. C. 1854. Agriculture and geology of Mississippi. Fauna. Published by State, Jackson, 310-338.

Walker, R. B. 1954. Ecology of serpentine soils (a symposium). II. Factors affecting plant growth on serpentine soils. Ecology 35:259-266.

Warming, E. 1909. Oecology of plants. An introduction to the study of plant communities. Clarendon Press, Oxford. xi + 422 pp.

Waterman, W. G. 1925. Plant communities of Alpine Park. Botan. Gaz. 80:188-202.

———. 1926. Ecological problems from the sphagnum bogs of Illinois. Ecology 8:255-272.

Watson, J. R. 1912. Plant geography of north central New Mexico. Botan. Gaz. 54:194-217.

———. 1926. Florida. In Shelford 1926, 427-444.

Weaver, J. E. 1917. A study of the vegetation of southeastern Washington and adjacent Idaho. Univ. Nebraska Stud. 17:1-131.

Weaver, J. E., and F. W. Albertson. 1939. Major changes in grassland as a result of continued drought. Botan. Gaz. 100:576-591.

———. 1940. Deterioration of grassland from stabilization to denudation with decrease in soil moisture. *Ibid.* 101:598-623.

Weaver, J. E., and F. E. Clements. 1929. Plant ecology. McGraw-Hill Book Company, Inc., New York. xx + 520 pp.

Weaver, J. E., and T. J. Fitzpatrick. 1932. Ecology and relative importance of the dominants of tall-grass prairie. Botan. Gaz. 93:113-150.

Weaver, J. E., H. C. Hanson, and J. M. Aikman. 1925. Transect method of studying woodland vegetation along streams. Botan. Gaz. 80:168-187.

Weaver, J E., and J. Kramer. 1932. Root system of *Quercus macrocarpa* in relation to the invasion of prairie. Botan. Gaz. 94:51-84.

Weaver, J. E., and A. F. Thiel. 1917. Ecological studies in the tension zone between prairie and woodland. Botan. Surv. Nebraska, n.s. 1:3-60.

Webster, J. D. 1950. Altitudinal zonation of birds in southeastern Alaska. Murrelet 31:22-26.

Weese, A. O. 1919. Environmental reactions of *Phrynosoma*. Amer. Nat. 53:33-54.

———. 1924. Animal ecology of an elm-maple forest. Illinois Biol. Monogr. 9:345-438.

———. 1939. The effect of overgrazing on insect population. Proc. Oklahoma Acad. Sci. 19:95-99.

WELCH, P. S. 1935. Limnology. McGraw-Hill Book Company, Inc., New York. xiv + 471 pp.

WELLS, B. W. 1928. Plant communities of the coastal plain of North Carolina and their successional relations. Ecology 9:230-243.

———. 1939. A new forest climax, saltspray climax of Smith Island, North Carolina. Bull. Botan. Cl. 66:629-634.

———. 1942. Ecological problems of the southeastern United States coastal plain. Botan. Rev. 8:533-561.

———. 1946. Vegetation of the Holly Shelter Wildlife Management Area, North Carolina. Dept. Conserv. Devel., Div. Game and Inland Fish. St. Bull. 2:1-40.

WELLS, B. W., and I. V. SHUNK. 1931. The vegetation and habitat factors of the coarser sands of the North Carolina coastal plain: an ecological study. Ecol. Monogr. 1:465-520.

WENT, F. W. 1948. Ecology of desert plants. I. Observations on germination in the Joshua Tree National Monument, California. Ecology 29:242, 253.

———. 1949. Ecology of desert plants. II. The effect of rain and temperature on germination and growth. *Ibid.* 30:1-13.

WENT, F. W., and M. WESTERGAARD. 1949. Ecology of desert plants. III. Development of plants in the Death Valley National Monument. Ecology 30:26-38.

WETMORE, A. 1941. Notes on birds of the Guatemalan highlands. Proc. U.S. Natl. Mus. 89:523-581.

WETZEL, R. M. 1949. Analysis of small mammal populations in the deciduous forest. Ph.D. Thesis, Univ. Illinois.

WHELAN, D. B. 1936. Coleoptera of an original prairie area in eastern Nebraska. J. Kansas Ent. Soc. 9:11-115.

———. 1938. Orthoptera of an eastern Nebraska prairie. *Ibid.* 11:3-6.

WHITFIELD, C. J. 1933. Ecology of the vegetation of the Pikes Peak area. Ecol. Monogr. 3:1-103.

WHITFIELD, C. J., and E. L. BEUTNER. 1938. Natural vegetation in the desert plains grassland. Ecology 19:26-37.

WHITTAKER, R. 1952. Some summer foliage insect communities in the Great Smoky Mountains. Ecol. Monogr. 22:1-42.

———. 1954. Ecology of serpentine soils. Vegetational response to serpentine soils. Ecology 35:274-288.

———. 1956. Vegetation of the Great Smoky Mountains. Ecol. Monogr. 26:1-80.

WHITTEMORE, W. L. 1937. Summer birds of Reelfoot Lake. J. Tennessee Acad. Sci. 12:114-128.

WIGGINS, DOROTHY, and I. LOREN. 1953. A winter journey to Point Barrow, with summer pictures of the tundra vegetation. Asa Gray Bull. 2:83-92.

WIGHT, H. M. 1926. In Shelford 1926, 181.

WILLIAMS, A. B. 1936. The composition and dynamics of a beach-maple climax community. Ecol. Monogr. 6:317-405.

WILLIAMS, E. C., JR. 1941. An ecological study of the floor fauna of the Panama rain forest. Bull. Chicago Acad. Sci. 6:63-124.

WILLIAMSON, A. W. 1913. Cottonwood in the Mississippi Valley. U.S. Dept. Agr. Bull. 24:1-62.

WILLISTON, S. W. 1908. Manual of North American Diptera. Hathaway, New Haven. 405 pp.

WING, L. 1943. Summary of detection distance and flock size reports St. Coll. Washington, Pullman [mimeographed].

WISE, H., and C. M. ZACARIAS. 1958. Analisis de Caida de Lluvia en cincuenta estaciones de Honduras. 1913-1957. Boletin Tecnico 9, Ser. E. A. 1, Ministerio de Recursos Naturales, STICA, Tegucigalpa, D.C., Honduras, C.A., 2-62.

Wolcott, A. B., and B. E. Montgomery. 1933. An ecological study of the coleopterous fauna of a tamarack swamp. Amer. Midl. Nat. 14:113-169.

Wolcott, G. N. 1937. An animal census of two pastures and a meadow in northern New York. Ecol. Monogr. 7:1-90.

Wolf, F. A. 1938. The fungi of Duke forest. Duke Univ. Sch. For. Bull. 2:1-122.

Woodbury, A. M. 1933. Biotic relationships of Zion Canyon, Utah with special reference to succession. A survey of the geological, botanical and zoological interrelationships within a part of Zion National Park, Utah. Ecol. Monogr. 3:147-246.

———. 1947. Distribution of pigmy conifers in Utah and northeastern Arizona. Ecology 28:113-126.

Woodbury, M., and A. M. Woodbury. 1945. Life history studies of the sage-brush lizard *Sceloporus g. graciosus* with special reference to cycles in reproduction. Herpetologica 2:175-196.

Wooster, L. D. 1935. Notes on the effects of drought on animal population in western Kansas. Trans. Kansas Acad. Sci. 38:351-353.

———. 1936. The contents of owl pellets as indicators of habitat preferences of small mammals. *Ibid.* 39:395-397.

———. 1938. An attempt at an ecological evaluation of predators on a mixed prairie area in western Kansas. *Ibid.* 41:387-394.

———. 1939. An ecological evaluation of predators on a mixed prairie area in western Kansas. *Ibid.* 42:515-517.

Wright, A. H. 1945. Our Georgia-Florida frontier. The Okefinokee Swamp, its history and cartography. Vol. I. Studies in history 9:1-40; 10:1-20; 11:1-46; 12:1-26; 13:1-44; 14:1-47. Published by author, Ithaca, New York.

Wright, A. H., and S. C. Bishop. 1915. The reptiles. II. Snakes. A biological reconnaissance of the Okefinokee Swamp in Georgia. Proc. Acad. Nat. Sci. Philadelphia, 139-192.

Wright, A. H., and W. D. Funkhouser. 1915. The reptiles. I. Turtles, lizards, and alligators. A biological reconnaissance of the Okefinokee Swamp in Georgia. Proc. Acad. Nat. Sci. Philadelphia, 107-139.

Wright, A. H., and A. A. Wright. 1932. The habitats and composition of the vegetation of Okefinokee Swamp, Georgia. Ecol. Monogr. 2:111-232.

———. 1949. Handbook of frogs and toads of the U.S. and Canada. Comstock Publishing Company, Inc., Ithaca, New York. 640 pp.

Wright, G. M., J. S. Dixon, and B. H. Thompson. 1933. Fauna of the national parks of the United States. Fauna Natl. Parks U.S., Fauna Ser. 1:1-157.

Wright, G. M., and B. H. Thompson. 1935. Wildlife management in the national parks. Fauna Natl. Parks U.S., Fauna Ser. 2:1-142.

Yeatter, R. E. 1943. The prairie chicken in Illinois. Illinois Nat. Hist. Surv. Bull. 22:375-416.

Yeatter, R. E., and D. H. Thompson. 1952. Tularemia, weather and rabbit populations. Illinois Nat. Hist. Surv. Bull. 25:351-382.

Young, F. N. 1951. Vanished and extinct colonies of tree snails, *Liguus fasciatus* in the vicinity of Miami, Florida. Occ. Pap. Mus. Zool. 531:1-21.

Young, S. P., and H. W. Dobyns. 1937. Den hunting as a means of coyote control. U.S. Dept. Agr. Leafl. 132.

Young, S. P., and E. A. Goldman. 1944. The wolves of North America. Amer. Wildl. Inst., Washington, D.C. 588 pp.

Yunker, T. G. 1938. A contribution to the flora of Honduras. Fld. Mus. Nat. Hist., Botan. Ser. 17:389-407.

Zetek, J. 1918. Mollusca of Piatt, Champaign, and Vermilion counties of Illinois. Trans. Illinois Acad. Sci. 11:151-182.

————. 1938. Rainfall, temperatures and relative humidity, 1937. 14th annual report, March 1, 1937 to Feb. 28, 1938. Barro Colorado Island Biological Station in the Panama Canal Zone. Natl. Res. Coun., Washington, D.C., 20-27 [mimeographed].

ZWOLFER, W. 1931. Studien zur Oecologie und Epidemologie der Insecten. I. Die Kieferneule (*Panolis flammea*). Zeit. f. Angew. Ent. 17:475-682.

Locality Index

533

Species Index

545

Wren, Carolina, *Thryothorus ludovicianus*, 75, 81, 97

Wren, long-billed marsh, *Telmatodytes palustris*, 145

Wren, rock, *Salpinctes obsoletus obsoletus*, 163

Wren, rufous-breasted, *Thryothorus rutilus*, 443

Wren, white-breasted wood, *Henicorhina leucosticta*, 415

Wren, winter, *Troglodytes troglodytes*, 133, 137, 146, 171, 219, 221

Wrentit, *Chamaea fasciata*, 253, 255

Wyeomyia smithii, mosquito, 55

Xanthippus corallipes, 303, 338, 366

Xanthippus lateritius, 270

Xanthocephalum gymnospermoides, 363

Xanthonia decemnotata, 36

Xanthoria, lichen, 197

Xanthosarus latimanus, leaf cutter bee, 141

Xanthosarus melanophaea, 141

Xenochalepus dorsalis, beetle, 36, 42

Xenops minutus, plain xenops, 414, 438

Xeracris minimus, 390

Xerophloea viridis, leafhopper, 355, 370

Xiphidium nigropleuroides, grasshopper, 84

Xiphomyrmex spinosus, ant, 295

Xolisma ferruginea, staggerbush, 73, 79

Xyele, sawfly, 181

Xyleborus affinis, bark beetle, 420

Xyleborus propinquus, 419

Xylopia frutescens, 410, 443

Xylosma hemsleyana, 409

Xylosma sylvicola, 405

Xylota curvaria, syrphid, 135

Xylotrechus colonus, cerambycid beetle, 61

Xysticus, spider, 153

Xysticus cunctator, 278, 295

Xysticus gulosus, 316

Xysticus punctatus, 53

Xysticus triangulosus, 148

Xystocheir, millipede, 253

Yarrow, *Achillea*, 251

  Achillea lanulosa, 176

  Achillea millefolium, 227

Yaupon. *See Ilex vomitoria*

Yellowthroat, *Geothlypis trichas*, 323, 386

Yellowthroat, gray-crowned, *Chamaethlypis poliocephala*, 491

Yerba bucca, *Micromeria chamissonis*, 245

Yersiniops solitarium, ground mantis, 365

Yobius haywardi, centipede, 304

Yucca, 80, 292, 294, 295, 305, 349, 350, 358, 361, 363, 364, 365, 371, 377, 391, 393, 458, 463, 473

Yucca baccata, 287, 364

Yucca brevifolia, Joshua tree, 337, 387, 388

Yucca elata, palmilla, 361, 365, 366, 391, 392

Yucca glauca, soapweed, 304, 335

Yucca periculosa, 452

Yucca schidigera, Mojave yucca, 387, 388, 390

Yucca treculeana, cenizo, 369

Yucca whipplei, whipple yucca, 256, 388, 390

Zannichellia palustris, horned pondweed, 108

Zanthoxylum, 49, 310, 409

Zanthoxylum caribaeum, 490

Zanthoxylum fagara, lime prickly-ash, 449, 475

Zanthoxylum flavum, yellowheart, 475

Zanthoxylum panamense, 407

Zanthoxylum spinifex, 488

Zapata salutator, 382

Zapote. *See Calocarpum sapota*

Zapus hudsonicus. See Mouse, meadow jumping

Zelmira subterminalis, myceteophilid, 135

Zelus exsanguis, assassin bug, 53

Zelus laevicollis, 253

Zenobia pulverulenta, 58, 85

Zerene eurydice, butterfly, 257

Zinnia acrosa, 365

Zinnia anomola, 367

Zizaniopsis miliacea, cut grass, watermillet, 70, 93, 105, 108, 109, 110, 111

Zizyphus sonorensis, 455, 457

Zollernia, tango, 404

Zonitoides, snail, 98

Zonitoides arboreus, 61, 69, 134, 257

Zoögenetes harpa, 134

Zootermopsis, termite, 169, 221

Zootermopsis angusticolles, 228

Zornia, 442

Zuelania guidonia, aiguane, 453

Zygadenus elegans, 163

Zygadenus venenosus, 278